D1498285

ITHDRAWN

AIChE Symposium Series No. 304
Volume 91, 1995

Fourth International Conference on

FOUNDATIONS of COMPUTER-AIDED PROCESS DESIGN

Proceedings of the Conference held at
Snowmass, Colorado, July 10-14, 1994

Lorenz T. Biegler
Carnegie Mellon University

Michael F. Doherty
University of Massachusetts

Volume Editors

CACHE
American Institute of Chemical Engineers

1995

Neither AIChE nor CACHE shall be responsible for statements or opinions advanced in their papers or printed in their publications.

Library of Congress Cataloging-in-Publication Data

International Conference on Foundations of Computer-Aided Process Design (4th : 1994)
Fourth International Conference on Foundations of Computer-Aided Process Design / Lorenz T. Biegler, Michael F. Doherty, volume editors.
p.cm. — (AIChE symposium series ; no 304)
Includes index.
ISBN 0-8169-0666-1
1. Chemical processes — Data processing — Congresses.
I. Biegler, Lorenz T. II. Doherty, Michael F. III. Title. IV. Series.
TP155.7.158 1994
660'.2815-dc20 95-15014
CIP

FOREWORD

Chemical process industries remain one of the strongest segments of the worldwide economy. This is due largely to the cost effectiveness of its processes as well as to its technological innovations. However, as we enter the next century, the industry faces major challenges through increased competition, greater regulatory pressures, and uncertain prices for energy, raw materials and products. Given these competitive concerns, there is an increasing focus on processes with subsystems that have a tighter level of integration and coordination, on tools that communicate with each other across design levels and on consideration of multiple design criteria including profitability, safety, operability and environmental concerns.

To provide perspective and direction for research in this area for the remainder of the decade, a conference on the Foundations of Computer Aided Process Design (FOCAPD '94) was held on July 10-15, 1994 in Snowmass, Colorado, under the sponsorship of the CACHE Corporation and the CAST Division of AIChE. This volume contains the proceedings of the conference. FOCAPD '94 was the fourth in a series of highly successful meetings on computer-aided chemical process design that have focused academic research and industrial practice over the past decade. It also complements other CAST/CACHE conferences on Foundations of Computer-Aided Process Operations (FOCAPO) in 1986 and 1993 and the Chemical Process Control (CPC) series, most recently held in 1991.

The conference brought together 144 engineers and scientists from universities, industry, and government laboratories from 17 countries in order to assess and critique the current status and future directions of computer-aided process and product engineering. The tone of the conference was process-oriented rather than methods oriented, as in past FOCAPD meetings. This led to presentations and discussions that were focused on important unsolved design problems in the process industries. Topics of the conference focused on the integration of process design across subsystems, downstream concerns such as operability, safety and the environment, and the opportunities and challenges in design posed by continuous advances in computer hardware and software environments. Twenty invited plenary papers were presented by internationally known speakers to survey the state-of-the-art and promote discussion on future research directions for chemical engineering design. In addition, thirty one contributed research papers were presented to discuss recent results in the areas of process synthesis and design. These papers describe leading edge research from design centers around the world.

The success of this conference was due to many individuals and several organizations. First, we are grateful to the National Science Foundation for providing travel support for invited US. speakers. For this we are especially thankful to Dr. Edgar O'Rear at NSF for his advice and encouragement. We are also grateful to six industrial sponsors, Air Products and Chemicals, Aspen Technology, DuPont, Eastman Chemicals, Hyprotech, and Weyerhaeuser, who supplied further financial support for this conference.

We gratefully acknowledge the guidance provided by the Technical Organizing Committee and the work of the invited session chairs, speakers and participants. The research impact of this conference is entirely due to their efforts. Most of these contributions represent a series of collaborative efforts and this hard work has led to fresh perspectives for many topics in the design area. In addition, we are very pleased by the number and quality of contributed papers at this conference. We gratefully acknowledge Prof. G. V. Reklaitis for the outstanding organization and coordination of the contributed session. Lastly, collecting the final papers in electronic form and reproducing them in a polished and professional format was facilitated by the selfless efforts of Prof. Brice Carnahan, assisted by Matthew Smart and Adam Thodey, undergraduate students at the University of Michigan.

We gratefully acknowledge our sponsoring organizations, the CAST Division of the American Institute of Chemical Engineers, chaired by Dr. W. D. Smith, and the CACHE Corporation. From this organization, we thank particularly Prof. David Himmelblau, Executive Officer, Prof. Michael Cutlip, President and Prof. Richard Mah, Chair of the Conferences Task Force, for their support and encouragement. We are especially grateful to Robin Craven of Smart Meetings, Inc., for her excellent conference facilitation and to Janet Sandy and Margaret Beam of the CACHE office for their outstanding help on a day-to-day basis. Their efforts in providing an excellent conference environment contributed much to the success of this event.

Lorenz T. Biegler
Carnegie Mellon University

Michael F. Doherty
University of Massachusetts

FOCAPD TECHNICAL ADVISORY COMMITTEE

TABLE OF CONTENTS

Design for Operations and Control

Process Engineering Software Tools and Environments

Methods Driven by Advanced Computing Environments

Integrating Design Criteria, Subsystems and Tools

Contributed Papers

IMPACT OF GLOBAL ECONOMY ON NEW DIRECTIONS FOR THE COMPETITIVENESS OF THE CHEMICAL INDUSTRY

J. A. Miller, Senior Vice President
Research and Development, DuPont
Wilmington, DE 19880-0328

Introduction

Thanks to Mike Doherty and Larry Biegler for their invitation to participate in this conference and speak on a subject which I know is of paramount interest to all of us: a healthy competitive chemical industry, and the role of chemical engineers in making it competitive. As shown on this first slide, I first plan to review the current research environment in the context of a longer term perspective on the chemical industry. Then I will cover the challenges we face, discuss what we have done at DuPont in the face of these challenges, and finally look at the new directions our industry must take to stay competitive.

With the end of the Cold War, it is obvious to all of us that we have entered an environment of rapid economic, political and organizational change and a period of great uncertainty. Changes which were developing in an evolutionary way seem to have accelerated to revolutionary or even warp speed. Businesses throughout the country and the world have reorganized, downsized and globalized; local economies have regionalized into trading blocs. Hiring of current graduates is down significantly and serious questions are being asked about both undergraduate and graduate education of chemists and chemical engineers. There is widespread concern about the health and continued viability of academia.

Major changes are also occurring at the federal level which impact R&D and the chemical industry. Government support for research is being constrained and the missions of the National Laboratories are being redirected to serve more civilian needs. Speaking at the Council of Scientific Society Presidents, Ed David, former Science Advisor to President Nixon, projected a worst case scenario in which large, centralized, corporate research labs would disappear completely, the national investment for R&D would shrink by 25-30% and the National Laboratory system would be reduced to 30% of its current size. (I suspect that there is a threat in those suggestions for just about everyone in this room.)

In this environment of heavy rhetoric, seeming chaos and the push for immediate action, it is essential that we stand back from the turmoil and develop a longer, more pragmatic perspective on our industry. By any measure, it has been one of the most successful enterprises in the country and still possesses an exceptionally strong intellectual and industrial base. Since its emergence as an industry in the latter part of the last century, it has grown to a world-wide one trillion dollar giant. In the United States, it accounts for 10% of all U.S. manufacturing with annual sales exceeding $100 billion

Our industry is also the only unsubsidized industrial sector which has sustained a long term positive balance of trade as shown in Fig. 1; all others including textiles, machinery, electronics and transportation continue to run negative balances.

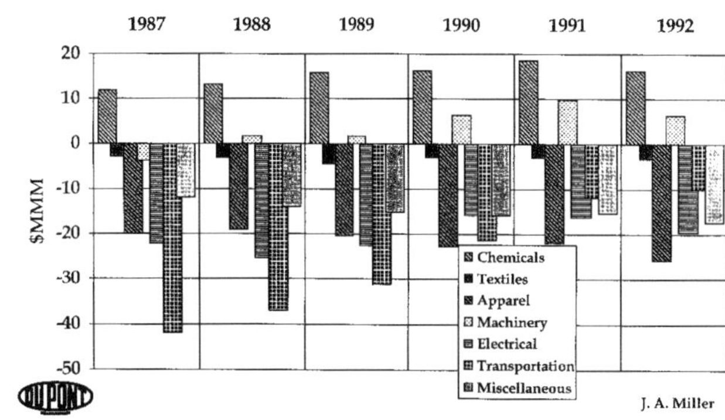

Figure 1. Trade balance by industry.

This positive balance, shown in Fig. 2, has been growing since 1987, and in 1992 added $15 billion to the United States trade balance. According to the latest C&E News, this trend continued in 1993. The size of this surplus and its growth is a direct measure of the strength of our competitive position. In addition, although it is not obvious in these statistics, the U.S. chemical industry has already extensively globalized, with over $20 billion of the 1992 export-import business involving inter-company transfers.

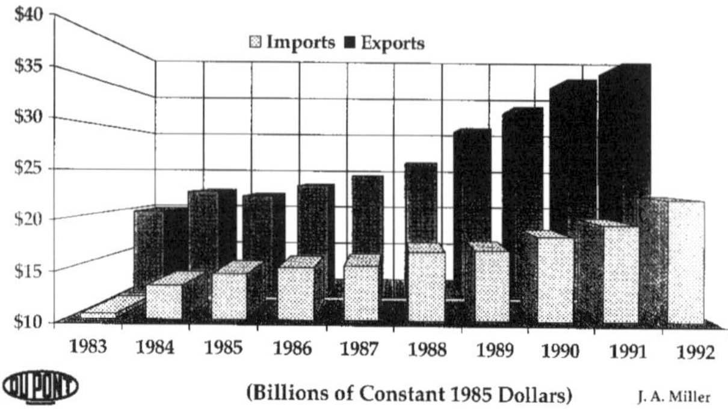

Figure 2. U.S. chemical industry imports and exports.

The chemical industry has always been and must continue to be technology driven. The process and product innovations from which the growth of the last half century has come are the direct result of aggressive development of new technology based on solid science and engineering. Since 1971, expenditures for R&D in the chemical industry have grown at a rate one and one-half times as fast as the national rate for all R&D, rising from $1.8 billion to $15 billion in 1993. During this period, the number of scientists and engineers employed in R&D more than doubled. While the most recent data show a marked flattening of this growth curve with little hiring for the last several years, this is unquestionably a real long-term success story.

However, any complacency we might have had about such a favorable situation certainly has been shaken in the last few years. Since 1989, return on net sales and on net assets has declined sharply and by 1992, these performance measures were below the lows reached in 1985. As shown in Fig. 3, net returns recovered slightly in 1993, and early 1994 data indicate that this trend is continuing. These returns are still well below levels which will sustain the industry over the long term especially when we consider that the competitive situation has become much tougher in the interim.

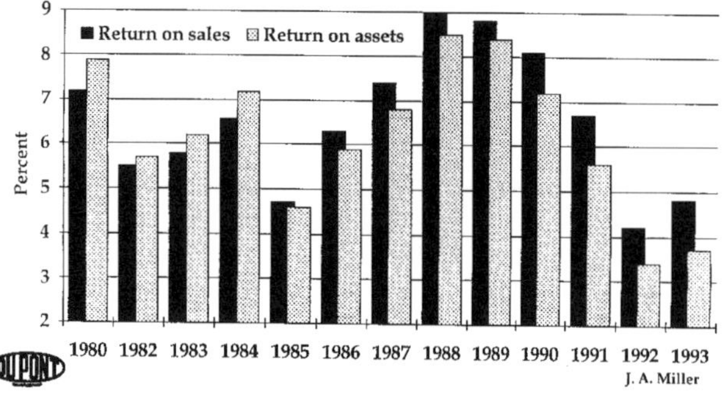

Figure 3. U.S. chemical industry margins and returns.

In the last fourteen years, our net returns have not been sufficient to allow the industry to attract the capital needed for any sustained growth. We seem to be trapped in a box defined both by major sociopolitical forces and by the way we develop and implement technology.

Our future depends on finding a way of breaking out of this box, and such a breakout I believe will require a commitment not only to the development of new technology, but also to implementing it more quickly and effectively than we do now. I am confident that, if we in industry, academia and government work together to grow and sustain fundamental research and develop the tools needed to facilitate the commercialization process, we will find a way of breaking out. The chemical industry is truly a national resource, but it will need such a commitment if we are to sustain and grow it.

With that as background let me consider in more detail the impact of these longer range forces on our industry. First, growth rates have been slowing down in the U.S. and also in other major developed countries. This removes an important demand-side driver for our products. Figure 4 shows Production Capacity Index for the U.S. Chemical Process Industries, a measure of installed manufacturing capacity. This grew steadily through the 1960s and 1970s, but levelled off dramatically in the 1980s as demand dropped.

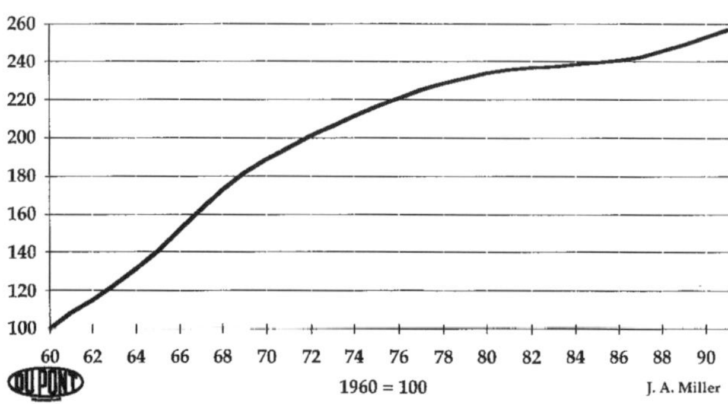

Figure 4. U.S. production capacity index.

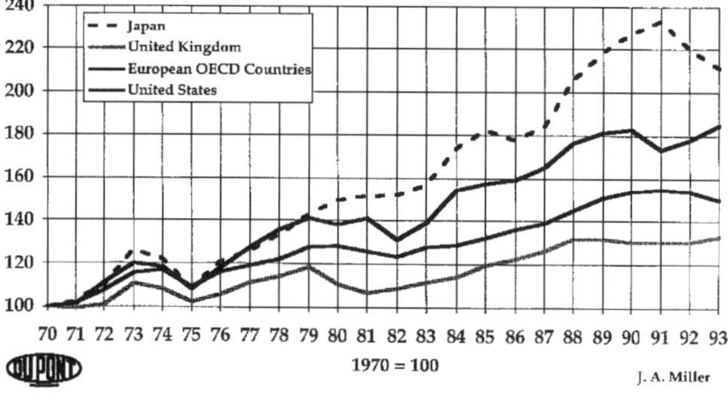

Figure 5. Growth of industrial production.

In contrast, as shown in Fig. 5, the growth rate in Japan (and also for most of Asia-Pacific) has been accelerating, exceeding that of the U.S through much of the 1980s. This provides a short-term opportunity for export growth but in the long-term will inevitably lead to increased competition for U.S. based industries.

Second, investment productivity has been declining in the U.S. both for total manufacturing and for the chemical industry, as shown in Fig. 6.

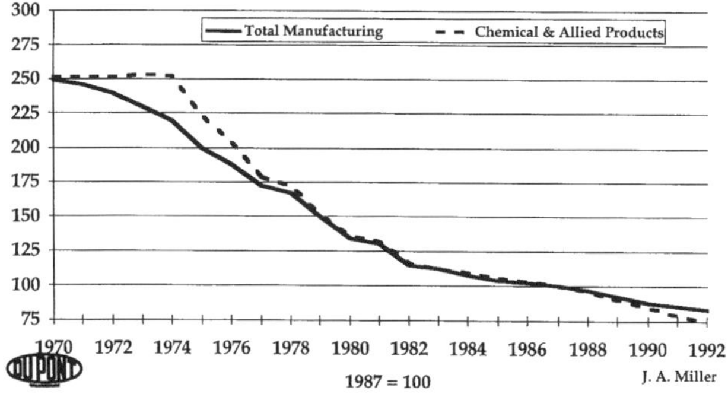

Figure 6. Capital productivity for all manufacturing and for chemicals.

In essence, the chemical industry has been increasing its capital investment at a much faster rate than it has been gaining increased revenue from this investment. In fact, since 1990, capital expenditures in the chemical industry have been about double industry profits, as shown in Fig. 7. Some of this is certainly related to investments required to meet environmental regulations, which bear more heavily on the chemical industry than other manufacturing. While such an excess of expenditures over income may be required and while they can be tolerated over the short run, they are untenable over the long run.

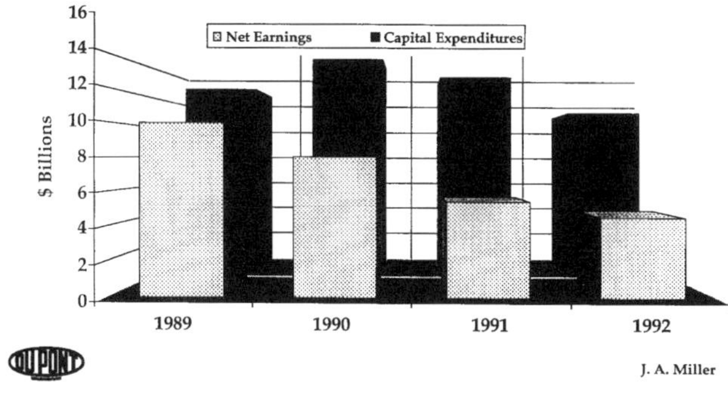

Figure 7. U.S. chemical industry net earnings and captial spending.

Third, our industry is faced with more sophisticated customers who have increased expectations in terms of higher quality, lower prices and more tailoring of the product for their specific needs. This demand to tailor products can have a significant adverse impact on the performance of existing manufacturing facilities which were typically built as very large, single-line plants. The necessity for reducing prices to stay competitive while the chemical construction price index is continuing to increase makes it difficult to impossible to justify new capital investments.

In this new environment, every chemical company has had to become more productive to survive. I will discuss primarily what we have done at DuPont, but I am sure it is representative of what has occurred in most companies.

We have narrowed our business focus, concentrating primarily on those businesses in which we can be globally competitive. That is, where we can compete and have a chance of becoming the best in the world. This has meant divesting certain businesses and acquiring others. In some cases, we have been able to swap businesses as we did in the recent ICI-DuPont nylon-for-acrylics deal. Such exchanges can turn out to be win-win propositions for both parties.

We have examined all of our internal activities, seeking to focus our efforts on those which bring value to our businesses. As a result we have stopped many activities we once considered essential and outsourced others which can be done better by other suppliers.

R&D has not been immune to these changes. Quite to the contrary, R&D has been specially scrutinized since we believe that R&D must play a key role in changes needed to keep our company competitive. The current DuPont Company with its many polymer-based businesses is a direct result of our commitment to basic research and to the development of new technology. We value research not only intrinsically, but because of what it has done in the past to create and renew our company, and how it has shaped the current chemical industry. We do not plan to walk away from research.

As a result of this scrutiny, we have reorganized R&D to eliminate organizational redundancies, thus concentrating our resources on critical business needs and strengthening our core competencies. The core competency of most interest to this audience is Process Science and Engineering, which is headed by Jim Trainham. Essential technologies in this competency cover the topics found in most chemical engineering curricula (e.g. reaction engineering, catalysis, transport phenomena, process modeling and control). We have substantially strengthened our work in Process Modeling and Process Control with a commitment to become world class in these technologies. We believe that they are critical to the success of our efforts to make significant improvements to current manufacturing processes.

We have also shifted our funding priorities to increase R&D in direct support of existing products. Our Central Science and Engineering Laboratory has been challenged to search for breakthrough opportunities, in addition to maintaining its current responsibilities for basic research.

We recognize that successful research programs are no guarantee of ultimate competitive advantage. That advantage is only achieved when the chemistry is embodied in a plant. We know that our competitors will not be standing still. To stay competitive, our pace of innovation has to increase and our target must be "best in the world" chemical plants. These plants will need to be environmentally benign, and they must have the flexibility to make the products customers want, when they want them, with the highest quality and at a price they are willing to pay.

Since the focus of this conference is design, let me share with you some specific challenges we in DuPont face in process design. While the problems are not totally separate, it is helpful to discuss the design of a new plant and the retrofit design problem separately.

First, we recognize that with current capital limitations and the need to improve our capital productivity, most of our capacity increases in the near term are going to come from the retrofits of existing processes. Many of our plants are over twenty-five years old. When they were built, energy and raw materials costs were low and there were not as many environmental constraints. Therefore, with a new objective function to optimize, new constraints, and improved process understanding, we are confident that there are large opportunities to improve existing processes.

There may be opportunities to increase capacity or reduce cost by process simplification or clever modification of the flow sheet while the chemistry remains unchanged. For example, the new Eastman Chemical methyl acetate process, which Jeff Siirola and his colleagues developed, combines the chemistry and complex separation processes into a single reactive distillation column. We would all like to have many successes like that. However, in reality there is probably a much greater opportunity to improve existing processes if, in addition to flow sheet optimization, we consider modifying some or all of the chemistry in the process. This will be especially important if we want to solve some of the environmental problems that we face.

To be successful, we believe it is imperative to get the chemist directly involved in the retrofit process. It is of little value just to give the chemist a detailed flow sheet of the process and ask for help. We need tools that will allow the chemist to visualize the process in a way that is meaningful to her or him. Good teamwork between the engineers and the chemists can have a profound effect on the outcome of such projects. Similarly, business managers need to have a still different view of the process, one which would show the financial impact of proposed changes. We have started working on some of these issues with Mike Malone from the University of Massachusetts but we do not profess to have all the answers. The retrofit problem is a significant challenge for the chemical industry and one for which adequate computer tools are not yet available.

The design of new plants is still essential to us even though we have made the decision to out-source the detailed mechanical, electrical and civil design activity. We still do all the preliminary design work and develop the process specifications that go to our full-service design partners.

In this area, it is critical that we find ways to shorten the cycle time of the total process design activity. Currently, the average industry time to go from the bench top to an on-line plant is about nine years. In that period, the market place and/or the economics for a particular product may change drastically. Yet we have cases inside DuPont, and I am sure in other companies, in which we have cut the process development time in half. We know it can be done but we haven't institutionalized the work processes throughout the corporation. One major problem continues to be the lack of a completely integrated set of computer tools which will support concurrent engineering design instead of doing almost everything serially. A second problem is that chemical plants, designed on the basis of steady-state operation, rarely run at steady state in practice and sometimes end up being very hard to control. Rather than trying to use process control to correct a poor design, we need better methods to evaluate the controllability and operability of competing process alternatives as early as possible in the design process.

Finally, our existing processes tend to be large, world-scale continuous plants that were designed to make one or a few products. Over time, the product line has increased, customers are demanding greater responsiveness, and business managers are being driven to reduce working capital. These are contradictory goals and present an excellent opportunity to apply optimization techniques, but the problems are usually more complicated. For example, it is quite common for a single business to have multiple plants located around the world with multiple warehouses and distribution centers. The critical problem for us is to be able to optimize the whole supply chain subject to inventory constraints while meeting service-level commitments to our customers. We are actively working on these types of optimization problems. When we need to build a new plant for such a business, we have to learn to do the design in the context of the whole supply chain.

Before I turn to recommendations, let me repeat the key challenges we face. Aggressive global competition and increased customer expectations combined with rising societal demands for environmentally benign plants will require new and better chemical processes for current and future products. With costs increasing and prices decreasing, productivity must increase to justify the required capital investments.

These productivity improvements can only be obtained with continuing advances in technology and through their more effective implementation. This is central to maintaining the health and competitiveness of the chemical enterprise. It is also closely coupled to improving our national competitiveness, a problem which extends beyond DuPont or any one company or any one organization.

What specifically then must we do to deal with this?

First, we must have a strong science base.

The technology supply chain is ultimately anchored in good science. Like any supply chain, all parts are necessary and must be healthy and interacting. If we weaken one part, we weaken the entire system.

Our universities have always been the major source of this knowledge and have given the United States a strong leadership position, but federal support for science in the universities is decreasing. This is the wrong action at this time, and for that matter at any time. The National Science Foundation has long sustained American Universities in their vital part of this work. Today there are significant pressures to change the well established NSF practice of supporting outstanding research in a wide variety of fields, a practice which recognizes the value of discipline driven research. This ignores the fact that there are already existing federal organizations, such as the National Institute of Standards and Technology (NIST) and the many national laboratories which are better suited to working on strategic technological issues.

It is imperative, now more than ever, that we work together to ensure that federal policy is committed to support fundamental, comprehensive, science and engineering research at our universities and national laboratories. This would both supplement and complement industrial research to the benefit of all. To capture this benefit, each of us must work to build stronger more effective partnerships between industry and academia, a point I will return to shortly.

Second, we must ensure a continuing supply of top quality engineers a and scientists.

Universities are the main source of new knowledge and, more importantly from industry's perspective, the source of high quality engineers and scientists.

We must ensure a continuing supply of people capable of understanding and contributing to scientific understanding and technology development. I am certainly aware that there are many concerns among engineers about future job prospects. While the last few years have been difficult, I expect that recruitment will increase next year. As I look at the future, I can confidently say that we will always need chemical engineers in DuPont, not only to run our plants, but to strengthen our core competencies and to manage our businesses.

Moreover, if we want to accelerate the innovation process, all engineers will need a broader range of skills than in the past, including a greater working knowledge of chemistry, the ability to work in teams and to communicate more effectively with chemists and biochemists. Engineers must be able to understand and work effectively with chemists; chemists must be able to understand and work effectively with engineers!

Third, we need to improve the effectiveness of our partnerships.

Collaboration in science and engineering must be designed both to complement and supplement industrial research in a way that benefits all. It will require all of us to work to build stronger, more effective partnerships. These partnerships must be based on a deeper understanding of each other's needs and strengths, and unite our efforts to bring understanding and common sense to our business processes and to the political processes in Washington.

Although education must stay as the prime objective of universities, the times and emerging national policies call for an expanded role in working with industry.

Universities and government laboratories are now facing the same market forces which industry has had to face in the past decade. Funding for research will be constrained, there will be fewer jobs for graduates and increased competition from foreign universities to supply the people and the services they currently offer, and corporate support for research will be more selective and come with more expectations. This is certainly true in DuPont where we are seeking to focus our research grants on building long term relations with fewer institutions. Universities will need to work harder at understanding their customer's needs and define how best to contribute to them.

The government also has an important role. The government labs, with their combination of basic research and outstanding engineering capabilities are well situated to contribute to pre-competitive research and technology. For example, under the leadership of Hazel O'Leary, industry is now actively involved with DOE in seeking more efficient uses of energy. We need more examples such as this.

Finally, we need to strengthen our K-12 education system.

Here I speak in my role as co-chair of Delaware's Science Frameworks Commission charged with establishing world-class science standards for our students. It was only through this work that I really learned how much we need to do to improve our schools.

While we have the best graduate and undergraduate education in the world, we have major weaknesses in K–12, especially in math, science and technology — all fields critical to our long-term scientific and engineering capabilities. K–12 students form an important part of the technology supply chain. As citizens and as members of the work force, they will have a significant role in our future competitiveness and in setting national science and engineering policy. Yet many of these students today are getting a marginal science and technology education. Many will be unprepared to fill jobs in an industrial society that is increasingly technology based. Even such traditional "blue collar" jobs as automotive assembly work now require people with more advanced academic preparation.

Never has the need been more urgent to improve the technological and scientific literacy of the public and our students, to strengthen their basic skills in mathematics, to develop their capabilities for thinking critically, and to inculcate in them the desire to learn continuously — for the rest of their lives. I personally would like to see more people who really understand and practice technology participating in solving this problem.

I know this conference deals primarily with design. In my talk today, I have asked you to consider the entire technology development supply chain from pre-college education to fundamental research through process design to commercialization. The chemical engineering design community has a critical role to play in improving the competitiveness of the chemical industry — to help it break out of the box. Each of us have a major stake in the outcome of this process. We must succeed.

SEPARATION SYSTEM SYNTHESIS AND DESIGN: INTRODUCTORY REMARKS

Warren D. Seider
University of Pennsylvania
Philadelphia, PA 19104-6393

Introduction

There are three topics, emphasized to varying extents, in the three papers presented in this session. These are process synthesis, process modeling and process algorithms. To introduce the session, in each of these areas, I identified several typical concerns and questions that the audience could expect the speakers to address.

Indeed, many of these concerns and questions were addressed during the presentations and the discussion that followed. As you read the papers, I recommend that you keep these issues in mind. For each area, Tables 1, 2 and 3 provide my lists of concerns and associated questions.

Table 1. Separation Process Synthesis.

Concerns	Questions
Role of experimentation	For non-ideal systems, is this needed during process synthesis?
Analysis and synthesis	For non-ideal systems, can these activities be separated?
Residue-curve maps and geometric methods	How useful are these approaches?
Liquid-phase splitting	Is this advantageous or disadvantageous in distillation operations?
Chemical reactions	How do they impact the analysis and synthesis strategies?
Batch distillation	Are batch strippers widely used? What is the role of optimal control?
Membranes and pressure-swing adsorption for bulk-gas separations	At what scale do these become advantageous?
Pressure-swing adsorption	How are the operating cycles selected and optimized?
Entrainers, membranes and adsorbents	How are these selected?

Table 2. Separation Process Modeling.

Concerns	Questions
Equilibrium-stage models	Are they adequate or are mass transfer models gaining importance?
Mass transfer models for multicomponent systems	How difficult is it to master multicomponent diffusion and mass transfer?
Tray hydraulics — flooding, entrainment, weeping	When are these important in dynamic models? Can they be accounted for properly?
Steady-state multiplicity	Is this an important issue in separation processes?
For reactive separations, transformations permit the usage of models without reactions	What are the limitations of this approach?
Experimental data for multistaged towers	Can this be made available to model and algorithm developers? Need industrial cooperation.
Membrane modules	Are permeabilities predicted well for multicomponent mixtures?
PSA models	When are equilibrium-based models appropriate? ... kinetically controlled models? How important is diffusion?

Table 3. Separation Process Algorithms.

Concerns	Questions
Local convergence methods	When do global convergence strategies gain importance?
Simulations in the complex domain	Will these mitigate the problems at phase boundaries?
For PSA models, PDEs are often solved	What are the preferred methods? Is stiffness an issue?
Some specifications lead to higher-index DAEs	Are the generalized integrators able to handle these systems?
Sensitive movements of steep fronts	Are specialized methods needed to track their movements?
For multiphase systems, stability analysis is needed	Can tangent-plane distance algorithms be improved?
Solution diagrams are often needed as parameters change	Are generalized systems, such as AUTO, effective for large systems?

SEPARATION SYSTEM SYNTHESIS FOR NONIDEAL LIQUID MIXTURES

M. F. Malone and M. F. Doherty
Department of Chemical Engineering
University of Massachusetts
Amherst, MA 01003-3110

Abstract

Computer aids for process synthesis and design which give both quantitative results and intuition are valuable. This paper describes such tools, currently developed for continuous distillation systems, as well as some examples based on similar ideas for other separation technologies applicable to nonideal liquid mixtures.

Keywords

Process synthesis, Conceptual design, Separation systems, Azeotrope, Azeotropic distillation, Extractive distillation, Reactive distillation, Catalytic distillation, Extraction.

Introduction

Many separation systems have been invented by experience, leaving open the questions of what potential there may be for improving existing separation systems as well as what methodology to use in efforts to devise new and improved systems. Furthermore, new technologies, along with environmental and economic conditions present problems for which there is a relatively small base of experience to draw upon. The recent development of systematic design procedures, especially for nonideal mixtures and startling decreases in the cost of computing for their implementation can be used to address these questions.

Some may have the impression that the sole purpose of process synthesis is to provide the optimal specification of equipment based on known physical properties and specifications. We believe otherwise — that the major benefits from computing tools come when they are used in conjunction with experimental studies. For separations, this is partly because there remains a significant uncertainty in thermodynamic predictions made for very nonideal systems, despite the availability of large databases for vapor-liquid and liquid-liquid equilibrium. The data available for reacting mixtures is even more sparse. Furthermore, synthesis results for separation systems are often coupled to effects on a higher level, connected to the basic chemistry and reactor systems that give rise to the separations.

We would also like to keep in mind not only the difference between analysis and synthesis but also the fact that they are complementary activities. Synthesis relies on results from analysis and much of the analysis is fueled by alternatives generated in the synthesis stage. Early work on process synthesis gives the impression that these two activities can be separated, but our experience suggests the contrary. Figure 1 portrays the idea that we need both synthesis and analysis for conceptual design.

A critical issue in any synthesis exercise is the number of alternative solutions. There may be many alternatives and there is no general rule to choose among them. Many may have comparable costs, or all but a few may be economically unattractive so some analysis for the ranking of alternatives is useful. Fortunately, this analysis need not be a rigorous design. When the differences among alternatives are small, high accuracy is required to identify the true optimum, but the choice among neighboring alternatives is unimportant because the differences are small. Conversely, larger differences among the alternatives mean that less accurate models will not lead to bad decisions. In other words, models need to be sufficiently accurate to discard poor alternatives and yield a smaller number of candidate designs worthy of further attention. A well-known example is the large number of alternative distillation sequences for ideal mixtures, but for very nonideal mixtures many of these alternative sequences are infeasible. This does not necessarily mean that there are no alternatives, but that more work is required to find feasible configurations.

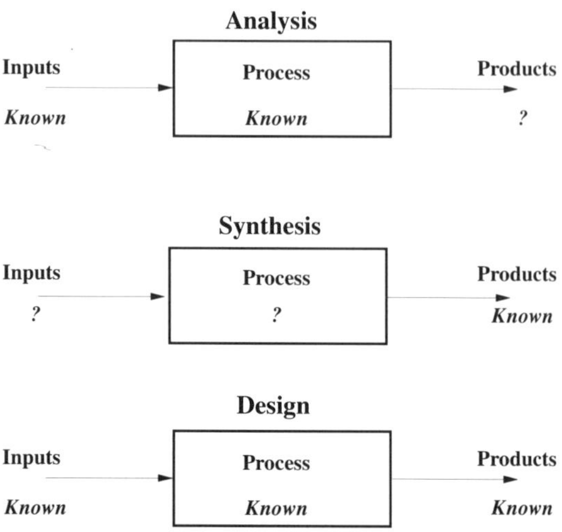

Figure 1. Synthesis + Analysis = Design.

Our initial focus is on continuous distillation processes for strongly nonideal mixtures, an area where useful tools for computer-aided conceptual design are now appearing. Extension of these ideas to other technologies such as liquid extraction and reactive distillation are discussed briefly later in the paper.

Computer-Aided Design Tools

During the last five years there has been an emergence of computer aids for flowsheet synthesis and for design for liquid separation systems. Barnicki and Fair (1990) developed a knowledge-based expert system for synthesizing commercially important methods for the separation of liquid mixtures. The procedure is hierarchical in nature and uses a rank-ordered set of heuristics for making decisions. In its original form, the expert system was not intended for treating highly nonideal systems such as azeotropic distillation, although research is continuing in this direction. More recently, Westerberg and co-workers at Carnegie Mellon University have embarked on an ambitious project to develop an automated synthesis procedure for hybrid liquid separation systems (called SPLIT) with the specific intention of treating highly nonideal liquid mixtures (Wahnschafft et al. 1991; Wahnschafft et al. 1992b; Wahnschafft et al. 1993). A decomposition approach is adopted in which streams are considered sequentially and the separation method is chosen before determining which splits are feasible and before selecting the operating conditions. The approach is capable of introducing recycles, and arrives at some interesting flowsheets for mixtures of industrial complexity. During the same period we have developed a computer aid (called Mayflower) that has the capability to check feasibility and to design individual separation units for quite complex mixtures. Separation system synthesis in Mayflower is currently user-driven as described below.

None of the above methods represents a complete solution for the conceptual design of separation systems for multicomponent liquid mixtures in which azeotropes, tangent pinches, and liquid immiscibilities are present in any combination. However, progress is being made rapidly and we can expect useful design tools to be commercially available in the next few years.

Approach

We describe a hybrid procedure that combines heuristics and models for conceptual design. The models are typically nonlinear and make use of *geometric methods* to find solutions. The models are intended to preserve the essential nonlinearities and to address synthesis as well as analysis. On the other hand, for conceptual design, a complete first principles model is often too demanding of data, engineering time, and perhaps computational resources so we will also use some heuristics to avoid expending resources to solve hard problems which are either not worth solving or which typically have good solutions that are not very different from one another. Our approach consists of the following steps.

1. Specifications

Choose a pressure, from available utility levels, reaction conditions, or constraints. Determine the feed composition(s) available and the product purities desired[1].

2. Feasibility

This requires *VLE* information (often forcing experiments) and consists of two parts.

a. Construct *residue curve maps* by solving for the phase plane portrait of

$$\dot{x} = x - y \tag{1}$$

where the liquid and vapor compositions, x and y are related by a *VLE* model. If the *VLE* model is incomplete, sketch the residue curves based on boiling temperatures and compositions (Foucher et al. 1991). If the data are not sufficient for a sketch, then do an experiment. Alternatively, use a guess or a group contribution method, *then* do an

[1] The pressure is a candidate for optimization, provided a reliable properties model is available for the range in question. Note that pressure changes can alter the number of azeotropes and other important features of the vapor-liquid equilibrium behavior.

experiment! Use the following heuristics and rules for sequencing.

i. The compositions of the desired products should lie in the same distillation region (Foucher et al. 1991; Laroche et al. 1992).
ii. Unstable nodes can be taken as distillate.
iii. Stable nodes can be taken as bottoms streams.
iv. Feeds and product flows and compositions must satisfy overall material balance e.g. as illustrated in Figure 3.
v. Recycle alone should not be used to cross distillation boundaries.
vi. Distillation boundaries can be crossed by mixing or, if the mixture forms multiple liquid phases, by decanting.
vii. Distillation boundaries can be crossed by reactions if the stoichiometry permits.

b. For each feasible alternative, estimate the particular range of product compositions available for simple columns (Wahnschafft et al. 1992a; Fidkowski et al. 1993a).

If no alternatives are feasible, we adjust the pressure or add another component as an entrainer or "mass separating agent" and re-start the feasibility step[2].

3. Flows and Theoretical Stages

a. Estimate the minimum reflux and, for extractive distillation, the minimum entrainer flow.
b. Find the number of theoretical stages, using the heuristics.
   i. The reflux should be 50% greater than the minimum.
   ii. The entrainer flow should be two to four times greater than the minimum (Knapp and Doherty, 1994).
   iii. In extractive distillation, the recycle purity of entrainer should be midway between the maximum and minimum values. The maximum value is either 100% purity or the composition of an azeotrope if one is used as the entrainer. The minimum value is determined from the amount of light impurity whose presence in the distillate from the azeotropic column would just meet the overhead purity specification.
c. Unspecified compositions should be 99.5% of the maximum fractional recovery.

[2] This procedure is not useful to choose candidate entrainers, but it is useful to screen candidates once they are selected. A useful short list of candidates can often be developed from components already present in the process. Computer-Aided Molecular Design Approaches for the selection of solvents have also been proposed, e.g. Pretel et al. (1994).

Often, alternatives can be eliminated at this point if the stage or vapor boilup requirements are excessive.

4. Cost Estimates

For comparison purposes among alternatives, first estimates are based on

a. A 50% stage efficiency and 0.5 m tray spacing for the column height.
b. A vapor rate at 60% of flooding.
c. Average overall heat transfer coefficients.
d. A cost correlation for the equipment, e.g. Guthrie, 1974.
e. Utility costs from the flows and site-specific information.

For the least expensive alternative and those with comparable costs, refine the design estimates and optimize pressure, reflux ratio, etc. The meaning of "comparable" depends on the accuracy of the models, but within 25% is not atypical.

5. Complex Columns

Sidestream columns are the most common, followed by sidestream strippers or reboilers. These configurations often lead to significant savings when volatility differences are large, or when sidestream purity requirements are not extreme. Relatively ideal mixtures can be handled easily (Tedder and Rudd, 1978; Glinos and Malone, 1988), but little is known about mixtures containing azeotropes.

There are certainly questions about the operability and control of complex columns, but our intent here is first to estimate the incentive to answer these questions. We believe that it is feasible to reach a side stream purity close to a saddle (middle-boiler) in the same distillation region as the other products. In many cases this will not be economical but if the sidestream purity requirements are not extreme, e.g. the sidestream is recycled, then the savings may be quite significant.

6. Sensitivity

It is particularly important to study sensitivities. These may be in the form of disturbances in process variables or parameters in physical property models.

Notes on the Approach

1. For feasibility, we ask for a picture suited to design, that also reflects the basic phase equilibrium information. For distillation, this is the well-known *residue curve map*; this is simply a picture of the composition changes in a simple open evaporation or in a continuous column at infinite reflux (Foucher et al. 1991). These maps have been well-studied theoretically (Doherty and Perkins, 1979; Rev, 1992) and have been measured for a few mixtures (Yamakita et al. 1983; Bushmakin

and Kish, 1957). The benefits of this visual representation should not be underestimated.

2. It is possible to violate Heuristic i and v when the distillation boundaries have large curvature (Levy, 1985; Laroche et al. 1992; Wahnschafft et al. 1992a) and such cases *can* be added to the list of alternatives. However, such schemes seem to be quite sensitive, e.g. to *VLE* and other process parameters.
3. Residue curves for three and four-component mixtures are straightforward to construct when a model is available. A more qualitative, but surprisingly useful, picture can often be deduced when only limited data on the boiling temperatures and approximate compositions of azeotropes is available (Foucher et al. 1991).
4. Even in mixtures with more components, it is quite useful to consider subsets of the components for feasibility. With five or more components, the full picture cannot be drawn, although it is straightforward to determine when particular compositions lie in the same distillation region. This can be done by integrating the residue curve equations forwards and backwards, starting from the compositions in question.
5. The minimum reflux can be estimated for nonideal and azeotropic mixtures quite closely from the "zero-volume" geometric construction (Julka and Doherty, 1990; Fidkowski et al. 1991).

Example 1: A Continuous Distillation System

A residue curve map is shown in Figure 2 for a mixture of diethoxymethane, water and ethanol. A ternary azeotrope and three binary azeotropes give rise to three distillation regions, each containing one of the pure components. There is a region of limited miscibility and, except for the fact the ternary azeotrope is outside of this region, traditional heterogeneous azeotropic distillation would be feasible. A sequence which removes stable nodes as bottoms products and which crosses distillation boundaries by mixing and decanting is shown in Figure 3. The addition of water moves the overall decanter composition inside the shaded triangle. The precise position depends on the amount of water added and the figure corresponds to an overall composition on the tie-line e-h. This sequence has not been optimized!

There are actually several other feasible sequences, two of which have been patented (US 4,740,273, 1988). The preferred sequence from this patent is reproduced in Figure 4; a distillation boundary is crossed by reacting part of the mixture to obtain a feed in the distillation region containing water. Note that there are at least three sequences for this separation in addition to the traditional sequence using hexane as an entrainer; feasibility alone is insufficient for making decisions.

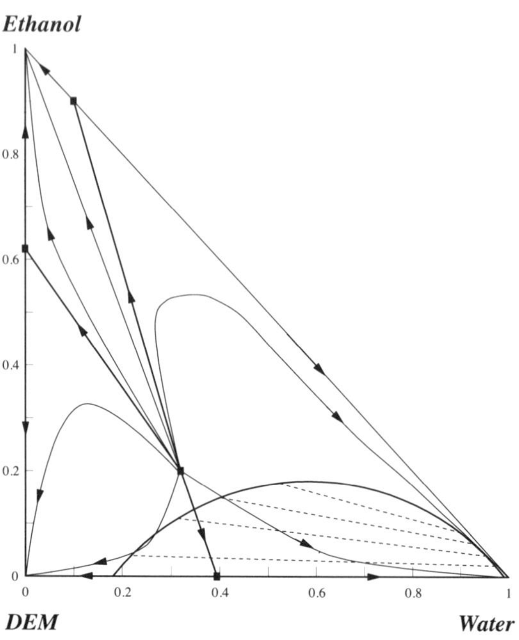

Figure 2. Residue curve map (schematic) for Example 1 (From US Patent 4,740,273).

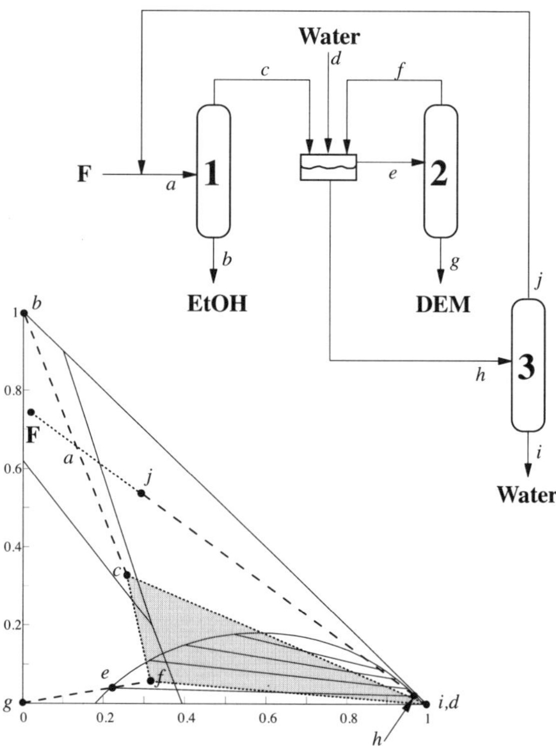

Figure 3. A sequence for separating the mixture ethanol, water and diethoxymethane at 1 atm.

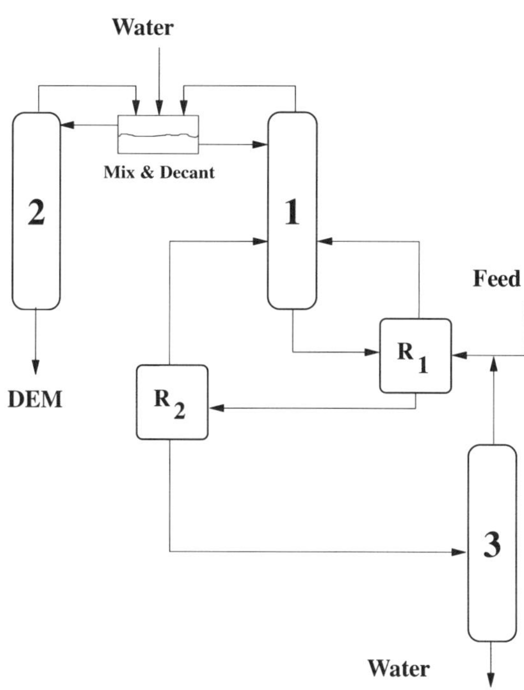

Figure 4. Alternative patented sequence for DEM recovery (US 4,740,273, 1988).

Example 2: Computing Azeotropes

Geometric methods are also useful to understand mixtures containing more components, although more work certainly remains to be done. An important consideration is the extent of non ideality and the first question is the number and the nature (stable node, unstable node or saddle) of any azeotropes predicted by a VLE model.

We have devised a homotopy continuation method to track azeotropes in any number of dimensions (Fidkowski et al. 1993b). The idea is to begin with a (purely artificial) ideal model and gradually deform this into one describing the real mixture, viz.,)

$$\tilde{y} = (1 - t)y^{ideal} + y^{real} \tag{2}$$

When the homotopy parameter, t, is zero the only solutions to Eq. (2) are the pure components. As t is increased, bifurcations occur in azeotropic mixtures. The number of branches at t = 1 beyond those for the pure components give the number of azeotropes. With Eq. (2) and for "j-g" models, the solution branches are connected, making the approach robust. Arc-length continuation (Keller, 1977) is effective for finding solutions and the eigenvalues of the Jacobian give the stability of the azeotropes and pure components. Figure 5 gives an example.

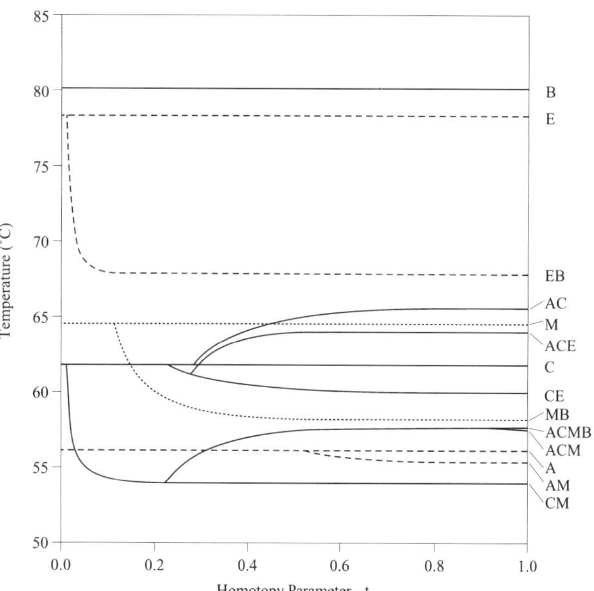

Figure 5. Azeotrope calculation for acetone (A), chloroform (C), methanol (M), ethanol (E) and benzene (B) at 1 atm (After Fidkowski et al. 1993b, Fig. 16).

Results like this can be used for feasibility (Baburina and Platonov, 1987; Baburina and Platonov, 1990) and as a starting point for tracking fixed points (pinches) which can be used to estimate minimum reflux and design (Julka and Doherty; 1990, 1993).

Batch Distillation

The product distributions and maximum amounts of each cut in batch distillation can also be studied with these geometric methods (Bernot et al. 1990). It turns out that several configurations beyond the usual "batch rectifier" are useful. One of these is the "inverted" configuration or "batch stripper" — a configuration known in the literature for a long while (Robinson and Gilliland, 1950). This configuration may be useful to recover certain components from azeotropic mixtures which cannot be produced in a batch rectifier (Bernot et al. 1991). More complex configurations, such as combinations of the rectifier and stripper, can also be analyzed with a geometric approach. (Davydian et al. 1994, for constant volatility mixtures).

Example 3: A Batch Distillation Sequence

Figure 6 shows the simple distillation residue curves for a mixture of methyl acetate, methanol, ethyl acetate and ethanol. These components appear in the product stream from a reactor in which the following liquid-phase transesterification reaction occurs.

$$\text{MeOH} + \text{EtOAc} \Leftrightarrow \text{EtOH} + \text{MeOAc} \tag{3}$$

There are five distillation regions, all of which intersect the stoichiometric plane, also shown in Figure 6. The stoichiometric plane is simply the mass balance constraint that results from Eq. 3 (c.f., Bernot et al. 1991).

There are many alternatives, but the simplest separations arise if the ratio of methanol to ethyl acetate in the reactor feed is kept above the ratio in the MeOH/EtOH binary azeotrope, resulting in a product stream in distillation region 5 which yields product cuts consisting of the pure alcohols along with the MeOH/EtOH and MeOH/MeOAc binary azeotropes.

Two possible alternatives are shown in Fig. 7. In the first alternative, a batch rectifier first removes the MeOAc/Me azeotrope which is the unstable node in region 5. Methanol and the MeOH/EtOAc azeotrope are saddles or middle-boilers and are produced in the next two cuts for recycle (these cuts need not be sharp). Ethanol is the stable node in region 5 and is either left in the pot or taken overhead. The second column is a batch stripper where methyl formate is used as an entrainer for batch extractive distillation to break the MeOAc/MeOH azeotrope.

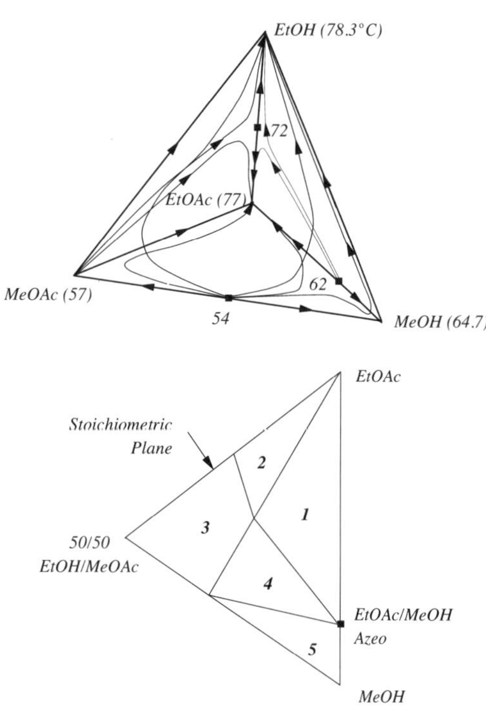

Figure 6. Residue curve map for methanol, methyl acetate, ethanol and ethyl acetate at 1 atm pressure (top) and stoichiometric plane (bottom) showing the five distillation regions.

This choice of entrainer was not optimized and it is worth noting that this entrainer actually exploits the effects of a curved distillation boundary as discussed by Bernot et al. 1991. The heuristics for continuous systems imply that these effects are not generally useful to devise a separation sequence, although this can occasionally be a good strategy. In batch systems, the effects of curvature in the distillation boundaries are more difficult to avoid, because the composition of the batch must change with time. Consequently, the behavior of at least parts of some batch distillation residue curves and the corresponding product compositions can be governed by the distillation boundary. In fact, a more detailed knowledge of the curvature may be valuable in batch distillation because the effects can be important even if they are not exploited to devise the separation scheme. For example, a strongly curved distillation boundary may mean that certain product fractions cannot be obtained in high purity no matter how large the reflux ratio and number of stages are chosen.

An alternative operates both columns alternately as a batch rectifier and then a batch stripper. The first cut in column 1 is the same as in the first alternative, but ethanol is removed in a second cut by operating as a batch stripper. All of the ethyl acetate and a significant amount of the methanol is left for recycle. We know of no columns built and operated in this manner, but the savings in time and energy could be significant.

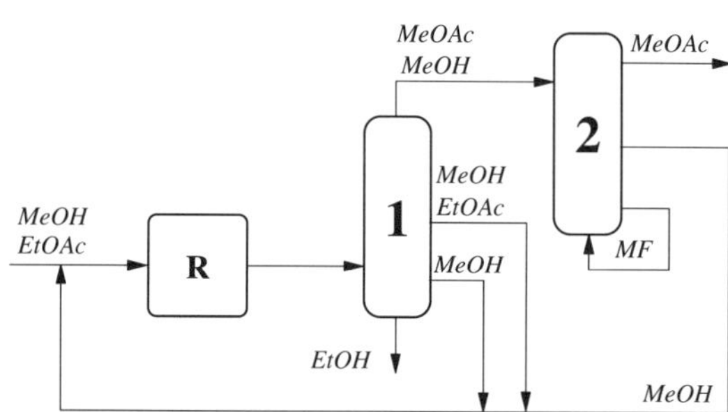

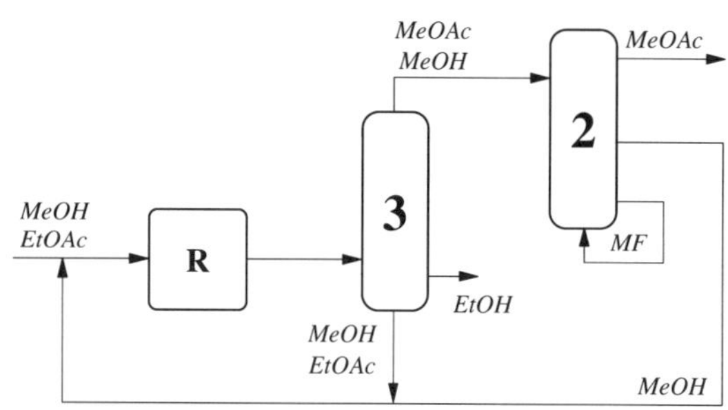

Figure 7. Alternatives for batch reaction and separation of ethanol and methyl acetate from methanol and ethyl acetate. The first alternative (top) uses a batch rectifier (1) and stripper (2); the second alternative (bottom) uses two units (3 and 4) that are operated as strippers or rectifiers.

This is another example of a nonideal separation with several alternatives; no one is obviously superior and some analysis is needed in order to rank them. With a bit more work, the reflux ratio policy for the batch rectifier and the reboil policy for the batch stripper as well as some costs can be estimated (Bernot et al. 1993).

In both alternatives, we seek a product from the reactor with a composition that is in a particular distillation region. While separations can be done in other regions, they are almost certainly more expensive. For example, if an equimolar mixture of methanol and ethyl acetate is fed to the reactor, the product may have a composition in a region where a second extractive distillation is needed (Bernot et al. 1991). Once again, we can think of the proper choice of reactor exit composition as "crossing" a distillation boundary. It would be extremely useful to integrate the geometric methods recently developed for assessing "attainable" reactor product compositions (Hildebrandt et al. 1990) with the geometric methods described here for assessing distillation feasibility.

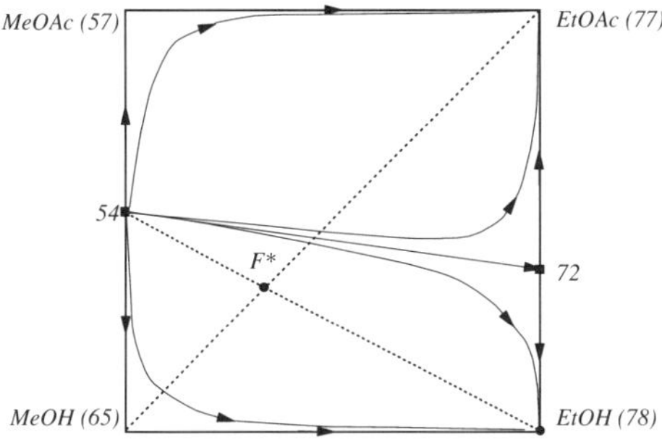

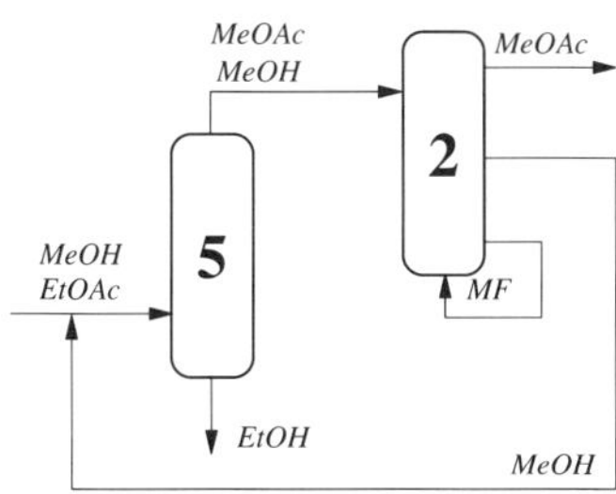

Figure 8. Residue curves (schematic) and sequence for reactive distillation in Example 4.

Example 4: Batch Reactive Distillation

The mixture in Example 3 is an obvious candidate for reactive distillation. Figure 8 shows a schematic of the residue curves for reactive distillation which be made from a knowledge of boiling temperatures along with two assumptions: (i) rapid reaction and (ii) no reactive azeotropes. Lack of accuracy in either assumption simply makes reactive distillation less attractive and investigation of either one requires experimental data. Two of the binary azeotropes survive, but the other two are not present. The low-boiling binary azeotrope between MeOH and MeOAc is an unstable node and is connected to the intermediate-boiling saddle azeotrope between EtOH and EtOAc. This gives rise to a distillation boundary dividing the alcohols from the acetates. Feeds of EtOAc and MeOH have compositions on the straight line connecting the upper right and lower left corners of the figure, and feeds on this line and in the lower region can produce high-purity ethanol as a bottoms product. If the average feed composition is at F^*, the distillate can be the MeOH/MeOAc azeotrope. A feed along the line joining F^* to the MeOH vertex gives a distillate with a higher concentration of MeOH than in the azeotrope. The optimum value will balance costs for MeOH recycle against the costs of the reactive distillation. Fig. 8 shows a sequence, which is similar to the second alternative for Example 3, but the reactor and first distillation have been combined. There is only one recycle stream of MeOH from the extractive distillation.

Liquid Extraction

Liquid extraction in combination with distillation can accomplish difficult separations, overcoming azeotropes and tangent pinches. The classic example is breaking the tangent pinch in the acetic acid-water mixture. Liquid extraction is used to isolate the pure component closest to the tangent pinch (water), and distillation us used to recover the other pure component (acetic acid) as well as the solvent for recycle. The structure of the distillation system depends on the choice of solvent and the resulting sequences can be very different than those encountered in more conventional distillation systems. There are often many alternatives and these can be synthesized, designed and ranked by the methods described earlier. Conceptual design techniques have not been completely worked out for combined extraction/distillation systems and the main research issues may be highlighted by means of an example.

Example 5: An Extraction/Distillation System

Consider chloroform as the solvent for separating acetic acid and water. This solvent is chosen deliberately because it is not used commercially and therefore our discussion does not infringe on any proprietary process. The liquid-liquid phase diagram for the ternary mixture at 25 °C is shown in Fig. 9 based on predictions made by the UNIQUAC equation with parameters taken from Sorenson and Arlt, 1980, p. 18. The liquid-liquid envelope is very close to the pure water vertex making it possible, in principle, to obtain high purity water in the raffinate stream from a liquid extraction column.

If the process feed contains 80 mol% water and 20 mol% acetic acid and the solvent feed is a mixture of 95.8 mol% chloroform and 4.2 mol% water (we will see why later), the first step in the design is to pick a ratio of solvent to feed for the extractor. This value must be above the minimum which can only be calculated using current equation-based methods if the pinch in the extractor is at one end or the other. However, for many systems of practical interest, a tangent pinch away from the ends of the extractor controls the minimum solvent rate. It is an open research problem to devise equation-based methods for detecting and calculating such points. In this example, the minimum solvent-to-feed ratio is 0.58 and is controlled by a tangent pinch. If we take an operating value higher than the minimum, say (S/F) = 1, an overall material balance gives the composition of the extract phase leaving the extractor (see Fig. 9). The extract phase is fed to a distillation system to produce the acetic acid product and recover the chloroform. The residue curves are shown in Fig. 10; this is a classical extractive map overlaid with a heterogeneous liquid region. That is, the mixture has a single unstable node (the minimum-boiling heterogeneous binary azeotrope) and a single stable node (acetic acid); the other two pure components are saddles.

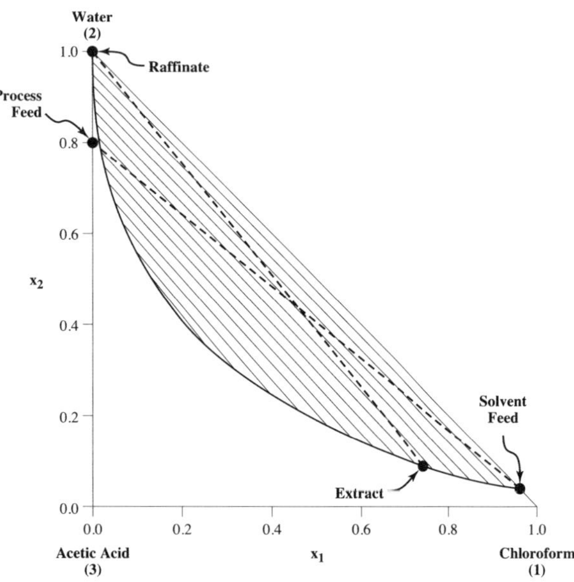

Figure 9. Liquid-liquid equilibrium diagram for acetic acid, water and chloroform at 25 °C, showing the compositions of feed and exit streams for the extractor.

Various distillation alternatives may be contemplated and the feasible regions for a single-feed, two-product column are shown in Fig. 11 (Wahnschafft et al. 1992a; Fidkowski et al.1993a).

One alternative is to take acetic acid as the bottom product from the distillation column and a mixture of water and chloroform as the distillate (Fig. 12). This arrangement requires condensation and separation of the overhead vapor into (well-mixed) two-phase reflux and distillate streams. The distillate is decanted into an aqueous product and a chloroform-rich solvent stream that is recycled to the extractor. The final sequence is shown in Fig. 13. This sequence is not the normal split associated with the extractive map, although the split shown here is a natural one since the bottom product is a stable node. The split shown in Fig. 12 can be achieved easily with quite modest conditions. The minimum reflux ratio is 0.13 and, with an operating value of 0.2, we find 13 theoretical stages; the column profile and remaining design parameters are as shown in the figure.

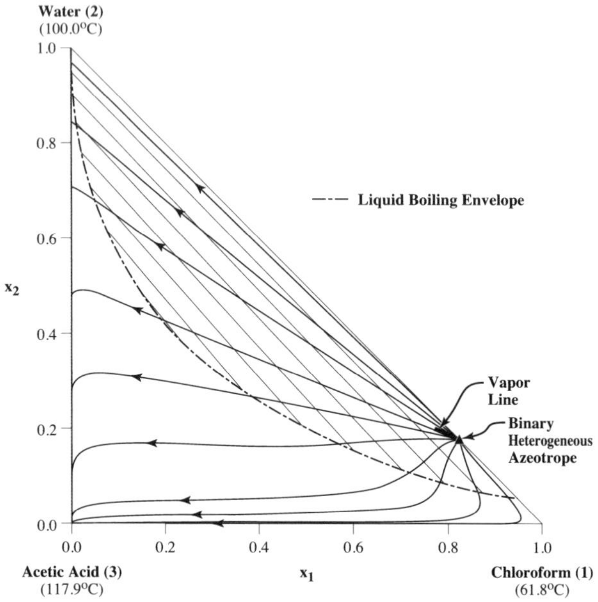

Figure 10. Residue curves at 1 atm pressure with the liquid-liquid-vapor boiling diagram overlaid.

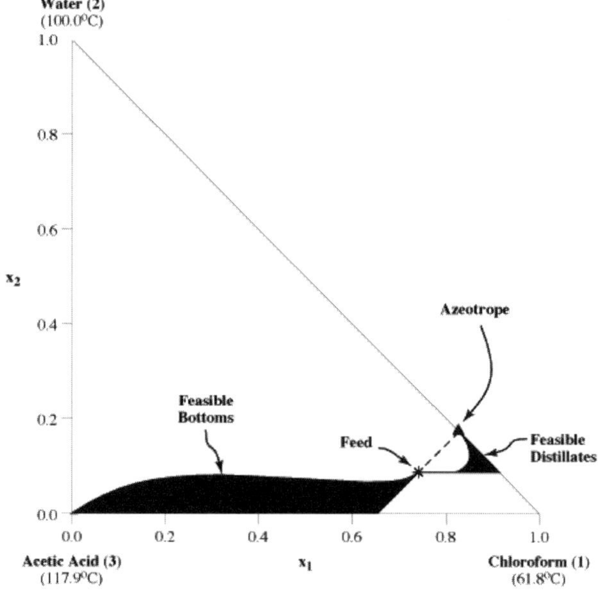

Figure 11. Product regions for Example 5.

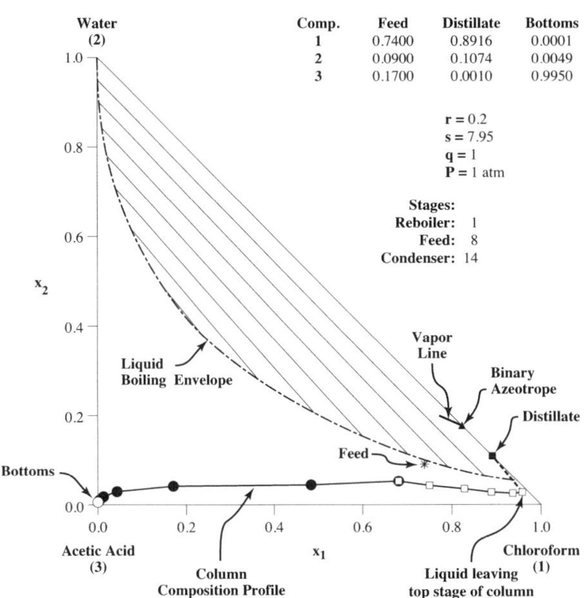

Figure 12. Liquid composition profile in the distillation column for Example 5.

As an alternative it may also be possible to feed the extract phase to an extractive distillation column. This would require a second feed to the distillation column consisting of pure acetic acid. The distillate is expected to be chloroform (which is recycled to the extractor) and the bottoms stream will be a binary mixture of acetic acid and water. This stream would be distilled into acetic acid product and a water-rich stream for recycle to the extractor feed. Without further feasibility studies it is not clear whether this alternative is viable. Such studies would consist of (i) checking the separation regions for the extractive column, (ii) finding the minimum entrainer-to-feed ratio for the extractive distillation column, (iii) checking whether chloroform or water will be the distillate from this column, and (iv) configuring the distillation sequence in accord with these results.

Methods have been developed for solving each of these steps for homogeneous mixtures but it not clear whether they are complete when multiple liquid phases occur.

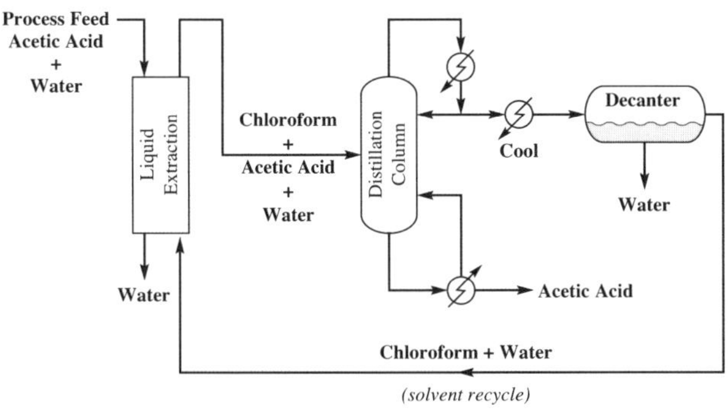

Figure 13. Extraction/distillation system for separating acetic acid and water.

Conclusions

Ten years ago there was very little information available on the systematic design of separation systems involving nonideal liquid mixtures. Since 1985 rapid progress has been made on this subject and it is now possible to invent and design such systems for interesting classes of mixtures. During the next five years we can expect to see progress on the following fronts:

- Development of computer aids for the conceptual design of separation systems for nonideal liquid mixtures without chemical reactions. This includes automated synthesis capability, user-driven synthesis capability, feasibility checking, design calculations, economic evaluation of alternatives and suggested experiments to verify the critical steps.
- Development of the basic methods and technology needed to incorporate chemical reactions into the separation system/total flowsheet synthesis problem. This area has enormous economic potential for inventing new classes of flowsheets where the chemistry and the separations are carried out in a much more integrated fashion than is currently thought possible.
- Education of the next generation of chemical engineers who are capable of understanding and using these tools once they leave the research stage of development. This is already occurring at some universities and with the advancement of new educational materials (e.g. textbooks, monographs, software, faculty interest and expertise, etc.) we can expect wider interest in teaching these new methods.

Acknowledgments

We are grateful to M. Minotti for extraction solutions, to S. Wasylkiewicz for phase equilibrium and column calculations, to R. E. Rooks for calculation of the separation regions, and to P. Stephan for figures. Financial support for parts of this work was provided by E. I. DuPont Co., Inc., Eastman Chemical Company, Union Carbide Chemicals and Plastics, Imperial Chemical Industries (UK), and NSF (Grant No. CTS-9113717). MFM is also grateful to the DuPont Company for sabbatical accommodations.

References

Baburina, L.V. and V.M. Platonov (1987). Analysis of Separating Manifolds in Multidimensional Simplices of Polyazeotropic Mixtures. *Khimicheskaya Promyshlennost* (English Translation), **19**, 62-67.

Baburina, L.V. and V.M. Platonov (1990). Application of Theory of Conjugate Tie Lines to Structural Analysis of Three and Four Component Mixtures. *Theor. Found. Chem. Engng.*, **24**, 287-291.

Barnicki, S.D. and J.R. Fair (1990). Separation System Synthesis: A Knowledge-Based Approach. 1. Liquid Mixture Separations. *Ind. Eng. Chem. Research*, **29**, 421-432.

Bernot, C., M.F. Doherty, and M.F. Malone (1990). Patterns of Composition Change in Multicomponent Batch Distillation. *Chem. Engng. Sci.*, **45**, 1207-1221.

Bernot, C., M.F. Doherty, and M.F. Malone (1991). Feasibility and Separation Sequencing in Multicomponent Batch Distillation. *Chem. Engng. Sci.*, **46**, 1311-1326.

Bernot, C., M.F. Doherty, and M.F. Malone (1993). Design and Operating Targets for Nonideal Multicomponent Batch Distillation. *Ind. Eng. Chem. Research*, **32**, 293-301.

Bushmakin, I.N. and I.N. Kish (1957). Isobaric Liquid-Vapor Equilibrium in a Ternary System with an Azeotrope of the Saddlepoint Type. *J. Appl. Chem. USSR* (Engl Trans.), **30**, 205-215.

Davydian, A.G., V.N. Kiva, G.A. Meski, and M. Morari (1994). Batch Distillation in a Column with a Middle Vessel. *Chem. Engng. Sci.*, in press.

Doherty, M.F. and J.D. Perkins (1978). On the Dynamics of Distillation Processes — I. The Simple Distillation of Multicomponent Non-Reacting Homogeneous Liquid Mixtures. *Chem. Engng. Sci.*, **33**, 281-301.

Doherty, M.F. and J.D. Perkins (1979). On the Dynamics of Distillation Processes — III. Topological Classification of Ternary Residue Curve Maps. *Chem. Engng. Sci.*, **34**, 1401-1414.

Fidkowski, Z.T., M.F. Doherty, and M.F. Malone (1993a). Feasibility of Separations for Distillation of Nonideal Ternary Mixtures. *AIChE J.*, **39**, 1303-1321.

Fidkowski, Z.T., M.F. Doherty, and M.F. Malone (1991). Nonideal Multicomponent Distillation: Use of Bifurcation Theory for Design. *AIChE J.*, **37**, 1761-1779.

Fidkowski, Z.T., M.F. Doherty, and M.F. Malone (1993b). Computing Azeotropes in Multicomponent Mixtures. *Computers Chem. Engng.* **17**, 1141-1155.

Foucher, E.R., M.F. Doherty, and M.F. Malone (1991). Automatic Screening of Entrainers in Homogeneous Azeotropic Distillation. *Ind. Eng. Chem. Research*, **30**, 760-772.

Glinos, K.N. and M.F. Malone (1988). Optimality Regions for Complex Column Alternatives in Distillation Systems. *Chem. Engng. Res. Design*, **66**, 229-240.

Guthrie, K.M. (1974). *Process Plant Estimating, Evaluation and Control*. Craftsman Book Company, Solana Beach, CA.

Hildebrandt, D., D. Glasser, and C.M. Crowe (1990). Geometry of the Attainable Region Generated by Reaction and Mixing: With and Without Constraints. *Ind. Eng. Chem. Research*, **29**, 49-58.

Julka, V. and M.F. Doherty (1990). Geometric Behavior and Minimum Flows for Nonideal Multicomponent Distillation. *Chem. Engng. Sci.*, **45**, 1801-1822.

Julka, V. and M.F. Doherty (1993). Geometric Nonlinear Analysis of Multicomponent Nonideal Distillation: A Simple Computer-Aided Design Procedure. *Chem. Engng. Sci.*, **48**, 1367-1391.

Keller, H.B. (1977). Numerical Solution of Bifurcation and Nonlinear Eigenvalue Problems. In P. H. Rabinowitz (Ed.), *Application of Bifurcation Theory*. Academic Press, New York, 359-384.

Knapp, J.P. and M.F. Doherty (1994). Minimum Entrainer Flows for Extractive Distillation. A Bifurcation Theoretic Approach. *AIChE J.*, **40**, 243-268.

Laroche, L., N. Bekaris, H.W. Andersen, and M. Morari (1992). Homogeneous Azeotropic Distillation: Separability and Flowsheet Synthesis. *Ind. Eng. Chem. Research*, **31**, 2190-2209.

Levy, S.G. (1985). Design of Homogeneous Azeotropic Distillations. Ph.D. thesis, University of Massachusetts, Amherst.

Pretel, E.J., P.A. Lopez, S.B. Bottini and E.A. Brignole (1994). Computer-Aided Molecular Design of Solvents for Separation Processes. *AIChE J.*, **40**, 1349-1360.

Rev, E. (1992). Crossing of Valleys, Ridges, and Simple Boundaries by Distillation in Homogeneous Ternary Mixtures. *Ind. Eng. Chem. Research*, **31**, 893-901.

Robinson, C.S. and E.R. Gilliland (1950). *Elements of Fractional Distillation*, 4th ed. McGraw Hill, New York. 387.

Sorenson, J.M. and W. Arlt (1980). Liquid-Liquid Equilibrium Data Collection: Ternary Systems. In *DECHEMA Chemistry Data Series*, Vol. V, part 2. DECHEMA, Frankfurt/Main.

Tedder, D.W. and D.F. Rudd (1978). Parametric Studies in Industrial Distillation. *AIChE J.*, **24**, 303-334.

US Patent 4,740,273 (1988). Process for the Purification of Diethoxymethane from a Mixture with Ethanol and Water. Daniel L. Martin and Peter W. Reynolds, assigned to Eastman Kodak Company.

Wahnschafft, O.M., T.P. Jurian, and A.W. Westerberg (1991). SPLIT: A Separation Process Designer. *Computers Chem. Engng.*, **15**, 565-581.

Wahnschafft, O.M., J.W. Koehler, E. Blass, and A.W. Westerberg (1992a). The Product Composition Regions for Single-Feed Azeotropic Distillation Columns. *Ind. Eng. Chem. Research*, **31**, 2345-2362.

Wahnschafft, O.M., J.P.L. Rudulier, P. Blania, and A.W. Westerberg (1992b). SPLIT: II. Automated Synthesis of Hybrid Liquid Separation Systems. *Computers Chem. Engng.*, **16**, 305-312.

Wahnschafft, O.M., J.P.L. Rudulier, and A.W. Westerberg (1993). A Problem Decomposition Approach for the Synthesis of Complex Separation Processes with Recycles. *Ind. Eng. Chem. Research*, **32**, 1121-1141.

Yamakita, Y., J. Shiozaki, and H. Matsuyama (1983). Consistency Test of Ternary Azeotropic Data by Use of Simple Distillation. *J. Chem. Eng. Japan*, **16**, 145-146.

Original: May 11, 1994. Revised after reviews: September 6, 1994.

MODELING AND ANALYSIS OF MULTI-COMPONENT SEPARATION PROCESSES

Ross Taylor and Angelo Lucia
Department of Chemical Engineering
Clarkson University
Potsdam, NY 13699-5705

Abstract

This paper looks at recent developments in the modeling and numerical and theoretical analysis of multicomponent separation processes. We look at the well known equilibrium-stage model before focusing on the newer nonequilibrium models (also known as rate-based models) now used for simulating operations such as distillation, absorption and extraction. Algorithms that are used to solve the model equations are discussed, with particular attention given to the use of Newton's method (including highlighting some of the pitfalls of this method).

Keywords

Equilibrium stage model, Nonequilibrium models, Rate-based models, Separation process modeling.

Introduction

The equilibrium stage model has been with us for almost exactly 100 years. It is the standard model for simulating distillation, absorption, stripping and extraction operations. The key assumption in the model is that the vapor and liquid streams leaving a stage are in equilibrium with each other. The equations that model equilibrium stages are known as the *MESH* equations. The *M* equations are the Material balance equations, the *E* equations are the equilibrium relations, the *S* equations are the mole fraction summation equations and the *H* equations are the enthalpy balance equations. The unknown variables determined by solving these equations are the mole fractions of both phases, the stage temperatures and the flow rate of each phase. Methods of solving these equations are discussed next.

Tearing Methods

Bubble point and sum-rates methods belong to a class of algorithms known as equation-tearing methods and have been widely used in industry for more than thirty years now (see Wang and Wang (1981) and Seader (1985) for brief details, an interesting history and literature citations). These methods pair model equations and variables in two loops and alternately solve the equations in each loop until convergence is reached. For example, sum-rates methods fix the temperature and pressure profile for the column and solve the mass balance and phase equilibrium equations simultaneously for the component flow rates in an inner loop. The temperature profile is then adjusted in an outer loop by solving the energy balance equations. Bubble point methods, on the other hand, use temperatures as inner loop variables and adjust the total vapor flow profile in an outer loop using the energy balances. Inside-out methods (see, Boston and Britt, 1978; Russell, 1983) also belong to the class of tearing methods and are the subject of ongoing industrial research in multistage separation process simulation (Venkataraman et al., 1990).

It is widely accepted that for relatively narrow boiling mixtures (i.e. small values of D_{DB}, where D_{DB} is the difference between the bubble and dew point temperatures of the primary feed), bubble point methods are preferred. For relatively wide boiling mixtures (large values of D_{DB}), sum-rates methods are preferable. Intermediate boiling mixtures, on the other hand, represent something of a dichotomy and have always been a source of difficulty for either method (see, Friday and Smith, 1964; Tomich, 1964).

Outside of the work on inside-out algorithms, there has been very little recent research on tearing algorithms, particularly in the academic community. However, Sridhar and Lucia (1990) have developed a modified sum-rates method based on insights provided by rigorous analysis and have reported improved numerical results. In particular, this analysis provides an analytical expression for the Newton-acceleration of the inner loop as well as the partial derivatives of the total vapor flow with respect to temperature needed in the outer loop and results in a modified sum-rates method that is capable of solving narrow, intermediate and wide boiling mixtures alike. Sridhar (1990) has also developed a modified bubble point method using the same principles and has obtained similar improvements in reliability when compared to traditional bubble point methods.

With the recent widespread interest in parallel processing, tearing methods for separation process simulation may once again come into vogue.

Newton's Method and its Relatives

To the best of our knowledge, a method to solve all the MESH equations for all stages at once using Newton's method was first described by Whitehouse (1964) (see, also, Stainthorp and Whitehouse, 1967). Among other things, Whitehouse's code allowed for specifications of purity, *T, V, L* or *Q* on any stage. Problems involving interlinked systems of columns and nonideal solutions could also be solved. Since then many others have used this approach (see, e.g. Naphtali and Sandholm, 1971; Goldstein and Stanfield, 1970) and the method has, at long last, been accorded a measure of acceptance by the engineering community; Newton's method has found use in some commercial simulation programs.

How should the Jacobian be Computed?

Newton's method requires the evaluation of the partial derivatives of all equations with respect to all variables. The partial derivatives of thermodynamic properties with respect to temperature, pressure and composition are most awkward to obtain (and the ones that have the most influence on the rate of convergence). It is rather a painful experience to differentiate, for example, the UNIQUAC equations with respect to temperature and composition, so, often, this differentiation is done numerically. However, neglect of these derivatives is not recommended except for nearly ideal solutions, since, to do so, will almost certainly lead to an increase in the required number of iterations or even to failure. Michelsen and Mollerup (1986) and Baden and Michelsen (1987) have shown that coding analytical derivatives of the complicated physical property models is well worth while as this leads to substantial reductions in the computer time required for this portion of the calculations. With the availability of computer algebra systems like Mathematica and Maple (Gonnet and Gruntz, 1991), there is no excuse for not using analytical derivatives in newly created programs.

Quasi-Newton methods (e.g. Broyden, 1965) sometimes are used in order to reduce the time required to calculate the necessary derivative information for the Jacobian. In these methods, approximations to the Jacobian are updated through the use of formulae derived to satisfy certain constraints that force the approximate Jacobian to mimic the behavior of the actual Jacobian. Experience with quasi-Newton methods has been reported by Gallun and Holland (1980), Lucia and coworkers (Lucia and Macchietto, 1983; Lucia and Westman, 1983; Westman, et al., 1984, Venkataraman and Lucia, 1987, 1988) and others. It is probably fair to say that pure quasi-Newton methods are not especially useful for separation process problems. Lucia and coworkers have developed what they call a hybrid approach to the solution of sets of model equations. In this hybrid approach, any derivative information which is easy (i.e. inexpensive) to calculate is included in a *computed* part of the Jacobian, C, while derivative information which is difficult or expensive to obtain or unavailable (perhaps because the simulation uses a proprietary physical properties package) is included in an *approximated* part of the Jacobian, A. In equilibrium stage calculations, A includes derivatives of activity and fugacity coefficients, and excess enthalpies with respect to composition, temperature and pressure. C is calculated exactly each iteration and A is updated using a quasi-Newton update that satisfies secant conditions, symmetry with respect to mole number derivatives and Gibbs-Duhem and Gibbs-Helmholtz equations. The result is a method that takes more or less the same number of iterations as Newton's method with analytical derivatives to solve a given problem, while taking significantly less computer time to converge and has nearly the same region of convergence as Newton's method (see Venkataraman and Lucia, 1988). Pantelides (1988) also provides some numerical results for a secant-only hybrid method implemented in the Speed-Up commercial simulation program.

How Should the Linearized Equations be Solved?

Newton's method requires the solution of a linear system of equations in each iteration. Since the vast majority of Jacobian elements are zero, it is absolutely essential to take account of the sparsity of this matrix when solving the linearized equations. Software for solving sparse linear systems is described by Seader (1986).

For simple columns the Jacobian matrix has a block-tridiagonal (BTD) structure. Linear BTD systems may efficiently be solved using a generalized form of the Thomas algorithm, one where the scalar arguments are replaced by matrices and divisions are replaced by premultiplication by inverse matrices. The steps of this algorithm are given by Henley and Seader (1981). Still further improvements in the block elimination algorithm can be obtained if we take advantage of the special structure of the submatrices

(Baden and Michelsen, 1987). Onana and Hikolo (1993) report on the use of iterative methods for solving BTD systems and conclude that SOR methods are more efficient than the Thomas algorithm for systems with large numbers of components.

How Should the Initial Guess be Obtained?

Newton's method requires initial estimates for all variables. It is obviously impractical to expect the user of a distillation code to guess what may be hundreds of numbers. Thus, the designer of a computer code must provide one or more methods of generating initial estimates of all the unknown variables from, at most, one or two user supplied values of end stage temperatures and flow rates (and, preferably, from no user supplied information). A good initial guess is of central importance to obtaining convergence with Newton's method but many papers do not even address the issue (see, however, Henley and Seader, 1981, p 605; Wayburn and Seader, 1983; Burton and Morton, 1987; and Venkataraman and Lucia, 1988).

Continuation Methods

Homotopy or continuation methods are one way to solve very difficult problems when Newton's method fails. In these methods, we start with a problem that is easy to solve. This easy problem is transformed through some parameterization into the problem that we wish to solve. This can be done by using an additional parameter or one that occurs naturally in the model (parametric continuation).

The homotopy method that, to date, has been used most often for solving equilibrium stage model problems (at least in literature articles) is the Newton homotopy (Wayburn and Seader, 1983, Chavez et al., 1986; Lin et al., 1987; Kovach and Seider, 1987; Ellis et al., 1986; Burton and Morton, 1987, and many others). Other methods are proprietary (e.g. Byrne and Baird, 1985).

Mathematical homotopies (such as the Newton homotopy) make no concessions to the model being solved. Consequently, difficulties can arise if variables take on inappropriate values at intermediate values of the continuation parameters. Negative mole fractions at intermediate parameter values (as encountered by Vickery and Taylor (1986) and by Burton and Morton, 1987) lead to difficulties in the physical properties routines which require at least nonnegative mole fractions in order to return realistic values. Ellis et al. (1986) report negative total flow rates (a much less serious problem) at intermediate values of the continuation parameter.

Vickery and Taylor (1986, 1987) designed a homotopy in which the thermodynamic quantities are modified, simplifying them for the initial problem so that the K-values and enthalpies are much easier to deal with. As the homotopy parameter is increased, the thermodynamic quantities are brought back to their original forms, until, finally, the original problem has been solved. The thermodynamic homotopy was claimed to be rather more effective than a simple implementation of the Newton homotopy for solving distillation problems involving strongly nonideal systems. Fidkowski et al. (1993) have used a related method for finding all the azeotropic points in multicomponent mixtures.

Parametric continuation methods are most often employed to investigate the sensitivity of a model to a particular quantity, for example, the reflux ratio or bottoms flow rate. Parametric continuation has also been used to detect multiple solutions of the MESH equations (Ellis et al., 1986; Kovach and Seider, 1987, and others).

A parameter occurring naturally in the MESH equations that makes a good continuation parameter is the stage efficiency. Müller (1979) appears to have been the first to do this and Sereno (1985) used this approach for solving liquid-liquid extraction problems. Vickery et al. (1988) used the efficiency as a continuation parameter to solve multicomponent distillation problems in single and interlinked systems of columns.

An efficiency-based continuation method is very effective at solving problems involving standard specifications but cannot, handle problems in which a product stream purity is specified. Woodman (1989, 1990) used efficiency-based continuation to solve a thermodynamically ideal problem involving standard specifications. Thermodynamic continuation was then used to bring in the solution nonidealities in a controlled way. Finally, parametric continuation was invoked to solve any non-standard specifications that were involved in the original problem.

Collocation Methods

Collocation methods are useful for solving systems of partial differential equations and have been adapted to solve both steady state and dynamic equilibrium stage model problems. The reader is referred to papers by Stewart et al. (1985) and Swartz and Stewart (1986). There does not appear to have been much use of these methods (see, however, Seferlis and Hrymak, 1994a,b).

Three Phase Distillation

The extension of the equilibrium stage model of two phase mixtures to three-phase systems is straightforward in principle. However, the numerical solution of these equations is a decidedly nontrivial task. There are a number of reasons for this:

- The very high nonlinearity of the thermodynamic quantities (the K-values and enthalpies) means that the numerical computations are unusually sensitive to estimates of the temperatures and compositions and convergence often is very difficult to obtain.
- Unlike two phase separation process problems, the number of equations and variables necessary to determine the operating condition of a column is not known until the problem has been solved. The reason for this is that even though the process is called three

phase distillation, there will almost certainly be a number (probably a majority) of stages with only one liquid phase present.

- The set of model equations listed above always admits a (trivial) two-phase solution; that is, one vapor and one liquid phase. Kingsley and Lucia (1988) show that the two-phase solutions lie underneath the three-phase solutions.

Most algorithms for solving three-phase distillation problems are, in one sense or another, extensions of methods that have been found useful for solving two-phase distillation problems. A review of these algorithms is provided by Cairns and Furzer (1990).

Dynamic Models

Dynamic simulation of chemical processes is becoming increasingly important in recent years with dynamic process simulation systems becoming available (see, e.g. Pantelides, 1988). Some recent contributions to the dynamic simulation of distillation-type operations are noted here.

Gallun and Holland (1982) used Gear's method to solve the equations involved in dynamic simulation. Holland and Liapis (1983) discuss the use of semi-implicit Runge-Kutta methods as well as the multi-step methods of Gear for the integration of the differential equations. Prokopakis and Seider (1983) simulated the dynamics of azeotropic distillation towers. Gani and coworkers (1986, 1987, 1989) and Cameron et al. (1988) have developed a comprehensive model for the dynamic simulation of continuous distillation columns. Gani et al. (1987) also discussed the optimization of the dynamics of start-up/shutdown operations and the hydraulics involved (flooding, entrainment, and weeping). They found that plate hydraulics play an important role in these kinds of simulations. Ranzi et al. (1988) looked at the effects of the energy balances and found that the energy balances must be evaluated properly in order to predict correct behavior. Dynamic equilibrium stage models of three-phase systems have been developed by Rovaglio and Doherty (1990), Widagdo et al. (1992), and by Bossen et al. (1993).

Distillation with Reaction

The past few years have seen considerable interest in distillation with reaction. Doherty and Buzad (1992) have recently reviewed the literature on kinetics, equilibria, and equilibrium stage modeling of this interesting area.

Theoretical Analysis and Multiple Solutions

Although there were several early studies concerned with various aspects of the analysis of multistage separation processes (Acrivos and Amundsen, 1955; Rosenbrock, 1960, 1961), recent numerical studies reporting multiple steady-state solutions (Shewchuk, 1974; Magnussen et al., 1979; Prokopakis and Seider, 1983; Lin et al., 1987; Venkataraman and Lucia, 1988; and Rovaglio and Doherty, 1990; Bossen et al., 1993; Gani and Jorgensen, 1993; and others) have rekindled interest in rigorously determining the number of steady-state solutions to multistage separations problems. Most of the more recent theoretical results (Doherty and Perkins, 1982; Sridhar and Lucia, 1989, 1991; Lucia and Li, 1992; Kienle and Marquardt, 1994) have been established using either a dynamical systems approach or classical fixed-point theory and algebraic techniques. The main thrust of this work has been to try and relate the occurrence of multiplicity to (physical property) model-independent attributes such as the homogeneity or heterogeneity of the mixture, column specifications and the presence of energy balances but only limited success has been achieved in this regard. Furthermore, while there are reports of unusual and extreme sensitivity, there is no definitive experimental evidence that shows that two or more steady-states can exist for the same operating conditions. In our opinion, industrial evidence of this is sorely needed.

Doherty and Perkins (1982) have used residue curves and linear stability theory to show that flash models involving multicomponent homogeneous mixtures and multistage separators involving binary homogeneous mixtures admit unique and stable steady-state solutions. Sridhar and Lucia (1989) and Lucia and Li (1992), on the other hand, use classical fixed-point theory and algebraic methods to establish the uniqueness of solutions to models for binary multistage columns for a variety of column specifications. They also show which sets of column specifications can lead to model multiplicity. Sridhar and Lucia (1991) prove that multicomponent homogeneous multistage separators with fixed temperature and pressure profiles have unique steady-state solutions and Lucia and Li (1994) show that multiple solutions to binary multistage separators imply multiple solutions to multicomponent columns under the same set of specifications.

Jacobsen and Skogestad (1991) show that ideal two-product binary distillations have multiple steady-states when specifications are provided in mass flow instead of molar flows and Bekiaris et al. (1993, 1994) give a numerical example of a column involving the homogeneous mixture acetone/methanol/chloroform that has three steady-state solutions for specifications of reflux and distillate. However, the proofs in these papers are not rigorous and therefore contain only anecdotal evidence similar to other numerical studies. In fact, Lucia and Li (1994) have shown that the arguments in Bekiaris et al. are inconsistent with overall mass balance and can lead to incorrect conclusions for columns with finite numbers of stages at finite reflux.

Kienle and Marquardt (1991) analyze an infinitely extended column using differential techniques and more recently Kienle et al. (1994) use continuation methods to find multiple steady-states for a single section of an azeotropic distillation involving a ternary homogeneous mixture. Both papers suggest that multiplicity is a consequence of repeated eigenvalues in the nonlinear operator associated with the differential mass balances and phase equilibrium equations that model the column section.

Kovach and Seider (1987), Kingsley and Lucia (1988), Widagdo et al. (1989), Cairns and Furzer (1990), Rovaglio et al. (1993), and others have investigated the multiplicity of solutions associated with three-phase distillation problems. Rovaglio et al. (1993) used a dynamic model and state that steady-state multiplicity in heterogeneous azeotropic columns is a consequence of both extreme sensitivity and small inaccuracies in column inventories due to the convergence criteria used by most steady-state solvers. They claim that some solutions reported by some authors are not really solutions at all. However, Lucia et al. (1990) solve a single feed, two-stage heterogeneous column to a two-norm value for the model equations of less than 10^{-9} (even for unscaled energy balances) and still find three two-phase solutions underneath the three-phase region. Furthermore, in Venkataraman and Lucia (1988), the material balances are linear and are solved to machine precision. Therefore, entrainer inventory problems are not possible.

The analysis of the number of solutions to models of multistage separators remains a difficult theoretical problem with many unanswered questions and much experimental, numerical and analytical work remaining to be done.

Distillation Calculations in the Complex Domain

Flash and distillation calculations have been carried out in the complex domain by Lucia and Taylor (1992) and by Taylor et al. (1994) who found that a complex code will converge to a real solution, even from a complex starting point if there is a real solution to be found. However, complex codes can solve certain problems far more easily than can programs that are confined to the real domain. Examples include separations carried out at high pressure where an equation of state must be used to model the thermodynamic properties. Complex solutions have been obtained with a variety of activity coefficient models including the Wilson equation. Complex solutions have been found by Lucia and Wang (1994) for the TV flash of a system whose thermodynamics were modeled using a cubic equation of state while Lucia and Sridhar (1994) have shown that TP flashes involving cubic equations of state must have real-valued solutions.

Complex solutions may simply be an artifact of the complicated algebraic structure of the model equations. However, they do show that certain (e.g. high purity) solutions are not feasible for some specifications. Moreover, the presence of complex solutions may also be an indication that the actual process is hard to control.

It is likely that performing three-phase distillation calculations in the complex domain will mitigate some of the convergence problems associated with moving from two-phase to three-phase stages. Not only will it be possible to converge problems where the specifications ensure that real domain solutions exist but the model discontinuities and one-sided differentiability difficulties associated with the appearance and disappearance of liquid phases will be lessened as well. There is a close analogy here with equation of state models where the bifurcation into the complex domain of the compressibility coincides with the disappearance of a real phase.

Since Newton's method can exhibit periodic and/or aperiodic behavior in both the real and complex domains (see Lucia et al., 1993) and because we have observed similar difficulties in solving difficult distillation problems, we recommend that some type of stabilization method, extended to the complex domain, be used to improve convergence. Lucia et al., (1993) have extended Powell's dogleg strategy to the complex domain to avoid both the periodic/aperiodic behavior of Newton's method and termination of traditional dogleg methods at singular points and have reported improved numerical performance on small multivariable problems. In a similar way, continuation methods with real/complex solutions and real (and perhaps complex) parameter values can be used to find desired solutions and/or clarify the feasibility of these solutions for any given set of parameters. In our opinion, it is a simple matter to extend existing arc-length and other homotopy/continuation methods to do this.

Nonequilibrium (Rate-based) Models

Up to this point we have focused our attention on the standard model used in separation process simulation — the equilibrium stage model. The model is so simple in conception, so elegant from the mathematical viewpoint, has inspired so many algorithms, and has been used to simulate and design so many real columns that it seems almost heretical to mention that the model is fundamentally flawed! The trays of an actual column are not equilibrium stages. What is more, this fact is well known.

The usual way of getting around the fact that real trays are not equilibrium stages is to use an efficiency factor of some kind. If engineers use an efficiency it is most likely to be either the overall efficiency or the Murphree efficiency. The latter is easily incorporated into the equilibrium equations. It is common practice to assume that the efficiencies are the same for all components on any given tray (and only slightly less common to assume that they are the same from tray to tray as well) even though there is abundant experimental and theoretical evidence to show that the efficiencies of different components are not the same (see Taylor and Krishna (1993) for a review of the data for two phase systems). An experimental study by Cairns and Furzer (1990) has shown that the component efficiencies in three-phase systems also could be very large, positive or negative numbers. Seader (1989) discusses some of the pitfalls of using an efficiency.

The building blocks of a nonequilibrium model include: material balances, energy balances, equilibrium relations, and mass and energy transfer models. The first three of these are also used in building equilibrium stage models; however, there is a crucial difference in the way in which the conservation and equilibrium equations are used in the two types of model. In equilibrium stage models the balance equations are written around the stage as a whole and the composition of the leaving streams related through an assumption that they are in equilibrium or by use of an efficiency equation. In a nonequilibrium model, separate balance equations are written for each phase. The conservation equations for each phase are linked by material balances around the interface; whatever material is lost by the vapor phase is gained by the liquid phase. The energy balance for the stage as a whole is treated in a similar way, split into two parts — one for each phase, each part containing a term for the rate of energy transfer across the phase interface. Efficiencies are not required by a nonequilibrium model (although they can be calculated following a simulation).

In equilibrium stage calculations the equilibrium equations are used to relate the composition of the streams leaving the stage. K-values are evaluated at the composition of the two exiting streams and the stage temperature (usually assumed to be the same in both phases). In a nonequilibrium model the equilibrium relations are used to relate the compositions on either side of the phase interface; the K-values being evaluated at the interface compositions and temperature. The interface composition and temperature must, therefore, be determined (by calculation) as part of a nonequilibrium column simulation. The assumption in the equilibrium model of thermal equilibrium forces the liquid and vapor leaving a stage to have the same temperature. In reality, heat transfer between the two phases is limited and the separate phases have their own temperatures.

Mass and energy are transferred across the phase boundary at rates that depend on the extent to which the phases are not in equilibrium with each other. These rates are calculated from models of mass transfer in multicomponent systems (see Taylor and Krishna (1993) for an extended discussion of such models). Mass and heat transfer coefficients and interfacial areas must be computed from empirical correlations or theoretical models. These coefficients depend on the column design as well as on its method of operation. It is essential to ensure that the flows and design parameters are consistent with the satisfactory operation of the column.

The nonequilibrium model contains the equilibrium stage models as a further limiting case. Equilibrium between the bulk phases is attained when the driving forces vanish. This will happen if the mass transfer coefficient interfacial area products are very high. In fact, if one does a nonequilibrium simulation with interfacial areas about 100 times larger than they really are, then the results are very close to those obtained with an equilibrium model (Powers et al., 1988).

Algorithms, Models and Applications

A nonequilibrium model has been developed in a series of papers by Taylor and coworkers (Krishnamurthy and Taylor (KT); 1985, 1986; Taylor et al., 1992, 1993). The KT model is formulated in way that clearly shows the relationship to the equilibrium stage model and it is possible to use the algorithms that are useful for solving equilibrium stage simulation problems for solving nonequilibrium problems. We are thinking, in particular, of Newton's method and the continuation methods discussed above. Analogous to the efficiency continuation method is to multiply the mass transfer coefficients by the homotopy parameter so that the first problems involve no, or very little, mass transfer and are, therefore, very easy to solve (Powers et al., 1988). Biardi and coworkers (Biardi and Grottoli, 1987; Grottoli et al., 1991) have presented alternative algorithms that belong to the class of tearing methods and that retain the use of efficiencies. Young and Stewart (1990) use collocation techniques to solve a boundary layer model of cross flow on a tray. Sivasubramanian and Boston (1988) have described a commercial implementation of a nonequilibrium model.

The KT model has been used to simulate simple distillation columns, absorbers, and extractive distillation operations) distillation and absorption columns) processes (Krishnamurthy and Taylor, 1985b, c, 1986; Taylor et al., 1987, 1992). Krishnamurthy and Andrecovich (1989) adapted the model for simulating cryogenic distillation operations. Nonequilibrium models of liquid-liquid extraction have been described by Lao et al. (1989) and by Zimmerman et al. (1992).

Nonequilibrium models are capable of better predictions of actual column performance than is possible with an efficiency-modified equilibrium stage model (see Taylor and Krishna, 1993). The model is more computationally involved but computational results obtained by Taylor et al. (1994) show that the penalty for solving a larger problem is not unacceptable.

Packed Columns

The equilibrium stage model is often used to simulate packed columns. The efficiency is replaced by the Height Equivalent to a Theoretical Plate (HETP). However, HETP's for multicomponent systems suffer from the same problems that plague tray efficiencies.

Packed columns are, of course, continuous contact devices and it is, perhaps, more common to use a differential mass transfer-based model. Gorak (1987, 1991) has used multicomponent mass transfer theory to model packed distillation columns. Standard numerical integration methods were used to solve the systems of ODEs that make up this model. Cho and Joseph (1983, 1984) used collocation methods for solving the partial differential equations that make up their packed column model. Unfortunately, their mass transfer model is flawed in that too many rate equations (of the pseudo-binary type) are employed.

Packed columns (of pilot plant to full size industrial scale) have been simulated using the KT model by Krishnamurthy and Taylor (1985d, 1986), Taylor et al. (1987), and Taylor et al. (1992). A nonequilibrium stage may represent either a single tray or a section of packing in a packed column. The same basic equations may be used to model both types of equipment and the only significant difference between these two models is that alternative expressions must be used for estimating the mass transfer coefficients and hydrodynamic parameters. Gorak et al. (1991) and Wozny et al. (1991) have presented a brief description of their use of a nonequilibrium stage model to simulate vacuum distillation of fatty alcohols in columns fitted with structured packing. They found that the compositions predicted by the nonequilibrium model to be closer to the experimental data than were the results of an equilibrium stage — HETP calculation. McNulty and Chatterjee (1992) discuss the use of nonequilibrium models to design packed bed pumparound zones of crude distillation towers.

Simultaneous Column Design and Simulation

A nonequilibrium simulation cannot proceed without some knowledge of the column type and the internals layout. Tray type and mechanical layout data, for example, are needed in order to calculate the mass transfer coefficients for each tray. For packed columns the packing type, size and material must be known.

The need for column design information might, on the face of things, limit the application of a nonequilibrium model to situations where the column design is already known. It does not seem possible to carry out flowsheet simulations and column design studies, for example, where the flows are determined only after material and energy balance calculations around other process units.

To get around this limitation on the use of nonequilibrium models Taylor et al. (1994) describe an implementation of the a second generation model that includes a design mode that eliminates the need for any hardware design information (other than the number of stages and the specific type of internal — i.e. whether or not the trays are sieve trays or another type of tray or packing). In this mode the program performs an equipment design calculation for each stage on each iteration. When the entire problem has converged, the design can be rationalized so that a more or less uniform design applies over a section of the column (usually between feeds and product streams) and the simulation repeated using the rationalized equipment design parameters (design mode being turned off as it were).

Three Phase Distillation

Nonequilibrium models of three-phase distillation can be devised in essentially the same way that models for two-phase systems are constructed; conservation equations are written separately for each phase. The material balances for the vapor phase include a term for mass transferred from the vapor phase to each of the liquid phases. The material balances for each liquid phase includes a term for mass transfer from the vapor phase and a term for mass transfer between the two liquid phases. The interfaces are assumed to offer no resistance to mass transfer; all of the usual conditions of phase equilibrium apply there.

The general framework described above reduces to the two phase model of Krishnamurthy and Taylor (1985) if the second liquid phase is not present. If the vapor phase is not present then the model simplifies to a model of liquid-liquid extraction (Lao et al., 1989).

A very important difference between equilibrium and nonequilibrium models must be noted here. In the mass transfer model we do not solve a three phase equilibrium problem for the interface (or anywhere else for that matter) because we do not believe that all three phases in equilibrium exist together anywhere on a distillation tray. Only two phases come into contact at any particular time and place. This means that all three sets of interface equilibrium equations are independent in the nonequilibrium model whereas only two of the three sets of the equilibrium relations are independent in the equilibrium model.

A consequence of this is that for the liquid-liquid equilibrium (LLE) computation we can use activity coefficient model parameters fitted to LLE data whereas we can use parameters fitted to vapor-liquid equilibrium (VLE) data for the VLE calculations. We may even use two different sets of interaction parameters for the two independent VLE computations at the VL interfaces. Thus, the inability of existing thermodynamic models to fit three phase, vapor-liquid-liquid, equilibrium (VLLE) data as well as they fit two phase VLE and LLE data is much less critical in a mass transfer model than it would be in an equilibrium stage model. In fact, it may not even be desirable to use three phase VLLE data in fitting the parameters used in the interface equilibrium computations. Furthermore, we do not even have to use the same thermodynamic models for each interface if there is a good reason not to. For example, we might use UNIQUAC to model the vapor-liquid equilibria and the NRTL equation to model the liquid-liquid equilibrium if that would provide better results. It should also be possible to use an activity coefficient model that does not admit two liquid phase (e.g. Wilson) in the VLE calculations.

Since the hydrodynamics of three phase distillation trays is still a subject for research, it is necessary, at least as a first step, to develop a model for vapor and liquid flow on a three phase tray. Lao and Taylor (1994) have put forward four simple models of flow on three phase distillation trays: the homogeneous liquid model, the segregated liquid model, the stratified liquid model, and the dispersed liquid model. Simulations of the experimental data of Cairns and Furzer showed that the dispersed liquid model was the best of these four. More data on three phase distillation needs to be madc available if we are to develop better models of this important operation.

Dynamic Models

If real columns do not consist of equilibrium stages at steady state, it seems unlikely that efficiency-modified models can be adequate for modeling column dynamics where composition transients imply changes in the component efficiencies over time. Further, the assumption of thermal equilibrium makes it difficult to model the dynamics of sections in a column that are purposely used for heat transfer.

Since the nonequilibrium model avoids the use of tray efficiencies it should be suitable as a basis for developing a dynamic column model. Some preliminary work to this end has been reported by Kooijman and Taylor (1992, 1993). Their model uses four distinct holdup terms: the liquid in the froth (spray/emulsion) on a tray, the vapor in the dispersion on a tray, the liquid in the downcomer below a tray, and the vapor above the froth/downcomer on a tray. Simplifications often made in other dynamic models such as constant holdups, neglecting energy derivatives, neglecting vapor holdups, ignoring the liquid in the downcomer, and constant (tray/component) efficiencies also are avoided.

Distillation with Reaction

If any distillation-type operation deserves to be modeled as a rate-process, it is distillation with reaction. A word of caution is in order here. Proper care needs to be exercised in the setting up of the mass (and heat) transfer sub-problems in reactive distillation operations. If the reaction occurs only in the bulk liquid then a simple modification of the KT model may be adequate. However, if the reaction is fast enough it will influence the mass transfer process and the liquid phase mass transfer rate equations will need to be modified for this effect. If the reaction involves a supported catalyst then we must account for diffusion to and from a solid surface as well as reaction at the surface. Finally, if the catalyst is porous it is necessary to account for diffusion and reaction in porous media. Berg and Harris (1993) have made some progress in this direction. Yuxiang and Xien (1992) have used a nonequilibrium model to simulate catalytic distillation operations. However, theirs is a homogeneous model whereas a heterogeneous model is more appropriate for this situation.

Conclusions

Since the late 1950s, hardly a year has gone by without the publication of at least one (and usually more than one) new algorithm for solving the equilibrium stage model equations. One of the incentives for the continued activity has always been (and remains) a desire to solve problems with which existing methods have trouble. With the development of improved algorithms some of the incentive for further developments has gone and we can detect a slowing down in the rate at which papers presenting new algorithms are published. We do not mean to imply, however, that we have reached that happy (for some) situation where all problems can be solved first time, every time, in only a little elapsed time. On the contrary, there remain many separations problems that are very hard indeed to solve.

Any proposed improvement to existing algorithms must be subjected to the most demanding tests before being granted an entry into the literature. Algorithms should be tested on the most difficult test cases and on more than 2 or 3 rather trivial problems. Test cases should have some industrial relevance. Industry can play an important role here by providing (or creating) examples that test algorithms to their limits.

Mass transfer rate based models will continue to be developed. Indeed, it does not require too much imagination to foresee a time when there will be no excuse for not using a nonequilibrium model; they will be particularly useful for multicomponent nonideal systems, reactive systems and low-efficiency systems. A thorough understanding of mass transfer (with reaction) in multicomponent systems will be essential for the engineer concerned with developing nonequilibrium models.

The needs of a nonequilibrium model simulator for thermodynamic properties are no less than those of conventional equilibrium stage simulation programs. In addition, a nonequilibrium simulation requires transport properties and mass and heat transfer coefficients, properties that are entirely absent from some simulation systems.

The greater emphasis on mass transfer rate based models in the future should act as a spur to equipment vendors to develop sound correlations of mass transfer coefficients as well as hydraulic performance. Too many existing correlations are limited to narrow ranges in some design or operation parameter. It is essential that performance models cover wide ranges in plant design and operating conditions and extrapolate sensibly outside normal limits.

In a similar vein we need better methods of predicting physical and transport properties of mixtures. This will prove most important in dynamic simulations but property models are important in steady state simulation as well. Better methods of predicting binary pair diffusivities in multicomponent systems are needed for liquids and dense gases. Existing methods cannot be generalized to multicomponent systems. Methods for estimating other properties like viscosity and density need to developed so that they are continuous over very wide ranges of temperature, pressure and composition.

The nonequilibrium models discussed in this article represent our first (and second) steps towards developing more realistic models of separation processes. It is possible to create much more sophisticated nonequilibrium models but it is not clear whether such developments are warranted at present. With the models available today it is possible to predict actual plant performance within the accuracy of plant data. However, most industrial plant data is not of very high quality. We will need more and better data if we are to develop better models.

Abbreviations

CEF 87 = Conf. on The Use of Computers in Chemical Engineering, Sicily, 1987

Distillation 79 = I. Chem. E. Symposium. Series No. 56, International Symposium on Distillation, 1979

Distillation 87 = I. Chem. E. Symposium Series No. 104, Distillation and Absorption, 1987

Distillation 92 = I. Chem. E. Symposium Series No. 128, Distillation and Absorption, 1992

ESCAPE-1 = European Symposium on Computer-Aided Process Engineering, Elsinore, Denmark, 1992

FOCAPD-1 = Proceedings of the First International Conference on Foundations on Computer-Aided Process Design, R.S.H. Mah and W.D. Seider eds., AIChE, 1981

FOCAPD-2 = Proceedings of the Second International Conference on Foundations on Computer-Aided Process Design, A.W. Westerberg and H.H. Chen, eds., CACHE, 1984

References

Acrivos, A. and N.R. Amundsen (1955). *Ind. Eng. Chem.*, **47**, 1533.

Baden, N. and M.L. Michelsen (1987). *Distillation* 87, A425-436.

Bekiaris, N., G. A. Meski, C. M. Radu, and M. Morari (1993). *Ind. Eng. Chem. Res.*, **32**, 2023-2038.

Berg, D. and T. Harris (1993). *Ind. Eng. Chem. Res.*, 32, 2147.

Biardi, G. and M.G. Grottoli (1989). *Comput. Chem. Engng.*, **13**, 441-449.

Boston, J.F. and H.I. Britt (1978). *Comput. Chem. Eng.*, 2, 109.

Bossen, B.S., S.B. Jorgensen, and R. Gani (1993). *Ind. Eng. Chem. Res.*, **32**, 620-633.

Broyden, C.G. (1965). *Math. Comp.*, **19**, 577.

Burton, P.J. and W. Morton (1987). *Proc. CEF* 87, 59.

Byrne, G.D. and L.A. Baird (1985). *Comput. Chem. Eng.*, 9, 593.

Cameron, I.T., C.A. Ruiz, R. Gani (1988). *Comput. Chem. Eng.*, **12**, 377-382.

Cairns, B.P. and I.A. Furzer (1990). *Ind. Eng. Chem. Res.*, **29**, 1349-1363, 1364-1382.

Chavez C, R., J.D. Seader and T.L. Wayburn (1986). *Ind. Eng. Chem. Fundam.*, **25**, 566.

Cho, Y.S. and B. Joseph (1983). *AIChE J.*, **29**, 261-269, 270-276.

Cho, Y.S. and B. Joseph (1984). *Comput. Chem. Eng.*, **8**, 81-90.

Doherty, M.F. and J.D. Perkins (1982). *Chem. Eng. Sci.*, **37**, 381-392.

Doherty, M.F. and G. Buzad (1992). *Chem. Eng. Res. and Design*, **70**, 448.

Ellis, M.F., R. Koshy, G. Mijares (1986). A. Gomez-Munoz and C.D. Holland, *Comput. Chem. Eng.*, **10**, 433.

Friday, J.R. and B.D. Smith (1964). *AIChE J.*, **10**, 698.

Gallun, S.E. and C.D. Holland (1980). *Comput. Chem. Eng.*, **4**, 93.

Gani, R., C.A. Ruiz, I.T. Cameron (1986). *Comput. Chem. Eng.*, **10**, 181-198.

Gani, R., C.A. Ruiz (1987a). I.T. Cameron, *Proc. CEF* 87.

Gani, R., C.A. Ruiz (1987b). *Distillation* 87, B39-B50.

Gani, R., I.T. Cameron (1989). *Comput. Chem. Eng.*, **13**, 271-280.

Gani, R., S.B. Jorgensen (1994). *Comput. Chem. Eng.*, **18S**, S55-S61.

Goldstein, R. and R. Stanfield (1970). *Ind. Eng. Chem. Proc. Des. Develop*, **9**, 78-84.

Gonnet, G.H., and D.W. Gruntz (1991). *Algebraic Manipulation Systems, in Encylcopedia of Computer Science and Engineering, 3rd Edition*, Van Nostrand Reinhold.

Gorak, A. (1987). *Distillation* 87, A413-424.

Gorak, A. (1991).*Habilitationschrift der RWTH Aachen* (Germany).

Gorak, A., Wozny, G. and Jeromin, L. (1991). *Proc. 4th World Congress Chem. Eng.*, Karlsruhe, Germany.

Grottoli, M.G., Biardi, G. and Pellegrini (1991). *L. Comput. Chem. Eng.*, **15**, 171-179.

Henley, E.J., and J.D. Seader (1981). *Equilibrium-Stage Separation Operations in Chemical Engineering*, Wiley.

Jacobsen, E.W. and S. Skogestad (1991). *AIChE J.*, **37**, 499.

Kienle, A. W. Marquardt (1991). *Chem. Eng. Sci.*, **46**, 1757-1769.

Kienle, A. E.D. Gilles and W. Marquardt (1994). *Comput. Chem. Eng.*, **18**, S37-S41.

Kingsley, J.P., and A. Lucia (1988). *Ind. Eng. Chem. Res.*, **27**, 1900.

Kooijman, H.A. and R. Taylor (1992). *Proc. ESCAPE*-1.

Kooijman, H.A. and R. Taylor (1993). *AIChE National Meeting*, St. Louis.

Kovach III, J.W., and W.D. Seider (1987). *AIChE J.*, **33**, 1300.

Kovach III, J.W., and W.D. Seider (1987). *Comput. Chem. Eng.*, **11**, 593.

Krishnamurthy, R. and Taylor, R. (1985a,b,c). *AIChE J.*, **31**, 449-456, 456-465, 1973-1985.

Krishnamurthy, R. and Taylor, R. (1985d). *Ind. Eng. Chem. Process Des. Dev.*, **24**, 513-524.

Krishnamurthy, R. and Taylor (1986). *R. Can. J. Chem. Eng.*, **64**, 96-105.

Krishnamurthy, R. and Andrecovich (1989). *M.J. Proc. International Symposium on Gas Separation Technology*, E.F. Vansant and R. Dewolfs (Editors), Antwerp, Belgium, Elsevier.

Lao, M., Kingsley, J.P., Krishnamurthy, R. and Taylor (1989). *R. Chem. Eng. Commun.*, **86**, 73-89.

Lao, M. and R. Taylor (1994). Paper under review.

Lin, W-J., J.D. Seader and T.L. Wayburn (1987). *AIChE J.*, **33**, 886.

Lucia, A. (1985). *AIChE J.*, **31**, 558-566.

Lucia, A., and S. Macchietto (1983). *AIChE J.*, **29**, 705.

Lucia, A. and K.R. Westman (1984). Proc. *FOCAPD*-2, 741.

Lucia, A., X. Guo, P. J. Richey and R. Derebail (1990). *AIChE J.*, **36**, 641-654.

Lucia, A. and R. Taylor (1992). *Comput. Chem. Eng.*, **16**, S387.

Lucia, A. and H. Li (1992). *Ind. Eng. Chem. Res.*, **31**, 2580.

Lucia, A. and H. Li (1994). Paper under review.

Lucia, A., X. Guo, and X. Wang (1993). *AIChE J.*, **39**, 461-470.

Lucia, A. and X. Wang (1994). *Ind. Eng. Chem. Res.*, in press.

Magnussen, T., M.L. Michelsen and A. Fredenslund (1979). *Distillation* 79, 4.2/1.

McNulty, K.J. and Chatterjee, S.G. (1992). *Chem. Eng. Res. and Design*, **70**, 479.

Michelsen, M.L. and J. Mollerup (1986). *AIChE J.*, **32**, 1389.

Müller, F.R. (1979). Ph.D. Thesis in Chem. Eng., ETH Zurich.

Naphtali, L.M. and D.P. Sandholm (1971). *AIChE J.*, **17**, 148-153.

Onana, A. and A.M. Hikolo (1993). *Comput. Chem. Eng.*, **17**, 799.

Pantelides, C.C. (1988). *Comput. Chem. Eng.*, **12**, 745-755.

Powers, M.F., Vickery, D.J., Arehole, A. and Taylor (1988). *R. Comput. Chem. Eng.*, **12**, 1229-1241.

Prokopakis, G.J., and W.D. Seider (1983a). *AIChE J.*, **29**, 49.

Prokopakis, G.J., and W.D. Seider (1983b). *AIChE J.*, **29**, 1017.

E. Ranzi, M. Rovaglio, T. Faravelli, G. Biardi (1988). *Comput. Chem. Eng*, **12**, 783-786.

Rosenbrock, H.H. (1960). *Trans. Inst. Chem. Engrs.*, **38**, 279-287.

Rosenbrock, H.H. (1962). *Automatica*, **1**, 31-53.

Rovaglio, M. and M.F. Doherty (1990) *AIChE J.*, **36**.

Rovaglio, M., T. Faravelli, G. Biardi, P. Gaffuri, S. Soccol (1993). *Comput. Chem. Eng.*, **17**, 535.

Ruiz, C.A., I.T. Cameron, R. Gani (1988). *Comput. Chem. Eng.*,

12, 1-14.
Russell, R.A. (1983). *Chem. Eng.*, **90**, 53-59.
Shewchuk, C. (1974). Ph.D. Thesis, Cambridge University.
Seader, J.D. (1985). *Chem. Eng. Educ.* **19(2)**, 88-103.
Seader, J.D. (1986). *AIChE Monograph Series*, Number 15, 81.
Seader, J.D. (Oct. 1989). *Chem. Eng. Progress*, 41-49.
Seferlis, P. and A. Hrymak (1994a). *Chem. Eng. Sci.*, **49**, 1369-1382.
Seferlis P. and A. Hrymak (1994b). *AIChE J.*, **40**, 813-825.
Sereno, A.M. (1985). Ph.D. Thesis, University of Porto, Portugal.
Sivasubramanian, M.S. and J.F. Boston (1990). *Proc. Com-Chem* **90**, 331, Elsevier.
Sridhar, L.N. (1990). Ph.D. thesis, Clarkson University.
Sridhar, L.N. and A. Lucia (1989). *Ind. Eng. Chem. Res.*, **29**, 1668-1675.
Sridhar, L.N. and A. Lucia (1990a) *Ind. Eng. Chem. Res.*, **29**, 793-803.
Sridhar, L.N. and A. Lucia (1990b) *Comput. Chem. Eng.*, **14**, 901-905.
Sridhar, L.N. and A. Lucia (1994). *AIChE J.*, in press.
Stainthorp, F.P., and P.A. Whitehouse (1967). *I. Chem. E. Symp. Series*, **23**, 181.
Stewart, W.E., K.L. Levien, and M. Morari (1985). *Chem. Eng. Sci.*, **40**, 409-421.
Swartz, C.L.E. and W.E. Stewart (1986). *AIChE J.*, 32, 1832-1838.
Taylor, R., Powers, M.F., Lao, M., and Arehole, A. (1987). *Distillation* 87, B321-329.
Taylor, R., Kooijman, H.A., and Woodman, M.R. (1992). *Distillation* 92, A415-427.
Taylor, R., R. Krishna (1993). *Multicomponent Mass Transfer*. Wiley NY.
Taylor, R., H.A. Kooijman, J-S. Hung (1994). *Comput. Chem. Eng.*, **18**, 205-217.
Taylor, R., K. Achuthan and A. Luica (1993). *AIChE National Meeting*, St. Louis.
Tomich, J.F. (1971). *AIChE J.*, **16**, 229.
Venkataraman, S., and A. Lucia (1986). *AIChE J.*, **32**, 1057.
Venkataraman, S., and A. Lucia (1988). *Comput. Chem. Eng.*, **12**, 55-69.
Venkataraman, S., W.K. Chan and J.F. Boston (1990). *Chem. Eng. Progress* , **8**, 45.
Vickery, D.J. and R. Taylor (1986). *AIChE J.*, **32**, 547.
Vickery, D.J., T.L. Wayburn, and R. Taylor (1987). *Distillation* 87, B305.
Vickery, D.J., J.J. Ferrari and R. Taylor (1988). *Comput. Chem. Eng.*, **12**, 99.
Wang, J.C. and Y.L. Wang (1981). Proc. *FOCAPD*-1 II, 121.
Wayburn, T.L. and J.D. Seader (1983). *FOCAPD*-2, 765.
Whitehouse, P.A.(Apr. 1964). Ph.D. Thesis, *UMIST* .
Westman, K.R., A. Lucia and D. Miller (1984). *Comput. Chem. Eng*, **8**, 219.
Widagdo, S., W.D. Seider, and D.H. Sebastian (1989). *AIChE J.*, **35**, 1457-1464.
Widagdo, S., W.D. Seider, and D.H. Sebastian (1992). *AIChE J.*, **38**, 1229-1242.
Woodman, M.R. (1989). Ph.D. Thesis, Cambridge University.
Woodman, M.R., W. Morton and W.R. Paterson (1990). *CHISA*, Prague.
Wozny, G., M. Neidert, and A. Gorak (1991). *Fat Sci. Technol.*, **93**, 576-581.
Young, T.C. and W.E. Stewart (1990). *AIChE J.*, **36**, 655-664.
Yuxiang, Z. and X. Xien (1992). *Chem. Eng. Res. and Design*, **70**, 459, 465.
Zimmermann, A., Gourdon, C., Joulia, X., Gorak, A., and Casamatta, G. (1992). *Comput. Chem. Eng.*, **16S**, S403-410.

DESIGN OF MEMBRANE AND PSA PROCESSES FOR BULK GAS SEPARATION

Douglas M. Ruthven
Department of Chemical Engineering
University of New Brunswick
Fredericton, NB, Canada

Shivaji Sircar
Air Products and Chemicals Inc.
Allentown, PA 18195

Abstract

The mathematical models used in the design of membrane and pressure swing adsorption (PSA) processes are briefly reviewed. Although these processes have similar overall performance characteristics their principles of operation are entirely different, and this is reflected in the design procedures. Issues such as parametric sensitivity and the stability of numerical schemes for the solution of differential equations are generally much more important for PSA processes, which are modelled by a set of coupled non-linear partial differential equations. Brief comments on overall process economics and the role played by CAD are also included.

Keywords

Membrane, Pressure swing adsorption, Gas separation.

Introduction

During the past fifteen years pressure swing adsorption (PSA) and membrane processes have experienced a rapid growth of market share and are now considered economically competitive options for bulk gas separations even for quite large scale operations (up to perhaps 200 tons/day product gas). This development has been driven largely by the reduction in capital costs, relative to the more traditional cryogenic distillation process.

Although they operate on entirely different principles, PSA and membrane processes are often bracketed together as non-cryogenic alternatives for smaller scale gas separation applications. In general economic terms this makes sense since both PSA and membrane processes are best suited to the production of a pure raffinate product (less strongly adsorbed or less permeable species). In PSA process the less strongly (or less rapidly) adsorbed species passes through the adsorbent bed and is recovered as the high pressure (raffinate) product. In a membrane process the less permeable species is rejected by the membrane and is recovered as the high pressure retentate product. The retentate from a membrane system is therefore analogous to the raffinate stream in a PSA process. In both processes, the major operating cost is the power required to compress the feed stream or to evacuate the desorbed product. However, there are also important differences; membrane processes operate under steady state conditions while PSA is, by its very nature, a transient process so the design equations are quite different. Furthermore, despite the similarity in operating costs, the scaling of capital costs with throughput for the two processes, is quite different. While the capital cost of a membrane system increases almost linearly, the capital cost of PSA process varies with a fractional power of the throughput (about 0.6). This has important consequences for the ranges of economic viability.

In this paper we present a brief summary of the basic approaches used in the design of both PSA and membrane processes for bulk gas separation with an indication of the role played by CAD in modern design procedures. Our discussion is limited to the process design aspects; CAD procedures are also widely used in detailed mechanical design and for optimization of plant layout as well as for final optimization of the operation of the process, but these aspects are not considered in any detail.

Membrane Processes

The principle of operation of a membrane separation process (Fig. 1) is deceptively simple. The separation depends on differences in permeation rate arising from differences in solubility and diffusivity between the different molecular species in the feed stream. An economic process requires both a membrane material with a sufficiently high selectivity between the key components and a module which gives a sufficiently high permeation flux at an acceptable operating pressure. These requirements are more restrictive than might be expected and it is only during the last decade that such processes have achieved widespread commercialization. Not surprisingly the first commercial membrane gas separation processes were for the recovery of H_2 or He since the permeation rates for these gases are substantially greater than for all other heavier species.

(a)

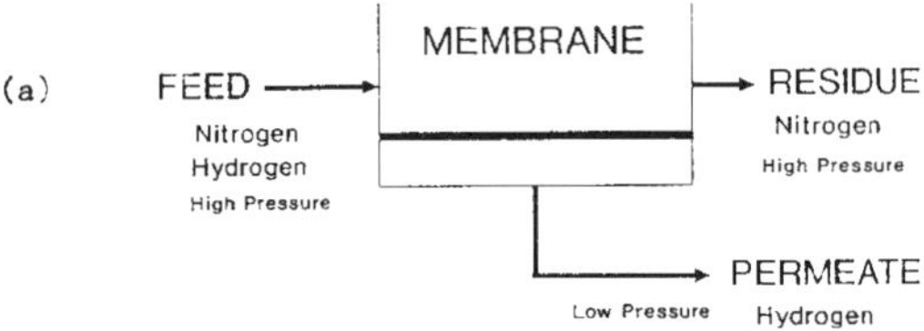

(b)

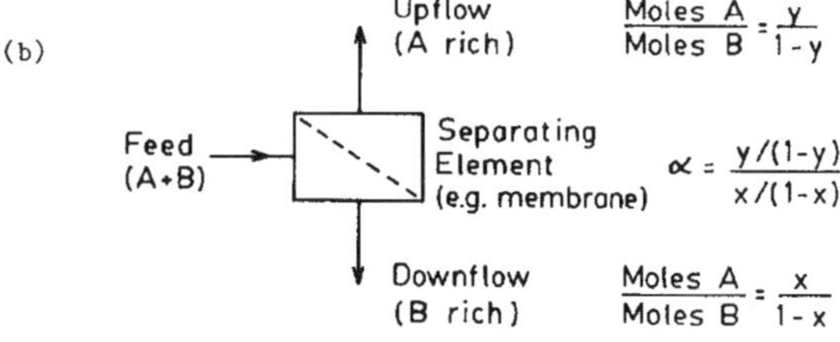

Figure 1. (a) Concept of membrane separation process and (b) definition of separation factor α.

Selectivity and Permeability

The permeability (π) for any species depends on the product of the diffusivity and solubility and the intrinsic selectivity (s) is defined by the permeability ratio:

$$N_A = \frac{K_A D_A}{\delta}(P_H x - P_L y); \quad s = \frac{\pi_A}{\pi_B} = \frac{K_A D_A}{K_B D_B} \tag{1}$$

In contrast to the intrinsic selectivity, the separation factor (α), defined in Fig. 1b, depends on the pressure ratio across the membrane. By considering the net flux for each species as proportional to the differences in partial pressures across the membrane one obtains:

$$\frac{y}{1-y} = \frac{N_A}{N_B} = s \cdot \frac{rx - y}{r(1-x) - (1-y)} \tag{2}$$

where $r = P_H/P_L$ is the pressure ratio. In the limit when r becomes very large, the effect of the back pressure becomes negligible and the separation factor approaches the intrinsic selectivity:

$$\alpha \equiv \frac{y/(1-y)}{x/(1-x)} \rightarrow s \ (\text{for } r \rightarrow \infty) \tag{3}$$

The more general relation for the separation factor when back-pressures are significant was first given by Naylor and Backer (1955):

$$\alpha = \left(\frac{1+s}{2}\right) - \frac{1}{2x} - \frac{s-1}{2xr} + \left[\left(\frac{s-1}{2}\right)^2 + \frac{r(s-1) - (s^2-1)}{2rx} + \left(\frac{s-1+r}{2rs}\right)^2\right]^{1/2} \tag{4}$$

In comparing the properties of different polymeric materials a compensation between solubility and diffusivity is commonly observed so that the selectivity is generally lower than might be expected simply from a comparison of the diffusivities or solubilities. This is illustrated in Table 1.

Table 1. Compensation of Diffusivity and Solubility in Polymer Membranes (H_2S at 30°).

Membrane	P (Torr)	D ($cm^2.s^{-1}$)	K (-)	π = KD (cm^3STP/ cm.s.atm)
Nylon	110	3×10^{-10}	7.9	2.4×10^{-9}
	620	4.9×10^{-10}	5.3	2.6×10^{-9}
Polyvinyl	244	5.5×10^{-9}	0.4	2.2×10^{-9}
Trifluoro	453	4.9×10^{-9}	0.4	2.0×10^{-9}
Acetate	751	6.8×10^{-9}	0.3	2.0×10^{-9}

An inverse relationship between selectivity and permeability is also commonly observed (Fig. 2) so that the choice of a membrane material involves a compromise between these properties. Since the flux also depends inversely on the membrane thickness it is possible to compensate for a low permeability by reducing the thickness. However, the membrane must have sufficient physical strength to withstand the applied pressure difference, which is often quite large, and this imposes a limit on the minimum thickness. The usual solution is the asymmetric membrane (Fig. 3) in which the active polymeric membrane film is supported on an inert macroporous inorganic support.

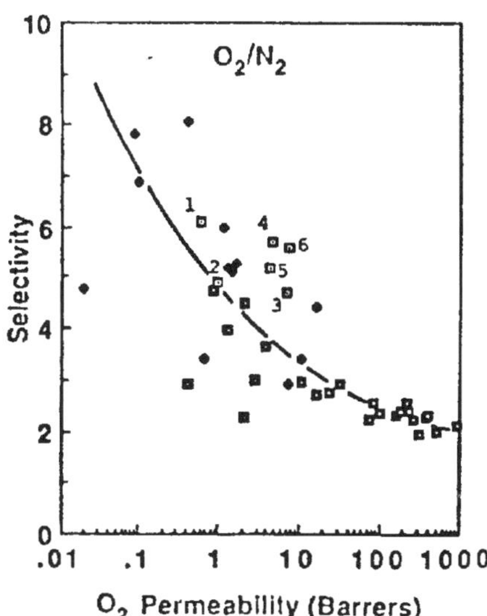

$$\left[\text{Barrers} = \frac{cm^3 STP}{cm^2 \sec (cm\ Hg\ /cm)} \right]$$

Figure 2. Variation of selectivity with permeability for O2/N2 separation on polymeric membranes (Koros et al. 1988).

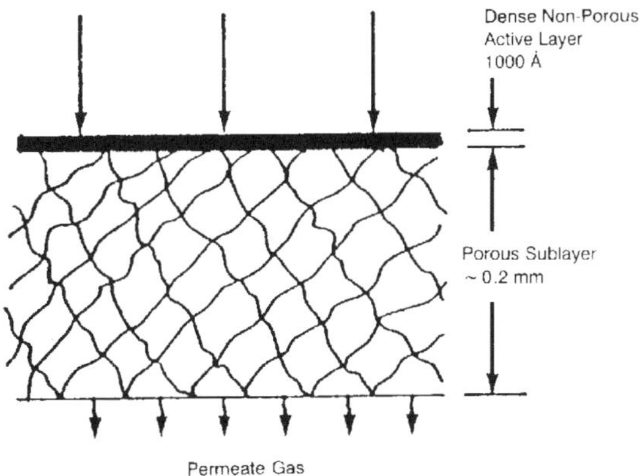

Figure 3. Schematic diagram solving construction of an asymmetric membrane.

Membrane Modules

Even with a thin polymer film (typically 0.1-0.5 μm) and a pressure ratio of several atmospheres, permeation fluxes are modest so, for large scale process, a large membrane area is required. The two most common designs of membrane module (spiral wound and hollow fibre) are shown in Fig. 4. In recent years the hollow fibre module has become the preferred choice.

The choice of flow direction in a hollow fibre membrane presents an interesting problem. It is physically easier to coat the outer surface of the hollow fibre with the active polymer film but, if the active membrane is on the outer surface, the system must be designed with this as the high pressure side. From the process standpoint it is more important to maintain a close approximation to plug flow on the high pressure side and this is more easily achieved if the high pressure is on the *inside*. Mechanical considerations also favour maintaining high pressure on the tube side. In practice both arrangements are found and the best choice seems to be system specific.

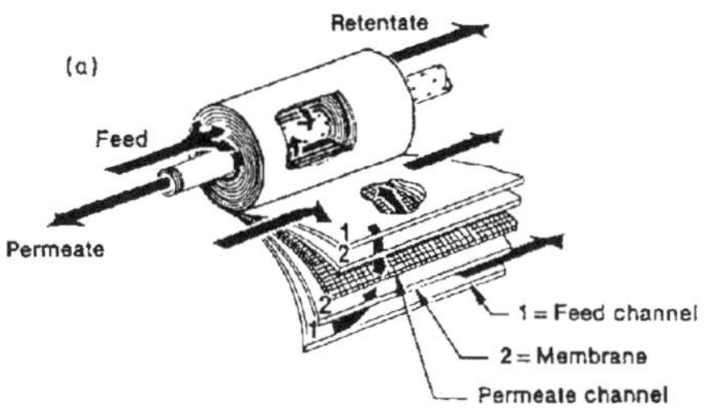

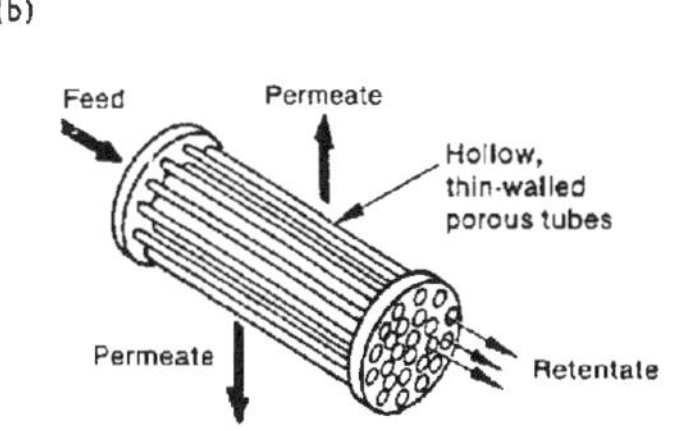

Figure 4. (a) Spiral wound and (b) hollow fiber membrane modules.

Flow Pattern and Membrane Cascades

Simple theoretical considerations suggest that, in the ideal system, counter-current plug flow should be maintained on both sides of the membrane (Fig. 5a). However, this is not easy to achieve in practice. Where a membrane is used to produce a pure raffinate (retentate) product, the flow on the low pressure (permeate) side, at the product end, will be very low. The large variation in flow rate makes it difficult to maintain a good approximation to plug flow. A cascade of several cross-flow elements in which plug flow is maintained on the high pressure side with partial mixing on the low pressure side (Fig. 6a) is therefore sometimes used to provide a practically reasonable alternative to a counter-current system. Somewhat different forms of cascade are used when the permeate stream has to be recycled or where a pure permeate is the required product (Figs. 6b and 6c). However, recompression adds greatly to the cost of the system so a membrane system tends to be most cost effective where a pure raffinate is the required product. This is the situation in air separation to produce nitrogen where a simple counter-current arrangement, or the equivalent cross-flow cascade, is normally used.

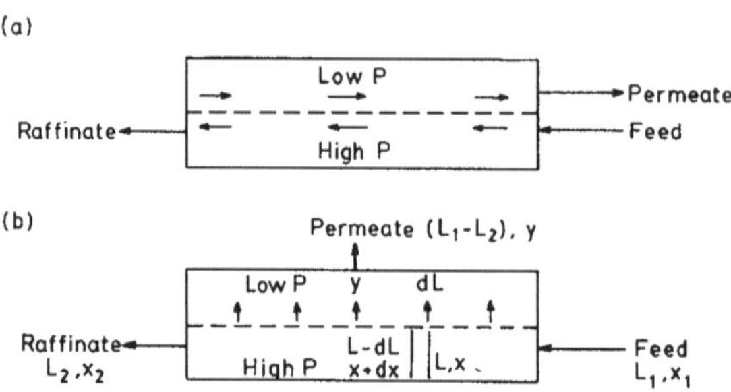

Figure 5. Schematic diagram showing the flow pattern in (a) counter-current and (b) cross-flow membrane element.

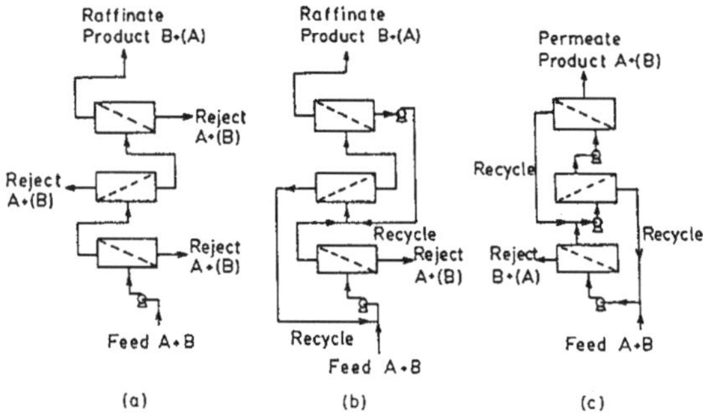

Figure 6. Cascades for membrane separation processes. (a) to produce pure raffinate product (permeate discharged) (b) to produce pure raffinate product (permeate recycled) (c) to produce a pure permeate product.

Design Calculations

Membrane separation processes are designed to operate under steady state conditions so the basic design calculations are relatively straightforward, involving integration of the differential mass balances to determine the concentration profile through the unit and hence to estimate the fractional recovery and required area for any defined feed and product specifications. The relevant equations are summarized in Tables 2 and 3 for cross-flow and counter-current flow systems. In the usual design problem the throughput, feed composition and product purity are specified and the properties of the available membrane materials are also known (permeability, selectivity and minimum allowable thickness). The main variables available for process optimization are the overall pressure ratio and, for a cross-current cascade, the number of series sections into which the unit is divided. Increasing the number of sections gives a closer approximation to true counter-current flow but the capital cost is, of course, increased. The main role of CAD is to arrive at the economic optimum, taking due account of both operating costs (mainly the power requirement) and the capital cost. It is also possible, by such procedures, to explore the trade-off between permeability and selectivity offered by the choice of different membrane materials.

Table 2. Design of Cross-Flow Membrane Module (Fig. 5b).

$$y\,dL = d(Lx) = L\,dx + x\,dL$$

$$\frac{dL}{L} = \frac{dx}{y-x} = \frac{dx}{x(1-x)(\alpha-1)} + \frac{dx}{1-x} \qquad \text{(A)}$$

$$\ln\left(\frac{L_2}{L_1}\right) = \ln\left(\frac{1-x_1}{1-x_2}\right) + \int_{x_1}^{x_2} \frac{dx}{(\alpha(x)-1)(1-x)x} \qquad \text{(B)}$$

Integration with $\alpha(x)$ given by Eq. 4 yields:

Recovery:

$$R \equiv \frac{L_2}{L_1}\left(\frac{1-x_2}{1-x_1}\right) = \exp\left[\int_{x_1}^{x_2} dx\,(\alpha-1)(1-x)x\right] \qquad \text{(C)}$$

Area:

$$dL_A = \frac{dA}{\delta}\pi_A P_H (x - y/r);$$

$$dG_B = dA\delta\frac{\pi_B P_B}{s}[(1-x)-(1-y)/r]$$

$$dL = dL_A + dL_B = \qquad \text{(D)}$$

$$\frac{dA\pi_A P_H}{\delta}\{(x-y)r + [(1-x)-(1-y)/r]/s\}$$

In crossflow system, y is constant but x varies from x_1 to x_2:

$$\frac{A\pi_A P_H}{\delta} = \int_{x_1}^{x_2} \frac{dL}{(x-y)r + [(1-x)-(1-y)/r]/s}$$

Substitute for dL from (A) with $L = L_2$, $x = x_2$ from (B) and integrate to find A for specified x_1, x_2.

Table 3. Design of Counter Current Flow Membrane Module (Fig. 5a).

$$-dL_A = -d(Lx) = \frac{dA\pi_A}{\delta}(P_H x - P_L y)$$

$$-dL_B = -d[L(1-x)] = \frac{dA\pi_A}{\delta}[P_H(1-x) - P_L(1-y)]$$

$$d(Lx) = d(L'y);\quad dL = dL'$$

$$\text{Hence: } -\frac{1}{L}\frac{dL}{dx} = \frac{1}{x + \dfrac{s}{(1-s)+(1-r)/(xr-y)}}$$

Mass balance around raffinate end:

$$L'y + L_2x_2 = Lx$$

$$L' = L - L_2 \qquad \text{(F)}$$

$$y = \frac{L_x - L_2x_2}{L'} = \frac{Lx - L_2x_2}{L - L_2}$$

To avoid trial and error substitute $g = L/L_2$ (in (E) and integrate from raffinate end ($g = 1$ at $x = x_2$):

$$-\frac{1}{g}\frac{dg}{dx} = \frac{1}{x + \dfrac{\alpha}{(1-s) + (1-r)/(xr-y)}}$$

with y from (F). At the raffinate end y_2 is given by Eq. 2 (with x_2 specified). Integration is continued to $x = x_1$ (the feed value) to obtain L(x) and hence L_1 (at the feed). Recovery follows from (C). Area is found from integral (D) with y from (F) and L from the integration described above.

PSA Processes

The selectivity in a PSA process comes from differences in either adsorption equilibrium or adsorption rate between the components to be separated. Differences in sorption kinetics large enough to provide a basis for a kinetically selective PSA process are obtained only for microporous adsorbents (zeolites and carbon molecular sieves) for which, in the sterically hindered diffusion regime, micropore diffusivities vary strongly with molecular size and shape. However, even for such systems, the sorption rate depends also on the equilibrium, so kinetic and equilibrium effects are closely coupled.

In contrast to a membrane process in which there are only a few different types of cascade, between which the optimal choice can generally be deduced by intuitive reasoning, for PSA processes a wide range of different operating cycles have been developed and the optimal choice of both the cycle and the number of adsorbent beds are generally not intuitively obvious. However, once the adsorbent is specified (assuming that the key kinetic and equilibrium parameters are known) a relatively simple simulation of the process can provide valuable guidance concerning the best choice of operating cycle. In the wider sense this can be regarded as an important (perhaps even the most important) role played by CAD in the design of a PSA process. Once an operating cycle has been selected CAD procedures are also used to "design" the process by optimizing such variables as the number of beds, the operating pressures and the times for the various steps. This step in the design procedure generally requires a more detailed and sophisticated dynamic model than that needed to guide the initial choice of operating cycle.

Dynamic Models for a PSA Process

The basic differential equations governing the performance of a PSA process come from transient differential mass, momentum and energy balances, together with appropriate expressions for the (multicomponent) adsorption equilibrium and the mass transfer rate expression (which may in fact be a coupled set of diffusion equations with associated boundary conditions). The models used to represent a PSA process differ mainly in the form of the mass transfer rate expression, the form of the (multicomponent) equilibrium isotherm and the accuracy with which thermal effects and pressure drop through the adsorbent bed are accounted for. Two representative models are summarized in Tables 4 and 5. In general, for an equilibrium-based process, a relatively simple kinetic model is adequate but thermal effects and even pressure drop can be very important. In contrast, a kinetic separation requires an accurate and detailed representation of the kinetics but heat effects are generally less significant and the accuracy of the equilibrium expression is also less critical. In both kinetic and equilibrium based processes the assumption of local thermal equilibrium is generally a valid approximation although a significant (time varying) axial temperature profile is quite common, particularly in bulk separations.

Table 4. Mathematical Model for Separation of CH_4-C_2H_4 on Activated Carbon Skarstrom Cycle — Equilibrium controlled process (i = CH_4 or C_2H_4).

Component Balance:

$$v\frac{\partial c_i}{\partial z} + c_i\frac{\partial v}{\partial z} + \frac{\partial c_i}{\partial t} + \left(\frac{1-\varepsilon}{\varepsilon}\right)\frac{\partial \bar{q}_i}{\partial t} = 0$$

Heat Balance:

$$\frac{\partial}{\partial z}(vT) + \frac{\partial T}{\partial t} + \frac{C_s}{C_g}\left(\frac{1-\varepsilon}{\varepsilon}\right)\frac{\partial T}{\partial t} + \sum_i \frac{\Delta H_i}{C_g}\cdot\frac{\partial \bar{q}_i}{\partial t} = 0$$

Momentum:

$$-\frac{\partial P}{\partial z} = \frac{150\mu(1-\varepsilon)^2}{\varepsilon^3 d^2}\cdot(\varepsilon v) + \frac{1.75\rho(1-\varepsilon)}{\varepsilon^3 d}(\varepsilon v)^2$$

Equilibrium (Toth):

$$\frac{q^*}{q_s} = \frac{b_i c_i}{\left[1 + \left(\sum_i b_i c_i\right)^k\right]^{1/k}}$$

Rate:

$$\frac{\partial q_i}{\partial t} = k_i[q_i^* - \bar{q}_i]$$

Boundary conditions through which these equations are coupled are not quoted.

In both cases the mathematical model consists of a set of coupled partial differential equations (PDEs) with associated initial and boundary conditions. To complete the simulation this set of equations must be solved repeatedly for each step of the cyclic process in sequence, using the final concentration profile for each step as the initial condition for the next step in the cycle. This procedure is repeated for a sufficiently large number of cycles until convergence to the cyclic steady state is achieved. The calculations are bulky and an efficient numerical scheme is therefore essential.

Table 5. Mathematical Model for Air Separation over CMS to Produce N_2 Self-Purging Cycle ($i = O_2$ or N_2).

Component Balance:

$$-D_L\frac{\partial^2 c_i}{\partial z} + v\frac{\partial c_i}{\partial z} + c_i\frac{\partial v}{\partial z} + \frac{\partial c_i}{\partial t} + \left[\frac{1-\varepsilon}{\varepsilon}\right]\frac{\partial \bar{q}_i}{\partial t} = 0$$

Continuity:

$$\sum_i c_i = c = \text{constant (negligible pressure drop)}$$

Overall balance:

$$c\frac{\partial v}{\partial z} + \left(\frac{1-\varepsilon}{\varepsilon}\right)\sum_i \frac{\partial \bar{q}_i}{\partial t} = 0$$

Equilibrium (Langmuir):

$$\frac{q_i^*}{q_i} = \frac{b_i c_i}{1 + \sum_i b_i c_i}$$

Rate:

$$\frac{\partial q_A}{\partial t} =$$

$$\frac{D_{AO}}{1-\theta_A-\theta_B}\cdot\left[(1-\theta_B)\left(\frac{\partial^2 q_A}{\partial r^2} + \frac{2}{r}\frac{\partial q_A}{\partial r}\right) + \theta_A\left(\frac{\partial^2 q_B}{\partial r^2} + \frac{2}{r}\frac{\partial q_B}{\partial r}\right)\right]$$

$$+\frac{D_{AO}}{(1-\theta_A-\theta_B)^2}\cdot\left[(1-\theta_B)\frac{\partial \theta_A}{\partial r} + \theta_A\frac{\partial \theta_B}{\partial r}\right]\left(\frac{\partial q_A}{\partial r} + \frac{\partial q_B}{\partial r}\right)$$

$$\theta_A = q_A/q_s, \theta_A = q_B/q_s;$$

-expression for $\delta q_B/\delta t$ is similar

Average concentration:

$$\bar{q}_i = \frac{3}{R}\int_0^R q_i r^2 dr$$

Boundary conditions through which these equations are coupled are not quoted.

Numerical Methods

The numerical procedure involves reducing the set of PDEs to a (much larger) set of ordinary differential equations (ODEs) which can then be integrated by standard routines. This reduction can be achieved either by finite difference methods or by the method of weighted residuals, of which orthogonal collocation is the most popular variant (Villadsen 1978). For a given accuracy, collocation methods are generally much faster than finite difference methods (Sun; Hassan et al. 1987). This advantage was crucial in the earlier days of PSA simulation but has become less significant with the increase in computing speed during the last decade. The efficiency and accuracy of the finite difference method can be improved significantly by moving the grid with the adsorption front (see, for example, the adaptive grid extension of the method of lines, proposed by Hu and Schiesser (1981)). A similar variant of the collocation method (collocation is finite elements) in which the collocation frame is moved with the adsorption front, has also been developed but has not been widely exploited, probably due to the additional complexity (Carey and Finlayson 1975).

With the plug flow approximation the key equations (the differential fluid phase mass balances for each component) are of hyperbolic form. The equations may be reduced to parabolic form by incorporating an axial dispersion term but this term is generally not large so the behaviour is still "quasi hyperbolic." If mass transfer is rapid and the equilibrium is favourable this can lead to very sharp concentration (or temperature) fronts in certain regions of the bed and this in turn may cause significant numerical problems since the resulting set of ODEs is then "stiff" and requires a numerically stable integration routine, such as Gear's "stiff" algorithm (Gear 1971; Hindmarsh 1985). Such routines are, however, relatively slow. The most effective ODE integration packages include a choice of integration routines that can be selected depending on the "stiffness" of the equations. One of the first such packages was FORSIM, developed in 1974 by M.B. Carver at AECL. In its updated version this is still superior to most of the more widely available commercial routines. A more detailed discussion of numerical methods has been given by Costa, Loureiro and Rodrigues (1989; 1984).

C_2H_4-CH_4 Separation

As an example of an equilibrium based PSA simulation we have summarized in Table 4 the system for methane-ethylene separation studied by Hartzog and Sircar (Villadsen 1978). With accurate model parameters and proper consideration of the heat balance including the variation of isosteric heat with loading, (the base case) the simulation provides a good representation of the experimentally observed behaviour. However, relatively modest perturbations to the parameters lead to large differences in predicted performance (Table 6). The profiles for the extreme assumption of an isothermal system are compared with the "base case" profiles in Fig. 7.

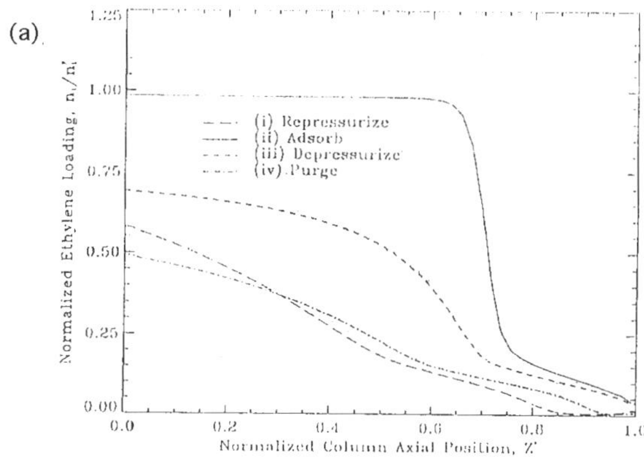

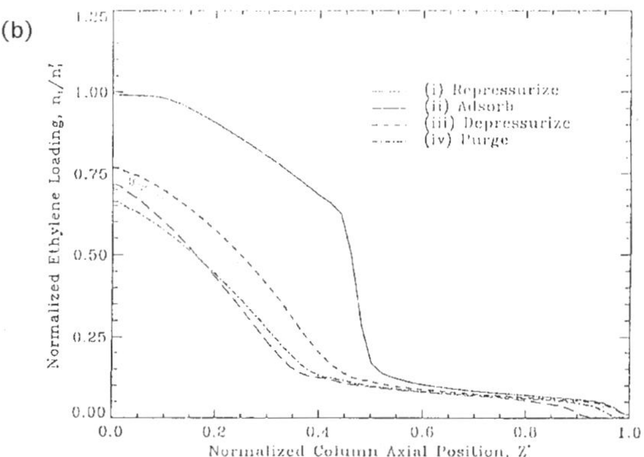

Figure 7. Profiles of C_2H_4 concentration at the ends of the four steps of the cycle. (a) system assumed isothermal; (b) 'base case' with realistic heat effects (adiabatic).

Table 6. Sensitivity of PSA Performance to Errors in Model Parameters.

CH_4-C_2H_4 separation over CMS to produce 99.9% CH_4 as Raffinate Product.

	CH_4 Recovery (%)	Productivity (g moles/g adsorbent day)	Purge / Feed (-)
Base Case	28.4	0.195	2.4
Case I	22.1	0.14	3.54
Case II	36.2	0.29	1.95
Case III	60	0.83	1,48
Case IV	18.9	0.125	5.0

Cases I and II — isotherm increased/decreased by 5% at constant selectivity by increasing /decreasing = -ΔH by 1 k cal/mole for both components.
Case III — system assumed isothermal.
Case IV — variation of ΔH with loading neglected.

The conclusion is that a numerical simulation can provide a good representation of an equilibrium based process but a realistic model is necessary and even modest errors in the equilibrium isotherm can lead to large errors in the predicted performance. Commonly used approximations such as isothermal operation are obviously inadequate for detailed design and optimization studies. Thus while CAD can play a valuable role in selecting and optimizing the design, final verification in a pilot system is still essential.

Air Separation to Produce N2

Air separation over carbon molecular sieve to produce nitrogen as the high pressure raffinate product provides a good example of a kinetically controlled PSA separation. Brief details are given in Table 5. Perhaps not surprisingly, the most stringent requirement for proper simulation of a kinetic PSA process is an accurate representation of the sorption kinetics. It appears that the cyclic nature of the process means that small errors are amplified so that the prediction of the cyclic steady state is very sensitive to small differences in rate. Commonly used approximations such as the linear driving force rate expression and even the use of a diffusion model with constant Fickian diffusivities lead to inaccurate predictions of performance, even though such models can provide an excellent representation of the breakthrough behaviour for a single step. To represent the behaviour adequately it is necessary to account for the concentration dependence of the diffusivities and to use chemical potential gradient rather than concentration gradient as the driving force in order to account for coupling between the component fluxes. This is illustrated in Fig. 8.

Process Economics

A breakdown of the capital and operating costs of a typical commercial PSA nitrogen process (CMS) is given in Table 7. The most important component is the power cost which is essentially determined by the pressure ratio and the "recovery." From the process standpoint, optimization of recovery subject to the required product purity is therefore the major aim of CAD optimization studies. The major component of the capital cost is attributable to the mechanical components, so it is clear that the contribution of CAD to the optimization of the mechanical design, including mundane aspects like component layout can be quite important.

Table 7. Comparative Economics of PSA/Membrane Processes for N_2 Production.*

	PSA	MEMBRANE
Operating Costs (Power)	50%	50%
Capital Costs		
- Adsorbent/Membrane	10	30
- Compressor	10	10
- Piping Valves, etc.	30	10

*Based on cost of gas for intermediate scale unit (10-20 tons/day)

The breakdown of costs for a membrane process (N_2 production) is broadly similar although the cost of the membrane element is relatively more important while the costs of piping, valves, etc. are relatively less important. Since, for both types of process, the major cost is the power, a simple analytic comparison can be drawn between the two processes, based simply on the fractional recovery at a given pressure ratio. Such a comparison is shown in Fig. 9 for conditions typical of current technology. With currently available adsorbents (diffusivity ratio ~ 45) and membranes (s ~ 5-6) the performance of PSA and membrane systems over the range of practical interest (> 98% N_2) is remarkably similar. The economic choice is therefore generally governed by the capital costs which, at low throughputs, are similar but become more favourable to PSA at higher throughputs.

More detailed discussion of PSA/membrane process economics have been given by Spillman and Thorogood.

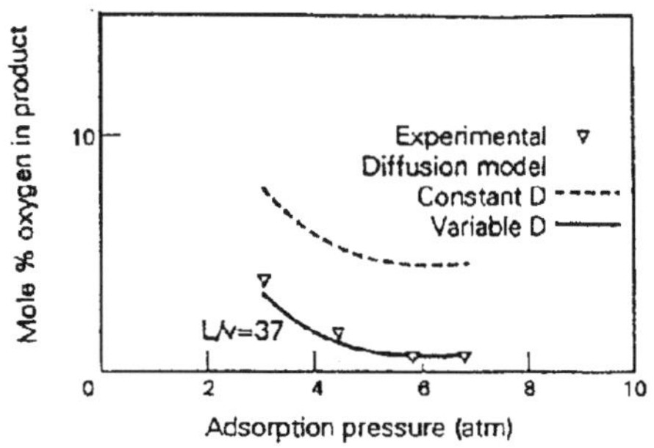

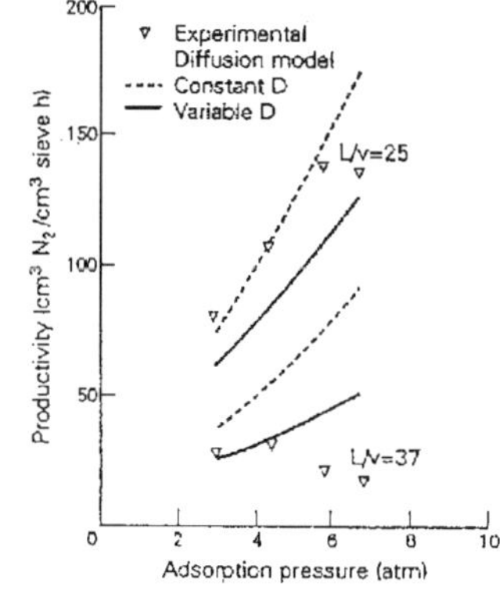

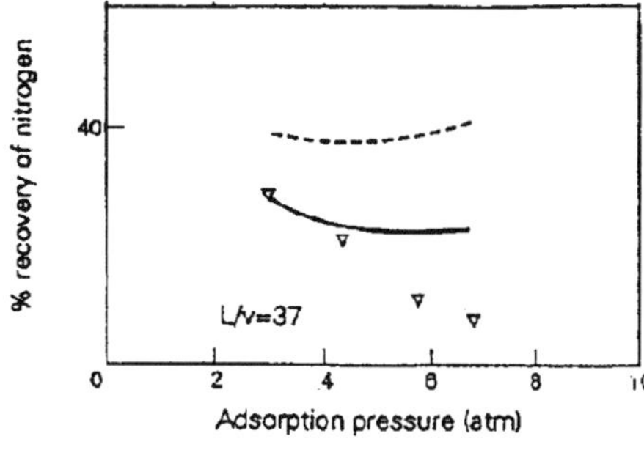

Figure 8. Purity, productivity and recovery for N_2 production over CMS as function of adsorption pressure showing comparison of simulation predictions and experimental data.

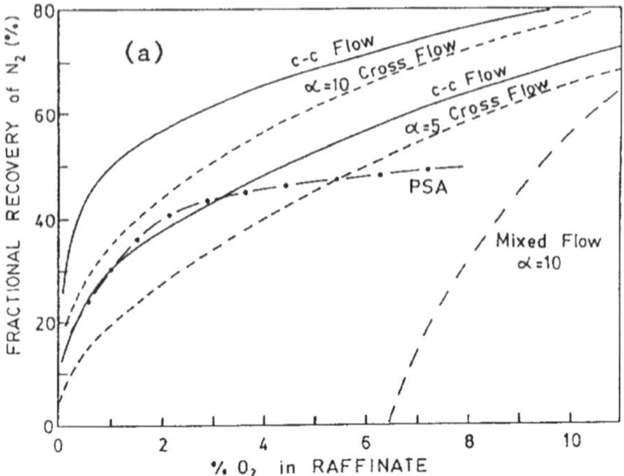

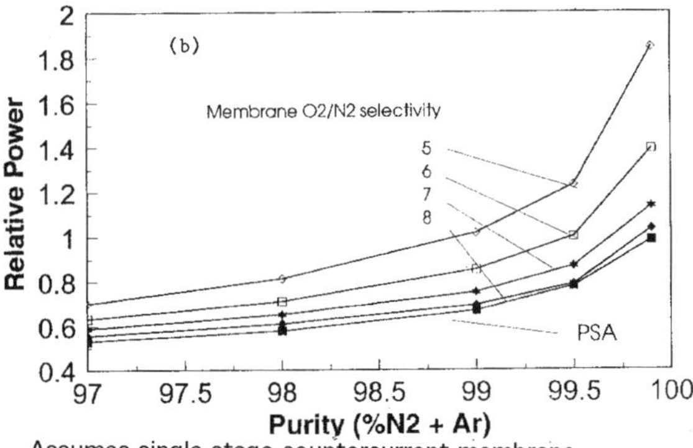

Figure 9. Comparison of performance of PSA and Membrane Processes to produce N_2 from air. (a) Recovery-Purity profiles; (b) Relative power requirements [Part (b) was kindly provided by Prof. R.M. Thorogood of North Carolina State University.

Nomenclature

A	= membrane area
b_i	= adsorption equilibrium constant
c_i	= sorbate concentration of species i in gas phase
c	= total molar density of gas phase
C_s	= volumetric heat capacity of adsorbent
C_g	= volumetric heat capacity of gas
d	= particle diameter
D_i	= diffusivity of species i
k	= mass transfer rate coefficient
L,L'	= molar flows on high and low pressure sides of membrane
K_i	= dimensionless equilibrium constant
N_i	= flux of component i
P	= total pressure
P_H,P_L	= high and low pressures
q_i	= adsorbed phase concentration of component i
q_s	= saturation limit
r	= pressure ratio P_H/P_L

R = fractional recovery
s = selectivity of membrane ($K_A D_A / K_B D_B$)
t = time
T = temperature
v = interstitial velocity
x = mole fraction of more permeable species on high pressure side
y = mole fraction of more permeable species on low pressure side
z = distance
α = separation factor (Eq 3)
δ = membrane thickness
ΔH = heat of adsorption
ε = voidage of adsorbent bed
π_i = permeability ($K_i D_i$)

References

Blaisdell, G.T. and K. Kammermeyer (1973). Counter-Current and Co-Current Gas Separations. *Chem. Eng. Sci.*, **28**, 1249.

Carey, G.F. and B.A. Finlayson (1975). Orthogonal Collocation on Finite Elements. *Chem. Eng. Sci.*, **30**, 587.

Carver, M.B. (1974, 1978). FORSIM: A Fortran Package for Automated Solution of Coupled Partial and/or Ordinary Differential Equations. AECL Report 4844, Nov. 1974. (See also AECL Report 5821 Feb. 1978 FORSIM VI Simulation Package.)

Costa, C. and A. Rodrigues (1989). Numerical Methods for Solution of Adsorption Models. *Adsorption: Science and Technology*. NATO ASI E158 A.E. Rodrigues, M.D. LeVan and D. Tondeur eds. Kluwer, Dordrecht.

Gear, C.W. (1971). Numerical Initial Value Problems in ODEs. Prentice Hall, Englewood Cliffs NJ.

Hassan, M.M., N.S. Raghavan and D.M. Ruthven (1987). Numerical Simulation of a PSA Air Separation System — Comparative Study of Finite Difference and Collocation Methods. *Can. J. Chem. Eng.*, **65**, 512.

Hindmarsh, A.C. (1985). Stiff Problems and their Solution at LLNL. *Stiff Computation*. R. Aikin Ed. Oxford Press, London.

Hu, S.S. and W. Schiesser (1981). Adaptive Grid Method in the Numerical Method of Lines Times. *Advances in Computer Methods for PDEs*. R. Vichnevetsky and R.S. Stapleman eds. IMACS.

Koros, W.J., G.K. Fleming, S.M. Jordan, T.A. Kim and H.A. Hoehn (1988). Polymeric Membrane Materials for Solution-Diffusion Based Permeation Separations. *Prog. Polymer Sci.*, **13**, 339.

Loureiro, J.M. and A.E. Rodrigues (1994). Two Solution Methods for Hyperbolic System of PDEs. *Chem. Eng. Sci.*, in press.

Naylor, R.W. and P.O. Backer (1955). Enrichment in Gaseous Diffusion Separations. *AIChE J.*, **1**, 95.

Sun, L.M. and M.D. LeVan. Numerical Solution of Diffusion Equations by Finite Difference Methods. *Chem. Eng. Sci.*, in press.

Villadsen, J. and M.L. Michelsen (1978). *Solution of Differential Equations by Polynomial Approximation*, Prentice-Hall, Englewood Cliffs, NJ.

REACTOR SYSTEM SYNTHESIS AND DESIGN

Jan J. Lerou
Central Research and Development Department
E. I. DuPont de Nemours & Company
Experimental Station
P.O. Box 80262
Wilmington, DE 19880

Introduction

As Joe Miller pointed out in his opening lecture, the pressure is on all of us to design and operate our chemical processes better than ever. How to do that is our job, and it is a very challenging one. If one looks at the complexity of the chemical reactions we are dealing with and the selectivity issues we have to address, one has to understand the chemistry in order to be able to design better reactions and better reactor networks.

The chart displays the evolution which has been going on for the last ten years and which is still going on how to handle homogeneous and catalytic reactions for complex chemistries. We used to do a lot of lumping to describe the global process chemistries; this was useful when the availability of computer power was limited. Gradually, we are moving to describe in more detail, on the molecular level, all the elementary steps which are taking place, be it on a catalyst or in a radical scheme. That is where the design of reaction paths occurs. The future will probably extend this approach to the description of the chemistry on the electronic level. Molecular modeling is already applied in some areas and I believe that we will see more of this in the future, also for complex reactions.

Complex reaction networks have been tackled in different ways in the last decade. Froment and co-workers (1984, 1985 and 1988) used a graph theory to generate reaction paths for thermo-cracking and hydrocracking. Froment's work in this area is guided by the desire to be able to achieve fundamental kinetic modeling of complex processes.

Klein and co-workers (1992, 1993) followed a different route to describe the kinetics of complex reactions. They used Monte Carlo simulations of the complex mixture to get to the kinetics. They applied this approach to the kinetics of polynuclear aromatic hydrocarbons, the pyrolysis of heavy oils, coal and lignite, and to model the fluid catalytic cracking of gas-oils.

The microkinetic analysis proposed by Dumesic and co-workers (1993) is a third way to get catalytic reaction path synthesis, which is related to the electronic structure of the species.

Another view and procedure to synthesize reaction paths and to prune them to an adequate level, if presented, is in the Mavrovouniotis and Bonvin paper. The procedure presented in this paper which includes chemometric techniques from a basic framework for molecular-level systems engineering of complex chemical processes.

The synthesis of reactor networks as the authors of the second paper state is much less advanced and is not yet able to take full advantage of all the information one can get out of the reaction pathway analysis. Hildebrandt and Biegler give a very clear description of the state of the art in the synthesis of chemical reactor network and of the effort which lead to this state.

In conclusion, we can state that both papers of this section contain building blocks which will ultimately lead to improved processes synthesis. One key element of success will be our ability to integrate these approaches and procedures.

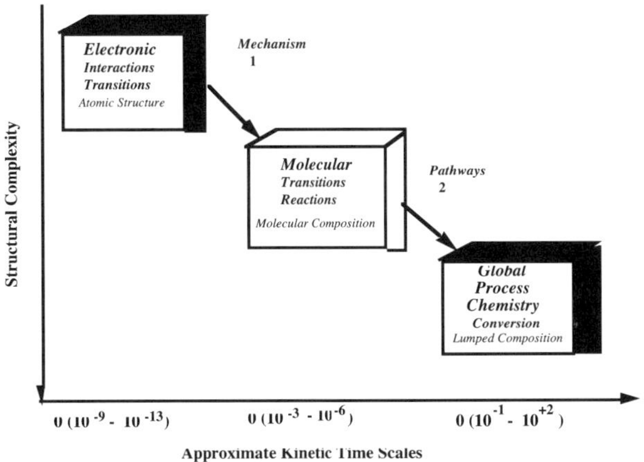

References

Clymans, P.J., and G.F. Froment (1984). *Computers and Chem. Engg.*, **8**, 137.

Baltanas, M.A., and G.F. Froment (1985). *Computers and Chem. Engg.*, **9**, 71.

Hillewaert, L.P., J.L. Dierickx, and G.F. Froment (1988). *AIChE Journal*, **34**, 17.

Liguras, D.K., M. Neurock, M.T. Klein, S. Stark, C. Libianati, A. Nigam, H.C. Foley, and K.B. Bischoff (1992). AIChE Symposium Series: *Advanced FCC Technology*, G. Young and R.M. Benslay, eds.

Klein, M.T., M. Neurock, L. Broadbelt, and H.C. Foley (1993). ACS Symposium Series: *Selectivity in Catalysis*, M.E. Davis and S.L. Suib, eds., 291.

Dumesic, J.A., D.F. Rudd, L.M. Aparicio, J.E. Rekosko, and A.A. Trevino (1993). *The Microkinetics of Heterogeneous Catalysis*, ACS Washington, D.C.

TOWARDS DESIGN OF REACTION PATHS

Michael L. Mavrovouniotis
Chemical Engineering Department
Northwestern University
Evanston, IL 60208-3120

Dominique Bonvin
Institut d'Automatique
Ecole Polytechnique Federale de Lausanne
CH-1015 Lausanne, Switzerland

Abstract

In the design, operation and optimization of chemical processes that involve complex chemical reaction systems, stricter specifications and increased competition necessitate detailed analysis and design of the chemical reaction network. This paper presents a set of techniques that carry out synthesis and analysis tasks in a complex chemical reaction network. The network itself can be generated from a formal description of the general types of reactions that take place in the process; this description is compiled into procedures which screen candidate compounds and automatically apply the reaction to compounds that meet the reaction type's restrictions. Recursive application of the reaction-generation to a small set of initial species can yield the complete, complex reaction network of the process. Within this network, only some compounds are stable enough to be overall raw materials or products of the process; a synthesis algorithm can construct paths, and derive all permissible stoichiometries, for the interconversion of stable compounds. This reaction set can be pruned by analyzing experimental results through chemometric techniques. The remaining stoichiometries provide a simplified system description which allows the statement of meaningful design objectives. Iterative improvement of the performance of the process can be achieved through selection of a promising set of process inputs, and application of chemometric techniques on experimental results to decipher how these inputs affect the dominant reactions in the process. These techniques form a promising basic framework for molecular-level systems engineering of complex chemical processes.

Keywords

Reaction paths, Factor analysis, Reaction identification, Singular value decomposition, Reaction generation, Pathway synthesis, Chemometrics, Stoichiometric models, Synthesis, Reaction processes.

Introduction

In recent years, the design and operation of chemical processes have been faced with increasing competitive pressure and stricter specifications. The latter can stem from product quality requirements or from environmental or other regulations. Tighter specifications arise even for bulk products that are not analyzed to the level of individual chemical compounds; for example, in fluid catalytic cracking (FCC) of oil fractions, one must cope with environmental restrictions on aromatics and performance requirements on octane numbers. We note that the quantitative changes required are often small on an absolute scale (this is certainly true for FCC) but nevertheless have a substantial economic impact.

These new challenges cannot be met with ordinary lumped models that are useful for determining equipment sizes. They require manipulations which change rates in the detailed chemical reaction network. The design and operation of the processes can no longer ignore the detailed molecular transformations, and must find ways to cope with the complexity of their description and modeling.

The goal of this paper is to pose questions, elucidate problems, and show the complementary relevance of different methodologies in the analysis and synthesis of reaction paths for complex processes.

Since there is no general framework cemented in place, it is hard to assess the impact and role of various isolated tools; we therefore profess that we will devote our attention to those methods whose role and interplay we understand best: methods developed in our own laboratories. We briefly note other relevant prior work in the area of generating and describing reaction paths. Quann and Jaffe (1992) have developed a method called structure-oriented lumping to describe the composition, reactions and properties of complex hydrocarbon mixtures. Froment (1991) automatically generates the detailed reaction networks of processes involving either carbenium ions (catalytic reforming, catalytic cracking and hydrocracking) or free radicals (olefins production), using boolean matrices to represent the hydrocarbons. Chevalier et al. (1990) generate the reaction network of oxidation reactions involving aliphatic hydrocarbons, using a similar philosophy. A computer generated model of pyrolysis recently developed by Broadbelt et al. (1994) also uses matrices to represent the reactants and reactions (the reaction matrices only contain the atoms involved in the reaction).

Complex Chemical Systems

It is useful to divide chemical reaction processes into three broad categories, based on the complexity of the reaction system that takes place:

1. Low reaction-system complexity. Many processes entail only a small number of *known* reactions (often just one or two main reactions and one or two undesired side reactions). The reaction engineer's task is mainly to arrange process conditions, heat transfer, mixing, and other transport phenomena to favor the desired reactions over the undesired ones.
2. Intermediate reaction-system complexity. In some processes we have a system with 5-15 reactions, most of which are known (some side-reactions might not be known).
3. High reaction-system complexity. The system involves a combinatorially formed reaction set; a few reaction types acting recursively give rise to a large, complex network. These processes are common in refinery operations aimed at upgrading the value of streams by reducing the molecular weight, increasing branched hydrocarbons, etc.

This classification is based on complexity resulting from the shear number of reactants, products, and molecular pathways — not complexity of mechanistic steps within simple overall pathways. Thus, if there are only two or three overall molecular reaction routes but numerous mechanistic steps (due to complexity of surface phenomena, for example), we will regard this system as low complexity.

This paper is not concerned with the first category of reaction systems, since the task of designing reaction paths is not relevant for such processes. The second category presents more interest, in that the selection of desired paths and the identification of reactions from data begin to interact. However, algorithms and computational techniques are not indispensable, and they often take a back seat to the qualitative analysis of an expert.

The complexity of the reaction system in the third category of processes makes computational treatment essential. The rest of the paper focuses on precisely this category of processes.

Raw materials and target products are usually not specified at the detailed, molecular level in these processes. They are defined in terms of a variety of coarse physical measures and sets of compounds (although occasionally key individual compounds are singled out). The general types of reactions taking place are usually known, in terms of the substructures or categories of compounds on which they act. However, the specific reactions of specific chemical compounds in the system are not enumerated.

Another characteristic of such systems is that some of the compounds involved are stable species that can be present in the feed and product streams, but others are unstable intermediates of the reaction system. The unstable intermediates occur in small amounts; they are difficult to measure and usually are not measured. The intermediates have high turnover; they are formed and consumed at rates similar to the stable compounds. But during most of the operation of the process their formation very nearly balances their consumption and their net rate of accumulation is close to zero. This leads to the common quasi-steady state or pseudo-steady state assumption for the unstable intermediates.

Spectrum of Problems

In this paper the focus is exclusively on reactions taking place in a single processing unit. We are not concerned with the synthesis of reactor networks, although some of the insights gained from analyzing the reaction system (through the techniques presented here) may be useful in configuring multiple reactors.

We want to be able to determine whether a transformation or reaction sequence can take place or not, or more generally, what set of transformations can take place in the system. We use the term pathway to describe reaction sequences that achieve interesting overall transformations.

The presence of suitable pathways is relevant during preliminary process conception and the iterative improvement of an initial design. It is also relevant in the improvement of an existing, operating process

Within the design of chemical reaction processes, we focus on the portion of the design activity that is concerned with the reaction paths: choosing appropriate paths, making the paths take place and analyzing which of the paths operated in experimental runs.

These tasks have implications in many practical decisions, such as the design or selection of catalysts, since the catalyst affects the predominance of pathways. A catalyst (its characteristics including preparation conditions) affects the rates of many reactions simultaneously, and *a priori* quantification of these effects is difficult. Thus, the design cannot be carried out in a purely synthetic manner; one must analyze the effects of changes on experimental measurements. Although we will not specifically discuss catalysts, we will be examining the analytical side of this activity.

The objective of the design of reaction paths is the production of specific target products from specific raw materials. Suppose, for example, that the objective is to increase the octane number of the product of fluid catalytic cracking. We have a general idea for what products lead to higher octane numbers (e.g. branched hydrocarbons) but which ones we target would depend on the particular set of transformations taking place; a product may be desirable in general, yet undesirable if it is accompanied by an undesirable byproduct. We are in fact more comfortable judging reaction stoichiometries between raw materials and products, rather than individual species. And we prefer selecting among specific candidate stoichiometries, rather than attempt to define favorable stoichiometries *a priori*. from a large space of possibilities. Since, in the end, we are looking at properties of mixtures, we will have to test our preferences and our design decisions to verify whether the overall product streams have met our objectives.

Reaction Generation

Complex reaction systems usually involve known types of reactions but unknown specific reactions. A general reaction type specifies what molecular substructures or categories of compounds the reaction applies to; and how it alters the structure of the compounds. This description is usually informal, verbal or schematic.

A specific reaction, on the other hand, involves individual chemical compounds with a specific stoichiometry.

In a complex chemical system, if the chemistry of the process is understood then the general reaction types are known. But the generation of the specific reactions requires considerable effort, because of the large number of chemical compounds likely to be present in the system.

Language for Reaction Generation

Efforts towards automated generation of reactions are maturing into a *language* for the description of generic types of chemical reactions (Prickett, 1994). Compared to prior work in the generation of reactions, this language provides a generic and extensible framework. The language includes descriptors of the actual changes caused by the reaction to the molecular structure, as well as preconditions that must be satisfied in order for the reaction to be applicable. The language has a LISP-like parenthesized syntax; alternate front-end command processors that modify the syntax (e.g. to remove the parentheses or change the order of the command options) are easy to develop.

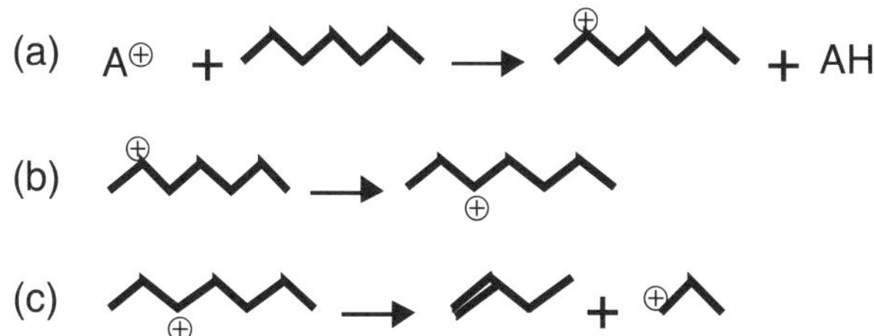

Figure 1. Examples of three reaction types: (a) Hydride abstraction from a neutral alkane, where A⊕ represents the catalyst site that picks up the hydride. (b) Hydride shift, leading to isomerization of an ion. (c) Scission of an ion to form an alkene and another ion.

Examples of Generic Reaction Types

The removal of a hydride (as in an acid catalyst for FCC) with formation of a cation (Fig. 1a) is described in this language as follows:

```
(Require hydrocarbon molecule)
(Require neutral molecule)
(Label C1 (Find carbon))
(Forbid (Find-exactly 3 hydrogens
   attached-to C1))
(Label H1 (Find hydrogen attached-to
   C1))
(Disconnect C1 H1)
(Add-charge C1)
(Set number-of-reactions (Symmetry-num-
   ber H1))
```

The last command is intended to reduce complexity by treating all equivalent atoms collectively. In a methyl group, the 3 hydrogens are equivalent; the procedure will examine only one and will store the symmetry number 3 in the generated reaction to account for the multiplicity. The line `(Forbid (Find-exactly 3 hydrogens attached-to C1))` is intended to exclude primary ions, on the assumption that they are not favored thermodynamically.

Another example, below, is α-hydride shift (Fig. 1b), in which a hydride that is in α position to a charge moves to that charge (while the positive charge moves, in effect, in the opposite direction).

```
(Label C1+ (Find positive-carbon))
(Label C2 (Find carbon attached-to C1+))
(Forbid (Primary C2))
(Forbid (Find double-bond attached-to
   C2))
(Label H1 (Find hydrogen attached-to
   C2))
(Disconnect C2 H1)
(Connect C1+ H1)
(Add-charge C2)
(Remove-charge C1+)
(Set number-of-reactions (Symmetry-
   number H1))
```

The line `(Forbid (Primary C2))` precludes the formation of a primary ion by this reaction. This line has the same function as the expression `(Forbid (Find-exactly 3 hydrogens attached-to C1))` used in the hydride-removal description. In fact, `(Primary C2)` is only a macro for `(Find-exactly 3 hydrogens attached-to C2)`. The language includes facilities for defining such macros; the above descriptions already use other macros.

Our last example will be β-scission of a alkyl cation (Fig. 1c):

```
(Label C1+ (Find positive-carbon))
(Label C2 (Find carbon attached-to C1+))
(Label C3 (Find carbon attached-to C2))
(Forbid (Find aromatic-ring belonging-
   to C3))
(Forbid (Find double-bond attached-to
   C3))
(Forbid (Find-exactly 3 hydrogens
  attached-to C3))
(Disconnect C2 C3)
(Increase-bond-order (Find (Connecting-
   bond-of C1+ C2)))
(Subtract-charge C1+)
(Add-charge C3)
(Set number-of-reactions (Symmetry-
   number C3))
```

The expression `(Find positive-carbon)` is another macro that could easily be replaced with a sequence of simpler expressions.

Example of Complete Network

The descriptions generated in this language are compiled into functions that screen candidate molecules and carry out the reaction. Thus, given an initial set of molecules, these procedures, recursively applied, will generate the complete reaction system. One such example was processed (Prickett, 1994), involving the following reaction types: Hydride-removal from a neutral alkane, and its reverse (hydride addition to an alkyl cation); hydride shift, described in the preceding subsection; β-scission and its reverse (polymerization of an olefin and cation). These reaction types do not exhaust FCC chemistry; they are not intended to generate a realistic example, but rather to illustrate the complexity of generated networks.

The example started with 2-methyl butane as the initial species, and restricted the formation of primary ions to those formed by β-scission and the two primary ions of 2-methyl butane. The size of molecules was also restricted, to no more than 8 carbons. Recursive application of the generic reaction types yielded a large and dense network of reactions. In all, 151 species (with 37 different skeletal arrangements) were involved. Of these, 35 were neutral alkanes, 23 were neutral alkenes, while 93 were ions. Classified by carbon number, there were 3 species of C_2; 4 species of C_3; 9 species of C_4; 16 species of C_5; 23 species of C_6; 29 species of C_7; and 67 species of C_8.

Throughout the generation of the system, only unique reaction sites and species were considered (through the symmetry number technique mentioned in the previous subsection, and through canonicalization of molecular structures). A total of 441 distinct specific reactions were constructed. We note that, despite the high complexity, this system does not include all possible isomers and reactions in this carbon-number range. There are many more than 29 C_7 compounds (even restricted to at most 1 charge and 1 double bond), but most of them are simply not accessible from the starting materials and the 5 reaction types of this example.

Simulated Example

We will not use the chemical reaction system produced earlier for the rest of our study. At this early stage, the real example does not leave us with enough freedom to adjust the size of the system, the sparseness of the reactions, the mix of reversible and irreversible reactions, etc., to have results that are interesting yet comprehensible. The difficulty does not lie in the formal applicability of the algorithms, but in the human judgment involved in guiding the application and analyzing the results. When the framework is completed, we will be able to deal with realistic examples throughout.

As a substitute, we have constructed a simulated example which retains the basic features of a realistic system, but gives us more control to adjust the attributes of the reaction system. The scheme used in the construction of this simulated example is described in Appendix A.

Overall Pathways and Reactions

What transformations are in principle possible, given the set of specific reactions? The reactions themselves provide one answer, but we are more interested in the net interconversion of neutral compounds, viewing the ionic intermediates only as bridges enabling that interconversion.

Many of our insights into the behavior of complex chemical systems stem from this kind of separation of the species into unstable intermediates and stable compounds; we will call the latter *terminal* species.

We noted in an earlier section that design objectives involve judging candidate reaction stoichiometries (between raw materials and products), rather than designating individual species as desirable products. This requires the construction of suitable paths whose overall (net) reaction stoichiometries involve only terminal species. We will do this with the algorithm of Mavrovouniotis (1992). The algorithm accomplishes the synthesis of such paths through iterative elimination of the intermediates. In each iteration, one intermediate is eliminated by combining all partial paths in which it participates, with the correct combination coefficients.

In our example (defined in Appendix A), we will consider the site C_0^{+1} as a terminal species and the hydride-loaded C_0 as an intermediate. Among other species, we will consider neutral compounds to be terminal species. Ions will be considered unstable intermediates, because:

- They are chemically unstable; they cannot be raw materials and final products of the process.
- They cannot be measured.
- After an initial transient, they will have fast turnovers but slow changes in concentration (i.e. they will be in a quasi-steady state).

A total of 213 paths interconverting terminal species were synthesized for the simulated example. The pathways show that the example actually retains a good measure of similarity with real chemical pathways.

One path constructed involves formation of an ion, β-scission of the ion (creating a new unsaturation) and finally neutralization of an ion. Its steps are shown below. The coefficient -1 displayed before a step indicates that the step is utilized in its reverse direction.

$$(-1) * C_0 + (C_4U^{+1})_1 \rightarrow C_0^{+1} + (C_4U)_3$$
$$(C_4U^{+1})_1 \rightarrow C_2U^{+1} + C_2U$$
$$C_0 + C_2U^{+1} \rightarrow C_0^{+1} + C_2U$$

The net reaction accomplished by the pathway is: $(C_4U)_3 \rightarrow 2 * C_2U$. The path is reversible, because all its constituent steps are reversible.

Another path accomplishes overall β-scission, with a net reaction $C_5U \rightarrow C_2U + (C_3U)_1$, but there is an intermediate isomerization of the ion $(C_5U^{+1})_4$ to the ion C_5U^{+1} which also takes place through β-scission. The steps are:

$$(-1) * C_0 + (C_5U^{+1})_4 \rightarrow C_0^{+1} + C_5U$$
$$(C_5U^{+1})_4 \rightarrow (C_3U^{+1})_2 + C_2U$$
$$(-1) * C_5U^{+1} \rightarrow (C_3U^{+1})_2 + C_2U$$
$$(-1) * (C_5U^{+1})_3 \rightarrow C_5U^{+1}$$
$$(-1) * (C_5U^{+1})_2 \rightarrow (C_5U^{+1})_3$$
$$(-1) * C_3U^{+1} + C_2U \rightarrow (C_5U^{+1})_2$$
$$C_0 + C_3U^{+1} \rightarrow C_0^{+1} + (C_3U)_1$$

Because we have already used reaction reversibilities (discussed in the next section), the construction of paths was efficient (Mavrovouniotis, 1992) and we were able to identify the reversibility of each path.

Reaction Pruning

Having generated a reaction system, one invariably finds it too complex for subsequent study. How could we reduce our initial system?

A first question is the assignment of reversibility which limits the space of permissible transformations and paths. If we know or can estimate suitable thermodynamic parameters (Gibbs energies or, as a first approximation, enthalpies) for the compounds, then we can use a thermodynamic criterion to decide whether a reaction is reversible, and if it is irreversible what direction it will follow. The unstable intermediates are likely to present difficulties, but for simple ions there are suitable estimation methods (Mavrovouniotis and Constantinou, 1994).

This is a viable option, except for one difficulty: Initiation reactions may not be favored thermodynamically but they play a crucial role in the operation of the system. In our example, initiation requires the formation of ions. One must ensure that initiation is not eliminated by thermodynamic arguments.

The strategy followed in our example was to use a different thermodynamic threshold for each reaction type. Thus, hydride removal was made reversible in many cases allowing initiation to take place (it stayed irreversible in the direction of destroying the ion in some specific reactions, accounting for the fact that extremely unfavored ions will not form). Other general reaction types were made reversible by a thermodynamic threshold that led to the presence of both reversible and irreversible reactions.

The derivation of an overall strategy for reaction reversibility decisions that would not disable key initiation reactions is an open problem.

Experimental Pruning

While it is reasonable to assume that thermodynamic data are available, kinetic data usually are not. Thus, we cannot use *a priori* kinetic models to further reduce the reaction space. Any further pruning must be based on experimental data.

Bonvin and Rippin (1990) and Prinz (1992) presented a set of chemometric techniques for analyzing experimental data to determine what reactions are taking place in a system. The techniques are based on factor analysis (FA) or principal component analysis (PCA) of a matrix of concentration changes. The matrix is decomposed into orthogonal factors or principal components using the singular value decomposition (SVD). The numerical magnitude of a singular value indicates the relative importance of the corresponding factor. A variety of tasks are covered, including:

- Determination of the number of linearly independent reactions, in the presence of measurement noise.
- Target factor analysis to test proposed reaction stoichiometries for consistency with experimental data.
- Incremental factor analysis which allows one to use already established reactions to improve subsequent analysis. Both the determination of the number of remaining reactions and the testing of subsequent candidate reactions are facilitated by this approach. A key assumption in the incremental procedures is that reactions are linearly independent.
- Methods for coping with incomplete composition measurements or incomplete target stoichiometries.
- Criteria for equilibrium and enforcement of atomic consistency of data.
- Global compatibility checks for entire sets of proposed reaction stoichiometries.

A necessary condition for the application of these techniques is $S > R$ where S is the number of measured species and R the number of independent reactions (i.e. linearly independent stoichiometries). Since all intermediates are unmeasured we have only 24 measured species in our example. Thus, the analysis is possible only if there are no more than 23 independent reactions. While the system description includes a higher number of reactions, we can proceed with the analysis, in the hope that many of these reactions only take place in negligible rates.

The fact that in each reaction stoichiometry there are unmeasured species is not an obstacle, because the techniques can, in principle, deal with incomplete measurements. The insurmountable difficulty here is that there will only be one measured species for each reaction (see Appendix A), leading to extremely weak results. Any step would look plausible, provided its measured species reacted or was produced at all. Thus, the chemometric techniques cannot be applied at the level of individual reactions in our example.

The approach that we will follow here examines the system at the level of basic paths synthesized in the previous section, rather than individual reactions in the initial system description. Since the paths do not involve any unmeasured intermediates in their overall reactions, we can obtain reliable results.

We can now see that the pathway synthesis algorithm naturally complements the chemometric analysis. Where unstable and unmeasured intermediates prevent the direct application of chemometric techniques, the synthesis of paths enables chemometric analysis for net reactions which do not involve these unmeasured species.

Given the absence of information on intermediates, a path's plausibility depends solely on its overall (net) reaction. Therefore, we need to assess only the net reactions accomplished by the paths, rather than their internal steps. There are only 27 different overall reactions (for the 213 paths); they are shown in Table 1. The number of candidate reactions is still higher than the number of measured species (24), but we expect that only a few of these reactions actually take place (satisfying the condition $S > R$).

Table 1. Overall Reactions Accomplished by the Synthesized Paths.

#	Reaction stoichiometry	Reversibility	No. of paths
1	null reaction, representing circular paths	1	26
2	$(C_5)_1 \rightarrow C_2U + C_3$	0	1
3	$(C_5)_1 \rightarrow C_2U + (C_3)_1$	0	1
4	$(C_5)_1 \rightarrow C_5$	0	2
5	$(C_5)_1 \rightarrow C + (C_4U)_2$	0	1
6	$(C_3U)_1 + (C_5)_1 \rightarrow (C_4)_2 + (C_4U)_2$	0	1
7	$(C_3U)_1 + (C_5)_1 \rightarrow C_4 + (C_4U)_2$	0	1
8	$(C_4U)_3 + (C_5)_1 \rightarrow (C_4U)_2 + C_5$	0	1
9	$(C_4U)_3+(C_5)_1\rightarrow C_2 +(C_3U)_1+(C_4U)_2$	0	1
10	$(C_5)_1 \rightarrow (C_5)_3$	0	1
11	$(C_4U)_3 \rightarrow 2{*}C_2U$	1	4
12	$(C_4U)_1 \rightarrow 2{*}C_2U$	0	1
13	$(C_5U)_1 \rightarrow C_2U + (C_3U)_1$	1	16
14	$(C_4U)_1 \rightarrow (C_4U)_3$	0	1
15	$C_2U + (C_5U)_1 \rightarrow (C_3U)_1 + (C_4U)_3$	0	5
16	$C_5U \rightarrow C_2U + (C_3U)_1$	1	39
17	$(C_5U)_4 \rightarrow C_2U + (C_3U)_1$	1	16
18	$C_5U \rightarrow (C_5U)_4$	1	12
19	$C_2U + (C_3U)_1 \rightarrow (C_5U)_2$	1	9
20	$(C_5U)_1 \rightarrow (C_5U)_2$	1	6
21	$C_5U \rightarrow (C_5U)_2$	1	15
22	$(C_5U)_4 \rightarrow (C_5U)_2$	1	7
23	$(C_5U)_1 \rightarrow C_5U$	1	18
24	$(C_5U)_1 \rightarrow (C_5U)_4$	1	8
25	$C_2U + C_5U \rightarrow (C_3U)_1 + (C_4U)_3$	0	12
26	$C_2U + (C_5U)_4 \rightarrow (C_3U)_1 + (C_4U)_3$	0	5
27	$C_2U + (C_5U)_2 \rightarrow (C_3U)_1 + (C_4U)_3$	0	3

A remaining question is the linear independence of the reaction stoichiometries, assumed by the factor analysis approach. In the initial examination of candidate reactions, the presence of dependent reactions does not, in itself, prevent the analysis. It may lead to false plausibility of some reactions, but it will not lead to false rejection of any candidate.

Linear dependence may lead to false rejection in incremental factor analysis. Incremental analysis is of tremendous help in dealing with noisy data (in our example we may regard the contributions of the less important reactions as noise). Incremental factor analysis modifies the data to remove the information that is due to any reactions already accepted; this has disastrous consequences for linearly dependent reactions.

Since linear independence of reactions cannot be guaranteed, we will avoid incremental analysis, and examine all the reactions separately and in parallel. We risk false acceptance of candidate reactions, but not false rejection.

Simulation of Reaction System

If we want the elimination of reactions and paths to have broad validity, we must make sure that the conditions of the experiment give all reactions an opportunity to take place. In our simulation, we start with a mixture of all the neutral compounds, in concentrations set to 1, and a much larger concentration of catalyst sites (here 10) to avoid tight competition for sites. We also seed the system with a smaller initial concentration (0.001) of all the intermediates. We assume that all reactions were adequately represented in this system. In order to carry this out as an experiment in practice, we need to make sure that all stable compounds are present, and all the intermediates have an opportunity to form. The latter objective may require the use of initiators or other precursor compounds. The formation of the intermediates is important, since they may act as homogeneous catalysts.

The selection of experimental conditions that allow adequate sampling of the reaction space is an open problem. Competition of reactions is the most difficult issue: Some reactions may be active primarily in the *absence* of competitors, and they may be overlooked in an experiment that samples a very broad set of reactions.

We simulated the example reaction system for 251 time points, producing a matrix of 250 × 24 changes in composition. The first 12 singular values of the matrix of concentration changes are shown in Table 2. As expected, there is a rapid decline in the magnitudes of the singular values (by roughly one order of magnitude for every 3 singular values).

Table 2. Singular Values of the Concentration Change Matrix, in the Reaction-space Reduction Experiment.

1-3 ($\times 10^0$)	4-6 ($\times 10^1$)	7-9 ($\times 10^2$)	10-12 ($\times 10^3$)
0.1607	0.0422	0.0597	0.0525
0.1184	0.0163	0.0218	0.0450
0.0261	0.0083	0.0138	0.0196

The identification of the number of dominant factors, i.e. the number of significant independent reactions active in the system, is a difficult problem. One approach is to look for a gap in the singular values (Table 2) or apply one of many empirical indicators (Malinowski, 1991). An F-test for the detection of the number of independent reactions (Malinowski, 1988; Prinz, 1992) involves a simple variance ratio on the singular values:

$$f_k = \frac{(n-k)\sigma_k^2}{\sum_{i=k+1}^{n} \sigma_i^2}$$

for each k ($1 \le k \le n$-1). We take the ratio of each f_k and the appropriate value from the F-distribution (depending on the number of degrees of freedom, n-k). The value of the ratio is shown in Fig. 2 as a function of k, for the 0.99 point of the F-distribution. A value less than 1 indicates that the corresponding factor can be explained by noise, i.e. the factor is not relevant. Fig. 2 also considers the reduced F-test (Malinowski, 1988, Prinz, 1992) which uses a slightly different formula for f_k:

$$f_k = \frac{\sigma_k^2 \sum_{i=k+1}^{n} (b-i+1)(n-i+1)}{(b-k+1)(n-k+1) \sum_{i=k+1}^{n} \sigma_i^2}$$

where b is the number of samples. Another alternative is the autocorrelation of singular vectors (Shrager and Hendler, 1982, Harmon et al., 1993), which will not be considered here.

The F-tests show a peak at 3 factors, but the value remains high because there are indeed many reactions in the system. The principal reason these tests do not give a clear answer is that the noise in our system is not random; what we call noise is the contribution of the less dominant reactions, not measurement error. For this reason, we will not commit ourselves to a specific number of reactions. Instead, we use a plausibility score for all the reactions in parallel. We examine, for a given number of factors from SVD, how close each reaction lies to the subspace of the factors, by comparing the reaction stoichiometry to its projection on this subspace (Bonvin and Rippin, 1990). We score the difference with an infinity norm. The results are shown in Table 3.

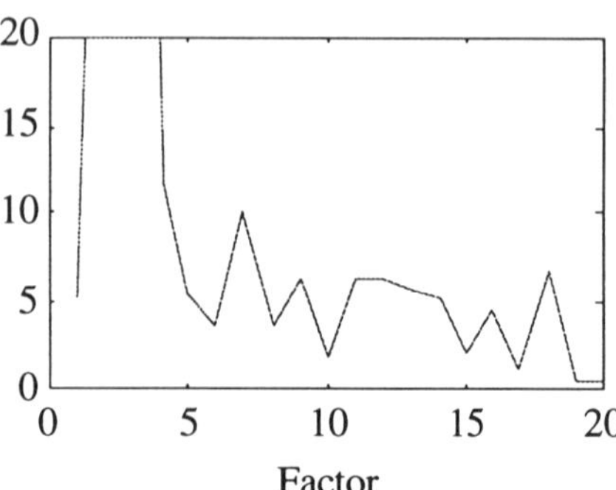

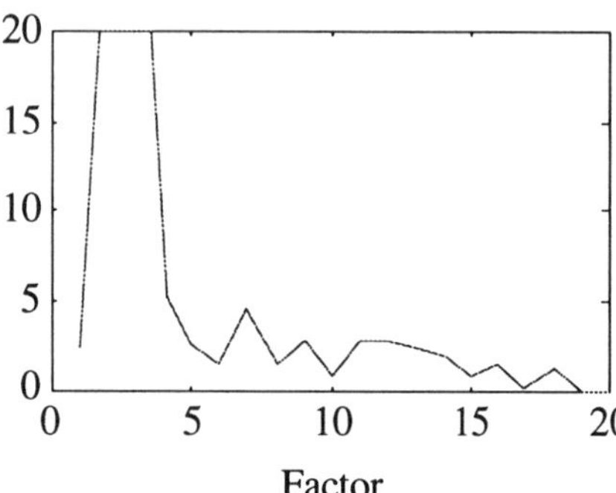

Figure 2. Tests for the number of dominant factors in the reaction-space reduction experiment. Top: F-test; Bottom: Reduced F-test.

If we select for consideration only those reactions which meet the criterion of distance < 0.3 for 9 factors, we obtain (excluding the null reaction) the net reactions 11, 13, 16, 19, 23, 25 and 27 (in Table 1). If we select for consideration only those reactions which meet the criterion of distance < 0.4 for 6 factors (excluding the null reaction) we obtain 11, 13, 15, 19 and 27. We take, as an aggressive strategy, the intersection of these two sets: 11, 13, 19, 27.

The above four net reactions include 32 of the original 213 paths. We can now prune individual steps from the system: If a step does not participate in one of the 32 remaining paths, it can be eliminated. Only 21 steps remain (Table 4) from the original set of Appendix A.

Table 3. Infinity-norm Test for Each of the Reactions of Table 1, for the Reaction-space Reduction Experiment. The Number of Factors Retained From SVD Varies From 3 to 18.

Rxn	3 factors	6 factors	9 factors	12 factors	15 factors	18 factors
1	0	0	0	0	0	0
2	1.0185	0.9419	0.8949	0.8260	0.6994	0.6421
3	1.0295	0.9856	0.9830	0.6086	0.5011	0.3303
4	0.9534	0.9021	0.8895	0.6581	0.4757	0.4359
5	1.0005	0.9996	1.0016	1.0286	1.0161	0.9979
6	1.0316	0.9995	1.0008	0.9795	0.9879	0.9714
7	1.0311	0.9998	1.0013	0.9842	0.9919	0.9745
8	1.0625	1.0665	1.0982	0.9842	0.9765	0.9811
9	1.0479	0.9968	0.9972	0.9926	0.9821	0.9773
10	0.9971	1.0049	1.0022	1.0123	0.7427	0.4407
11	0.4600	0.3029	0.2291	0.1784	0.1661	0.1654
12	0.9447	0.8287	0.7511	0.5498	0.5358	0.4677
13	0.5944	0.2805	0.2435	0.1327	0.0684	0.0132
14	1.0215	0.8866	0.8746	0.7283	0.7019	0.6331
15	0.8746	0.3675	0.3993	0.1820	0.1360	0.1567
16	0.9659	0.8890	0.2840	0.2295	0.1188	0.0119
17	0.9372	0.7772	0.6051	0.5646	0.3110	0.0223
18	1.0059	0.9552	0.6122	0.5909	0.3578	0.0307
19	0.4867	0.2153	0.1764	0.0748	0.0348	0.0072
20	1.0226	0.4958	0.4199	0.2075	0.1032	0.0195
21	0.9869	0.9764	0.3602	0.2712	0.1453	0.0052
22	1.0009	0.7346	0.5428	0.5128	0.2780	0.0260
23	0.9572	0.7967	0.2619	0.1820	0.0726	0.0241
24	0.9746	0.8940	0.7315	0.6736	0.3765	0.0284
25	0.6310	0.5861	0.2798	0.2399	0.1922	0.1773
26	0.6586	0.5681	0.4697	0.4916	0.2620	0.1466
27	0.6749	0.3903	0.2951	0.2197	0.1863	0.1726

Path Selection

At this point we have accomplished the generation of the reaction system from known general reaction types; the identification of overall reactions and paths among stable species; and the pruning of this space through analysis of its experimental behavior, and indirectly through reaction reversibility arguments.

Table 4. Surviving Steps (from Table 1) with the Number of Paths in Which They Now Participate.

Step	No. of paths	Step	No. of paths
3	9	39	10
6	7	40	6
8	13	43	7
13	7	49	1
21	12	50	14
22	7	51	1
27	9	52	6
35	16	53	8
36	14	55	13
37	6	56	5
38	9		

Our position has improved substantially. From knowing only the general reaction types involved, we now have a specific system of the most dominant reactions, assembled into paths that interconvert stable species. In the usual case of loosely defined objectives, the initial state of our problem did not permit us to arrive at practical specifications or criteria. With the space of transformations reduced to a manageable size (and involving only terminal species), we can now pursue our objectives by selecting reactions and paths from the list of remaining ones.

Table 5. Set of Paths After One Synthesis Round.

Category and reaction	Rev.	Origin
From basic set of synthesized paths		
$(C_4U)_3 \rightarrow 2 * C_2U$	1	11
$(C_5U)_1 \rightarrow C_2U + (C_3U)_1$	1	13
$C_2U + (C_3U)_1 \rightarrow (C_5U)_2$	1	19
$C_2U + (C_5U)_2 \rightarrow (C_3U)_1 + (C_4U)_3$	0	27
Elimination of $(C_3U)_1$		
$(C_5U)_1 \rightarrow (C_5U)_2$	1	13+19
$2 * C_2U + (C_5U)_2 \rightarrow (C_5U)_1 + (C_4U)_3$	0	27-13
$2 * C_2U \rightarrow (C_4U)_3$	0	19+27
Elimination of $(C_5U)_2$		
$2 * C_2U \rightarrow (C_4U)_3$	0	19+27
Elimination of C_4U_3		
$(C_5U)_2 \rightarrow (C_3U)_1 + C_2U$	0	11+27
Elimination of C_2U		
$2 * (C_5U)_1 \rightarrow (C_4U)_3 + 2 * (C_3U)_1$	1	2*13-11
$(C_4U)_3 + 2 * (C_3U)_1 \rightarrow 2 * (C_5U)_2$	1	2*19+11
$2 * (C_5U)_2 \rightarrow 2 * (C_3U)_1 + (C_4U)_3$	0	2*27+11

In effect, rather than having to specify what products are desirable (within a large set) before even knowing what is *feasible* in the system, we are now in a position to scan a list and express our preferences for very specific transformations.

An extra synthesis step is helpful in the selection: We can combine basic paths to uncover new useful transformations. These can be constructed using the algorithm of Mavrovouniotis (1992); one round of such longer paths is shown in Table 5 along with the original surviving ones. We note that the table shows for $2 * C_2U \rightarrow (C_4U)_3$ a new path which is irreversible.

The selection of good paths to promote and bad paths to suppress in the system depends on the specific objectives and constraints in the system.

Manipulation of Concentrations

We will assume that, after considering the possible transformations of Table 5, we select $2 * C_2U \rightarrow (C_4U)_3$ as a favorable transformation to promote. We carry out a new experimental run (in our case, of course, simulated) in which the concentrations were selected to favor this reaction. Specifically, we have a low baseline concentration (0.005) for all species (other than the catalyst site) and an enhanced concentration of 1 for the intended reactant, C_2U.

Table 6. Change in Concentration Over the Course of the Reaction, for the Reaction-manipulation Experiment.

Species	Conc. Change	Species	Conc. Change	Species	Conc. Change
C	0.0026	$(C_4)_1$	-0.0001	$(C_5)_2$	0
C_2	0.0015	$(C_4)_2$	0.0009	$(C_5)_3$	0.0014
C_2U	-0.4639	C_4U	0.0017	C_5U	0.0002
C_3	0.0016	$(C_4U)_1$	0.0396	$(C_5U)_1$	-0.0001
$(C_3)_1$	0.0011	$(C_4U)_2$	0.0003	$(C_5U)_2$	-0.0004
C_3U	0.0007	$(C_4U)_3$	0.1629	$(C_5U)_3$	0
$(C_3U)_1$	0.0101	C_5	0.0011	$(C_5U)_4$	0.0001
C_4	0.0011	$(C_5)_1$	0.0000		

Table 7. Infinity-norm Test for each of the Reactions of Table 1, for the Reaction-manipulation Experiment. The Number of Factors Retained From SVD Varies From 3 to 15.

Rxn	3 factors	6 factors	9 factors	12 factors	15 factors
1	0	0	0	0	0
2	1.0293	0.9973	0.9336	0.7733	0.5308
3	1.0238	0.9969	1.0071	0.8450	0.3166
4	1.0018	0.9853	0.9256	0.8951	0.5027
5	1.0031	0.9919	0.8938	0.8378	0.8733
6	1.0009	1.0165	0.9319	0.9169	0.9701
7	1.0011	1.0170	0.9679	0.9021	0.8206
8	1.0183	1.0207	0.9477	0.9190	0.8729
9	1.0177	0.9892	0.9519	0.8820	0.8912
10	1.0017	0.9939	0.9269	0.9170	0.9583
11	0.2236	0.0709	0.0641	0.0576	0.0419
12	0.9183	0.3168	0.2746	0.2498	0.1678
13	1.0170	1.0089	0.9288	0.6685	0.1052
14	1.1420	0.3877	0.3387	0.3073	0.2097
15	0.9852	1.0027	0.9186	0.6682	0.1098
16	1.0168	0.9332	0.8725	0.4066	0.3601
17	1.0169	0.9716	0.8122	0.6113	0.5409
18	1.0022	0.9260	0.8390	0.4218	0.3864
19	1.0169	0.8218	0.5446	0.3953	0.1504
20	1.0006	1.0083	1.0614	0.7360	0.1182
21	1.0022	0.9893	0.9520	0.2901	0.2443
22	1.0000	0.8404	0.8735	0.7120	0.6307
23	1.0020	1.0179	0.9636	0.6839	0.3277
24	1.0006	0.9959	1.0203	0.9450	0.6088
25	0.9851	0.9318	0.8623	0.4094	0.3571
26	0.9851	0.9141	0.7551	0.5652	0.4990
27	0.9851	0.8928	0.6087	0.4529	0.1791

The above choice represents the simplest way to favor the desired net reaction. The selection of better initial conditions for the concentrations, through analysis of the internal reactions of each path, is an open question.

Table 6 shows the changes in concentrations for all the species. It is easy to verify that the chosen reaction was overwhelmingly favored. This can be confirmed by an infinity-norm projection test (the same as that carried out earlier), shown in Table 7. If we use the same criterion as before to select the dominant reactions, we obtain (beyond the null reaction) only two reactions. Reaction 11 is $(C_4U)_3 \rightarrow 2 * C_2U$, which was the targeted reaction (except for its formal direction). Reaction 12 is $(C_4U)_1 \rightarrow 2 * C_2U$. Note that the isomerization of the product $(C_4U)_3$ is possible (reaction 14 in Table 1), but it does not take place. Instead, reaction 12 is in competition with the desired reaction for the raw material.

The only process manipulation we carried out here involved concentrations. However, if qualitative knowledge is available on how catalyst components or preparation conditions affect different types of reactions, one could proceed to manipulate catalyst selectivity to favor the desired transformations and inhibit the undesired ones.

Conclusions

In this paper we targeted the problem of design and improvement of complex chemical reaction processes. A convergence of circumstances makes the problem immensely important. Materials must be used more wisely because of economic and environmental restrictions, necessitating tighter design, analysis, and control of complex chemical processes. Although quantitative models are still hard to derive, high resolution measurement devices, such as GC-MS (Gas Chromatography — Mass Spectroscopy) allow the acquisition of detailed information from experiments. Developments in computer hardware and algorithms facilitate the processing of such information. Thus, both the need and some basic enabling technologies are in place for tighter design, analysis and control of complex chemical systems.

The paper discussed a spectrum of algorithmic techniques for the case when full quantitative models are either unavailable (because kinetic constants are not known) or would be too complex. The techniques form an initial framework for attacking this problem. The grand integration of these and other ideas will require much work, but a lot can be accomplished through careful application and customization of current algorithms.

References

Bonvin, D., and Rippin, D.W.T. (1990). Target Factor Analysis for the Identification of Stoichiometric Models. *Chem. Eng. Sci.*, **45**, 3417-3426.

Broadbelt, L.J., Stark, S.M., and Klein, M.T. (1994). Computer Generated Pyrolysis Modeling: On-the-Fly Generation of Species, Reactions, and Rates. *Ind. Eng. Chem. Res.*, **33**, 790-799.

Chevalier, C., Warnatz, J., and Melenk, H. (1990). Automatic Generation of Reaction Mechanisms for the Description of the Oxidation of Higher Hydrocarbons. *Ber. Bunsenges. Phys. Chem.*, **94**, 1362-1367.

Froment, G.F. (1991). Fundamental Kinetic Modeling of Complex Processes, *Chemical Reactions in Complex Systems: the Mobil Workshop*, A.V. Sapre and F.J. Krambeck, eds., Van Nostrand Reinhold: New York.

Harmon, J.L., Duboc, P., and Bonvin, D. (1993). Factor Analytical Modeling of Biochemical Data: A Tutorial. *Comp. Chem. Engng.*, submitted for publication.

Malinowski, E. R. (1988). Statistical F-tests for abstract factor analysis and target testing. *J. Chemometrics*, **3**, 49-60.

Malinowski, E. R. (1991). *Factor Analysis in Chemistry*. Wiley, New York.

Mavrovouniotis, M.L. (1992). Synthesis of Reaction Mechanisms Consisting of Reversible and Irreversible Steps: II. Formalization and Analysis of the Synthesis Algorithm. *Ind. Eng. Chem. Res.*, **31**, 1637-1653.

Mavrovouniotis, M.L., and Constantinou, L. (1994). Estimation of Heterolytic Bond Dissociation Energies of Alkanes. *J. Phys. Chem.*, **98**, 404-407.

Prickett, S.E. (1994). Object-Oriented Generation of Complex Reaction Systems for Chemical Processes. Ph.D. Thesis (advisor: M.L. Mavrovouniotis), University of Maryland at College Park.

Prinz, O. (1992). Chemometric Methods for Investigating Chemical Reaction Systems. Dissertation No. 9708 (examiners: D.W.T. Rippin, D. Bonvin, W. Simon), Swiss Federal Institute of Technology (ETH), Zurich.

Quann, R.J., and Jaffe, S.B. (1992). Structure-Oriented Lumping: Describing the Chemistry of Complex Hydrocarbon Mixtures. *Ind. Eng. Chem. Res.*, **31**, 2483-2497.

Shrager, R.L., and Hendler, R.W. (1982). Titration of Individual Components in a Mixture with Resolution of Different Spectra, pKs, and Redox Transitions. *Anal. Chem.*, **54**, 1147-1152.

Appendix A

We present here an illustrative example which is used in the body of the paper. This artificial reaction system is generated as follows. Molecules are constructed randomly. Each molecule is described by a number of carbons c, a number of unsaturations or double bonds u (where $u < c-1$), a number of positive charges p (which is restricted based on c and u), and a final random index i to account for isomers. We assign a Gibbs energy to each isomer, as a linear function of the indices c, u and p, and an extra random amount to account for isomers. The formula $(C_cU_u{}^{+p})_i$ will denote the simulated species with indices c, u, p and i; any indices that have trivial values (c = 1, u = 0, p = 0 or i = 0) may be omitted from the formula. This notation is not shorthand for real compounds; some isomers, such $(C_3U_1)_1$, as may have no real-compound counterpart.

Catalyst sites are explicitly represented as a species, $C_0{}^{+1}$, while the site with a hydride attached becomes C_0. We generate reactions from essentially the same reaction types as before, adjusted to the fact that we are no longer dealing with fully defined molecular structures. Hydride removal is interpreted as conversion of a catalyst site and a compound of charge p into a compound of charge $p + 1$; hydride shift is interpreted as isomerization of an ion (which must have charge $p > 1$); scission is interpreted as conversion of an ion into two smaller compounds, preserving the total number of carbons and charges but increasing the unsaturations by 1.

We adjust the density of the system by generating different numbers of isomers of compounds and generating reaction types with different probabilities. We also adjust reaction rates and the random variations in Gibbs energy to have an interesting mix of reversible and irreversible reactions. The reactions' direction and reversibility were determined through Gibbs energy thresholds for each reaction type. This general scheme allows us to select a system of suitable complexity and density, with 52 species and 56 reactions. The system is analyzed in the body of the paper.

SYNTHESIS OF CHEMICAL REACTOR NETWORKS

Diane Hildebrandt
Chemical Engineering Department
University of Witwatersrand
Johannesburg, RSA

Lorenz T. Biegler
Engineering Design Research Center
Carnegie Mellon University
Pittsburgh, PA 15213

Abstract

An interesting problem in chemical reactor theory is finding bounds or targets on a given performance index in a reacting system. Moreover, performance of the reactor subsystem has a key impact on the design of other processing subsystems. It determines the recycle structure of the process, the separation sequence and has a strong influence on the energy and environmental considerations. However, this area of process synthesis has seen relatively little development when compared to heat integration and separation synthesis. As with the design of heat exchanger networks, this approach has evolved into the (discrete and continuous) *optimization of network superstructures* as well as the *performance targeting* of the optimal network prior to its construction. In this study we review both methods for reactor network synthesis but concentrate on advances with the latter approach.

The targeting approach is based on geometric interpretations of reaction and mixing. It uses a constructive approach to find the attainable region; that is, it effectively captures all possible reactor structures and finds the bounds on the performance of a reacting system. The approach also generates reactor structures which are candidates for the optimal system. It is however severely limited by the dimensionality of the problem and in practice only 2 and 3 dimensional problems have been solved. Nevertheless, insights gained from this geometric approach have led to an understanding of more general properties of optimal reactor structures. In particular the reactors that make up optimal structures are parallel-series systems of plug flow reactors, CSTR's and differential side stream reactors. Furthermore, the number of parallel structures is related to the dimensionality of the problem. In addition, these properties can be embedded within optimization formulations in order to deal with more complex problems. In particular, we describe several formulations that incorporate simpler properties derived from attainable region concepts. At this point, this optimization approach is not as rigorous as the geometric, but readily extends to higher dimensional reaction systems. In addition, it can be integrated with other process subsystems and allows for simultaneous approaches for heat integration, separation structures and reactor network design. In this way, trade-offs resulting from different parts of the process are properly taken into account in the optimization. All of these concepts will be illustrated with numerous examples. Finally, future work will concentrate on the extension of geometric concepts to more general reactor systems as well as to separation systems. These will also lead to more compact optimization formulations and the consideration of larger and more complex process problems.

Keywords

Attainable region, Geometric properties, Nonlinear programming, Process integration, Process optimization, Process synthesis, Reactor networks.

Introduction

Over the last thirty years the field of process synthesis has matured into an established research area. Significant progress has been made particularly in the synthesis of homogeneous systems related to energy and separation. On the other hand, the synthesis of reactor systems has not developed to the same degree, despite the fact that the reaction subsystem is the central focus of most chemical processes and its performance (yield, selectivity, energy requirements and byproducts) has a direct impact on the synthesis of all of the other subsystems.

There are a number of reasons that explain the lack of powerful tools for reactor networks. First, reactor design has a strong experimental component that is driven by the exploitation of new chemistries. As discussed in the previous paper (Mavrovouniotis and Bonvin, 1994) the exploration of new reaction paths is often the key to advancing the competitiveness of a process. However, given the competitive nature of the process industries, the primary goal of an experimental program is frequently not to obtain a detailed kinetic model, but rather to provide the data necessary to design a scaled-up reactor. Consequently, the lack of a quantitative predictive model makes the derivation of systematic synthesis tools difficult.

The objective of reactor network synthesis is therefore to provide a scoping tool to aid in the design and scale-up of the reaction subsystem. This approach requires a predictive model, though not necessarily a mechanistic one. Moreover, this approach must incorporate the interactions of other process subsystems in order to exploit the synergy of a process effectively. However, even with a predictive kinetic model, the synthesis of reactor networks becomes difficult. First, there are numerous trade-offs to be made due to competing reaction and transport mechanisms. The choice of flow and mixing patterns as well as the addition and removal of heat at appropriate points is often impossible to evaluate entirely in an experimental program. Much more can be done with a predictive model, but even here these phenomena can lead to very difficult modeling and optimization problem formulations. Consequently, some idealization of the process is required.

Perhaps the most common idealizations of reactor networks occur in the choice of simple reactor types, plug flow reactors (PFRs), continuous stirred tank reactors (CSTRs), recycle reactors (RRs), etc. that are common to undergraduate textbooks (e.g. Levenspiel, 1962; Fogler, 1992). Here several well-known rules have been derived based on geometric and monotonicity concepts which apply to simple reaction systems (single reactions, series/parallel reactions, simple endothermic and exothermic reactions). However, while these concepts are especially useful for single reactions they often cannot be generalized, or lead to conflicting advice when extended to more general systems.

A straightforward extension of this approach is to postulate a network of idealized reactors and perform a structural optimization on this enlarged network or "superstructure." This concept was investigated by Horn and Tsai (1967), Jackson (1968) and Ravimohan (1971) through the application of optimal control policies. Chitra and Govind (1985) exploited the extreme limits of recycle reactors and optimized serial structures of these reactors. The optimal control approach was again revisited by Achenie and Biegler (1986, 1990) by treating a network of axial dispersion reactors. The same authors also explored a serial network of recycle reactors with bypass. A more general approach to the optimization of reactor superstructures was taken by Kokossis and Floudas (1990, 1991, 1993). Here the problem was formulated as a mixed integer nonlinear programming problem (MINLP) and a very rich superstructure of CSTRs and PFRs (actually serially linked CSTRs) was postulated and the formulation was solved with generalized Benders decomposition. These authors also extended the formulation to include nonisothermal systems, interactions with separation systems and the consideration of stability in the synthesized reactor network.

While the superstructure approach can lead to an effective synthesis strategy, there are a number of drawbacks. First, because of the nonlinear nature of reaction processes, it is difficult to determine when a given superstructure is "rich enough" to deal with general reaction systems. Second, the resulting problem formulation contains many nonconvexities with the possibility of numerous local optima. As a result, global optimization tools, still under development, need to be applied here. Finally, the optimal network frequently has a nonunique structure; i.e. several networks can have the same yield or selectivity characteristics. As a result, an alternate approach of bounding or targeting in the concentration space is extremely useful. Reactor targeting has an intuitive analog with targets employed in heat exchanger networks (HENs). In both cases, strong bounds on network performance can be derived (in terms of concentrations for reactor networks and energy consumption for heat exchanger networks) without the explicit construction of a network. Generally the targeting information gives useful insights about the global solution (although not complete information) and is much easier to obtain.

A powerful concept for reactor network targeting is that of an attainable region (AR). The notion of an attainable region stems from Horn (1964) who noted that once an AR is identified in concentration space for a particular reaction system, the task of finding an optimal reactor network is greatly simplified. In particular, by exploiting geometric properties of attainable regions, a constructive approach is developed to find a region that is closed to the operations of mixing and reaction. As a result, the performance of a reactor network can be targeted and the network itself can be derived from boundaries of the attainable region.

This paper reports on the success of AR approaches for reactor network synthesis and develops a number of extensions to this approach. In the next section, geometric concepts for attainable regions are reviewed and a constructive approach for its approach is outlined. Moreover, while the constructive approach is most easily illustrated in two or three dimensions, general properties for any number of dimensions will be summarized. Section three extends these concepts to deal with more complex geometric aspects in AR approach. Of particular interest here are the incorporation of additional rate processes due to catalyst mixing and separation. In the fourth section optimization formulations will also be explored that build on the concepts of attainable regions. Here we will see that while these are not as rigorous as the geometric concepts, they allow us to "see" in higher dimensions in order to expand an attainable region. The fifth section further explores reactor network synthesis through the integration of the reaction subsystem to the rest of the process. Here optimization formulations are particularly useful to model the interactions between the reaction, energy and separation subsystems. Finally, section six summarizes and concludes the paper.

Geometric Concepts of Attainable Regions for Reaction and Mixing

Definition and Geometric Properties of the Attainable Region

For a given system of reactions with given kinetics and given feed(s), the attainable region **A** for reaction and mixing is defined as the set of all possible outcomes from all physically realizable steady state reactors in which the only processes occurring are reaction and mixing.

Consider a homogeneous, isothermal, constant density system with species i = 1, ... , n participating in the reactions and where the objective function that we wish to optimize is only a function of output concentrations of the various species C_i. The AR will lie in the space $C = \{C_1, C_2, \ldots, C_n\}$ and we define the reaction vector $R(C) = \{r_1(C), r_2(C), \ldots, r_n(C)\}$, where the rate of formation of a species j, r_j, is defined in terms of the concentrations of the various species C_i, i.e. $r_j(C)$.

Now consider the geometric interpretation of the two processes, namely reaction and mixing that we are considering. If we have a mixture of composition C and we allow a differential amount of reaction, then the change in composition dC will be in the direction of R(C) i.e.

$$dC = R(C)\, d\tau \text{ where } d\tau > 0 \quad (1)$$

If we have a mixture of composition C and mix with material of composition C^0, then the composition of the resulting mixture C^* lies on the line between C and C^0, i.e.

$$C^* = \alpha\, C + (1-\alpha)\, C^0 \text{ where } 0 \le \alpha \le 1 \quad (2)$$

Let us look first at the geometry of two ideal reactors: the PFR and CSTR, where C_{feed} is the feed concentration and τ is the residence time of the reactor. The PFR is described by:

$$dC/d\tau = R(C) \text{ where } C = C_{feed} \text{ at } \tau = 0 \quad (3)$$

which describes a trajectory in the space with the reaction vector R(C) tangent everywhere along the curve. The CSTR is described by:

$$C_{feed} - C = -R(C)\, \tau \quad (4)$$

which has the property that the reaction vector R(C) is linear with the mixing vector $(C_{feed} - C)$ and the two vectors point in opposite directions along the CSTR locus. Another reactor that will be of interest in the subsequent discussion is the differential sidestream reactor (DSR). In this reactor we have plug flow of material along the reactor with addition of sidestream of composition C^0. The DSR is described by:

$$dC/d\tau = R(C) + \alpha(C)\,(C^0 - C) \text{ where } \alpha(C) \ge 0 \quad (5)$$

Thus the change in composition at any point along the DSR must lie between the reaction vector R(C) and the mixing vector $(C^0 - C)$. Note that the limiting behavior of the DSR is either a PFR ($\alpha = 0$) or a CSTR ($dC/d\tau = 0$). These are just a few examples of how we can describe a reactor in terms of the reaction and mixing occurring in the reactor and from this devise a geometric interpretation of the reactor. By considering the individual processes of reaction and mixing, we can show that **A** must satisfy the following necessary conditions:

1. All reaction vectors R on the boundary of **A**, ∂**A**, must be tangent, point inwards or zero. This follows from the PFR equation, because if, at some point on ∂**A**, the reaction vector pointed outwards, then by reaction we could extend the region.
2. **A** must be convex. This follows if we had a concavity in **A**, we could fill in the concavity by mixing.
3. No reaction vector R in the complement of **A** can point backwards into **A**. This follows because if, at some point C_1 in the complement of **A**, the vector $R(C_1)$ could be extrapolated backwards into **A**, then a CSTR operating with a feed in **A** could achieve C_1.

A region that satisfies these necessary conditions is a candidate for the attainable region. Unfortunately we do not yet have a sufficiency condition for **A**; however a region that satisfies the necessary conditions is closed with respect to differential reaction and mixing, PFRs, CSTRs and DSRs.

One can construct **A** in a subset C' of the full concentration space when the objective function, bound or target depends only on the concentration of the species defining C' and when the rates of formation of the species defining C' also only depend on C'. The space can also be extended to incorporate variables other than concentration variables provided the new variables obey linear mixing laws, and can be incorporated in the definition of R. Examples of such variables are residence time in constant density systems and specific enthalpy. We can also extend this approach to non-constant density systems by using mass concentration variables as discussed by Hildebrandt et. al. (1990).

Once we have found **A**, an optimization problem can be solved relatively easily by searching over **A** to find where the objective function is optimized. The optimum can either lie on $\partial\mathbf{A}$ or in the interior of **A**. If the optimum lies in the interior of **A**, we can achieve this point in infinitely many different ways and in particular by mixing between appropriate points on $\partial\mathbf{A}$. We first look at the geometric properties of $\partial\mathbf{A}$ and how we can interpret these properties in terms of the combination reaction and mixing occurring in $\partial\mathbf{A}$. By understanding the geometry, we will be able to translate this combination of reaction and mixing to determine the reactor structures that make up $\partial\mathbf{A}$.

The Geometry of the Boundary of Attainable Region

Results of the AR concepts are summarized below for $\Re^n$. These are developed and proved in Hildebrandt and Feinberg (1992). Firstly, $\partial\mathbf{A}$ is the union of straight lines and surfaces along which R is tangent. We interpret the surfaces as the union of PFR trajectories. This tells us that the structure of the boundary is rather simple and that the complexity of the reactors that make up the boundary is in fact fairly limited. If the objective function is optimized on a curved section of $\partial\mathbf{A}$ the optimal reactor structure that would produce this material will have a PFR as the last unit in the structure.

We next look at how the straight lines intersect the surfaces made up of PFR trajectories; we will refer to these intersections as **connectors**. When there is no unique tangent support hyperplane along the connector (i.e. PFR trajectories and straight lines do not intersect smoothly), the connector is itself a surface along which R is tangent and is thus a union of PFR trajectories.

If the PFR trajectories and the straight line sections do intersect smoothly, the tangent support hyperplane is uniquely defined along the connector and the mixing vectors and reaction vectors lie in the support hyperplane. We are really only interested in when the connector corresponds to *feed points* to the PFR trajectories, which will occur if the reaction vectors point away from the connector. This geometry implies that the connector is the union of CSTR operating points and DSR trajectories. We now look at the construction of **A** and at what these results imply in 2 and 3 dimensional space.

General Results in $\Re^2$

In $\Re^2$, ∂A is the union of straight lines, PFR trajectories and equilibrium and feed points. Consider the sections of ∂A made up of alternating PFR trajectories and straight lines. When one end of a straight line is a feed point to a PFR, this point, a connector, is achieved by a CSTR with its feed point being the other end of the straight line. Thus in $\Re^2$, the reactors that lie in ∂A consist of alternating PFRs and CSTRs in series-parallel arrangements. We need at most 2 parallel structures to achieve any point in the boundary of the AR and at most 3 parallel structures to achieve any point in the interior of the region. No DSR is found to lie in the boundary of the AR in 2-dimensional examples. The construction of the AR is particularly easy in 2-dimensional space and a general construction algorithm can be given.

1. Start from the feed point and work toward equilibrium or endpoint by drawing a PFR from the feed point.
2. If there is a concavity in the PFR trajectory, then straight lines would be drawn to fill in the concavities and find the convex hull of the PFR trajectory. If there is no concavity then we have found a candidate for ∂A and stop.
3. Else, we check along the straight line sections of the convex hull to see if reaction vectors point outwards. If no reaction vectors point outwards then we have a candidate for ∂A and stop.
4. Else, there exists a CSTR locus, starting from the PFR trajectory, that intersects the straight line at the point where the reaction vector becomes tangent. We then draw in the CSTR locus, with feed on the PFR trajectory, that extends the region the most. (Be sure to include all solutions (branches) if the CSTR can exhibit multiple steady states.) We next find the convex hull of the new extended region by filling in concavities in the CSTR locus. (The straight line that fills in the concavity from the feed point on the CSTR locus should not have reaction vectors pointing outwards if we have chosen the feed point to the CSTR correctly.)
5. Next, draw in a PFR trajectory from the end of the straight line filling in the CSTR concavity. If the trajectory is convex, then we have a candidate for ∂A. Otherwise, repeat from step 3 until all the concavities are filled in and have reached the equilibrium point.

Note that this algorithm can also be applied to higher dimensional problems that can be projected into a two dimensional space. For example, Omtveit and Lien (1993) applied the principle of reaction invariants (Fjeld et al, 1974) to reduce the size of a steam reforming problem to two dimensions and then construct the AR.

We illustrate this approach by means of an example based on van de Vusse kinetics.

$$A \underset{k_{1r}}{\overset{k_{1f}}{\Leftrightarrow}} B \xrightarrow{k_2} C \text{ and } 2A \xrightarrow{k_3} D \tag{6}$$

The reactions are elementary and the rate constraints are as follows: $k_{1f} = 0.01$, $k_{1r} = 5$, $k_2 = 10$ and $k_3 = 100$. We assume that the feed is pure A where $C^0_A = 1$ and we define $C = (C_A, C_B)$ where $R = (-0.01C_A + 5C_B - 100C_A^2, 0.01C_A - 5C_B - 10C_B)$. Applying the above procedure, both **A** as well as the reactors that make up ∂A are shown in Fig. 1. We can see that although we have 4 different reactor structures lying in ∂A, the individual structures are simple combinations of CSTRs and PFRs. An advantage of the constructive approach is that we can give geometric conditions for the critical operating points in the boundary, in this case points F and H. Point F is defined where the reaction vector, the tangent vector to the CSTR locus with feed A and the line AF are all collinear. Point H is defined as the point where the reaction vector on the PFR trajectory with feed F is collinear with the line from the origin.

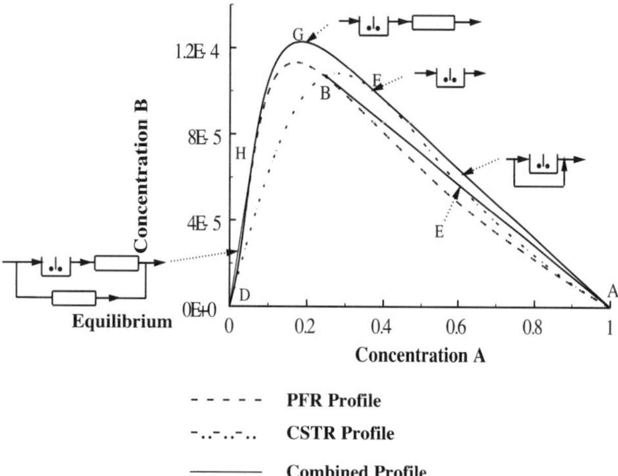

Figure 1. Isothermal van de Vusse example.

Once we have determined **A**, we are in a position to solve any optimization problem where the objective function is a function of the concentration of A and B only. Thus for example if we wanted to maximize the concentration of B at some specified conversion of A, we could read the answer off from Figure 1 and we could also determine the optimal reactor structure as well as the operating conditions of the various reactors in the structure.

General Results in $\Re^3$

The reactors that lie in ∂A in $\Re^3$ are a series-parallel arrangement of PFRs, CSTRs and DSRs. At most 3 parallel structures are needed to produce a point that lies in ∂A while at most 4 parallel structures are required to achieve a point in the interior of **A**. The most common side stream addition arrangement in the DSR will be the addition of either an equilibrium or feed point. The DSR that lies in the boundary of the AR also lies in the surface described by

$$R(C) \times (C^0\text{-}C) \cdot dR(C)\,(C^0\text{-}C) = \phi(C) = 0 \tag{7}$$

This property stems from a lengthy derivation of the connector relations and states that, at the connector, the change in the reaction vector projected along the mixing vector must lie in the plane spanned by the reaction and mixing vectors. Furthermore, from (7) we determine the sidestream addition policy so as to keep the DSR in the above surface. Thus α can be determined by:

$$\begin{aligned} d\phi/d\tau &= \nabla\phi(C)\cdot dC/d\tau \\ &= \nabla\phi(C)\cdot(R(C) + \alpha(C^0 - C)) = 0 \end{aligned} \tag{8}$$

At present we have only a trial and error construction method for 3-dimensional examples, but we do know that if we propose a region, we can test whether the region satisfies the necessary conditions.

An illustration of a typical 3-dimensional attainable region that can be found from constructing the AR geometrically is the following. Consider an exothermic, reversible reaction: A $\Leftrightarrow$ B, where $r_A = 5\text{x}10^5$ X exp(-4000/T) + $5\text{x}10^8$(1-X) exp(-8000/T), $X = C_A / C^0_A$, C^0_A is the feed concentration of A (pure A) and T is temperature. Let us look at the problem of finding the minimum volume of reactor for a given conversion of A. We have a feed of pure A at a temperature of 300 K and in addition to reaction and mixing, we are allowed to preheat the feed or a portion of the feed up to 400 K. We can choose how much of the feed to preheat and its preheat temperature. We assume constant density, constant heat capacity with ideal mixing. The energy balance for an adiabatic reaction is: $T = T^0_b + T_{ad}(1\text{-}X)$, where T is the temperature in K, T_{ad} is the adiabatic temperature rise (200 K) and T^0_b is the basis temperature if the mixture were adiabatically reacted to form pure A. The AR can be constructed in $\Re^3$, where $C = \{C_A, T, \tau\}$ and τ is the residence time. These variables (because of the assumptions used) follow linear mixing laws. We can also define a reaction vector $R = \{r_A, -T_{ad}r_A, 1\}$. Note that the new variables T and τ follow mixing laws and can be incorporated in R for the construction of **A**. The three dimensional region in temperature (T), conversion (X) and residence time (τ) for this example was constructed in Glasser et al (1992) and is shown in Figure 2. The reactor structures that make up the boundary of the region are also shown on the figure. Using the feed point, I, we see the effect of heating (e.g. line IE), the use of CSTRs (e.g. the surface IJG) and PFRs (e.g. the surface KJGH). The DSR is given by the curve GJ and lies in a surface defined by $\phi(C)$ which simplifies in this case to:

$$\phi(C) = \partial r_A/\partial X(X\text{-}1) + \partial r_A/\partial T\ (T\text{-}300) = 0 \qquad (9)$$

Again notice from Figure 2 that there are a great number of different optimal reactor structures that form ∂A but all of these structures are very simple series-parallel combinations of the 3 basic reactor units.

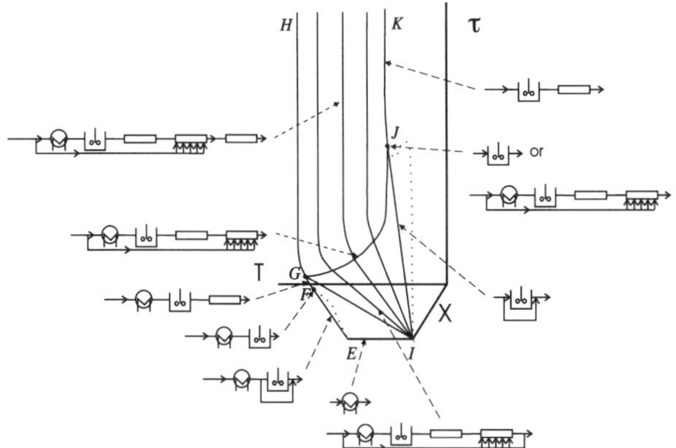

Figure 2. AR and optimal reactor structures for an exothermic reversible reaction (... CSTR loci).

We have constructed **A** for various 2 and 3 dimensional examples. We do not yet know, however, how to construct the region in higher dimensional spaces. The above results, on the other hand, have important implications as to the types of reactor structures that should be considered in optimization or targeting approaches in higher dimensions.

The reactor structures that need to be considered are only series-parallel arrangements of PFRs, CSTRs and DSRs. This means that we need not consider recycle reactors and other complex types of reactors and we can thus discard a very large number of possible structures immediately. We also do not need to include recycle within the structure itself, which also adds considerably to the simplicity of the structure.

Generally, the maximum number of parallel branches needed in the structure is related to the dimensionality of the problem. A point in the interior of **A** can be achieved by infinitely many different reactor structures; we can however achieve the point by mixing between (n+1) points on ∂A (as is consistent with our two and three dimensional observations). This means we can achieve any point by mixing the output of at most (n+1) parallel optimal structures where each parallel structure consists only of PFRs, CSTRs and DSRs. Moreover, the equilibrium points, corresponding to $dC/d\tau = 0$, of DSR trajectories are CSTR operating points. Thus, by using only CSTRs and PFRs a reasonable approximation of **A** can be expected. This again reduces the complexity of the reactor superstructure that we need to consider for a particular problem.

Multirate Processes and Geometric Concepts

We can extend the concept of the AR for reaction and mixing to include more processes. The processes we can include must be described by the vector field, P(C), in the space of the variables C. The field P(C) must be such that if we have a mixture of composition C and we allow the process to occur differentially, then the change in composition dC is in the direction of P(C) i.e., $dC = P(C)\ d\tau$, where $d\tau > 0$. Examples of such processes are separation by boiling and condensing, heating, cooling and allowing more than one reaction, such as when there is a choice of catalyst. An outline of these ideas is presented in Godorr et.al. (1994).

For simplicity in the following discussion we will regard mixing as a vector process described by $P_1 = (C - C^0)$, where both C and $C^0 \in$ **A**. The set of processes that are allowed are thus mixing P_1, reaction $P_2 = R$, and processes $P_3, P_4, \ldots, P_m$. We can incorporate all of these processes into $P = \{P_i\}$ where i = 1, ..., m. Geometrically we can say that at every point C in the space, there is a set of vectors defined such that the vectors point in the directions of change in C that can be achieved locally by allowing the individual processes P_i to occur differentially.

For a system of reactions with given kinetics and feeds, the AR is defined as the set of all possible outcomes from all physically realizable steady state systems in which only the processes defined in P are occurring. Thus, the necessary conditions and results given for reaction and mixing can be extended to incorporate the processes P. Firstly, along ∂A the components of P must not point outwards, that is they must be tangent, point inwards or zero. (This also implies that if $P_1 \in P$, then **A** must be convex if P_1 is not to point outwards over ∂A). Other necessary conditions could be added including those that cover processes which could have multiple steady states.

Furthermore, we can make the following assertions about the properties of ∂A. For example, ∂A would be the union of trajectories tangent to single vector processes P_i. The equipment needed to achieve a point on a trajectory of ∂A will have a unit in which only a single process is occurring as the last unit in the structure. Thus the equipment could have a section which can be heating only, reaction only or boiling only before the material exits the equipment.

Where do these trajectories originate from? Consider **A** in $\Re^n$ where we have k(n-1)-dimensional hypersurfaces

S_k that lie in ∂A, where $k \le m$ and which intersect. Let the surfaces be such that each surface S_i, $i = 1, \ldots, k$, is tangent to one of the elements of P and that all the other elements of P point into **A**. Furthermore, let each surface S_i be tangent to a different element of P. For simplicity let us assume that surface S_i is tangent to P_i. Surface S_i can thus be regarded as the union of trajectories which are tangent to P_i.

Consider now that $k = 2$ and suppose that S_1 and S_2 intersect smoothly and, furthermore, that P_1 and P_2 point away from the intersection. The intersection will be an (n-2)-dimensional hypersurface in ∂A and we can again call the intersection a **connector**. The support hyperplane to this connector will be tangent to both P_1 and P_2. It follows that this connector is the union of trajectories described by a differential equation which is a linear combination of P_1 and P_2. When $k = 3$ and S_1, S_2 and S_3 intersect smoothly, this connector/intersection will be an (n-3)-dimensional hypersurface in ∂A, and P_1, P_2 and P_3 will be tangent to the support hyperplane along the connector. If we consider only connectors that are feed points to the trajectories, i.e. connectors where P_1, P_2 and P_3 point away from the connector, then this connector is the union of trajectories which correspond to a process/operation defined by a differential vector equation which is a linear combination of P_1, P_2 and P_3.

We can further generalize these ideas. Firstly, n processes will operate simultaneously at isolated points in ∂A. Similarly, (n-1) processes will operate simultaneously along a 1-dimensional curve in the boundary of ∂A. In general, (n-m) processes will operate simultaneously along an (m)-dimensional hypersurface in ∂A, where $0 < m < n$. We should then be able to translate this geometry into equipment or a unit process. Notice that the processes making up ∂A will again come out of the construction of **A** and do not have to be specified.

These results, together with the previous results regarding the number of parallel structures needed, can be used to propose a suitable candidate for the targeting approach. We could propose that the optimal structure would be series-parallel arrangements of units described by differential equations that are linear combinations of the individual elements of P. We should be able to relate these structures to unit operations or process equipment. If we allow mixing, condensing and boiling, the type of equipment that should be used will come out of the construction and it may turn out, for example, that distillation columns are not optimal and some other combination of flashes and mixing is better. (However, if the distillation column is described by a differential equation, as in van Dongen and Doherty (1985), then the resulting equation is a linear combination of a mixing vector and a vector describing boiling and thus can be a trajectory in ∂A.)

These results reduce the complexity of the required structure as well as the types of units that need to be considered in a proposed structure. A limitation of the approach is that it currently works only for single input single output problems, and many practical problems, for example, separation have multiple outputs. We are currently looking at the implications of multiple outputs on the geometry. Moreover, the constructive approach to finding **A** in these types of problems is also limited by constructions in $\Re^2$ and $\Re^3$. However, it is of interest to look at a few of these examples to illustrate the ideas and the implications of the geometry.

Reaction and Mixing with Catalyst Profile Optimization

The following problem has been looked at by various researchers. Suppose we are given two different catalysts that catalyze different sets of reactions. We wish to choose the reactor as well as the catalyst profile in the reactor in order to minimize the total catalyst volume used to produce some specified product. Given the following kinetics, catalyst 1 catalyzes the following two reactions:

Reaction 1: $A \Leftrightarrow B$ where $r_{A1} =$
$(-k_1C_A + k_2C_B)/(1 + k_pC^2{}_A)$

Reaction 2: $A + C \Leftrightarrow B + C$ where $r_{A2} =$
$(-k_3C_AC_C + k_4C_BC_C)/(1+k_pC^2{}_A)$

and catalyst 2 catalyzes two different reactions:

Reaction 3: $A \Leftrightarrow C$ where $r_{A4} =$
$(-k_5C_A + k_6C_C)/(1 + k_pC^2{}_A)$

Reaction 4: $A + B \Leftrightarrow C + B$ where $r_{A4} =$
$(-k_7C_AC_B + k_8C_BC_C)/(1 + k_pC^2{}_A)$

We are given that the feed is pure A ($C^2{}_A = 1$) and we are allowed to use 3 processes: reaction with catalyst 1 which we will describe by reaction vector R_1, reaction with catalyst 2 which we will describe by reaction vector R_2 and mixing. The reaction constants are $k_1=k_5=2$, $k_2=k_6=1$, $k_3=k_7=60$, $k_4=k_8=10$ and $k_p=20$. We can construct **A** in $\Re^3$ space where $C = (C_B, C_C, \tau)$. The reaction vectors are thus given by $R_1 = (r_{A1} + r_{A2}, 0, 1)$ and $R_2 = (0, r_{A3} + r_{A4}, 1)$. We can thus see that R_1 is parallel to the C_B axis and R_2 is parallel to the C_C axis. The solution to this problem is given in Godorr et. al. (1994) and the results for this example are shown in Figure 3. The projection of **A** onto C_B - C_C space together with the various optimal reactor structures are given in Figure 3. Note that LSRQT is the equilibrium for the system while O is the feed point.

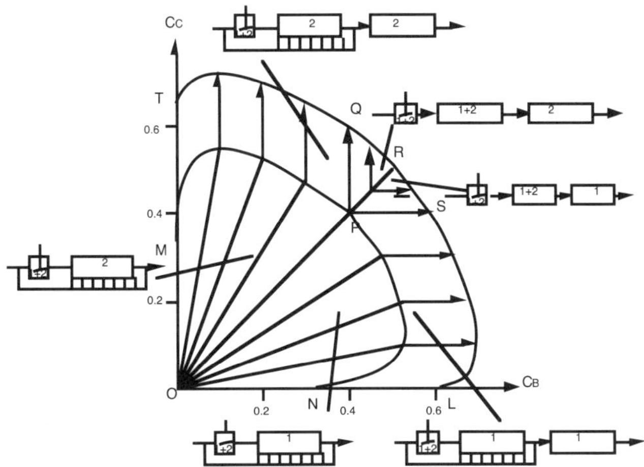

Additional AR boundary structures.

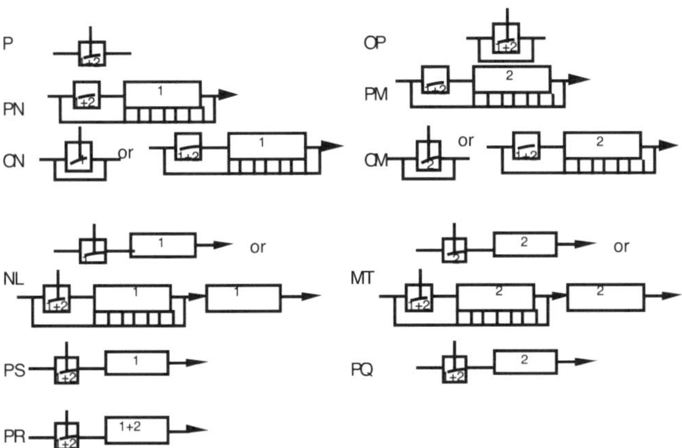

Figure 3. AR and optimal reactor structures for two catalyst problem.

Connectors PN and PM correspond to DSRs and lie in the 2-dimensional surfaces defined by equation (7), where the R refers to the relevant reaction vector. The mixing policy is described by $d\phi/d\tau = 0$. Connector PR, on the other hand, is a new kind of connector. The PFR with mixed catalyst is described by:

$$dC/d\tau = (1-\beta)\, R_1(C) + \beta\, R_2(C) \text{ where } 0 \leq \beta \leq 1 \quad (10)$$

and the connector lies in the 2-dimensional surface defined by:

$$\Phi(C) = R_1(C) \times R_2(C) \cdot \{dR_1(C)\, R_2(C) - dR_2(C)\, R_1(C)\} = 0 \quad (11)$$

Finally β, the fraction of catalyst 2 along the PFR, must be chosen to keep the mixed catalyst trajectory in the surface described by $\Phi(C) = 0$. Thus from $d\Phi/d\tau = 0$:

$$\nabla\Phi(C) \cdot dC/d\tau = \nabla\Phi(C) \cdot \{(1-\beta)R_1(C) + \beta\, R_2(C)\} = 0 \quad (12)$$

Lastly, all 3 connectors intersect at point P, which corresponds to a point where all 3 processes occur simultaneously. Thus we interpret this as a CSTR with mixed catalyst. This example demonstrates all the types of connectors that were described earlier in the discussion.

Reaction, Mixing and Separation by Evaporation

We also consider the contribution of boiling to our multirate processes. If we have a liquid mixture with species i = 1, ..., n, where the mole fraction of species i is X_i we can describe the composition of the liquid by $X = (X_1, \ldots, X_{n-1})$ and assume that the liquid molar density is constant. If we allow simple boiling (at constant pressure) to occur such that the vapor removed is in equilibrium with the liquid, we can describe the change in the composition of the liquid by:

$$dX/d\tau = N(X-Y(X)) = NS \quad (13)$$

where N is the molar rate of vapor removal per unit volume, Y is the mole fraction vector describing the vapor composition in equilibrium with the liquid of composition X and τ is a scalar parameter. We will refer to $S = (X - Y(X))$ as the separation vector. We can also define a reaction vector R(X) at every point in X.

Consider now an example in $\Re^2$, where we have $A \Leftrightarrow B \Leftrightarrow C$ and $X = (X_A, X_B)$. Let us suppose that the reactions are first order and that $R = (-k_{1f}X_A + k_{1r}X_B, k_{1f}X_A - k_{1r}X_B - k_{2f}X_B + k_{2r}(1 - X_A - X_B))$. We wish to find the maximum mole fraction of B that can be obtained for some specified mole fraction of A using reaction, mixing and separation by boiling only. Note that we are assuming that the vapor that is boiled off is "lost" and that it is not condensed and returned to the system. We have a feed of pure A and we adjust the temperature of the mixture to keep it at its bubble point assuming that the liquid and vapor behave ideally i.e. $Y_iP^0 = X_iP_i^{Vap}$, where P^0 is the total pressure (1 bar) and P_i^{Vap} is the vapor pressure of pure i. The temperature dependence of the vapor pressures is given by $a + b\,T$ (in ^oC), with a = 0.4, 0.5 and 0.3 for components A, B and C, respectively, and b = 0.005 for all components. Thus $S = (X_A - X_AP_A^{Vap}(T)/P^0, X_B - X_B P_B^{Vap}(T)/P^0)$ and T is defined implicitly by: $\Sigma(X_iP_i^{Vap}(T)/P^0) = 1$. We can find **A** for this problem and we would expect that at most 2 processes operate simultaneously at a point in ∂A and that nowhere in ∂A can all 3 processes occur simultaneously. Thus we would not expect to find, for example, a CSTR with simultaneous reaction and boiling (3 processes occurring simultaneously) or even a PFR trajectory with simultaneous reaction and boiling in ∂A. If we have simultaneous mixing and boiling occurring at a point (i.e. a flash) we would find that the feed composition X_{feed} would lie between X and Y on S. As the magnitude of S is generally not large, this means that the flash will not extend the region very much, if at all. We subsequently would not expect the flash to play an important role in extending ∂A.

The structures that make up ∂A are in fact very simple and are shown in Figure 4. A PFR trajectory from the feed point operates between AB, the PFR trajectory is convex and S points inwards along the trajectory. B corresponds to reaction equilibrium. At B separation by boiling moves us along BCD which is concave. We can fill in that concavity by mixing B and D. Both R and S point inwards along line BD and so we claim that this region is fact **A**.

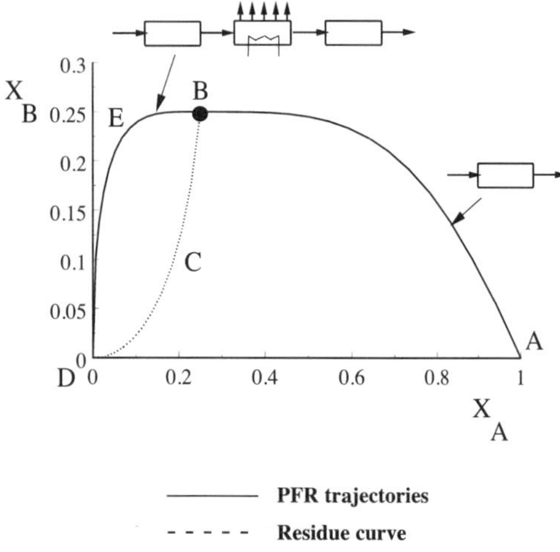

Figure 4. AR and optimal synthesis structure for reaction and separation example.

Note that one of the differences between this example and all the previous ones is that it is not clear from the construction how much material can be produced, as once we allow boiling, the quantity of material varies depending on how much vapor we have removed. Thus although all compositions along line BD can be achieved, the points along line BD correspond to different quantities of product.

Again, notice that in order to achieve a point on ∂A, we at most need 2 parallel structures with mixing at the outlet of the two structures. The ∂A is made up of curves along which only single processes are occurring, in this case reaction along AB and mixing along BD. In this example we do not have any points in ∂A where two processes operate simultaneously.

If we changed the vapor pressure relationships such that A had the lowest boiling point of the three components, then we would find that the whole mole fraction space was achievable in the limit by separation alone. Finally, it appears that we would have to look at $\Re^3$ or higher to find interesting examples where reaction, mixing and separation occurred simultaneously.

Optimization Formulations for Higher Dimensions

The previous two sections demonstrated the effectiveness of geometric concepts and constructions to a wide variety of synthesis problems. While geometric concepts lead to powerful tools for visualizing and constructing an AR in concentration space, obtaining this region can be much more difficult in higher dimensions. In this section, on the other hand, we explore an optimization-based formulations for reactor network targeting. This approach applies many of the concepts of attainable regions from the previous section and poses them as optimization problems. This allows the designer to probe in higher dimensional spaces (in principle, without limitation) without the need of visualization. As developed so far, optimization-based formulations consist of small nonlinear programming problems (NLPs) that describe the performance of PFRs and CSTRs and lead to a analogous approach for determining the attainable region. This NLP approach has a number of advantages as well as shortcomings.

In particular, it should be noted that NLP formulations do not entirely replace insight gained from the construction of an attainable region. With geometric constructions, one obtains a family of reactors that is complete in concentration space. The NLP approach rather finds the family of reactors within an attainable region that improve a given objective. Here one assumes that steady improvement can be found for each NLP extension of the attainable region. This is not always possible and, as a result, the NLP-based procedure can terminate in suboptimal networks. Moreover, the optimization-based targeting approach has only been tackled with local optimization methods, and solutions are obtained without a guarantee of a global optimum.

On the plus side, however, nonlinear programs can be formulated for arbitrarily large problems without restriction as to the features of the kinetics. As will be seen below, these formulations are quite easy to solve even for demanding kinetic problems. Moreover, while simple constraints can be incorporated into geometric constructions, the NLP approach offers greater flexibility in posing and synthesizing constrained reactor networks. This characteristic has its greatest advantage when integrating the reactor network within other process subsystems, as described in section 5. This integration step is done quite naturally with optimization based formulations as links from other parts of the process are treated directly through equality and inequality constraints.

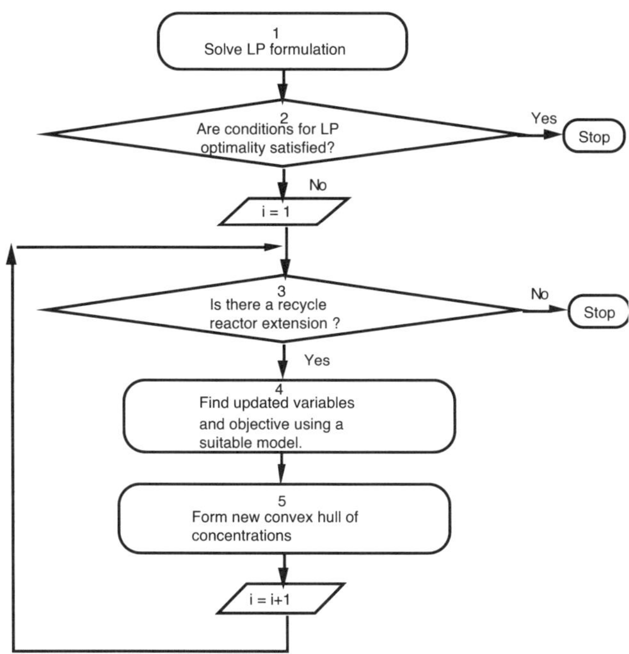

Figure 5. Flowchart for stagewise synthesis.

The NLP-based targeting strategy is summarized in Figure 5. Again, the basic properties of the attainable region are exploited and a constructive approach is developed in order to determine whether the best objective in the attainable region (but not the entire region itself) has been found. Here we modify the AR description slightly and first consider ARs in segregated flow, rather than simple PFRs. For isothermal problems, the segregated flow or PFR profiles are generated off-line by solving the rate equations:

$$dX_{seg}/dt = R(X_{seg})$$

$$X_{seg}(0) = X_0$$

and form the data for the problem given below:

$$\max_{f(t)} \quad J(X_{exit}, \tau) \tag{14}$$

$$Xexit = \int_0^\infty f(t)\, X_{seg}(t)\, dt$$

$$\int_0^\infty t\, f(t)dt = \tau \quad \int_0^\infty f(t)dt = 1$$

Here, f(t) and $X_{seg}(t)$ correspond to the residence time distribution and the dimensionless concentration vector (e.g. $C(t)/C_{feed}$), respectively. It follows from section 2 that if the problem can be represented in two dimensions and if the PFR profiles are convex, then solution of (14) yields the optimal network and the attainable region is given by at most two plug flow reactors. Moreover, discretization of the integrals yields a simple linearly constrained nonlinear program (NLP). In fact, when the objective is yield or selectivity, (14) is easily solved as a linear programming problem. Balakrishna and Biegler (1992) also considered convex two dimensional projections for multidimensional problems and derived more general conditions under which PFR profiles remain sufficient for constructing the attainable region.

If PFR trajectories are insufficient for an AR, however, the region can be enhanced by NLP formulations that describe CSTR or RR extensions. Strictly speaking, any attainable region that is closed to CSTRs is also closed to recycle reactors, and vice versa. However, an advantage to the recycle reactor formulation is that it can model both plug flow and CSTR extensions. From any point within the segregated flow region, recycle reactors are sought to improve the objective at points outside of the attainable region, as shown in the formulation below:

$$\max_{R_e,\, f(i,j),\, f_{model(k)}} \quad J(X_{exit})$$

$$dX_{rr}/dt = R(X_{rr}) \tag{15}$$

$$X_{rr}(t=0) = (R_e X_{exit} + X_{update}) / (R_e + 1)$$

$$X_{update} = \int_0^\infty f(t)\, X_{seg}(t)\, dt$$

(or X_{update} = previous point if further reactor extensions),

$$X_{exit} = \int_0^\infty f_{rr}(t)\, X_{rr}(t)\, dt$$

$$\int_0^\infty f(t)\, dt = 1, \quad \int_0^\infty f_{rr}(t)\, dt = 1$$

Problems (14) and (15) are used in the algorithm of Figure 5. (15) is augmented by any additional extensions and an optimal network is claimed when no further improvement can be found by these formulations. As mentioned above, an important limitation is that an improvement in the objective is sought with every extension to the attainable region. However, there are cases where the attainable region can be extended without improving the objective — and from these extensions further improvements could still be found. This approach has been applied by Balakrishna and Biegler (1992a) to several isothermal examples, with as many as seven independent reactions. Their results are at least as good as or better than previous literature results.

Finally, the NLP formulation can be extended naturally to systems with variable feeds compositions and to nonisothermal systems with arbitrary temperature profiles.

To deal with unknown or variable feed compositions, the formulation in (14) can be solved recursively in an inner loop with PFR profiles generated off-line for different feed conditions. This mimics the geometric approach where different attainable regions are constructed for a variety of feeds. However, a more direct approach involves a simultaneous NLP formulation that incorporates changes in segregated flow behavior, a differential equation model, with changing initial conditions. To reflect these in a nonlinear programming formulation, the differential and integral equations need to be discretized to algebraic equations. A natural way to do this is through collocation on finite elements. As a result, formulation (14) can be extended to:

$$\text{Max} \quad J(X_{exit})$$

$$dX_{seg}/dt = R(X_{seg})$$

$$h(X_{seg}(t=0), X_{exit}, y) = 0$$

$$X_{exit} = \int_0^\infty f(t)\, X_{seg}(t)\, dt \qquad (16)$$

$$\int_0^\infty t\, f(t)\, dt = \tau, \qquad \int_0^\infty f(t)\, dt = 1$$

$$\tau < \tau_{max}, \qquad l \le X_{exit} \le u$$

and solved as a discretized NLP. To generate extensions to the attainable region, the constraints related to recycle reactor behavior in (15) can be added as before and the algorithm in Figure 5 can be executed in the same manner. As will be seen in section 5, the simultaneous formulation allows for a natural integration with other flowsheet subsystems. In particular, trade-offs are established among these subsystems in a direct manner.

To deal with nonisothermal systems, formulation (16) can be extended to add decision variables which describe optimal temperature profiles. In a similar manner as with (15), recycle reactor and CSTR extensions can also be added to expand the attainable region. Here each of the reactor extensions is allowed different temperatures as well. Finally, for purposes of heat transfer, and by extension, energy integration with other process subsystems, an effective means of controlling the temperature profile is through cold-shot cooling profiles $\alpha(t)$, with the feed composition. This introduces only a slight modification of the nonisothermal formulation. Instead of an initial scheme based on reaction in segregated flow, the more general cross flow reactor model can be introduced as shown in Figure 6.

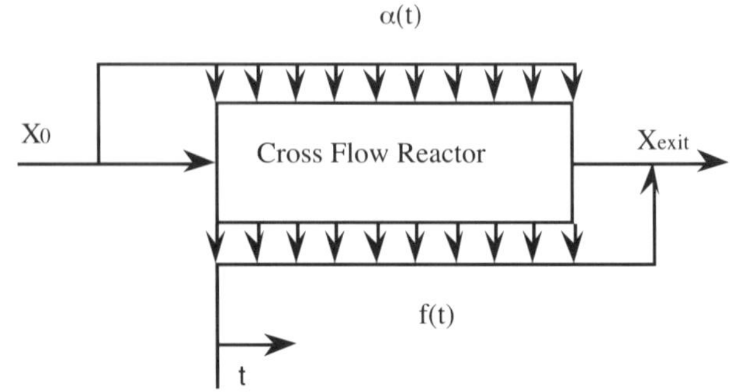

Figure 6. General cross flow reactor model.

Discretization of this model with the introduction of sidestreams at discrete points leads to the following NLP formulation:

$$\text{Max} \quad J(X_{exit}, \tau)$$

$$\alpha(t), f(t), T(t),$$

$$dX/dt = R(T(t), X) + \alpha(t) Q_0/Q(t) \cdot (X0 - X(t)),$$

$$X(0) = X_0$$

$$X_{exit} = \int_0^\infty f(t)\, X(t)\, dt \qquad (17)$$

$$\int_0^\infty f(t)\, dt = 1, \qquad \int_0^\infty \alpha(t)\, dt = 1$$

$$Q(t)/Q_0 = \int_0^t [\alpha(t') - f(t')]\, dt'$$

$$\int_0^\infty \int_0^t [\alpha(t') - f(t')]\, dt'\, dt = \tau$$

Here, $\alpha(t)$ is the feed addition profile for the cross flow model. Note that when f(t) becomes a Dirac delta function, the cross flow model becomes the DSR model described in the previous section with the sidestream set to the feed composition. Consequently, the targeting model is compatible with the geometric properties presented above. Balakrishna and Biegler (1992b) discretized (17) using collocation on finite elements and thus solved this problem as an NLP.

Again, RR or CSTR extensions similar to (15), or even further cross flow extensions can be applied to the attainable region constructed from (17). However, in contrast to the simplest approaches in (14) and (15) which often lead to LP formulations, (16) and (17) are more likely to lead to local optima. Clearly, the use of global optimization methods would be advantageous (see Floudas and Grossmann, 1994) but even with the use of faster local methods, good solutions to (17) can be obtained through incremental solutions of simpler problems. For example, solution of (17) with a fixed feed and temperature profile, and no sidestreams amounts to solving (14). With that solution, the feed compositions can be included as decisions (if desired) as in (16). Later with the solution from (16) or (14), temperature profiles can be varied and finally sidestreams can be allowed. While this approach does not guarantee global optimality, it builds on simple problem formulations and efficient solution procedures. In particular, more complex problem formulations are introduced only as they are needed. Numerous nonisothermal problems have been solved with this approach (Balakrishna and Biegler, 1992b). In the next section we will also see how this approach has advantages with flowsheet and heat integration.

Integration with other Process Subsystems

Reactor networks are rarely designed in isolation, but rather form an important part of an overall flowsheet. Moreover, since feed preparation, product recovery and recycle steps in a process are directly influenced by the reactor network, the synergy among these subsystems is a key factor in establishing an optimum process. Because of reactant recycling, overall conversion to product is influenced by selectivity to desired products rather than reactor yield, as noted by Conti and Patterson (1985). Douglas (1988) extends this notion of process synthesis by establishing trade-offs among conversion of raw materials, capital costs and operating costs. Here, although selectivity maximimization leads to optimum overall conversion to product, capital and operating costs affected by high recycles can improve if reactor yield is maximized instead. Hence, to balance these trade-offs, Douglas suggests a reactor network that operates between maximum yield and maximum selectivity.

A geometric approach to reactor/flowsheet integration was developed by Omtveit and Lien (1994) where separations and recycles were incorporated into the construction of the attainable region. Here, geometric constructions need to be performed iteratively as the reactor feed is unknown in the optimum flowsheet. Omtveit and Lien (1994) therefore constructed a family of attainable regions and used constraints due to reaction limitations to represent this problem in only two dimensions. This approach was demonstrated on the HDA process (Douglas, 1988) as well as methanol synthesis. In both problems the optimal reactor turned out to be a plug flow reactor and quantitative trade-offs were established between the purge fraction, reactor yield and economic potential.

While the qualitative concepts mentioned above yield useful insights for process integration, many quantitative aspects along with discrete and continuous decisions still have to be made. A natural way to account quantitatively for process trade-offs and to represent the interactions of process subsystems is to develop targeting models based on NLP and MINLP formulations. Again, as with reactor network targeting the goal of these formulations is to predict process performance without explicitly developing the network itself. Consequently, AR concepts are extremely useful here and dimensionality limitations can be overcome through the NLP formulations presented above.

For example, with an isothermal network, Balakrishna and Biegler (1992a) demonstrated the effectiveness of NLP formulations for flowsheet integration on the Williams-Otto (1960) process. Originally, the reactor was represented as a stirred tank with an optimal return on investment of about 130%. Application of (16) and (15) shows that much better performance can be obtained with a single plug flow reactor and the return on investment more than doubles to 278%.

For nonisothermal reaction systems and flowsheet integration, energy integration of the reactor network with separation units and process streams is a key consideration. Energy integration tools such as pinch technology have been extremely effective in reducing process operating costs, especially for existing processes, although often only the "sequential" energy integration problem is addressed. Here a heat exchanger network is targeted and synthesized only after the process is "optimized." On the other hand, numerous studies (Duran and Grossmann, 1986, Terrill and Douglas, 1987) have indicated the need for simultaneous heat integration and process optimization. This is especially important in order to reflect the correct "costs" of energy resulting for integration and therefore to establish an accurate balance of the trade-offs between energy, capital and raw material costs.

To include the integration of energy into the synthesis of the reactor network a related simultaneous strategy can also be developed. Here the nonisothermal NLP formulation in (17) is adapted to deal with heat exchange from other process streams. Within the discretized cross flow reactor, temperature segments within the reactor are identified as either hot streams (if system is exothermic) or cold streams (if endothermic) and additional sources of heating or cooling (heat exchangers) are included with the possibility of intermediate feedstreams along the reactor. As seen in the sketch in Figure 7, this formulation allows us to achieve an arbitrary temperature profile in the reactor along with the ability to evaluate the heat duties needed to achieve it.

With this framework and the identification of hot and cold reactor streams, we can now augment (17) with the Duran and Grossmann formulation for heat integration. In this approach additional constraints are included that reflect minimum utility consumption as a function of the flowrates and temperatures of the integrated system. Therefore, given a set of hot and cold streams, the minimum heating utility consumption is given by $Q_H = \max(z_H^p(y))$, where, z_H^p is the difference between the heat sources and sinks above the pinch point for pinch candidate p. For hot and cold streams for the reactor as well as the rest of the process, with inlet temperatures given by

T_h^{in} and t_c^{in}; and outlet temperatures T_h^{out} and t_c^{out} respectively, $z_H^p(y)$ is given by,

$$z_H^p(y) = \Sigma_c \left(w_c[\max\{0; t_c^{out} - T^p + \Delta T_m\}\right.$$

$$\left. - \max\{0; t_c^{in} - T^p + \Delta T_m\}]\right) -$$

$$\Sigma_h \left(W_h[\max\{0; T_h^{in} - T^p\} - \max\{0; T_h^{out} - T^p\}]\right)$$

for $p = 1, N_p$; where N_p is the total number of heat exchange streams. Here, T^p corresponds to all the candidate pinch points; these are given by the inlet temperatures for all hot streams and the inlet temperature added to ΔT_m for the cold streams. W_h and w_c are the heat capacity flows for the hot and cold streams and the vector y represents the set of all variables (temperatures, flowrates and compositions) in the reactor and energy network. The minimum cooling utility is given by the energy balance, $Q_C = Q_H + \Omega(y)$; where, $\Omega(y)$ is the difference in the heat content between the hot and the cold process streams, given by:

$$\Omega(y) = \Sigma_h W_h(T_h^{in} - T_h^{out}) - \Sigma_c w_c(T_c^{out} - T_c^{in})$$

These relations can be incorporated with the flowsheet model and the targeting formulation of (17) in order to develop the following simultaneous formulation.

$$\text{Max } \Phi(\omega, y, Q_H, Q_C) = J(\omega, y) - c_H Q_H - c_C Q_C$$

$$\text{s.t. } dX/dt = R(T(t), X) + \alpha(t)Q_0/Q(t) \cdot (X_0 - X(t)),$$

$$X(0) = X_0(\omega, y)$$

$$X(0) = X_0$$

$$X_{exit} = \int_0^\infty f(t)\, X(t)\, dt, \qquad \int_0^\infty f(t)\, dt = 1$$

$$\int_0^\infty \alpha(t)\, dt = 1, \qquad \int_0^\infty \int_0^t [\alpha(t') - f(t')]\, dt'\, dt = \tau$$

$$Q(t)/Q_0 = \int_0^t [\alpha(t') - f(t')]\, dt'$$

$$W_h = C_{ph}(T_h^{in})\, F_h, \qquad w_c = C_{pc}(t_c^{in}) F_c$$

$$Q_C = Q_H - \Sigma_h W_h[T_h^{in} - T_h^{out}] - \Sigma_c w_c[t_c^{out} - t_c^{in}]$$

$$Q_H \geq z_H^p(y) \quad p \in P$$

$$h(\omega, y) = 0, \qquad g(\omega, y) \leq 0$$

Here F_h and F_c are the flowrates for all process and reactor streams, ω is the set of flowsheet parameters and Q_H and Q_C represent hot and cold utility requirements. This optimization problem is expressed in a general abstract form, which is discretized in the same manner as (17). A discrete representation of this reactor targeting model is illustrated in Figure 7. In addition to discrete feed stream additions, any temperature profile can be determined through heating and cooling units at these points. Moreover, heats of reaction are directly incorporated through heat capacity flowrates of the reacting streams. Here in a discretized reacting segment, if Q_R is the exothermic (endothermic) heat of reaction to be removed (added) in order to maintain an isothermal segment, the equivalent W_h (or w_c) is equated to Q_R (we assume a 1 K temperature difference) for this reacting stream. Note that in addition to the performance of the reactor network, other process units and the energy network are captured in this compact formulation. Once this problem is solved, a formulation similar to (15) is employed to check for improvements in the network by extending the attainable region. Finally, we note that no assumptions were made as to the structure of the reactor or heat exchanger networks.

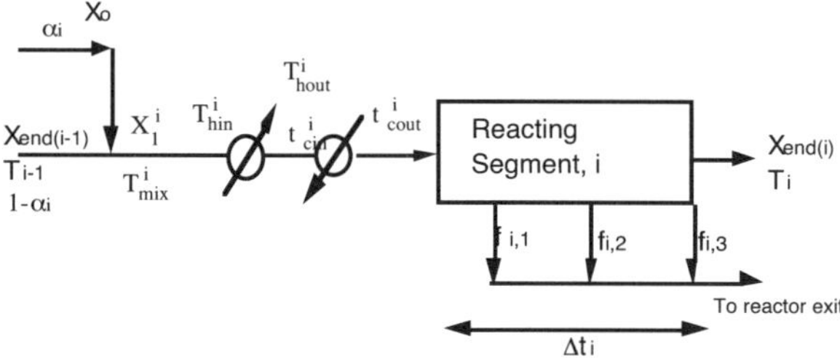

Figure 7. Reacting segment for heat integration.

This approach was applied to moderately sized process with van de Vusse kinetics, A —> B —> C, A —> D as shown in Figure 8. In this case, the reaction is highly exothermic and numerous opportunities exist for the reactor to generate steam for the reboilers and thus reduce the overall energy load. In addition, the process produces a valuable main product (B) and potentially harmful byproducts, C and D. Consequently, reactor selectivity is a key component of this process in order to maximize the process profit. In order to demonstrate the synergy of the flowsheet subsystems we present two cases for comparison. In the first instance, the sequential case, a reactor network is synthesized within a flowsheet and energy integration is performed for this design. For the second, simultaneous case, the energy target is determined together with the reactor target.

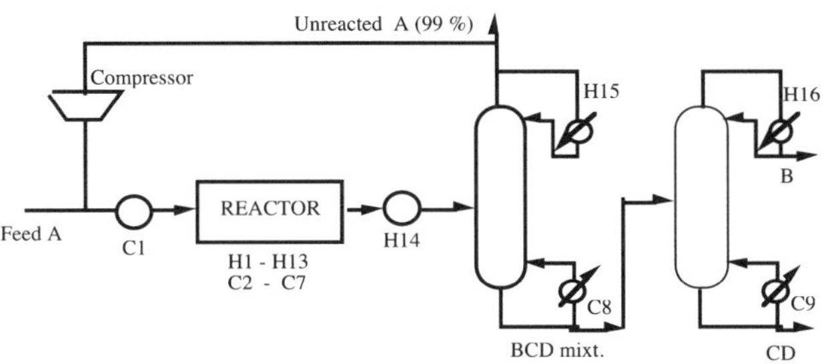

Figure 8. Flowsheet for reactor-energy network synthesis, "van de Vusse" process.

Interestingly, in both cases, a single plug flow reactor is chosen with similar residence times and both reactors have falling temperature profiles, with the simultaneous profile about 10-20 K lower than for the sequential case. Because the side reactions are more exothermic than the main reaction (A —> B), the simultaneous result has a much better selectivity and a higher overall conversion of raw material A to product B (61.5% vs. 49.6%). As listed below, the simultaneous case has a lower conversion per pass (77% vs. 87%), a higher recycle rate and requires about 20% less raw material for the process. This occurs simply because cheaper energy costs due to heat integration allow more emphasis to be placed on raw material conversion in the optimization.

	Seq'l	Simult
Overall Profit (10^5 \$/yr)	38.98	74.02
Overall Conversion to B (%)	49.6	61.55
Hot Utility Load (10^5 Btu/h)	3.101	2.801
Cold Utility Load (10^6 Btu/h)	252.2	168.5
Fresh Feed A (10^4 lb/h)	8.057	6.466
Byproducts C/D (10^4 lb/h)	4.045	2.44
Recycled A (10^4 lb/h)	1.22	1.963

A similar approach can also be adopted for the integration of reaction and separation systems. In previous studies, the separation sequence is considered to be a downstream process to recycle reactants, remove unwanted byproducts and purify the desired products. This approach allows for an easy decomposition of subsystems and this has been used to advantage in hierarchical decomposition (Douglas, 1988) and in MINLP synthesis (Kokossis and Floudas, 1991). In order to improve the synergy of these subsystems, strategies need to be developed that incorporate simultaneous reaction and separation. This topic is currently the focus of considerable research activity (Barbosa and Doherty, 1987, 1988; Omtveit and Lien, 1994; Balakrishna and Biegler, 1993). However, the synthesis of reaction-separation systems is still in its infancy. For instance, the targeting work described in Balakrishna and Biegler (1993) represents an idealized system that requires further development and generalization. Nevertheless, this topic has been spurred by significant industrial successes, where in lieu of clumsy conventional flowsheets, complex reactions and separations can be incorporated into a single reactor/separator (Agreda et al., 1990).

Finally, as new processes are invented and existing ones are revamped, the scope for minimizing waste and hazardous by-products becomes an important consideration. This aspect is directly focused on reactor performance. In addition to choosing new chemistries and reaction paths, insights gathered from attainable regions with knowledge of side reactions is a key element for waste-minimizing processes. As with any of the objectives used above for process synthesis, the application of geometric targeting approaches and the NLP extensions can be applied in a straightforward manner to waste minimization. Here constructive approaches can be applied to find an attainable region and subsequently to determine reactor networks with maximum selectivity, or minimum waste with a specified product yield. These results can then be embedded directly into an AR or NLP approach as well. Here, a useful decision-making tool, especially in dealing with uncertain waste treatment costs, is the use of multicriterion optimization and the generation of Pareto optimal or noninferior surfaces. Note here that on the noninferior curve, no objective can be improved without sacrifice to the values of the other objectives and trade-offs can be established clearly. In dealing with process profit vs. waste generation, Lakshmanan and Biegler (1994) adapted the NLP targeting approaches described in the previous section to develop these surfaces as well.

Conclusions

In this study, we have reviewed reactor network synthesis strategies based on attainable region concepts. These concepts have rigorous geometric foundations which can be used in complementary strategies; through direct construction of the attainable region (AR) or by embedding AR concepts within optimization formulations. At present we can construct attainable regions for two and three dimensional problems. This allows us to solve bounds on reactor performance problems and then specify the reactor network and operating parameters. These AR approaches can be extended to include other processes such as heat exchange and separation. The theory behind the attainable region still needs to be developed but there are many postulates that can be made by extension from reaction and mixing. These are outlined along with examples to demonstrate the constructive approach

The constructive approach has not been extended to higher dimensional (>3) problems as there are problems with the visualization of the region to higher dimensions.

Furthermore as the approach effectively finds all possible outputs from all possible reactor networks, and then searches over this set of solutions for the optimal one, it is not a feasible approach in higher dimensional problems. However, by understanding the geometry of individual processes of reaction and mixing we are able to predict the way in which individual processes make up the boundary of the attainable region. This can then be interpreted in terms of the types of reactors and ways in which the reactors are interconnected. Hence we are able to make some suggestions on how to synthesize a reactor superstructure rich enough to produce all possible output material, and then build these suggestions into the optimization formulations for reactor network targeting.

Optimization-based approaches, on the other hand, have other characteristics that must be carefully considered. While geometric constructions yield a global picture of a given reactor system, optimization formulations frequently rely on local tools. Clearly, the development of convex problem representations (as in segregated flow) or the application of global tools (Floudas and Grossmann, 1994) is an important topic for future work. At present, the application of small NLPs based on PFRs and CSTRs can be used as an approximation of the construction process. As a sequential approach, it ensures good, if not globally optimal solutions for higher dimensional problems.

In addition, NLP formulations have advantages as they can handle higher dimensional systems as well as problems in which the objective functions have parameters which cannot be incorporated in the constructive approach. Thus for example, constraints can be placed on the structure complexity and operating costs which cannot always be incorporated into the AR approach. Moreover, these formulations lead to straightforward integration of the reactor synthesis problem with other process subsystems in order to exploit their synergy. This was illustrated in a small process example.

In summary, this paper has explored ways in which the two approaches can be combined to exploit their strengths and advantages in reactor network synthesis, and ultimately for improved synthesis of process flowsheets. What we are aiming at is a method in which we can define the individual processes (i.e. reaction, mixing, separation, etc.), the process equipment and the flowsheet that can be predicted, either by the construction approach currently used or through further understanding of the geometric properties, and subsequent optimization. Through this approach we ultimately aim for a synthesis approach based on the process phenomena and not on the traditional unit operations. This will serve as a significant motivator for novel future processes.

Acknowledgments

The authors gratefully acknowledge the help and advice from Profs. David Glasser and Martin Feinberg. Research support for LTB from the US Department of Energy is also gratefully acknowledged.

References

Achenie, L.E.K. and L.T. Biegler (1986). Algorithmic Synthesis of Chemical Reactor Networks Using Mathematical Programming. *I & EC Fund.*, **25**, 621.

Achenie, L. and L.T. Biegler (1990). Algorithmic Reactor Network Synthesis with Recycle Reactor Models. *Computers and Chemical Engineering*, **14**, 1, 23.

Agreda, V.H., Partin, L.R. and Heise, W.H. (1990). High Purity Methyl Acetate via Reactive Distillation. *Chem. Eng. Prog.*, **86**(2).

Balakrishna., S., and Biegler, L.T. (1992). A Constructive Targeting Approach for the Synthesis of Isothermal Reactor Networks. *Ind. Eng. Chem. Research,* **31**, 300.

Balakrishna., S., and Biegler, L.T. (1992b). Targeting Strategies for the Synthesis and Heat Integration of Nonisothermal Reactor Networks. *Ind. Eng. Chem. Research,* **31**, 2152.

Balakrishna., S., and Biegler, L.T. (1993). A Unified Approach for the Simultaneous Synthesis of Reaction, Energy and Separation Systems. *Ind. Eng. Chem. Research,* **32**, 1372.

Barbosa, D. and M.F. Doherty (1987). The Theory of Phase Diagrams and Azeotropic Conditions for Two-Phase Reactive Systems. *Proc. R. Soc. (London),* **A413**, 443.

Barbosa, D. and M.F. Doherty (1988). The Simple Distillation of Homogeneous Reactive Mixtures. *Chem. Engg. Sci.,* **43**, 541.

Chitra, S.P, and R. Govind (1985). Synthesis of Optimal Reactor Structures for Homogenous Reactions. *AIChE J.*, **31**(2), 177.

Conti, G.A. and W.R. Paterson (1985). Chemical Reactors in Process Synthesis. *ICheme Symposium Series No. 92*, 391.

Douglas, J.M. (1988). *Conceptual Design of Chemical Processes*. McGraw-Hill Book Company, New York, NY.

Duran, M.A., and Grossmann, I.E. (1986). Simultaneous Optimization and Heat Integration of Chemical Processes. *AIChE J.*, **32**, 123.

Fjeld, M., O.A. Asbjornsen, K.J. Astrom (1974). Reaction invariants and the importance of in the analysis of eigenvectors, stability and controllability of CSTRs. *Chem. Eng. Science*, **30**, 1917.

Floudas, C.A. and I.E. Grossmann (1994). Global Optimization for Process Synthesis and Design. *FOCAPD '94*.

Fogler, H.S. (1992). *Elements of Chemical Reaction Engineering*. Prentice-Hall, Englewood Cliffs, NJ.

Glasser, B., Hildebrandt, D., Glasser, D. (1992). Optimal Mixing for Exothermic Reversible Reactions. *I & EC Research,* **31**, 6, 1541.

Glasser, D., Crowe, C.M. and Jackson, R. (1986). Zwietering's Maximum-Mixed Reactor Model and the Existence of Multiple Steady States. *Chem. Eng. Comm.*, **40**, 41.

Glasser, D., C. Crowe, and D. Hildebrandt (1987). A Geometric Approach to Steady Flow Reactors: The Attainable Region and Optimization in Concentration Space. *I & EC Research*. **26**(9), 1803.

Glasser, D., D. Hildebrandt, S. Godorr and M. Jobson (1993). A Geometric Approach to Variational Optimization. Presented at IFAC Conf., Sydney, Australia.

Godorr, S., D. Hildebrandt and D. Glasser (1994). The Attainable Region for Mixing and Multiple Rate Processes. Submitted to *Chemical Engineering J.*

Hildebrandt, D. and M. Feinberg (1992). Optimal Reactor Design from a Geometric Viewpoint. Paper 142c, presented at AIChE Annual Meeting, Miami Beach, FL.

Hildebrandt, D., D. Glasser, and C. Crowe (1990). The Geometry of the Attainable Region Generated by Reaction and Mixing: with and without constraints. *I & EC Research*. **29**(1), 49.

Hildebrandt, D., and D. Glasser (1990). The Attainable and Non-attainable Region and Optimal Reactor Structures. Proc. ISCRE Meeting, Toronto.

Horn, F. (1964). Attainable Regions in Chemical Reaction Technique. The *Third European Symposium on Chemical Reaction Engg.*, Pergamon, London.

Horn, F.J.M., and M. J. Tsai (1967). The Use of Adjoint Variables in the Development of Improvement Criteria for Chemical Reactors. *J. opt. Theory and applns.* **1**(2), 131.

Jackson, R. (1968) Optimization of Chemical Reactors with Respect to Flow Configuration. *J. opt. Theory and applns.* **2**(4), 240.

Kokossis, A.C. and C.A. Floudas (1990). Optimization of Complex Reactor Networks — I. Isothermal Operation. *Chem. Engg. Sci.*, **45** (3), 595.

Kokossis, A.C and C.A.Floudas (1989). Synthesis of Isothermal Reactor-Separator-Recycle Systems. 1989 *Annual AIChE meeting*, San Francisco, CA.

Kokossis, A.C. and C.A. Floudas (1991). Synthesis of Non-isothermal Reactor Networks. 1991 *Annual AIChE meeting*, Los Angeles, CA.

Lakshmanan, A. and L.T. Biegler (1994). Reactor Network Targeting for Waste Minimization. Presented at National AIChE Meeting, Atlanta, GA.

Levenspiel, O. (1962). *Chemical Reaction Engineering*, John Wiley, New York.

Mavrovouniotis, M.L. and D. Bonvin (1994). Synthesis of Reaction Paths. *FOCAPD '94*.

Omtveit, T. and K. Lien (1993). Graphical Targeting Procedures for Reactor Systems. Proc. ESCAPE-3, Graz, Austria.

Omtveit, T. and K. Lien (1994). Graphical Conceptualization of Reactor/Separation/Recycle Systems. *Proc. PSE'94*, Kyongju, Korea.

Pibouleau, L. Floquet, and S.Domenech (1988). Optimal synthesis of reactor separator systems by Nonlinear Programming Method. *AIChE Journal*, **34**, 163.

Ravimohan, A. (1971). Optimization of Chemical Reactor Networks with Respect to Flow Configuration. *JOTA*, **8**, 3, 204.

Terrill, D. and J.M. Douglas (1987). *I & EC Research*, **26**, 685.

Van Dongen, D.B., and M.F. Doherty (1985). Design and Synthesis of Homogeneous Azeotropic Distillations, 1. Problem Formulation for a Single Column.*I & EC Fund.*, **24**, 454.

INTRODUCTION TO GREEN TRENDS IN DESIGN

Gary E. Blau
DowElanco
Indianapolis, IN 46268

Introduction

The chemical industry is actively striving to restore its image and build credibility in the eyes of the public by supporting continuing efforts to improve the extent of responsibility it takes for the products and processes it develops. Toward this end a Responsible Care® initiative was adopted by the Chemical Manufacturers Association whose purpose is "to make health, safety and environmental protection an integral part of designing, manufacturing, marketing, and disposing of products." Many companies are approaching this "green" issue by using procedures to ensure that both new and existing products and processes are toxicologically and environmentally friendly. For example, life-cycle analysis has developed into a serious discipline for measuring the impact of products and processes on the environment and the financial implications of this impact to a company. This decision support tool places "system boundaries" around the whole life cycle of the material and energy flows. It then balances the upstream inputs of energy and materials with the downstream emissions and residues of a product as well as the environmental effects directly associated with its manufacturer. In his paper, Boustead demonstrates the life-cycle analysis methodology and shows how it can be used in weighing options for critical process issues such as recycling.

What has the impact of the Responsible Care® initiative and its trend toward green processes been on the process design function? These introductory remarks will attempt to shed some light on this question by addressing changes being experienced in the process research environment, the plant infrastructure and the process designs themselves.

Process Research Environment

Consider the new product development process in the specialty chemicals business. Once a new product has been discovered and a preliminary synthesis route identified by the chemist, it is the responsibility of the process research function to translate the laboratory route into a scalable process. A large amount of preliminary toxicology and environmental testing is done at the earliest stage of development to ensure that the product itself is benign. However, many of today's new products involve complex chemistry with many processing steps so that one or more of the intermediates, as well as the solvents selected by the chemist, may be toxic. Extensive recycling, advanced separation methods and solvent substitution programs are being developed to eliminate any toxic waste streams and minimize the potential for worker exposure.

In the last five years the laboratory and pilot plant environment in which new processes are being developed has changed dramatically. The productivity of the laboratory has been forced to increase dramatically to accommodate the challenges of developing green processes while facing downsizing issues which have characterized the industry. The laboratory chemist must now involve himself with extensive waste handling procedures previously deemed unnecessary. Government regulations require the preparation of premanufacturing notification (PMN) documents for all intermediates and products. Fortunately, advances in laboratory information systems have helped to keep materials handling and tracking tasks under control. Entire laboratory research programs are devoted to eliminate what historically were considered to be acceptable waste streams. Alternative process routes previously deemed economically infeasible are being considered. The process chemist is being forced to rely on a very limited variety of "green" solvents to replace any toxic solvents selected by the synthesis chemist. In some ways, this has made the engineers tasks simpler since reducing the number of solvents has always been one of the design engineers primary concerns.

Another positive aspect of the changing process environment has been the stimulus to better understand the chemistry and kinetics of new processes. It is no longer an option for a company to build a plant and then deal with Responsible Care® issues during the so called "optimization" or "learning" phase. This understanding, captured in mathematical models, is used to determine plant design and operating conditions which will minimize the formation of waste streams. Generally, this approach to plant design ensures that the plant is economically designed with Responsible Care® issues forming the constraint set. Other benefits of such an analysis are smoother start-ups and the ability to minimize operating costs in both off-line and online configurations.

The process research function has also been asked to examine existing processes. Many of the issues are the same as for new processes although economics dictate that process design changes be minimal. Quite frequently, elucidation of reaction kinetics of all the products and intermediates gives insights into how the process can be operated with minimal modifications to the existing plant superstructure to meet Responsible Care® guidelines.

Finally, Responsible Care® imposes the onus on plants to demonstrate to the public and government agencies that their processes are in compliance with government regulations. Retrievable process validation data in concert with a fundamental mathematical model to answer various "what if" questions are essential to help establish compliance and assure high product quality.

Plant Infrastructure

Although the efforts to generate green processes have significantly reduced the amount of by-products from new processes, there are still some waste streams which must be handled. Plant infrastructure considerations are different for new grassroots facilities as compared to existing plants. It is safe to say that every effort is made to avoid building grass roots facilities because of the expense associated with new waste handling capabilities. A few general observations can be made. There is a strong move to waterless plant sites to insure that local waterways are not contaminated. Incineration is preferred over landfill to eliminate the potential for future litigation. Storage tanks are being located above ground so that leaks can be identified quickly. Extensive monitoring programs are being put in place to check ground water under the plant and at the battery limits.

In a tangential move, companies are negotiating waste handling alliances with their suppliers. For example, spent catalyst is returned to the supplier for treatment where they have the expertise, infrastructure and economies of scale to perform the regeneration and handle the catalytic poisons. Many manufacturers are supplying material in returnable containers. Finally, companies are making decisions on selecting third party manufacturers based as much on their waste handling knowledge as their abilities to produce the desired product.

The Designs Themselves

From all indications 50% of the capital for new processes are devoted to handling wastes. As a result, considerable efforts are being devoted to incorporate waste handling in process design and synthesis efforts. The paper by Allen and Manousiouthakis is an example of one approach to accomplishing this task. As the capital for grass roots plants becomes prohibitive, new processes must share existing waste handling systems to spread infrastructure costs. This is a natural extension of the attention being devoted in the literature to the design and operation of multi-product plants. Although cross-contamination issues must be considered, sharing of infrastructure driven by the desire for "green" processes will make multi-product plants the wave of the future, especially for batch plants. Associated with such plants will be the need to develop faster sequencing and scheduling tools to operate these plants in an efficient manner.

Another important consideration that must be introduced into the design of new processes is risk management. The Responsible Care® codes are pushing industry to use probability-based procedures to estimate risk, factoring in information about health, safety and the environment. For example, many companies use a risk characterization system for each process that factors in a product's flammability, explosivity, reactivity, toxicity, as well as the potential for a safety incident. In his paper, Powers provides a state of the art methodology for characterizing the risk associated with a specific combustion process and showing how alternative designs can be generated. At the present time, procedures of this type are generally data limited. However, once this data becomes more readily available it will be necessary to develop more sophisticated software to generate meaningful probability distributions.

Additional Opportunities for Computer-Aided Design Tools

Application of the Responsible Care® initiative to process design has created some additional challenging research areas for computer-aided process design One such challenge is handling uncertainties imposed by dynamic changes in environmental regulations superimposed upon the normal uncertainties in product demand forecasting. The trend in environmental regulations toward insuring increasing lower exposure levels parallels the development in analytical chemistry techniques which enable the detection of extremely low impurity levels. Computational methods need to be developed to generate designs based on anticipated demand and environmental restrictions over the life of the project. Formally, the problem is to determine the design parameters $x(t)$ which

$$\text{maximize } f(x(t), u(t))$$

subject to the constraints:

$$g(x(t), a(t), u(t)) \leq b(t)$$

where $u(t)$ are demand and expected environmental concentration estimates and $a(t)$ are technical estimates such as kinetic parameters. A simple objective function, f(), such as net present value can be used as an objective function. However, a more realistic approach is to use a variety of performance criteria and extract an acceptable optimal policy. The problem constraints g() are conventional mass and energy balances from a steady state or discrete simulator and b(t) are expected environmental concentrations imposed by government regulations. Here, t is a discretized time dimension over the life of the project.

The output from this mathematical program is a design "profile" which specifies the number, size and time during the life of a project in which equipment is purchased. Since demand forecasting is generally made on an annual basis the number of time increments is roughly the same as the life of the project (15 - 20 years).

In most cases the optimal policy is to put off purchases as long as possible until the uncertainties in long range forecasts (greater than three years) have been reduced. Delaying purchases also allows for processing breakthroughs particularly in handling waste streams. The conceptually simplest way to handle the uncertainties in u(t), b(t) and a(t) is with a Monte Carlo approach. If the constraints can be evaluated rapidly, the number of uncertain parameters is small, and only a few of the discrete sizes and types of equipment are considered, this is a viable approach especially with the availability of high speed parallel computers. For larger problems, alternatives are needed. One approach is to use a "design of experiments" or "scenario" approach to select profiles of u(t) for evaluation. Another is to use classical mathematical programming under uncertainty concepts promulgated in the 1960's. It is safe to say that several orders of magnitude improvement in computational speed will be needed before we can accommodate all meaningful uncertainties in the design problem.

Conclusions

The Responsible Care® initiative has stimulated the trend toward green designs. This trend has had a profound effect on how designs are developed as well as the designs themselves. This session will illustrate how various computer-aided design techniques can be used to develop green designs. It is still apparent that significant advances in the computational sciences will be needed before uncertainty, the key element in the design process, can be successfully incorporated.

PROCESS SYNTHESIS FOR WASTE MINIMIZATION

Vasilios Manousiouthakis and David Allen
Department of Chemical Engineering
University of California, Los Angelos
Los Angeles, CA 90024

Abstract

The thesis of this paper is that waste minimization is a process synthesis activity. In turn, this position suggests that: process synthesis concepts can help guide and accelerate the development of systematic waste minimization procedures; waste minimization needs can help identify novel process synthesis problems. To support this position and its ramifications, several process synthesis concepts are briefly reviewed and their importance for waste minimization is outlined. Particular attention is given to the notion of mass exchange network synthesis whose conception was inspired by waste minimization needs.

Keywords

Waste minimization, Process synthesis, Mass exchange networks.

Introduction

> "Solid wastes are the discarded left-overs of our advanced consumer society ... If we are ever truly to gain control of the problem our goal must be ... to reduce the volume of wastes and the difficulty of their disposal and to encourage their constructive re-use"

These lines, pointing out the need for waste reduction, could have been part of a public officer's speech in the 1990's. They are not. They are part of President Nixon's 1970 Environmental Message. They indicate that waste reduction has long been the waste management option of choice in the United States. It is however in this decade that increased environmental awareness, tighter environmental regulations and the ever growing view of wastes as resources are all converging to make waste minimization a top agenda item for our society.

Large companies have received this waste reduction message and have instituted a variety of waste reduction programs (Pollution Prevention Pays, Save Money And Reduce Toxics, etc.). Undoubtedly progress is being made but it is also becoming apparent that the current waste reduction rate is inadequate. A step change is needed. But, what are the technical elements that will bring about this step change? It is our contention that one of the most important ones is chemical process synthesis (from here on simply referred to as process synthesis).

Definitions of Process Synthesis and Waste Minimization

Prior to demonstrating the importance of process synthesis for waste minimization it is desirable to understand its beginnings. Analysis and synthesis are composite words of Greek origin (*ana-lysis*, *syn-thesis*). *Ana* is a preposition implying distribution while *lysis* is a noun denoting resolution. Thus analysis is the resolution of a complex whole into its constituent elements. On the other hand, *syn* is the preposition with, while *thesis* is the noun position. Thus synthesis is the composition of elements into a complex whole.

Every synthesis activity is initiated once a need has been identified. Thus the first step in synthesizing an engineering system is to establish objectives and performance criteria which, if met, will address the identified need. Subsequently, a set of technologies is selected (or developed) and the optimal type, number, size and interconnection of technological units is determined. Process synthesis attempts to capture and quantify the decision making that takes place whenever an engineering system, consisting of chemical processing units, is put together. Therefore, process synthesis can be defined as follows:

Process Synthesis Definition

Process synthesis is the composition of chemical process elements into a complex whole that meets given objectives, according to predetermined performance criteria.

Having outlined the notion of process synthesis, we can now proceed to define the notion of waste minimization. The notion may seem simple, but its definition has long been the subject of controversy. Analyzing the reasons for this controversy is not an endeavor we wish to pursue and therefore, in the sequel, we define waste minimization in the spirit of the U.S. Office of Technology Assessment (1986).

Waste Minimization Definition

Waste minimization is the composition of in-plant practices that reduce, avoid or eliminate the generation of waste.

This definition, when examined in the light of the aforementioned notion of process synthesis, suggests that *waste minimization is a process synthesis activity*.(Manousiouthakis, 1989a, 1989b). Indeed waste minimization aims at composing chemical process elements ("in-plant practices") into a complex whole that meets certain objectives ("reduce, avoid or eliminate the generation of waste") according to predetermined performance criteria (process cost, waste volume and toxicity).

Waste minimization can be considered either as a grassroots or as a retrofit process synthesis activity. Historically, waste considerations have not been included in the development of design alternatives for the process industries. As a result waste effluent regulations were primarily met through waste treatment (often destruction) methods. As the cost of such methods escalated through the years it was realized that waste minimization, in its retrofit form, was both a viable and desirable alternative to waste treatment. It is now accepted that waste minimization should also be thought of as a grassroots synthesis activity and that waste considerations should be included at the early stages of process development.

Waste minimization can be classified in two major categories: process modification and on-site recycling. There is a hierarchical relation between these two activities since process modification efforts (changes in process technology and equipment, in plant operations, in process inputs and in end products) have the potential of completely eliminating the use of substances that lead to waste generation while on-site recycling can only prevent the emission of waste from the plant site. On the other hand, the development of systematic waste minimization activities is more difficult when it aims at process modification, than when it aims at on-site recycling.

The purpose of this paper is to suggest that waste minimization is a process synthesis activity, that process synthesis concepts can help guide waste minimization activities and that waste minimization needs can help identify novel process synthesis problems. This position is supported by examining in detail the relations of various process synthesis problems to waste minimization. Particular attention is given to the notion of mass exchange network synthesis whose conception was inspired by waste minimization needs.

Process Synthesis Classification and its Relations to Waste Minimization

The scope of process synthesis, as defined above, is so broad that a general process synthesis procedure has not yet been developed and is unlikely to be forthcoming in the near future. Two thorough reviews on the subject support this viewpoint (Nishida et al., 1981; Hlavacek, 1978). There have been, however, several successes when the problem scope was narrowed and special problem characteristics were capitalized upon. As a result, the following classification of process synthesis has evolved over the years:

1. Material Synthesis
2. Reaction Path Synthesis
3. Reactor Network Synthesis
4. Separator Network Synthesis
5. Heat Exchanger Network Synthesis
6. Mass Exchanger Network Synthesis
7. Total Flowsheet Synthesis

Next, we briefly review these process synthesis problems and examine their relations to waste minimization.

Material Synthesis

This synthesis task can be stated as follows: "Given a set of desirable properties, identify a material that possesses these properties." The first step in addressing this task is to establish quantitative (or at least semi-quantitative) structure-property relations. Progress has occurred in this direction, primarily through group contribution methods, for both physical and chemical properties. For example, Joback (1984) has demonstrated, that nine physical properties (e.g. heat of vaporization, critical temperature and pressure, etc.) can be expressed as linear combinations of three "factors" that assume unique values for each molecule. This result has two important ramifications for material synthesis.

- It suggests that if one is given an arbitrary set of desirable properties *there may not exist any material* that can possibly possess these properties. Indeed if desired values are given for more than three properties then an overdetermined set of linear equations would have to be solved. In this case only a least squares type solution for the "factor" values would be possible.

- It suggests the following evolutionary approach to material synthesis (Stephanopoulos and Townsend, 1986). First, a target point in the three dimensional factor space is identified based on the desired physical property values. Then a supply point is chosen, in the same space, that represents a seed molecule whose chemical structure is to be appropriately modified. One may evolve, in factor space, from the supply to the target point by inserting or deleting chemical groups. The intermediate points in factor space are identified through group contribution methods.

Similar ideas are applicable to the synthesis of mass separating agents (MSAs) for extractive separations. In this case the relevant properties of the MSA are (Hayes, 1988; Cohen, 1988):

- Partition coefficient of the solute among the MSA and the original solvent (if the solvents are immiscible this information can be recovered from the activity coefficients of the solute in each solvent). This property determines how the solute is distributed between the MSA and the original solvent.
- Selectivity for the considered solute over the original solvent or other solutes. This property relates the amounts of solute and solvent recovered by the MSA.
- Capacity for the considered solute. This property quantifies the ratio of solute recovered over MSA used.
- Normal freezing and boiling points and heat of vaporization. Thcsc properties may determine the ease of subsequent MSA recovery.
- Viscosity, surface tension and liquid-solid contact angle of a liquid MSA. These properties determine the rate at which the liquid MSA can penetrate into capillary spaces (Washburn, 1921). Such penetration is necessary for the removal of solutes from solid surfaces.
- Vapor pressure and density of a liquid MSA. These properties determine fugitive emissions of the MSA.
- Ozone depletion capability, photochemical reactivity, toxicity, flammability, corrosivity, chemical stability. These properties determine the risks involved with the MSAs use. Some risk levels may be unacceptable and may prevent the use of otherwise acceptable MSAs.

The synthesis of MSAs that possess all of these properties is an overwhelming task. Therefore process synthesis procedures have focused on a small number of these properties, namely the first two. Lo, Baird and Hanson (1983) have used solubility parameter theory to quantitatively express these two properties in terms of the solubility parameters (Hansen, 1967) of the original solvent, the solute and the MSA. The general premise of this theory is that when two substances have similar solubility parameters then they are mutually soluble. Thus, the solubility characteristics of a given substance (i) are determined by three solubility parameters: $\delta_{d,i}$: a dispersive (non-polar) term; $\delta_{p,i}$: a polar term; $\delta_{h,i}$: a hydrogen bonding term.

The solubility characteristics of a solute (s)-MSA (m) system can then be quantified in terms of a so-called radius of interaction R_{sm} defined as (Cabelka and Archer, 1987)

$$R_{sm}^2 \equiv \left(\delta_{d,s} - \delta_{d,m}\right)^2 + \left(\delta_{p,s} - \delta_{p,m}\right)^2 + \left(\delta_{h,s} - \delta_{h,m}\right)^2$$

Then $R_{sm} \rightarrow 0$ implies that s and m are soluble while $R_{sm} \rightarrow \infty$ implies that s and m are insoluble. To identify (or estimate) a substance's solubility parameters one may either consult solubility handbooks (Lo et al., 1983; Barton, 1983), or may employ group contribution methods (Hoy, 1970). Armed with this knowledge base, Stephanopoulos and Townsend (1986) developed an evolutionary MSA synthesis procedure in solubility parameter (δ) space. First, the original solvent and solute points are identified. Then a supply point is chosen that represents a seed MSA whose chemical structure is to be appropriately modified. Then one evolves in δ-space by inserting or deleting chemical groups from the supply point towards the solute and away from the original solvent. The intermediate points in factor space are identified through group contribution methods.

Within the context of waste minimization the material synthesis problem is often referred to as material substitution. This name alone clearly indicates that material synthesis aimed at waste reduction has been traditionally performed in a retrofit mode (as a process modification step) and is only now being considered as a grassroots synthesis activity (a process development step). In either case, the selection of raw materials, catalysts, mass separating agents, solvents, etc. has a large impact on all aspects of the process and therefore on the waste that the process generates. Close examination of past material substitution efforts reveals that they are mostly disconnected and case study specific (Huisingh et al., 1985). This fact does not negate the value of these efforts but does suggest that they are unlikely to decrease the development time for successful substitute materials. Next, we demonstrate how process synthesis ideas may help bring about change.

Consider for example the problem of substituting chlorofluorocarbons (CFCs) as cleaning agents. The production/use of these substances has been restricted by the Montreal protocol so as not to exceed the 1986 levels of production/consumption (2.1 billion pounds/year). Of particular concern to the printed circuit board manufacturing industry are the restrictions on CFC 113. Indeed, the 1986 production of CFC 113 was 360 million pounds, 45% of which was used in the electronics industry for flux removal from printed circuit boards (Shabecoff, 1988). An industry-wide effort has generated two alternative cleaning agents; water and terpenes. How did the development of these solvents come about? The problem at hand was to identify a MSA that can dissolve flux which is attached on a solid surface. As pointed out earlier, a fluid's ability to clean surfaces depends on its penetration rate into capillary spaces and on its ability to dissolve the flux. Typically both water and terpenes have higher surface penetration rates than CFC 113 (except when the contact angle of the water on the surface is higher than 70 degrees). Therefore it is their ability to dissolve flux that primarily determines their success or failure as CFC alternatives. It is not surprising therefore that water has been successful in cleaning ionic fluxes or in removing organic fluxes from surfaces that do not have excessively minute features (e.g. disc drive parts; Drier, 1988), but has failed to remove rosin fluxes from surfaces with minute features (e.g. silicon wafers). The selection of terpenes as a rosin flux remover was based on the same scientific tools we outlined earlier in MSA synthesis for extractive separations. Abietic acid (a major constituent of rosin flux) was chosen as the solute and its interaction radius R_{sm} was evaluated for several solvents. The abietic acid-terpene interaction radius was found to be in the 0-2 range which was small compared to the abietic acid-CFC 113/methylene chloride interaction radius which was found to be about 4.5. This prediction is confirmed in Fig. 1 which shows the mole fraction of abietic acid at saturation versus the interaction radius of abietic acid with various solvents. These experimental data (Hayes, 1988; Cabelka and Archer, 1987) confirm the trend high solubility <-> low R_{sm}, low solubility <-> high R_{sm}.

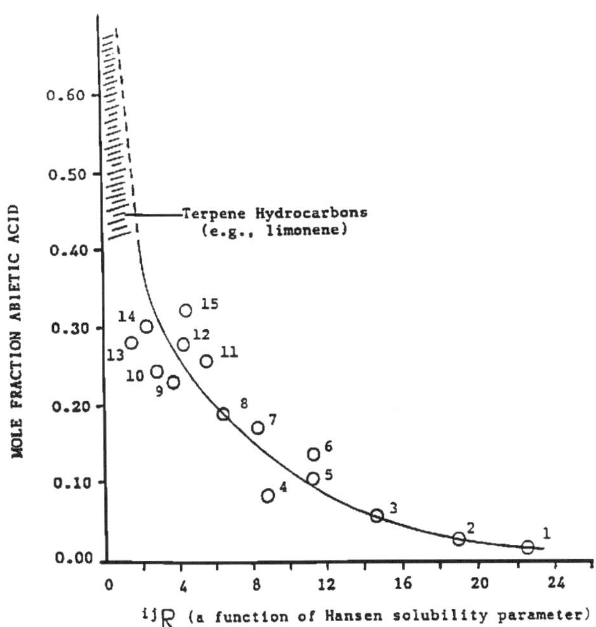

1. Acetonitrile
2. Methanol
3. Ethanol
4. CFC-113/Ethanol
5. Isoproponal
6. MEK
7. CFC-113/Methanol
8. 2-ethyl hexanol
9. Perchloroethylene
10. Methylene Chloride
11. 1,1,1-trichloroethane
12. CFC-113/Methyl Chloride
13. Trichloroethylene
14. Cyclopentanone
15. o-dichlorobenzene

* Data (except for limonene) from T.D. Cabelka and W. Archer, *Surface Mount Technology*, December, 1987, 18-21.

Figure 1. Solubility of abietic acid in selected solvents.

Reaction Path Synthesis

This synthesis task can be stated as follows: "Identify a reaction path that employs substances from a set of permissible chemicals to yield a desired product." As a research field, reaction path synthesis was initiated by organic synthesis chemists and was focused on the synthesis of large organic molecules. More recently it attracted the attention of chemical engineers and found applications in inorganic synthesis as well.

There are several considerations that must be taken into account during reaction path synthesis. The most important ones are as follows:

a. *Economic*: The cost of the involved chemicals and the reaction stoichiometry, yield and equilibrium conversion determine, to a large extent, the economic feasibility of the considered reaction.

b. *Thermodynamic*: The reaction Gibbs free energy change (ΔG) and the heat of reaction (ΔH) are two thermodynamic properties that critically influence the reaction selection process. ΔG determines reaction equilibrium conversion ($\Delta G = -RT\, ln\, K_{eq}$) and is, therefore employed in the thermodynamic criteria for reaction feasibility ($\Delta G < 10$ kcal/mole). ΔH determines reaction exothermicity ($\Delta H < 0$ <=> exothermic reaction) as well as the effect of temperature changes on equilibrium conversion ($\Delta H = RT^2 (d\, ln\, K_{eq} / dT)_p$ which suggests that $\Delta H < 0$ => a temperature increase causes equilibrium conversion decrease). Clearly these two properties determine, to a large extent, not only the reaction operating conditions but also the extent of byproduct formation, the types and quantities of utilities needed, and the difficulty of subsequent separations.

c. *Kinetic*: Reaction kinetics also play a critical role in the reaction and operating conditions. Unfortunately the theoretical prediction of reaction kinetics is still at its infancy and thus incorporating reaction kinetics considerations in reaction path synthesis may be thought of as one of the field's bottlenecks.

In the last two decades several researchers have successfully incorporated thermodynamic (ΔG) considerations in reaction path synthesis. Most notably, May and Rudd (1976) proposed synthesis methods for the creation of thermodynamically feasible closed-cycle chemical reaction sequences. A few years later Rotstein et al. (1982) and Fornari et al (1989) developed properties of the ΔG function, in the space of permissible chemicals, which they subsequently used in synthesizing single-step and multistep reactions. Starting from species classified as terminal (raw materials, products) and intermediates, an algorithm for the development of reaction mechanisms that incorporates reversible elementary reactions was proposed (Mavrovouniotis and Stephanopoulos, 1992; Mavrovouniotis, 1992). This algorithm is based on another algorithm proposed for the synthesis of biochemical pathways (Mavrovouniotis et al., 1990). Multistep reaction schemes can be classified into closed-cycle (Solvay clusters) and open-cycle sequences of reactions. A closed-cycle sequence consists of thermodynamically feasible reactions whose overall stoichiometry is identical to that of a given thermodynamically infeasible reaction. As an example (May and Rudd, 1976), the infeasible main reaction A + B -> Z may be realized through the closed-cycle sequence

$$\begin{array}{l} A + L \rightarrow N \\ N + B \rightarrow Z + L \\ \hline A + B \rightarrow Z \end{array}$$

Open-cycle sequences on the other hand have an overall stoichiometry which is different from that of the main reaction of interest (they may require additional reactants and/or yield additional products). For example, for the same main reaction A + B -> Z an open cycle sequence may be:

$$\begin{array}{lcl} A + Y & \rightarrow & N + X \\ N + B & \rightarrow & Z + L \\ \hline A + B + Y & \rightarrow & Z \end{array}$$

The significance of these ideas for waste minimization is paramount. As a general principle, *the use of closed cycle reaction sequences leads inherently to waste minimization* since no unnecessary reactants and/or byproducts are involved. As a result, byproduct and/or unreacted raw material separation steps are minimized. If a closed cycle sequence cannot be identified then open-cycle sequences with minimal number of reactants and byproducts should be considered. Douglas (1989, 1992) discussed the importance of alternative open-cycle sequences for the minimization of wastes in the production of the herbicide alachlor. The two multistep reaction schemes considered were both open-cycle sequences and were as follows:

1.
$$\begin{array}{lcl} A + B & \rightarrow & K + HCl \\ K + NaNH_2 + C & \rightarrow & Z + NaCl + NH_3 \\ \hline A + B + NaNH_2 + C & \rightarrow & Z + NaCl + NH_3 + HCl \end{array}$$

2.
$$\begin{array}{lcl} A + HCHO & \rightarrow & K + HCl \\ L & \rightarrow & M + H_2O \\ M + B & \rightarrow & N \\ N + CH_3OH & \rightarrow & Z + HCl \\ \hline A + B + HCHO + CH_3OH & \rightarrow & Z + H_2O + HCl \end{array}$$

where

A $= C_6H_3(C_2H_5)_2NH_2$
B $= ClCH_2COCl$
C $= ClCH_2OCH_3$
K $= C_6H_3(C_2H_5)_2NHCOCH_2Cl$
L $= C_6H_3(C_2H_5)_2NHCH_2OH$
M $= C_6H_3(C_2H_5)_2NCH_2$
N $= C_6H_3(C_2H_5)_2N(COCH_2Cl)(CH_2Cl)$
Z $= C_6H_3(C_2H_5)_2N(COCH_2Cl)(CH_2OCH_3)$ (alachlor)

The first reaction scheme employs only two reaction steps but creates a solid waste byproduct (NaCl) which is difficult to separate since it may be contaminated with A, B, K, C, Z. The second reaction scheme employs four reaction steps (thus requiring more reactors than the first scheme) and generates wastewater contaminated with A, HCHO, L, M and possibly B, N, CH_3OH and Z. It has however the advantage of not generating a solid byproduct. Clearly both alternatives have some shortcomings. Let us now consider how reaction path synthesis may be of help in this regard.

Consider, for a moment, that the reaction

$$NaCl + NH_3 \rightarrow NaNH_2 + HCl$$

were feasible. Then the first reaction scheme can be modified so that its overall reaction becomes

$$A + B + C \rightarrow Z + 2HCl$$

This overall reaction is clearly desirable from both an economic and a waste minimization viewpoint since it generates two valuable products that can be easily separated. Thus the waste minimization task has been translated into the following reaction path synthesis task: "Identify a closed-cycle sequence of feasible reactions for the overall reaction

$$NaCl + NH_3 \rightarrow NaNH_2 + HCl.$$

Reactor Network Synthesis

The task of synthesizing reactor networks is stated as follows: "Given a reaction mechanism identify a network of reactors in which these reactions transform raw materials to products at optimum cost." Typically the venture cost of the system need be minimized. Once this task is accomplished, the type, the number, the size, the interconnections and the operating conditions of all units are specified.

Several formulations of the reactor network synthesis problem have been employed by researchers. The maximization of the yield for simple reaction mechanisms through single unit design was proposed by Levenspiel (1962, 1984). Dynamic programming techniques were employed by Aris (1961) to identify the feed distribution that maximizes the yield for reactor networks. Dyson and Horn (1967) employed a variational formulation of the same problem and evaluated the feed distribution that maximizes the yield. Variational techniques have also been employed in the problem of identifying the optimal catalyst blend for tubular plug flow reactors (Jackson, 1968a. Horn (1964) proposed the notion of an attainable region in concentration space as a tool for reactor network synthesis. Hildebrandt and Feinberg (1989) further developed this idea and were able to establish a number of properties for the attainable region and its boundary. Another approach involves the study of the optimal flow configuration in a reactor network (Horn and Tsai, 1967, Jackson, 1968b, Ravimohan, 1971). Chitra and Govind (1985) proposed a technique that identifies the temperature profile of a serial structure of nonisothermal reactors. Campbell and Hopper (1989) and Hopper and Muninnimit (1989) examined the impact of reactor parameters and catalyst selection on waste generation. Balakrishna and Biegler (1992) utilized a segregation model in their nonisothermal reactor network representation and proposed an iterative procedure that identifies lower bounds on the achievable maximum of a general performance index. Kokossis and Floudas (1994) employed a superstructure parametrization of a nonisothermal reactor network which led to an MINLP problem formulation.

The impact of reactor network synthesis to waste minimization depends on the reaction mechanism at hand . For closed cycle mechanisms that do not produce any undesirable byproducts, optimal reactor network synthesis does not significantly impact waste minimization. If the formation of byproducts is encountered within a closed cycle mechanism, product yield maximization is an indirect manner of addressing waste minimization concerns. In the case of open cycle reaction mechanisms though, reactor network synthesis may have a significant impact in achieving waste minimization goals.

As an example consider the production of acrylonitrile from ammonia, oxygen (or air) and propane (or propylene), using ammoxidation (Jacob (1991), Catani et al. (1992)). This is a complex, catalytic open cycle reaction sequence that leads to a variety of byproducts (such as hydrogen cyanide, acetonitrile and carbon oxides) and involves a large number of phenomenological reaction steps (e.g. fourteen steps for the direct ammoxidation of propane).

To minimize waste generation for this process, one need only perform a reactor network synthesis study aiming at reducing the formation of hydrogen cyanide and acetonitrile (which are often incinerated). The scope of the study should be examined The scope of the study should be extended beyond the configuration and operating conditions of the network to include selection of the catalyst and of the network feed conditions. The latter in particular have a major impact on byproduct formation though in reactor network synthesis studies they are often considered to be known.

Separator Network Synthesis

This synthesis task can be stated as follows: "Given a set of multicomponent feed streams, identify a network of separators that can yield a set of desired product streams at a minimum venture cost." The successful completion of this task involves the selection of the type, number and size of the separator units as well as their interconnections and operating conditions. Thus information on separator modeling/costing and on phase equilibria is necessary. The overwhelming number of separator network alternatives that one may employ has led to a divide and conquer approach to the problem. As a result the following major classifications have arisen:

- The separating agent used to accomplish a given separation task is either energy or mass (King, 1980).
- The separations employed are simple (i.e. separate one feed stream into two product streams) or not simple; sharp (i.e. the recovery of the key components is unity) or not sharp.
- Ideal or nonideal phase equilibria relations are involved (Knapp and Doherty, 1989).
- Single or multiple separation methods are employed.
- A heuristic, evolutionary or algorithmic network synthesis strategy is employed.

A brief review of previous work on the subject can do no justice to the evergrowing literature. Thus we direct the interested reader to the review papers (Nishida et al., 1981; Hlavacek, 1978) and to a review paper on the synthesis of distillation networks (Westerberg, 1985). As a general overview, one may state that although no universal solutions are available, the methods available in the literature can significantly accelerate the development of alternative separation networks.

In the context of waste minimization, separation network synthesis plays a very important role. Consider for example the task of synthesizing a recycle/reuse network: "Given a set of multicomponent waste streams, identify a network of separators that allows the recycle of these streams at minimum venture cost." The waste streams can be thought of as the feed streams to a separator network. Their recycle will be allowed only if they meet some predetermined specifications which can help define a set of desired product streams. Thus the recycle/reuse network synthesis problem can be thought of as a separator network synthesis problem. Consider as an example the recycling of used solvents (Higgins, 1989) N-methyl-2-pyrrolidone (NMP) is used to remove fluxes from semiconductor substrates. Deionized waster (DI) is subsequently employed for rinsing purposes. The feed and product streams for the considered separator network are as follows:

Composition	Stream			
	Feed 1 45gal/m	Feed 2 15gal/m	Product 1	Product 2
NMP	99%	1%	100%	<500ppm
H2O	1%	99%	<500ppm	100%
Flux	100ppm	Trace	<5ppm	Trace
Metals	Trace	Trace	Trace	Trace
Oil	20ppm	Trace	Trace	Trace

Although several separation methods were considered for this task, most of them were rejected *not due to economic considerations* but rather based on ease of operation and maintenance and on lack of technical data. The chosen technology was vacuum distillation and the number of products was more than two, although concentration specifications were given only for two products.

From this and many other case studies it becomes apparent that, in waste minimization, technological choices are often based on an engineer's familiarity with a given separation method. In this context, *the development of design/modeling environments for hazardous waste problems is essential for waste minimization* (Westerberg, 1989a). Already existing environments such as ASCEND (Westerberg 1989b) and MODEL.LA (Henning et al., 1989) may serve as paradigms for such a development. Another realization that stems from these case studies is that *distillation network synthesis is of extreme importance to waste minimization*. This is the case not only for ideal mixtures but also for mixtures which form azeotropes. A prevalent example is the case of chlorofluorocarbon (CFC) recycling. Indeed CFCs are often used in conjunction with alcohols with which they form azeotropic mixtures. Therefore recycling of such contaminated solvent mixtures may require the use of azeotropic or extractive distillation networks (Knapp and Doherty, 1989).

Another separator network synthesis area with relevance to waste minimization is *reactor/separator network synthesis*. Consider for example the production of ethyl acetate from acetic acid and ethanol. The traditional production scheme employs a reactor and a distillation column and uses entrainer fluid to break up the resulting acetate-alcohol azeotrope. By studying the overall (reaction and phase) equilibria characteristics of the alcohol/acetic acid/water/acetate system, Doherty (1989) was able to demonstrate that a single distillation column with an appropriate feeding pattern and a reactive zone loaded with catalyst can transform acetic acid and alcohol into pure acetate and pure water. The new design is better than the old one, both in terms of economic profitability and waste generation.

Heat Exchanger Network Synthesis

This synthesis task can be stated as follows: "Given a set of hot and a set of cold streams identify a network of heat exchanger units that can transfer heat from the hot to the cold streams at minimum venture cost." Heat exchanger network synthesis flourished as a problem of interest during the 70's and 80's due to the energy crisis. The result has been an industry-wide application of heat exchange networks and a large number of journal publications. A thorough and recent review of these applications has been given (Gundersen and Naess, 1989). The prevalent design methods developed through the years are thermodynamic (Umeda et al., 1979; Linnhoff and Hindmarsh, 1983) and algorithmic (Cerda et al., 1983; Papoulias and Grossmann, 1983) in nature.

It has become apparent during these years that heat exchange network synthesis has had a significant impact on the reuse of wasted energy. Although it is not as relevant, it does have some impact on the generation of waste material as well. First of all, it has been shown that the proper integration of an overall plant from an energy viewpoint can reduce the consumption of raw materials (Duran and Grossmann, 1986) and thus indirectly lead to waste minimization. In addition, the reduction of process cooling and heating needs is associated with an analogous reduction in blowdown quantities (blowdown is typically 0.5% of the utility flowrates). Blowdown is known to contain toxic chemicals such as bactericides and fungicides and is classified as a hazardous waste (U.S. Office of Technology Assessment, 1986). Finally the energy savings resulting from the more efficient use of waste heat can be expected to reduce demand on power generation plants, thus reducing environmental emissions (NO_X, etc.).

Mass Exchanger Network Synthesis

This synthesis task (depicted in Fig. 2) was defined (El-Halwagi and Manousiouthakis, 1989a) in response to waste minimization concerns and can be stated as follows: "Given a set of rich streams and a set of lean streams, synthesize a network of mass exchange units that can transfer certain species from the rich streams to the lean streams at minimum venture cost." A mass exchange unit can be any countercurrent mass transfer operation, such as liquid-liquid extraction, adsorption, desorption, absorption and others. Further, each mass transfer operation may employ several types of mass separating agents (e.g. for adsorption one may use activated carbon, polymer resins, etc.).

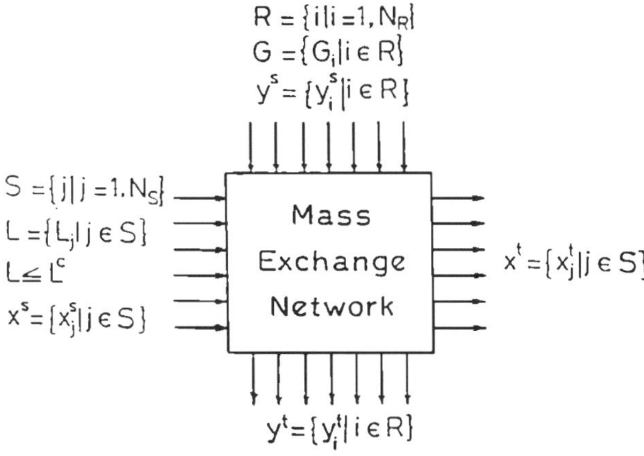

Figure 2. Schematic representation of the MEN problem.

Let us now consider a waste minimization problem that can be addressed through mass exchange network (MEN) synthesis.

Acrolein is a feedstock for the manufacture of acrylic acid, acrylic resins (latex), allyl alcohol, and methylethyl ketone. It is manufactured in petrochemical plants through high temperature catalytic oxidation of propylene with steam and air. In this process, a number of separation steps result in water streams containing acrolein (Weigert, 1973). Since acrolein is a polar-organic pollutant on the Priority Pollutants List of the U.S. EPA (Wise and Fahrenthold, 1981), these streams are characterized as waste streams. By removing acrolein from these water streams using a variety of solvents, one may satisfy environmental regulations and attain waste reduction.

Two acrolein-rich (R_1, R_2) and one acrolein-lean (S_4) wastewater streams are considered, as well as three possible acrolein-lean solvent streams (methyl isobutyl ketone: S_1, butyl acetate: S_2, toluene: S_3). The data for the various rich and lean streams are presented below, with $y^u{}_i$, $x^u{}_j$ representing upper bounds on acrolein target compositions.

Rich Streams				
Stream	G_i	$y^s{}_i$	$y^u{}_i$	
	kg/s	kg/kg	kg/kg	
R_1	14.085	0.0002	0.000002	
R_2	6.75	0.0018	0.000005	
Lean Streams				
Stream	L_i	$x^s{}_j$	$x^u{}_j$	c_j
	kg/s	kg/kg	kg/kg	$/kg
S_1	—	0.000005	0.0085	0.61
S_2	—	0.000001	0.000975	0.05
S_3	—	0.00002	0.0002	0.019
S_4	68.0	0.0	0.000005	0.0

Linear equilibrium relations, independent of the solute-solvent system, govern acrolein transfer from rich to lean streams (Joshi and King, 1981; Joshi, 1981):

(S1) $y = 0.2041(x_1 + \varepsilon)$
(S2) $y = 0.3846(x_2 + \varepsilon)$
(S3) $y = 0.4545(x_3 + \varepsilon)$
(S4) $y = 1.0x_4$

with $\varepsilon = 1.0\text{x}10^{-7}$ kg/kg denoting the minimum concentration driving force.

The MEN synthesis task at hand is to identify the type, number, size and interconnection of units that optimize mass transfer. In analogy to HEN synthesis, this task is decomposed in two subtasks: identifying the minimum possible cost of mass separating agents and then identifying a minimum utility network that features a minimum number of exchanger units. In accomplishing these subtasks the range of thermodynamically feasible operation for a mass exchanger is first quantified. Indeed in an equilibrium diagram practically feasible operation results if the operating line is some distance away from the equilibrium line. This realization allows one to establish a scale correspondence among the compositions of the rich and the lean streams. In turn, this correspondence permits the representation of all rich streams as a "composite rich stream" and of all the lean streams as a "composite lean stream." Then, in an exchanged mass vs. composition diagram, thermodynamic feasibility of mass exchange can be readily assessed. This problem representation not only quantifies a "mass exchange based waste minimization potential" but also leads to an LP based automatic MEN synthesis procedure (El-Halwagi and Manousiouthakis, 1990b).

Considering all stream target compositions to be fixed at their upper bounds permits application of this procedure to the above example. The resulting minimum utility cost is $0.3523/s, and the optimum solution features three lean streams, with flowrates: $L_1^* = 1.4361$ kg/s, $L_3^* = 13.9316$ kg/s, and $L_4^* = 68.0$ kg/s.

On the other hand, if one allows the target compositions of the streams to vary, then the above procedure is no longer applicable. MINLP and LP based solution methodologies for the resulting variable target MEN synthesis problem have been presented in Gupta and Manousiouthakis (1993; 1995b).

Their application to our example yields a minimum utility cost of \$0.2445/s, and an optimum solution that features two lean streams, with flowrates of $L_1^* = 4.0086$ kg/s and $L_4^* = 68.0$ kg/s. The cost reduction obtained in the variable target over the fixed target problem is significant (30.6%). This reduction can be attributed to the elimination of the thermodynamic bottlenecks imposed on S_1 when its target composition is fixed at its upper bound. Indeed, in the variable target problem the exit composition of S_1 is 0.003674 kg/kg as compared to 0.0085 kg/kg for the fixed target problem.

Two other important classifications of the MEN synthesis problem stem from the need to consider lean stream regeneration and transfer of multiple components. If regeneration is needed, the mass exchange/regeneration network synthesis problem should be considered (El-Halwagi and Manousiouthakis, 1990a). If, on the other hand, separation targets are imposed on multiple components, then the multicomponent MEN synthesis problem should be considered (El-Halwagi and Manousiouthakis, 1989b; Gupta and Manousiouthakis, 1994).

El-Halwagi and Srinivas (1992) developed minimum utility cost solutions for reactive mass exchange networks. Later, Srinivas and El-Halwagi (1994) generalized these results to include nonlinear equilibrium functions. El-Halwagi ct al. (1992) also applied MEN synthesis to the problem of phenol minimization in petroleum refinery wastewaters demonstrating significant savings over conventional waste treatment approaches.

Finally, Wang and Smith (1994) focused their attention on a mass exchange network synthesis problem in which the only lean utility is fresh water. Their problem formulation avoids the explicit use of the "rich composite curve" by employing the notion of a "limiting water profile" which accounts for both rich stream and driving force information.

MEN Synthesis Approach To Distillation

Bagajewicz and Manousiouthakis (1992) introduced the state-space approach to process synthesis as a comprehensive conceptual framework for the generation of all possible design alternatives. The main idea, inspired by the theory of dynamical systems, is that any process network can be represented as a distribution network that interacts with a process operator. The distribution network determines all the flows in the overall network while the operator determines how each flow is processed. Clearly, the process operator can itself admit a variety of representations. Some of these representations are based on diagonally ordered process units giving rise to superstructure type models. Others focus on aggregate models of process subsystems such as the pinch-based representation of an HEN operator. Employing these ideas, it was shown in the above work that one could represent a complex distillation network as an MEN and an HEN that interact through a distribution network. The MEN and HEN operators can be modelled either through unit based or pinch-based representations. In fact, the latter representation allows one to pursue the solution of the minimum utility cost problem for distillation networks without prior commitment to a network structure.

Application of the technique to propane-propylene splitting and solvent recovery problems resulted in 10% to 30% savings in utility cost.

To apply this methodology to a multicomponent distillation network synthesis problem one requires a model for a mass exchange unit. In previous studies on the design of complex distillation networks (Glinos and Malone, 1985; Carlberg and Westerberg, 1989a,b), Underwood-like models for distillation column sections were employed. Development of a similar Underwood-like model for a mass exchanger is therefore of interest. Let z_j be the net flow-normalized flux of component j in the V-direction, that is

$$z \equiv y_{j,N} - \frac{L}{V} x_{j,N+1} = y_{j,0} - \frac{L}{V} x_{j,1}, j = 1, \ldots, C$$

Then the conditions at one end of a mass exchanger can be related to those at the other end through the following equations:

$$\frac{\sum \frac{\alpha_j x_{j,N+1}}{\alpha_j - \varphi_p}}{\sum \frac{\alpha_j x_{j,N+1}}{\alpha_j - \varphi_q}} = \left(\frac{\varphi_p}{\varphi_q}\right)^N \frac{\sum \frac{\alpha_j x_{j,1}}{\alpha_j - \varphi_p}}{\sum \frac{\alpha_j x_{j,1}}{\alpha_j - \varphi_q}}$$

where, φ_p, φ_q, satisfy

$$f(\varphi) \equiv \sum \frac{\alpha_j z_j}{\alpha_j - \varphi} - 1 = 0$$

This mass exchanger Underwood-like model is similar to its distillation counterpart. It does possess, however, a unique characteristic. Unlike its distillation counterpart, in which φ is guaranteed to be real, in this model the φ-defining equation, $f(\varphi) = 0$ *may possess complex roots* (Gupta and Manousiouthakis, 1995a). In fact, as the following example demonstrates, there exist real mass exchangers for which this phenomenon occurs.

Example

Consider two three-component streams with compositions and relative volatilities as in Table 1 and an *L/V* ratio of 2.0. After being placed in contact with one another in a mass exchanger with one stage, their compositions are altered. The resulting outlet compositions and fluxes are summarized in Table 2.

Table 1. Data for the Three Components of Example.

Component	Rel. Volatility	Inlet Composition	
	α_j	$x_{j,N+1}$	$y_{j,0}$
1	4	0.2500	0.8500
2	2	0.4750	0.1000
3	1	0.2750	0.0500

Let us now identify the roots of $f(\varphi)$ for these z_j values. The roots of $f(\varphi)$ are: $\{1.18364, 3.58318 \pm 0.82367i\}$, *i.e., two of the roots are complex.*

Table 2. Results for the Three Components of Example.

Component	Outlet Composition		Flux
	$x_{j,1}$	$y_{j,N}$	z_j
1	0.2500	0.8500	0.100
2	0.4750	0.1000	-0.650
3	0.2750	0.0500	-0.450

Nonisothermal MEN Synthesis

The mass exchange network synthesis problems considered earlier employ implicitly the assumption that energy considerations need not enter in the design procedure. This is the case when all mass exchange operations take place at a single temperature and the only utilities employed are mass separating agents.

MEN synthesis is often called upon to address problems in which both mass and energy are used as separating agents. An optimal mass transfer policy may require the nonisothermal operation of the mass exchange network. In most mass transfer situations where the equilibrium distribution of the key component(s) depends on the temperature, use of heating or cooling utility to alter rich-lean contact temperatures may decrease the cost of MSAs needed. Besides, streams within a process exist at various enthalpy levels and for isothermal mass exchange, heating/cooling utility demands may be exorbitant. Instead, nonisothermal MEN synthesis can take advantage of this non-uniform stream temperature availability.

Heat exchange can take place in a nonisothermal MEN either within a mass exchanger itself or in heat exchangers placed throughout the network. In either case it is desirable that the minimum utility (both mass and energy) cost be identified. The problem can be formally stated as follows:

> Consider a number of streams with given flowrates and given inlet and outlet temperatures and concentrations. Consider also a number of mass and energy utilities with variable flowrates, given inlet and outlet temperatures and concentrations, and known costs per unit flow. Identify the minimum utility cost over all mass/heat exchange networks that meet these targets.

To perform this search over all networks, the state-space framework is employed. Each stream is given access to a family of junctions, some of which are connected to an HEN and others to an MEN. The number of junctions, distribution network flows, temperatures and concentrations and the HEN-MEN operator variables are all optimized to achieve minimum utility cost. The resulting optimization problem is significantly simplified for *separable nonisothermal MENs* in which heat and mass are not exchanged within the same unit.

Employing pinch operators for the MEN, HEN representation, Roxenby and Manousiouthakis (1993) proved that the minimum utility cost over all networks can be attained within a state-space representation which employs only three HEN junctions and one MEN junction per external stream.

In the same work, it is also established that under certain conditions the HEN flow path can be set to zero without increasing the minimum utility cost. These conditions are:

a. If the MEN operating temperature is between the supply and target temperatures of an external stream, then the HEN part of that stream can be set to zero.
b. If the minimum temperature difference (ΔT_{min}) of the HEN is zero then the HEN path of all streams can be set to zero. Furthermore, if the supply temperature of each lean stream is equal to its target temperature, then one can decompose the nonisothermal MEN synthesis problem into an HEN synthesis problem and a series of isothermal MEN synthesis problems.

These ideas are illustrated with the following example.

Example

Consider a process plant that has two wastewater streams, R_1 and R_2, from which a contaminant is to be removed. The flowrates, inlet and outlet contaminant compositions, and temperatures are as in Table 3. Also available are two lean utility (mass-separating agent) streams, L_1 and L_2. Their inlet and outlet contaminant compositions and temperatures, and per unit costs, are also given in Table 3. To help achieve the exit temperature specifications on these streams, a hot and a cold utility, with given inlet and exit temperatures and per unit costs, shown in Table 4, may be utilized. Specific heat information for these streams is also provided in Table 4.

Table 3. Stream Data for the MEN in the Non-isothermal Example.

Rich Streams			
Stream	F_i	y_i^s	y_i^t
	kg/s	g/kg	g/kg
R1	5.0	0.5	0.1
R2	10.0	0.25	0.05
Lean Streams			
Stream	x_j^s	x_j^t	c_j
	g/kg	g/kg	$/kg
L1	0.05	0.7	8.2
L2	0.04	0.6	0.15

The equilibrium governing the contaminant transfer between the rich streams and each of the MSA streams is linearly dependent on the temperature, T, of the mass exchange operation, and is also linearly dependent on the contaminant composition in the lean stream. These relations are:

$$y = (0.027 + 0.00128T)(x_1 + \varepsilon) + (0.016 - 0.00014T) \quad (1)$$

$$y = (-0.2204 + 0.0112T)(x_2 + \varepsilon) + (0.0622 - 0.000512T) \quad (2)$$

Table 4. Stream Data for the HEN in the Non-isothermal Example.

Hot Streams				
Stream	F_iC_p	$T^s_{H,i}$	$T^t_{H,i}$	c_i
	kW/C	^oC	^oC	$/kg
R_2	41.84	75	25	0.0
HU	? × 1547	271	270	0.2202
Cold Streams				
Stream	F_jC_p	$T^s_{C,j}$	$T^t_{C,j}$	c_j
	kW/C	^oC	^oC	$/kg
R_1	20.92	35	85	0.0
CU	? × 4.184	15	20	0.011

where, y is the composition of the contaminant in the rich stream, x_1 is the composition of the contaminant in lean stream L_1, x_2 is the composition of the contaminant in lean stream L_2, T is the temperature at which mass exchange takes place, and ε is the minimum composition driving force in the mass exchange network (fixed at $\varepsilon = 0.0001$). The relations are valid between 20°C and 200°C, and for a y composition greater than or equal to 0.05. We assume that the minimum temperature driving force (ΔT_{min}) for the HEN network is zero.

Since $\Delta T_{min} = 0$ and the supply and target temperatures of L_1 and L_2 are equal, the decomposition principle outlined earlier suggests that the minimum utility cost of the nonisothermal MEN is equal to the sum of the minimum utility cost for the HEN and the minimum utility cost for the MEN. In other words, only two pinch problems, one for the MEN and one for the HEN, need be solved.

The HEN problem is solved first. Only streams R_1 (cold) and R_2 (hot) are involved in the HEN, and have known flowrates. Streams L_1 and L_2 enter and exit at the same temperatures, and hence, may not be included in the HEN without any loss of optimality (provided $\Delta T_{min} = 0$). The solution is pinched at 60°C, and employs 0.135 kg/s of the hot utility (steam at 56 bar) and 60.0 kg/s of the cold utility (cooling water). The HEN minimum utility cost is $0.69/s. The temperature-enthalpy diagram is shown in Fig. 3.

The MEN problem is solved next. The cost of the MEN is determined by the lean utility consumption, which depends on the equilibrium between the rich and lean streams. This equilibrium, in turn, is determined by the operating temperature of the MEN. Hence, the MEN cost is dependent on the operating temperature. For a fixed MEN operating temperature one can solve a standard isothermal MEN pinch problem, and determine the MEN minimum utility cost for that temperature. This cost as a function of temperature is plotted in Fig. 4.

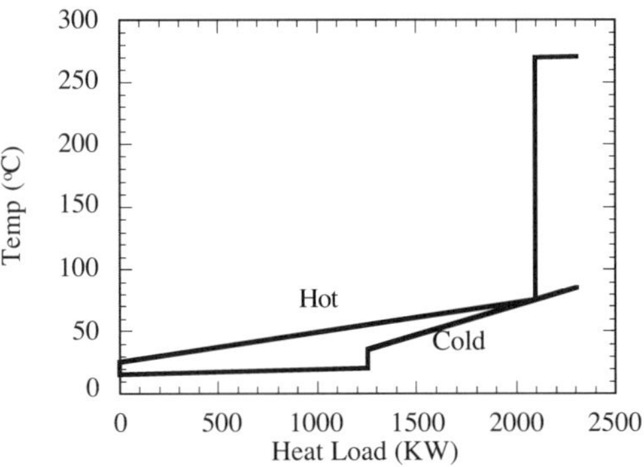

Figure 3. The T-H diagram for the optimal HEN. The pinch is located at 75°C.

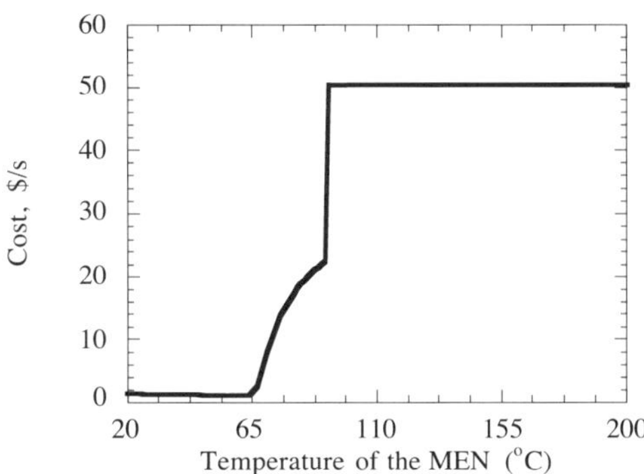

Figure 4. The optimal cost for the MEN as a function of the operating temperature of the MEN.

The middle portion of the curve is shown in Fig. 5. The MEN minimum utility cost over all temperatures is \$1.07/s, and is achievable for a range of temperatures between 53°C and 66°C. At these temperatures, the only employed lean utility is L2 which has a lower cost coefficient (\$0.15/kg) than L1 (\$8.20/kg). However, since t he equilibrium depends on temperature, the feasibility of employing L2 as an MSA varies with the operating temperature. In the range 53-66°C, it can remove the solute from both R1 and R2, and is the only MSA used, leading to a flat, minimum cost section of the plot. As the temperature is varied beyond this range, on either end, the exit composition of L2, x2t, moves higher in relation to the inlet compositions of R1 and R2. This imposes second law based limitations on the flowrate of L2. When stream L2 becomes infeasible, the cost goes to the upper limit that corresponds to the cost of employing L1 alone (\$50.46/s). The hot end temperature for this jump in the cost is 91.8°C, and is the temperature at which x_2^t is greater than y_1^s on a CID scale.

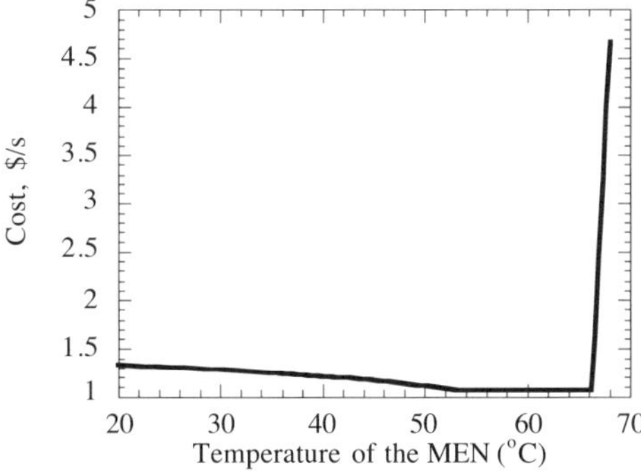

Figure 5. The optimal cost for the MEN versus the operating temperature of the MEN, in a smaller temperature range.

The composition-mass load diagram for an optimal MEN operating at 60°C is shown in Fig. 6. Only stream L_2 is employed with a flowrate of 7.14 kg/s.

The overall minimum utility cost for the nonisothermal MEN is, thus, \$1.07/s + \$0.69/s = \$1.76/s.

Total Flowsheet Synthesis

This synthesis task can be stated as follows: "Given a reaction path that transforms new materials to desired products identify a network of process units that accomplishes this transformation at minimum venture cost." As stated, the above synthesis problem encompasses all network synthesis problems mentioned earlier (reactor-separator-heat exchanger-mass exchanger). This problem generality creates an explosive number of alternatives and limits the designer's ability to address the problem in a comprehensive way. There have been however several attempts to bring some order into the problem. Most notably, the flowsheet synthesis programs AIDES (Siirola et al., 1971) and BALTAZAR (Mahalec and Motard, 1977a, 1977b), the hierarchical synthesis procedure of Douglas (1985), the heuristic-evolutionary approach of Lu and Motard (1985), the mixed-integer programming approach of Grossmann (1985) and the model based hierarchical design methodology of Kritikos and Stephanopoulos (1989). Ciric and Jia (1994) formulated waste reduction either as a multiobjective optimization problem or as a profit maximization problem with parametric dependence on the waste treatment cost coefficients. A sequential approximation procedure is then employed to over and underestimate the noninferior set.

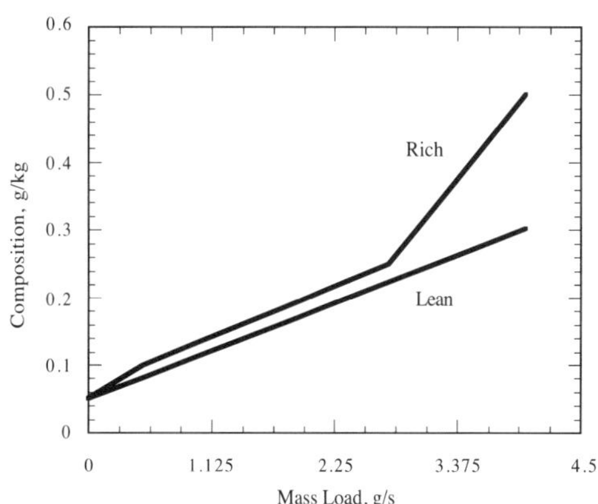

Figure 6. The composition-mass load diagram for an optimal MEN at 60°C. The pinch is located at y = 0.05.

Total Flowsheet Synthesis has an important role to play in waste minimization in that it forces the designer to consider the plant as a whole and to examine overall material and energy balances at the process development stage. As a result, the amounts and types of waste generated in a particular flowsheet can be easily assessed. Furthermore, a "waste audit" mechanism could be incorporated in the hierarchical and in the evolutionary flowsheet synthesis procedures so that it gears alternative design evolution towards minimum waste generation. In a similar vein, toxicology considerations can also be incorporated in the mathematical programming approach (Grossmann, 1989), so that noninferior (in terms of toxicological impact and economic benefit) configurations of a chemical complex can be identified.

Conclusions

In this paper we have tried to establish links between process synthesis and waste minimization. We have demonstrated that several process synthesis concepts have direct implications for waste minimization and that waste minimization concerns can lead to novel process synthesis problems. We believe that in the next few years the interplay between waste minimization and process synthesis will intensify. We only hope that this paper will further catalyze this interplay.

References

Aris, R. (1961). *The Optimal Design of Chemical Reactors. A Study in Dynamic Programming*. Academic Press, New York.

Bagajewicz, M.J., and V. Manousiouthakis (1992). Mass heat-exchange network representation of distillation networks. *AIChE J.*, 38, No. 11, 1769-1800.

Balakrishna, S., Biegler, L.T. (1992). Targeting Strategies for the Synthesis and Energy Integration of Nonisothermal reactor Netwroks. *Ind. Eng. Chem. Res.*, **31**, 2152-2164.

Barton, A.F.M. (1983). *Handbook of Solubility Parameters and Other Cohesion Parameters*, CRC Press, Florida.

Cabelka, T.D., Archer, W.L. (1987). *Surface Mount Technology*, 18-21.

Campbell Smyth, L., and J.R. Hopper (1989). Waste control by reactor design and catalyst X, Y, and Z in plug flow and semi-batch reactors. AIChE Conf., Pollution Prevention for the 1990's: A Chemical Engineering Challenge, Washington D.C.

Carlberg, N.A., and A.W. Westerberg (1989a). Temperature-heat diagrams for complex columns. 2. Underwood's method for the Petlyuk configuration. *Ind. Eng. Chem. Res.*, **28**, 1386-1397.

Carlberg, N.A., and A.W. Westerberg (1989b). Temperature-heat diagrams for complex columns. 3. Underwood's method for side strippers and enrichers. *Ind. Eng. Chem. Res.*, **28**, 1379-1386.

Catani, R., Centi, G., Trifiro, F., and Grasselli, R. (1992). Kinetics and Reaction Network in Propane Ammoxidation to Acrylonitrile on V-Sb-Al Based Mixed Oxides. *Ind. Eng. Chem. Res.*, **31**, 107-119.

Cerda, J., Westerberg, A.W., Mason, D., Linnhoff, B. (1983). Minimum Utility Usage in Heat Exchanger Network Synthesis — A Transportation Problem. *Chem. Eng. Sci.*, **38**, 373-387.

Ciric, A.R., Jia, T. (1994). Economic Sensitivity Analysis of waste treatment costs in source reduction projects: continuous optimization problems. *Comp. Chem. Eng.*, **18**, 481-496.

Chitra, S.P., Govind, R. (1985). Synthesis of Optimal Serial Reactor Structure for Homogeneous Reactions. Part II: Nonisothermal Reactors. *AIChE J.*, **31**, 185.

Cohen, R. (1988). Cleaning Agent Requirements. *Personal Communication*.

Doherty, M. (1989). Thermodynamics of Complex Separation Systems. UCLA Workshop on Pollution Prevention Design: Challenge for the 1990's, Los Angeles, California.

Douglas, J. (1989). Waste Minimization. UCLA Workshop on Pollution Prevention Design: Challenge for the 1990's, Los Angeles, California.

Douglas, J.M. (1985). A Hierarchical Decision Procedure for Process Synthesis. *AIChE J.*, **31**, No. 3, 353-362.

Douglas, J.M. (1992). Process Synthesis for Waste Minimization. *Ind. Eng. Chem. Res.*, **31**, 238-243.

Drier, J.S. (1989). Water as a Replacement for CFC's. UCLA Workshop on Pollution Prevention Design: Challenge for the 1990's, Los Angeles, California.

Duran, M.A., Grossmann, I.E. (1986). A Mixed-Integer Nonlinear Programming Approach for Process Systems Synthesis. *AIChE J.*, **32**, No. 4, 592-606.

Dyson, D.C., Horn, F.J. (1967). Optimum Distributed Feed Reactors for Exothermic Reversible Reactions. *Journal of Optimization Theory and Applications*, **1**, No. 1, 40-52.

El-Halwagi, M.M., Manousiouthakis, V. (1989a). Synthesis of Mass Exchange Networks. *AIChE J.*, **35**, No. 8, 1233-1244.

El-Halwagi, M.M., Manousiouthakis, V. (1989b). Design and Analysis of Mass Exchange Networks with Multicomponent Targets. 1989 Annual AIChE Meeting, paper 137f, San Francisco, California.

El-Halwagi, M.M., Manousiouthakis, V. (1990a). Simultaneous Synthesis of Mass-Exchange and Regeneration Networks. *AIChE J.*, **36**, 1209-1219.

El-Halwagi, M.M., Manousiouthakis, V. (1990b). Automatic Synthesis of Mass Exchange Networks with Single Component Targets. *Chem. Eng. Sci.*, **45**, 2813-2831.

El-Halwagi, M.M., and B.K. Srinivas (1992). Synthesis of reactive mass-exchange networks. *Chem. Eng. Sci.*, **47**, No. 8, 2113-2119.

El-Halwagi, M.M., El-Halwagi, A.M., Manousiouthakis, V. (1992). Optimal Design of Dephenolization Networks for Petroleum Refinery Wastes. *Process Safety and Environmental Protection*, **70**, 131-139.

Fornari, T., Rotstein, E., Stephanopoulos, G. (1989). Studies on the Synthesis of Chemical Reaction Paths. 2. Reaction Schemes with 2 Degrees of Freedom. *Chem. Eng. Sci.*, **44**, 1569-1579.

Glinos, K., and M.F. Malone (1985). Minimum vapor flows in a distillation column with a sidestream stripper. *Ind. Eng. Chem. Proc. Des. Dev.*, **24**, 1087-1090.

Grossmann, I. (1989). Mathematical Programming and Process Synergism. UCLA Workshop on Pollution Prevention Design: Challenge for the 1990's, Los Angeles, California.

Grossmann, I.,(1985). Mixed-Integer Programming Approach for the Synthesis of Integrated Process Flowsheets. *Computers and Chemical Engineering*, **9**, No. 5, 463-482.

Gundersen, T., Naess, L. (1989). The Synthesis of Cost Optimal Heat Exchanger Networks. An Industrial Review of the State of the Art. *Comp. Chem. Eng.*, **12**, No. 6, 503-530.

Gupta, A., and V. Manousiouthakis (1993). Minimum utility cost of mass exchange networks with variable single component supplies and targets. *Ind. Eng. Chem. Res.*, **32**, No. 9, 1937-1950.

Gupta, A., and V. Manousiouthakis (1994). Waste reduction through multicomponent mass exchange network synthesis. *Comp. Chem. Eng.*, **18**, No. S, S585-S590.

Gupta, A., and Manousiouthakis, V. (1995a). An Underwood-like Mass Exchanger Model (in press).

Gupta, A., and V. Manousiouthakis (1995b). Mass Exchange Networks With Variable Single Component Targets: Minimum Utility Cost Through Linear Programming. *AIChE J.* (accepted).

Hansen, C.M. (1967). The Three Dimensional Solubility Parameter and Solvent Diffusion Coefficient. Danish Technical Press.

Hayes, M.E. (1988). Chlorinated and CFC Solvent Replacement in the Electronics Industry: The Terpene Hydrocarbon Alternative. UCLA Workshop on Replacing CFC 113 in Critical Mechanical Cleaning Applications, November.

Henning, G., Leone, H., Stephanopoulos, G. (1989). MODEL.LA. A Modeling Language for Process Engineering. 1989 Annual AIChE Meeting, paper 137d, San Francisco, California.

Higgins, T. (1989). Hazardous Waste Minimization Handbook. Lewis Publishers, Chelsea, Michigan.

Hildebrandt, D. and Feinberg, M. (1992). Optimal Reactor Design from a Geometric Viewpoint. Paper #142c, AIChE Annual Meeting, Miami, Florida, November.

Hlavacek, V. (1978). Synthesis in the Design of Chemical Processes. *Computers and Chemical Engineering*, **2**, 67-75.

Hopper, J.R., and A. Muninnimit (1989). Waste minimization by modification of chemical reaction parameters. AIChE Conf., Pollution Prevention for the 1990's: A Chemical Engineering Challenge, Washington D. C.

Horn, F. (1964). Attainable and Nonattainable Regions in Chemical Reaction Technique. The Third European Symposium on Chemical Reaction Engineering, Pergamon, London.

Horn, F., Tsai, M. (1967). The Use of the Adjoint Variables in the Development of Improvement Criteria for Chemical Reactors. *Journal of Optimization Theory and Applications*, **1**, No. 2, 131-145.

Hoy, R.L. (1970). *J. Paint Technology*, **42**, 541, 76.

Huisingh, D., Hilger, H., Thesen, S., Martin, L. (1985). Profits of Pollution Prevention: A Compendium of North Carolina Case Studies in Resource Conservation and Waste Reduction. Report to North Carolina Board of Science and Technology.

Jackson, R. (1968a). Optimal Use of Mixed Catalysts for Two Successive Chemical Reactions. *Journal of Optimization Theory and Applications*, **1**, 27-39.

Jackson, R. (1968b). Optimization of Chemical Reactors with Respect to Flow Configuration. *Journal of Optimization Theory and Applications*, **2**, 240.

Jacob, A. (1991). Recycling technology in acrylonitrile production. *The Chem. Engr.*, January 31.

Joback, K., M.S. thesis (1984). Massachusetts Institute of Technology, Cambridge, MA.

Joshi, D.K., Ph.D. thesis (1983). Extraction of Polar Organic Chemicals. University of California, Berkeley, CA.

Joshi, D.K., Senetar, J.J., King, C. J. (1984). Solvent Extraction for Removal of Polar-Organic Pollutants from Water. *I&EC Proc. Des. Dev.*, **23**, 748-754.

King, C.J. (1980). Separation Processes. 2nd Edition, McGraw-Hill, New York.

Knapp, J.P., Doherty, M. (1989). Synthesis of Thermally-Integrated Homogeneous Azeotropic Distillation Sequences. 1989 Annual AIChE Meeting, paper 139b, San Francisco, California.

Kokossis, A.C., Floudas, C.A. (1994). Optimization of Complex Reactor Networks: II Nonisothermal Operation. *Chem. Eng. Sci.*, **49**, 1037-1051.

Kritikos, T., Stephanopoulos, G. (1989). Synthesis of Chemical Process Flowsheets — A Model Based Hierarchical Design Methodology. 1989 Annual AIChE Meeting, Paper 138h, San Francisco, California.

Levenspiel, O. (1962). Chemical Reaction Engineering: An Introduction to the Design of Chemical Reactors. John Wiley, New York.

Levenspiel, O. (1984). The Chemical Reaction Omnibook. Oregon State University Book Stores, Inc., Corvallis, Oregon.

Linnhoff, B., Hindmarsh, E. (1983). The Pinch Design Method for Heat Exchanger Networks. *Chem. Eng. Sci.*, 38, 745-763.

Lo, T.C., Baird, M.H.I., Hanson, C. (1983). *Handbook of Solvent Extraction*. Wiley & Sons, NY.

Lu, M.D., Motard, R.L. (1985). Computer-Aided Total Flowsheet Synthesis. *Computers and Chemical Engineering*, **9**, No. 5, 431-451.

Mahalec, V., Motard, R.L. (1977a. Procedures for the Initial Design of Chemical Processing Systems. *Computers and Chemical Engineering*, **1**, 57.

Mahalec, V., Motard, R.L. (1977b). Evolutionary Search for an Optimal Limiting Process Flowsheet. *Computers and Chemical Engineering*, **1**, 149.

Manousiouthakis, V. (1989a). Waste Minimization through Process Synthesis. UCLA Workshop on Pollution Prevention Design: A Challenge for the 1990's, Los Angeles, California.

Manousiouthakis, V. (1989b). Waste Minimization through Process Synthesis. AIChE Conference, Pollution Prevention for 1990's: A Chemical Engineering Challenge, Washington, D.C.

Mavrovouniotis, M.L., Stephanopoulos, G., Stephanopoulos, G. (1990). Computer-Aided Synthesis of Biochemical Pathways. *Biotechnology and Bioengineering*, **36**, 119-132

Mavrovouniotis, M.L., Stephanopoulos, G. (1992). Synthesis of Reactions Mechanisms Consisting of Reversible and Irreversible Steps. 1 A Synthesis Approach in the Context of Simple Examples. *Ind. Eng. Chem. Res.*, **31**, 1625-1637

Mavrovouniotis, M.L. (1992). Synthesis of Reactions Mechanisms Consisting of Reversible and Irreversible Steps. 2 Formulation and Analysis of the Synthesis Algorithm. *Ind. Eng. Chem. Res.*, **31**, 1637-1653.

May, D., Rudd, D. (1976). Development of Solvay Clusters of Chemical Reactions. *Chemical Engineering Science*, **31**, 59-69, 1976.

Nishida, N., Stephanopoulos, G., Westerberg, A. (1981). A Review of Process Synthesis. *AIChE J.*, **27**, No. 3, 321-351.

Papoulias, S.A., Grossmann, I.E. (1983). A Structural Optimization Approach in Process Synthesis — II Heat Recovery Networks. *Comp. Chem. Eng.*, **7**, 707-721.

Ravimohan, A.C. (1971). Optimization of Chemical Reactor Networks with Respect to Flow Configuration. *Journal of Optimization Theory and Applications*, **8**, 204.

Rotstein, E., Rasasco, D., Stephanopoulos, G. (1982). Studies of the Synthesis of Chemical Reaction Paths. I. Reaction Characteristics in the (ΔG,T) Space and a Primitive Synthesis Procedure. *Chemical Engineering Science*, **37**, No. 9, 1337-1352.

Roxenby, S., and V. Manousiouthakis (1993). Nonisothermal separable mass exchange networks: Minimum utility cost through the state-space approach. UCLA Internal Report (submitted).

Shabecoff, Philip (1988). New Compound is hailed as boon to ozone shield. *New York Times*, A. 15.

Siirola, J.J., Powers, G.J., Rudd, F.F. (1971). Synthesis of System Designs. III. Toward a Process Concept Generator. *AIChE J.*, **17**, 677.

Srinivas, B.K., El-Halwagi, M.M. (1993). Optimal Design of Pervaporation Systems for Waste reduction. *Comp. Chem. Eng.*, **17**, 957-970.

Stephanopoulos, G., Townsend, D. (1986). Synthesis in Process Development. *Chemical Engineering Research and Design*, **64**, No. 3, 160-174.

Umeda, T., Harada, T., Shiroko, K. (1979). A Thermodynamic Approach to the Synthesis of Heat Integration Systems in Chemical Processes. *Comp. Chem. Eng.*, **3**, 273-282.

Wang, Y.P., Smith, R. (1994). Wastewater minimization. *Chem. Eng. Sci.*, **49**, 981-1006.

Washburn, E.W. (1921). The Dynamics of Capillary Flow. *Physical Review Second Series*, **17**, No. 3, 273-283, March .

Westerberg, A.W. (1985). The Synthesis of Distillation-Based Separation Systems. *Computers and Chemical Engineering*, **9**, No. 5, 421-429.

Westerberg, A. (1989a). Conceptual Aspects of Design Systems. UCLA Workshop on Pollution Prevention Design: Challenge for the 1990's, Los Angeles, California.

Westerberg, A. (1989b). The Design Process. 41st Annual Institute Lecture, 1989 Annual AIChE Meeting, San Francisco, California.

Weigert, W.M. (1973). Acrolein-From-Propylene Aided By New Catalyst. *Chem. Eng.,* June 25, 68-69.

Wise, Jr., H.E., and Fahrenthold, P.D. (1981). Predicting priority pollutants from petrochemical processes. *Env. Sci. & Tech.*, **15**(11), 1292-1304.

U.S. Congress, Office of Technology Assessment (1986). Serious Reduction of Hazardous Waste: For Pollution Prevention and Industrial Efficiency. OTA-ITE 317 (Washington, D.C.: U.S. Government Printing Office) September.

DESIGN OF PROCESSES FOR COMBUSTION SYSTEM SAFETY

J. B. Gorss and M. J. Kinosz
Alcoa Technical Center
Alcoa Center, PA 15219

Gary J. Powers
Department of Chemical Engineering
Carnegie Mellon University
Pittsburgh, PA 15213

Abstract

The design of processing control systems for safety is described in this paper. An example of the application of probabilistic risk assessment via fault tree analysis and formal software and operating procedure verification via symbolic model checking is presented for an existing aluminum melting system.

Keywords

Design, Safety, Probabilistic, Risk, Verification, Combustion.

Introduction

The design of processing systems must accommodate the needs for productivity and safety. These needs are often in conflict and an explicit method for comparing the risks for various designs is illustrated in this paper. The method is based on probabilistic risk assessment based on fault tree analysis and on a new method (Model Checking) for the verification of the systems control system and operating procedures. The method identified several important cases for design improvements and the results have been included in the corporate engineering standards for this type of combustion systems. This is one of the first published examples of engineering standards being set based on the results of probabilistic risk assessments. These types of engineering standards are expected to be useful in addressing the requirements being developed by OSHA and EPA for the risk based assessment of processing systems.

As part of an ongoing combustion safety improvement program within Alcoa, a detailed analysis has been performed on a conventional, gas-fired door charged aluminum reverbatory melting furnace in order to establish a baseline of information with respect to the level of safety attained in the furnaces combustion system design. The furnace selected for the analysis met today's standards, as specified in the National Fire Protection Association (NFPA) 86 standard. The analysis included the potential for failures in the furnace hardware and instruments, failures of programmable logic controllers (PLC) including software and failures of human operating procedures, including human error during the maintenance time periods. Specifically, of interest was the potential for fatalities caused by natural gas explosions both during furnace start-up and shutdown and the maintenance time period. Other risks associated with molten metal explosions, asphyxiation and electrocution were recognized as important but were not developed in detail in this study.

The purpose of the Combustion Safety Improvement Program was to evaluate furnace safety and to determine whether today's standards are adequate. Our goal was to maximize furnace and personnel safety while minimizing both the cost and the complexity of control and safety equipment utilized.

The furnace selected for this audit was a 40,000 lb capacity, natural gas-fired, two (2) burner aluminum melter located at Alcoa's research center outside of Pittsburgh, PA. The furnace was originally built in the early 1980s and was designed to meet the existing standards of the day. These include National Fire Protection Association (NFPA), Factory Mutual Engineering (FM) and Alcoa engineering standards. Prior to performing the risk assessment, the furnace was upgraded to meet NFPA86-1990 edition standards and was submitted to and approved by Factory Mutual Engineering. As part of the approval process and in conformance with recent OSHA regulations, detailed startup and shutdown procedures complete with detailed lock, tag and try procedures were posted at the furnace in full view of the operating personnel. We also conducted a combustion training class for furnace operators. This class included startup/shutdown and emergency procedures. The following is a listing of the combustion system equipment and design following upgrading to NFPA86-1990 standards and prior to the Fault Tree Analysis:

- Double block valves on the main gas line with proof of position switches.
- Single block valve on the pilot gas line.
- Interrupted pilots – 15 mm burner; spark ignited – 6 mm burner.
- Dynamic self-checking UV flame detection system.
- PLC for furnace logic and PID control.
- Watch-dog timer circuit for PLC failure scan protection.
- Redundant contacts to close the main latch valve and blocking valves in the event of a PLC output card failure.
- Purge timer to ensure four (4) air changes.
- High and low gas pressure switches on both the main gas and pilot gas lines.
- Combustion air failure and proof of purge pressure switches.
- Gas line filters on both primary building supply and on the furnace main gas supply line.
- Flue gas thermocouple for over-temperature protection.
- Staged combustion control for the 6 mm BTU/hr and 15 mm BTU/hr burners.
- Lead/lag air/fuel ratio PID control logic.
- Enforced low fire start.

Normal maintenance has been performed on this furnace as part of standard procedures. A new more comprehensive preventative maintenance procedure was issued prior to requesting Factory Mutual acceptance and includes the following tests performed on a quarterly basis:

- Leak testing of the main gas blocking and latch valves, pilot shutoff valves and lockable main gas supply hand valve.
- Purge timer circuit and proof of purge switch.
- Air/fuel ratio shutdown circuits.
- All combustion air and natural gas pressure switches.
- Air and fuel flow transmitters.
- Over-temperature protection shutdown device.
- Dynamic self-checking flame supervisory system.
- Trial for ignition timer.

In addition, a second pilot gas shutoff valve was installed. All operators were trained to be familiar with startup and shutdown procedures and OSHA lock, tag and try procedures. These instructions were posted at the equipment in full view of the operators. Protection was provided against device output failures such as:

- PLC output cards.
- Transmitters.
- Pressure switches.
- Input devices.

All safety devices were hardwired fail safe and a watch-dog timer circuit for PLC controller protection was utilized. Verification of the PLC software for the furnace system was performed to check if logic errors existed in the software that would cause safety (explosion) or operability (won't start or spurious shutdown) problems. A model checking technique used for the verification of VLSI circuits was used for the verification of the furnace PLC logic (Moon, 1992).

Fault Tree Analysis

Fault Tree Analysis is a systems analysis technique for facilitating the design and operation of safer and more reliable production facilities. A Fault Tree consists of a logic diagram which identifies event sequences which can lead to a specified failure event (fatality, injury, equipment damage, explosion, etc.). The methodology involved in the analysis is the following:

1. A failure event of interest (top event) is specified, i.e. fatality at 40K furnace.
2. The consequences of the top event are estimated.
3. Within the context of the processing system and its environs, the immediate precursor events and their failure logic are identified.
4. The generation of precursor events is continued until primal (initiating) events are encountered.
5. Probabilities and failure rates are assigned to primal events and through the algebra of the logic model, the failure rate of the top event is computed.
6. The minimal cut sets for the system are computed from the logic of the Fault Tree and are presented in order of decreasing importance.
7. The top event rate is compared with a target for risk and changes to the system are made to reduce the risk.

The improvements to the system as identified by the fault tree analysis are combined to represent a proposed case for the design and is compared to a base case that represents the current operation of the equipment and operating procedures prior to system changes. Case studies can be prepared by removing features from the proposed case to test the sensitivity of the top event rate to each of the design changes.

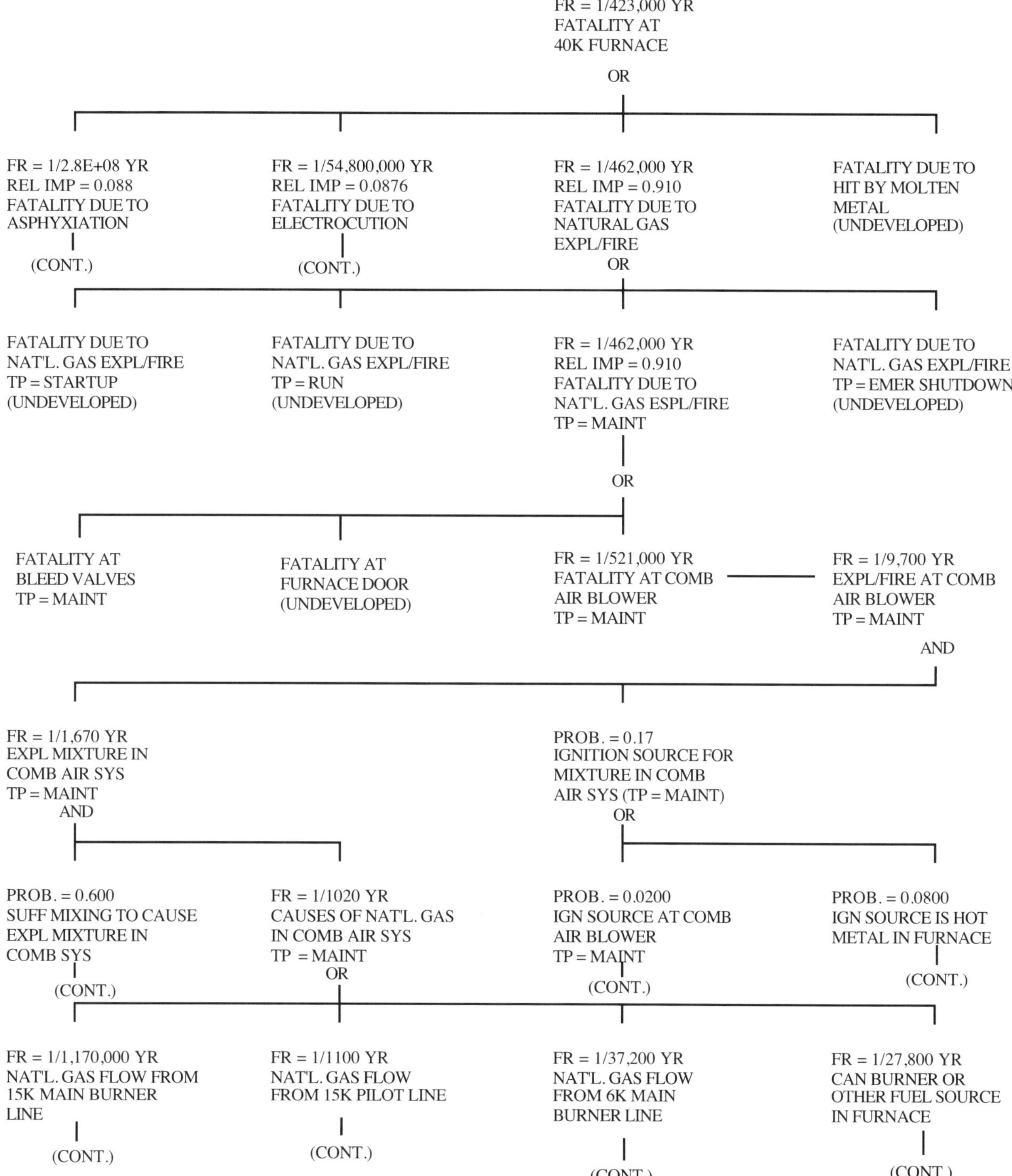

Figure 1. Part of fault tree for the proposed case 40,000 lb furnace; continued (cont.) means the fault tree was developed in more detail to predict the failure rate or probability.

Table 1. Minimal Cut Sets.

Top Event Rate of Occurrence = 1 event/423000 years

M.C.S.	No.	1	Prob.	Rate = 1/4160000 Years
EVENT	1		(0.100)	People present at combustion air blower TP = maint
EVENT	2		(0.200)	Probability of fatality (expl/fire) – no syn clothing
EVENT	3		(0.500)	Hit by expl/fire at combustion air blower (TP = maint)
EVENT	15		(0.500)	Suff air drawn into comb air sys by fuel leaking thru burner
EVENT	18		(0.400)	Hot metal in furnace TP = maint
EVENT	19		(0.200)	Combustible mixture finds path to hot metal
EVENT	45		(1/50 yrs.)	Can burner or other fuel source used in fce
EVENT	46		(0.0300)	Flame out or leak releases fuel into combustion air system

M.C.S.	No.	2	Prob.	Rate = 1/5480000 Years
EVENT	40		(4/yr.)	Electrician or others doing electrical maint
EVENT	41		(0.0100)	No or inad lock, tag & try for fce elec maint
EVENT	43		(0.500)	Prob of fatality electrocution
EVENT	44		(1/200 yrs.)	Sneak circuit or cross connect feeds power to maint circuit

M.C.S.	No.	3	Prob.	Rate = 1/7990000 Years
EVENT	2		(0.200)	Probability of fatality (expl/fire) – no syn clothing
EVENT	9		(1/6 yrs.)	Main gas filter bypass opened and injects dirt into lines
EVENT	21		(0.0500)	Dirt at valve BS11/LS12 causes leak across
EVENT	24		(0.0300)	Operator does not close LBV8 at TP = maint (with valve signal)
EVENT	61		(0.500)	Insuff mixing causes flam mixture in fce
EVENT	62		(0.100)	People present at 40K fce TP = emerg sd
EVENT	63		(0.500)	Hit by expl/fire TP = emerg shutdown
EVENT	64		(0.100)	Delayed ignition of flammable mixture in fce

M.C.S.	No.	4	Prob.	Rate = 1/10000000 Years
EVENT	1		(0.100)	People present at combustion air blower TP = maint
EVENT	2		(0.200)	Probability of fatality (expl/fire) – no syn clothing
EVENT	3		(0.500)	Hit by expl/fire at combustion air blower (TP = maint)
EVENT	9		(1/6 yrs.)	Main gas filter bypass opened and injects dirt into lines
EVENT	15		(0.500)	Suff air drawn into comb air sys by fuel leaking thru burner
EVENT	18		(0.400)	Hot metal in furnace TP = maint
EVENT	19		(0.200)	Combustible mixture finds path to hot metal
EVENT	21		(0.0500)	Dirt at valve BS11/LS12 causes leak across
EVENT	24		(0.0300)	Operator does not close LBV8 at TP = maint (with valve signal)

M.C.S.	No.	5	Prob.	Rate = 1/10400000 Years
EVENT	1		(0.100)	People present at combustion air blower TP = maint

Table 1. Minimal Cut Sets (cont.).

EVENT	2		(0.200)	Probability of fatality (expl/fire) – no syn clothing
EVENT	3		(0.500)	Hit by expl/fire at combustion air blower (TP = maint)
EVENT	16		(0.200)	Suff air drawn into comb air sys by natl conv
EVENT	18		(0.400)	Hot metal in furnace TP = maint
EVENT	19		(0.200)	Combustible mixture finds path to hot metal
EVENT	45		(1/50 yrs.)	Can burner or other fuel source used in fce
EVENT	46		(0.0300)	Flame out or leak releases fuel into combustion air system
M.C.S.	No.	6	Prob.	Rate = 1/13300000 Years
EVENT	2		(0.200)	Probability of fatality (expl/fire) – no syn clothing
EVENT	9		(1/6 yrs.)	Main gas filter bypass opened and injects dirt into lines
EVENT	21		(0.0500)	Dirt at valve BS11/LS12 causes leak across
EVENT	24		(0.0300)	Operator does not close LBV8 at TP = maint (with valve signal)
EVENT	58		(0.200)	People present at fce door TP = maint
EVENT	59		(0.500)	Hit by fire/expl at fce door
EVENT	60		(0.0300)	Ignition source ignites fuel in fce TP = maint
EVENT	61		(0.500)	Insuff mixing causes flam mixture in fce

Notes:

1. The abbreviation M.C.S. stands for Minimal Cut Set.
2. In numbers, E designates the power of ten, e.g. 1.2E+02 = 120 and 1.2E-02 = 0.012.

Table 2. Results of the Risk Assessment

The major findings of the analysis are presented below and the recommendations are based on examination of the Fault Tree "cut sets" to determine the primary risk scenarios.

- The top event rate for the **base case**, which assumes that <u>none</u> of the recommendations listed below are implemented is **1/7,260 yr.**
- The top event for the **proposed case**, which assumes that <u>all</u> the recommendations listed below are implemented, including the upgrade to meet NFPA 86-1990 standards, is **1/423,000 yr.**

CASE STUDIES	Fatality Freq.
• Base Case + PLC Logic Verification	1/12,100 yr
• Base Case + PLC Logic Verification + Second Block Valve in the Pilot Gas Line	1/17,300 yr
• Base Case + PLC Logic Verification + Second Pilot Block Valve + Leak and Function Testing of Solenoid Shutoff Valves and Hand Shutoff Valves Four (4) Times/Year vs. Once in Five (5) Years	1/23,500 yr
• Base Case + PLC Logic Verification + Second Block Valve in the Pilot Gas Line + Leak Testing Four (4) Times/Yr + Main Gas Hand Valve With Position Indicator and Alarm Signal Within 3 min of Loss of Combustion Air Flow	1/423,000 yr

A portion of the proposed case Fault Tree is shown in Fig. 1. The most likely cut sets for the Proposed Case Fault Tree are given in Table 1.

The results of the risk assessment given in Table 2 showed that a **base case fatality frequency rate** from a natural gas explosion would be expected to occur **once in 7,260 years (1/7,260 yr)** based on operating the furnace as originally built. Following the upgrade to meet NFPA86-1990 standards and acceptance by Factory Mutual Engineering, the fatality frequency rate improved to **1/23,500 yr**.

The normally accepted fatality frequency rate typical in the chemical and metals industries is **1/20,000 yr**, and can be compared with the risks of accidental deaths given in Table 3. The uncertainty associated with the primal event data was assessed and found to be approximately plus or minus a factor of 1.5 from the average used in the fault tree.

Table 3. Individual Risk of Acute Fatality by Various Causes.

Accident Type	Total Number in the U.S. – 1985	Average Fatality Rate for Individual (years)
Motor Vehicle	45,600	1/5,000
Falls	11,300	1/20,000
Drowning	5,700	1/40,000
Poison	5,000	1/40,000
Fires / Hot Substances	4,900	1/40,000
Machinery	1,850	1/100,000
Water Transport	1,650	1/100,000
Firearms	1,600	1/100,000
Air Travel	1,575	1/100,000
Falling Objects	1,270	1/125,000
Electrocution	1,050	1/200,000
Railway	870	1/200,000
Lightning	160	1/1,000,000
Tornadoes (1953-85 avg.)	91	1/2,000,000
Hurricanes (1901-85 avg.)	93	1/2,000,000
All Others	9,621	1/30,000
All Accidents	92,500	1/2,500

Normally accepted top event rate is 1/20,000 years based on experience in the chemical, petroleum and metals industries. These targets mean that the risk of fatality from fire, explosion and asphyxiation is **one-third** the average risk of fatality by automobile accident.

We were astonished to find that with all the safety modifications and increased maintenance attention that a gain to only 1/23,500 yr was realized. It was felt that after meeting NFPA 86-1990 standards and FM acceptance that the furnace was "state-of-the-art" as far as combustion system design was concerned. The value of the Fault Tree Analysis was that it identified several deficiencies in the furnace design and maintenance procedures that, when implemented, could significantly lower the fatality frequency rate to **1/423,000 yr**.

To put this frequency rate in proper perspective, fatality frequency rates for various accidents in the United States is listed in Table 3.

Detailed analysis of the Fault Tree showed that one specific event stood out as the primary and critical event in determining if a natural gas explosion would occur. The question was, what was the position (open or closed) of the furnace primary main gas supply manual shutoff valve and how quickly was this valve closed after the loss of combustion air flow. The Fault Tree clearly showed that if a visual and audible warning device and operating procedure were installed on the valve to indicate the valve was not closed within three (3) min of a loss in combustion air flow and that would result in the furnace operator manually closing the valve, the fatality frequency improved to 1/423,000 yr. The highest potential for fatality occurs not during furnace startup/shutdown or run time periods but when the furnace has been shutdown for maintenance. It is extremely important that the personnel responsible for operation and maintenance of the equipment know that furnace shutdown is as dangerous as startup. Explosions have been documented long after a furnace has been shutdown. These four (4) case studies given in Table 2 highlight the importance of:

1. Leak testing the shutoff valves even in a system that has double block valves on the main gas and pilot gas lines.
2. Having double block valves on <u>all</u> gas lines.
3. Giving the operator and other maintenance people a high integrity indication of the need to close the main gas hand valve during the time after loss of combustion air flow.
4. Need to verify the PLC logic to insure the program would provide safe operation of the furnace.

The improvement in fatality frequency rate was substantial, especially considering the investment of approximately $3,000 to purchase and install the lockable, position indicating hand valve and the necessary revisions to the control logic and human operating instructions. As a result of the risk assessment process, we feel our goals have been realized with the installation of this type valve and system on all our combustion equipment at the Alcoa Technical Center, regardless of BTU capacity.

The development of logic models to support the risk assessment of processes was a key to the success of this project. The relatively low cost for process risk reduction was due to the detailed and causal nature of the risk assessment. Our experience has indicated that process improvements made at primal causal levels are often the most cost effective. The challenge is to continue the development of detailed models and failure data that can be defended with process simulations and experience.

References

National Fire Protection Association Standard 86 - 1990 edition.

National Safety Council, Table 3.

Automatic Verification of Sequential Control Systems Using Temporal Logic, I. Moon, G. Powers, E. Clarke, J. Burch (Jan. 1992). *AIChE J.,* **38**, 1, 67-75.

Powers, G. and S. Lapp (1977). Computer-Aided Synthesis of Fault Trees. *IEEE Trans. on Reliability*, April, 2-13.

Fusillo R. H. and G. J. Powers (1988). Computer-Aided planning of purge operations. *AIChE J.*, **34**, 558-566.

Aelion V. and G. J. Powers (1991). A unified strategy for the retrofit synthesis of flowsheet structures for attaining or improving operating procedures. *Computer and Chemical Engineering*, **15**, 5, 349-360.

LIFE CYCLE ANALYSIS

Ian Boustead
BCL (UK)
West Grinstead, Horsham
GB-RH13 7BD

Abstract

This paper is intended as an introduction to life-cycle analysis. It does purport to be a complete description but presents to more important features and provides a selected bibliography for those who wish to read further into the subject.

Keywords

Life-cycle analysis, Resources, Industrial system, LCA, LCI, Inventory.

Introduction

This paper is intended to provide a some background information on resource analysis and the way it is carried out. It is not intended as a comprehensive coverage of the subject but aims to highlight the principal features of the methodology and raise some of the questions that need to be addressed when interpreting the results.

Resource analysis is concerned with the impact of industrial operations on the environment. Essentially there are two main groups of potential problems; conservation of resources and minimising environmental damage. Before solutions to these problems can be found, information is needed about the scale of the problems, if indeed there are any. The analysis of any process is therefore the first step to understanding these problems.

Such calculations for large systems are not new even though they have only just been "discovered" by some industrial sectors. The initial spur in the 1960's was the effect of groups pressing for conservation of resources, especially fossil fuels. An added incentive was the oil embargo in the mid-1970's. Since then, a relatively small number of individuals and groups have continued their work in this area, especially in the fields of energy and raw materials, which are interlinked, as well as solid waste generation. More recently, air and water pollution have been included in the calculations.

Basic Ideas

Resource analysis is concerned with the performance of systems and with the measurement or calculation of inputs and outputs. Essentially the starting definitions are the same as those used in thermodynamics. Any group of operations which perform some defined function can be enclosed by a boundary. The boundary encloses the system and the region surrounding the system is called the system environment (shown as the shaded area in Fig. 1).

In thermodynamics, the system environment represents an unlimited source of inputs for the system and a sink for its outputs. The analysis is concerned with measuring the flows across the system boundary. Flows of energy and raw materials across the boundary and into the system are positive; all flows across the boundary and out of the system are negative. These are standard thermodynamic definitions and should be applied rigorously.

Note that some confusion can arise over the use of the term environment. In resource analysis, environment refers to the system environment and, depending upon how the system is defined, this will not necessarily be the same as the natural environment used by ecologists. It is seldom the same as the loose usage by journalists of the word environment.

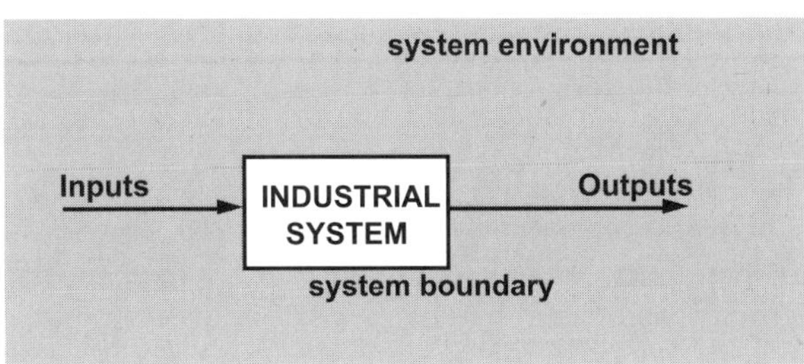

Figure 1. Schematic diagram of a simple industrial system.

The function of resource analysis is therefore to calculate all of the inputs and outputs for any defined system and this provides the raw information needed to discuss the effects of the system on the environment.

There is frequently some confusion about the nature of the subject itself. It aims to define the performance of industrial systems as they exist. It is not intended to solve the world's perceived environmental problems nor does it replace the work done in many other, different academic and engineering disciplines. A simple example illustrates this. Fig. 2 shows an industrial system for generating and distributing steam. The boxes represent some of the systems that can be defined and identify the specialists who are concerned with the different systems.

One important feature highlighted by Fig. 2 is that a specialist examining his own system must take into account the results of those systems which lie within it because they are sub-systems of his own system. He can however ignore those systems which external to his own because they form part of the system environment and, by definition, it is the behaviour of the system and not the system environment that is being measured. Thus the plant engineer must accommodate the work of the thermodynamicist but the climatological effects of his operation do not affect its performance and hence the analysis.

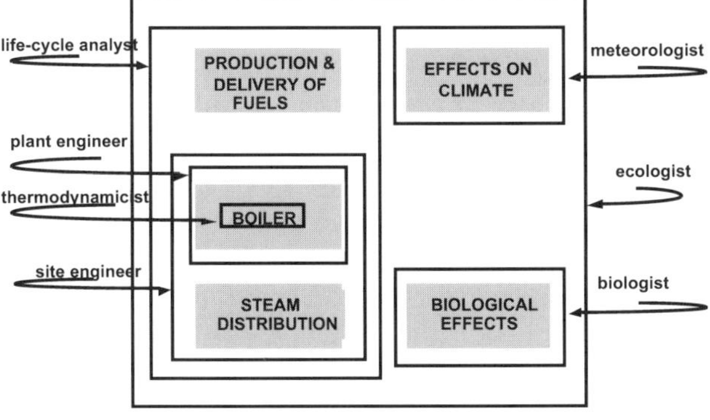

Figure 2. Simple industrial system showing the relationship of related systems and fields of study.

The resource analyst need not therefore be a specialist in meteorology or one of the biological sciences in order to carry out his analysis. His primary function is to provide performance data. However, when the implications of the results are discussed, inputs are needed from all disciplines shown in Fig. 2.

Fig. 3 shows the type of overall system that is analysed with the inputs and outputs broken down under the major headings of interest. The measurement or calculation of the inputs and outputs determines the impact of the system on its environment. When all inputs are derived from raw materials in the earth and all outputs are returned to the earth, including the usable products at the end of their useful lives, the results of such calculations are frequently referred to as ecobalances.

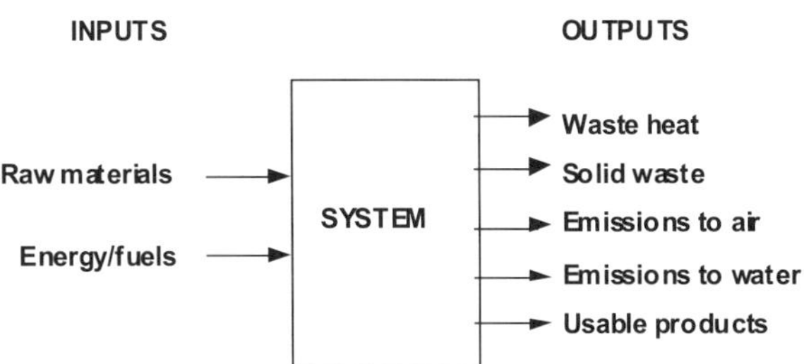

Figure 3. Simple industrial system showing inputs and outputs.

Fig. 3 highlights a number of important points.

1. Waste heat is seldom, if ever, measured directly. It can however be estimated from the input energy, since, with one exception, all of the energy input will appear as waste heat. The one exception is when a chemical change occurs during processing so that energy is either absorbed or liberated. Usually, however, the reaction energy change is small compared with the total process energy and so seldom invalidates this method of assessing waste heat.
2. The calculation of input energy and raw materials indicates the demand for primary inputs to the system and these parameters are important in conservation arguments because they are a measure of the resources that must be extracted from the earth in order to support the system.
3. Calculation of the outputs is an indication of the potential pollution effects of the system. Note that the analysis is concerned with quantifying the emissions. It does not assign any deleterious or beneficial qualitative properties to the emissions.
4. Because the inputs and outputs depend upon the definition of the system, any changes to the components of the system means that all

of the inputs and outputs will change because they are inter-related. One common misconception is that it is possible to change a single input or output whilst leaving the other parameters unchanged. In fact the reverse is true; because a new system has been defined by changing one component, all of the inputs and outputs are expected to change. If they happen to remain unchanged, it is a coincidence.

5. For a system where all inputs are traced back to the extraction of raw materials from the earth, those operations concerned with the extraction, processing and delivery of fuels form part of the system. Thus it is impossible to define identical systems for different countries because of the differing practices in the fuel producing industries.
6. Because any group of operations can be included within the system, there is no such thing as a "correct" system. Many of the apparent contradictions in the literature arise from comparisons of the results of analyses of similar but different systems. Usually such comparisons are made because of ignorance of the nature of the analysis or because analysts frequently define their system inadequately.

Source Data

In the early 1970's when this type of work was just beginning, there was a grave shortage of reliable primary information and considerable reliable had to be placed on published information. Today however, there is a considerable body of primary knowledge especially about energy and raw materials requirements yet reports are still appearing using pre-1980 published work, often of uncertain validity.

Reliable primary data for air and water emissions is still difficult to obtain. Usually organisations monitor air and water emissions as concentrations because this is the way that the statutory limits, with which they must comply, are expressed. However, in this type of work, information is needed on the total mass of each pollutant emitted per unit output of product and this is seldom monitored.

The Historical Legacy

The calculation of environmental impact parameters involves a considerable amount of tedious arithmetical work. In the early days of the development of the subject, when many calculations were carried out manually, attempts were made to simplify the system to minimise the amount of calculation. Two of the commonest simplifications which persist today in many models are given below.

Linear Systems Versus Networks

The calculations associated with linear sequences of operations are relatively easy to carry out and large segments of such sequences occur frequently in materials processing operations. However, it is important to recognise that networks also occur. Fig. 4, for example, shows a simplified instance. Here coal production requires inputs of electricity and steel but electricity production requires coal and steel as well as transport. Until recently, such networks were converted into pseudo-linear systems by assigning notional values to the downstream processes so that the circular links could be broken and hence the calculations simplified.

The assumption underlying this approach was that such networks were a relatively infrequent occurrence and that by choosing reasonably accurate notional values, no significant errors would be introduced into the calculations. Unfortunately such networks are more common than was originally thought, especially as the analysis of industrial systems becomes more detailed and sophisticated. The cumulative effects of a large number of approximations can introduce significant errors into the overall results.

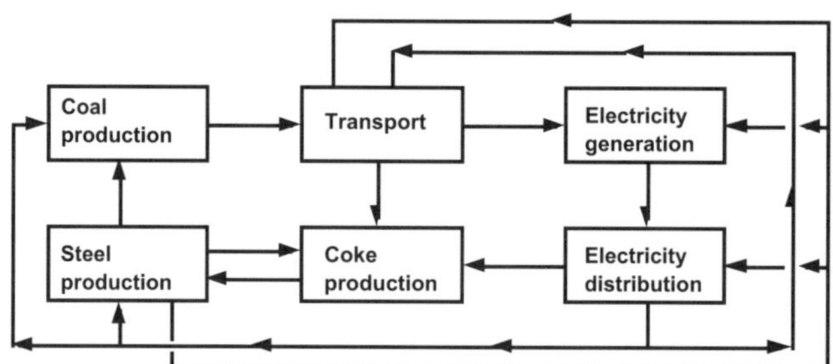

Figure 4. Simple network of unit operations requiring iteration in the calculation of resource parameters.

The solution to the problem is iteration; that is, notional values are chosen initially for the down-stream processes and the system evaluated. The calculated values are now substituted for the notional values and the calculation is repeated. This process is repeated until any changes in the recalculated values lie below the level of accuracy required. Whilst such iterative calculations are extremely time consuming using a hand calculator, they present no problems to a computer.

There is one area where iteration is vital and the use of pseudo-linear systems can lead to serious errors; this is in electricity production. The production efficiency for electricity in the UK is approximately 30% when measured as the ratio of electrical energy delivered to the consumer compared to the energy extracted as primary fuel from the earth. However, because of the large number of different inputs to electricity production and the primary fuels which feed it, iterative calculations show that the true production efficiency is less than 27%. Viewed as a change of only 3% in the overall production efficiency, the change does not seem particularly significant but in fact it represents a 10% change in the production efficiency — an error that can lead to significant under-estimates of resource requirements and emissions.

Separating Out the Fuel Producing Industries

The system can be greatly simplified if the operations concerned with the fuel producing industries are removed from the main system and treated separately. However, it must be remembered that the fuel producing industries consume materials from other parts of the industrial processing system. Divorcing them from the main system meant that often fuel production efficiencies were not amended when the detail in the main system was changed. With the advent of computing power, it is no longer necessary to treat the fuel producing industries in any special way and they can revert to what they are; simply industrial operations.

What Goes Into a System

The operations that are included within a system arise from the definition of the system function and the more closely this is defined, the less the likelihood of misinterpretation of the results. There are essentially three main groups of operations that always appear within industrial systems; these are shown in Fig. 5.

The main production sequence represents the group of operations which are the main focus of attention. For example, a liquid delivery system might be defined as the group of operations shown in Fig. 6. In addition a number of other parameters which affect the performance of the system would also need to be defined. Such additional parameters might be:

- Container mass
- Distribution system
- Container material
- Country
- Container size
- Year
- Container type
- Bulk or single-serve
- Liquid packed
- Returnable or one-trip
- Type of packaging for empty containers
- Type of outer packaging for full containers

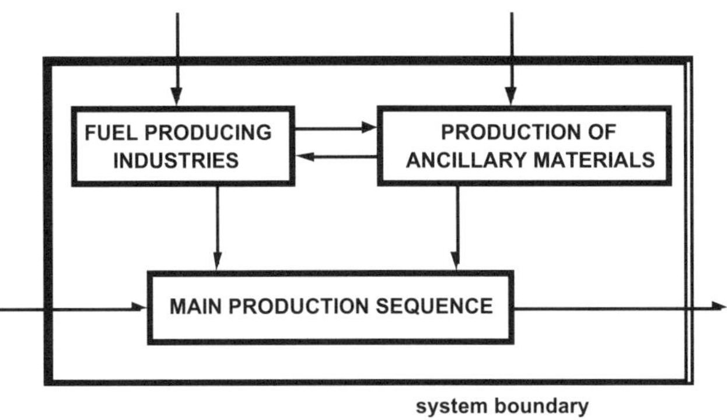

Figure 5. Main groups of operations included within any industrial system.

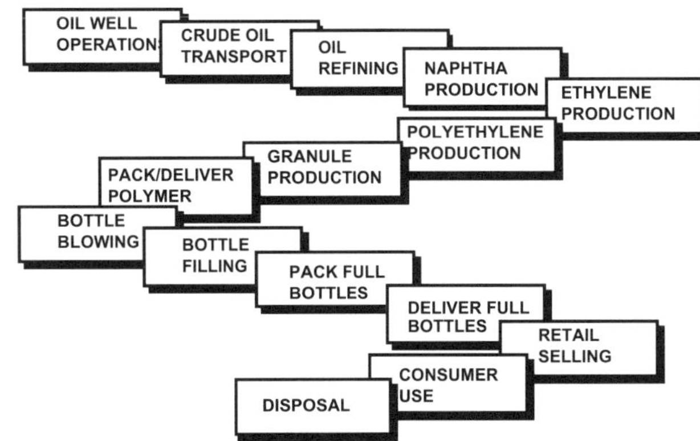

Figure 6. Main sequence of operations used in the production and use of PE bottles.

Most of the component operations shown in Fig. 6 require inputs of other materials. Packing full bottles ready for despatch, for example, requires an input of pallets, cardboard layer pads and shrink-wrap or stretch film. The additional inputs must also be produced from raw materials and are shown in Fig. 5 as ancillary inputs. For each of these ancillary inputs a sequence of operations similar to that shown in Fig. 6 can also be constructed.

Both the main production sequence and the production of ancillary materials demand a supply of energy and so all complete industrial systems must include the fuel producing industries. These industries are responsible for the extraction of primary fuels from the earth, converting them into a form suitable for use and then delivering them to the user. As a consequence of the processing that is carried out by them, they too use energy. Thus the energy that is finally delivered to the consumer represents only a part of the primary fuel that was originally extracted from the earth.

If the production of a fuel is represented as a single industrial system as shown in Fig. 7, then the overall efficiency of the fuel producing industries can be expressed as a production efficiency as shown in Fig. 7.

The important feature of this table is that it highlights the relative inefficiency of producing electricity; of the total primary fuel taken into the electricity production industry only about one quarter is finally delivered to the consumer. Whilst the industry may be able to improve its overall efficiency by 3 to 4%, the major losses are due to the physical impossibility of converting heat to other forms of energy; this is a natural phenomenon which will never be overcome.

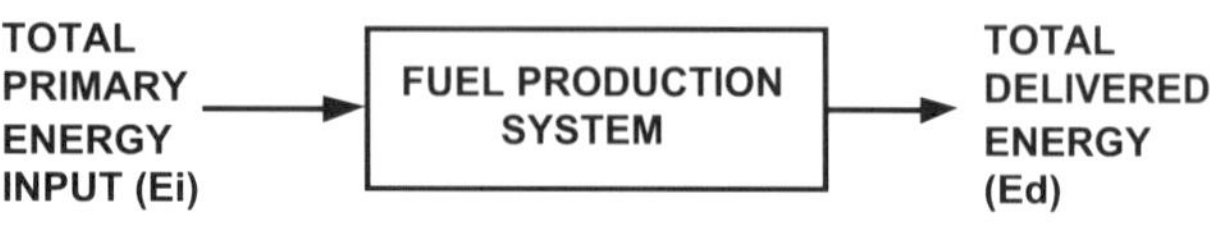

EFFICIENCY = Ed/Ei

FUEL	PRODUCTION EFFICIENCY/%
ELECTRICITY	26.0
COAL	95.3
COKE	82.2
NATURAL GAS	92.3
OIL FUELS	87.7

Figure 7. Typical averaged fuel production efficiencies (UK data).

The first step therefore, before any calculations can be made, is to analyse the overall system into a sequence of component operations, or sub-systems, such that the operators of these individual sub-systems can provide information about their performance.

Humans

No reference is made in the discussion so far about human beings. Yet they operate the machines and the whole system is set up for their benefit. So can they be ignored?

Within these systems, humans can be regarded as performing two types of function. First, if a human is regarded as an industrial machine, as for example when operating say an injection moulding machine, then his efficiency must be examined in comparison to the machine he is operating. The average energy extracted per day from food by humans is of the order of 0.144 MJ per kg of body weight. So for an adult of mass 70kg, this is equivalent to 10MJ per day and most of this is used to maintain the human as a living organism. By comparison, the gross energies required to produce 1 kg of tin-plate is 34 MJ and that to produce 1 kg of aluminium foil is 166 MJ. Consequently the energy associated with a human is negligible in comparison with his potential daily output of industrial goods and so can be ignored.

There is however another aspect to human behaviour which is especially important in delivery systems. Most distribution systems rely on the ultimate consumer to act as the final link in the delivery chain transporting goods from the retail outlet to home. In addition, some recycling processes rely on the consumer delivering materials to some collection point. It is difficult to calculate precisely the energy used by humans involved in these activities, but it is important to recognise that they are not insignificant. A journey of 1 km in a typical car will give rise to a gross energy consumption of the order of 2.8 MJ (or 4.5 MJ per mile). The total energy required to produce and use beverage containers, excluding the contents, lies in the range 5 to 20 MJ/container and so consumer transport is not negligible, even though it is difficult to measure with any precision.

How to Calculate Results

Once a system has been analysed into its component sub-systems, data for the performance of each sub-system are obtained from operators. The information is usually chosen to cover a twelve month period. Apart from being the period for which most firms maintain records, it also smoothes out any atypical behaviour such as machine breakdowns, or start-ups while being sufficiently short so that genuine improvements are not masked.

It must however be recognised that the results of such calculations represent a "snap-shot" of the industry for the year examined. For systems which have not previously been examined, there is no way of knowing whether the performance was good or bad in the year chosen. For this reason, the analysis should be repeated for more than one year until a pattern is established.

Dependence On Throughput

The demand for raw materials and energy and the outputs of materials obviously depend upon throughput. This dependence can be eliminated by normalising with respect to output of usable or saleable product as shown in Fig. 8.

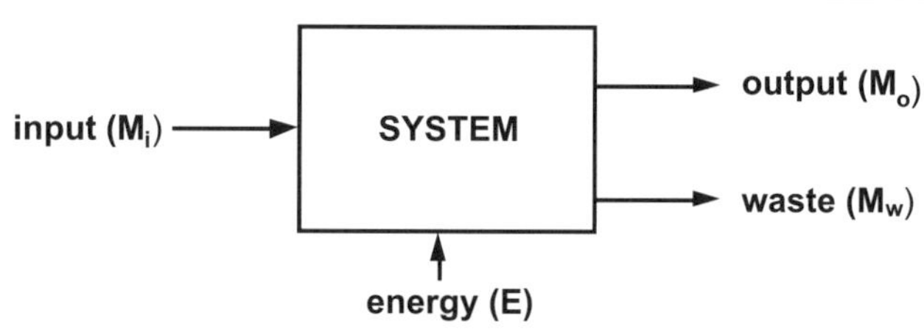

Mass balance: $M_i = M_o + M_w$

Normalised system energy = E/M_o

Normalised waste output = M_w/M_o

Normalised raw materials input = M_i/M_0

Figure 8. Normalising system inputs and outputs with respect to usable output.

Once information is available on the inputs and outputs per unit product from all of the sub-systems, the overall system can be re-assembled by balancing the mass flows. Fig. 9, for example, shows a simple linear sequence of three operations and, assuming that the masses are balanced, then it is a relatively easy matter to calculate the energy requirement of the combination. The same procedure is used for the other parameters.

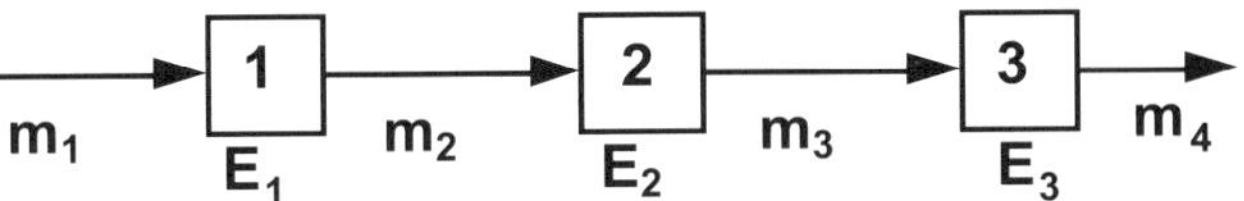

E_1, E_2 and E_3 are energy per unit output from the sub-systems.
Total system energy, E_s, is given by

$$E_s = m_4.E_3 + m_3.E_2 + m_2.E_1$$

Normalised system energy = E_s/m_4

Figure 9. Calculation of normalised energy requirement for a simple linear sequence of operations.

Extended Systems

Frequently this type of analysis is carried out on extended systems where the inputs are raw materials in the earth and the outputs are waste materials returned to the earth or atmosphere. The calculations are identical to those for smaller systems but there is now no useful or saleable output and the system inputs and output must be normalised with respect to some internal flow as shown in Fig. 10.

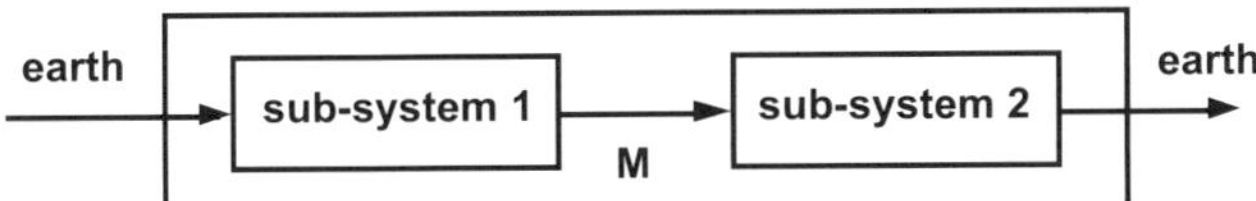

Energy associated with sub-system 1 = E_1
Energy associated with sub-system 2 = E_2
Total system energy requirement = (E_1+E_2)
Normalised system energy requirement = $(E_1+E_2)/M$

Figure 10. Normalising system energy when the system yields no usable or saleable output.

The choice of this internal normalising parameter is of some importance. In liquid delivery systems, for example, the choice of parameter can be different for small and large containers. For small containers, where the aim is to deliver a single "drink," the number of containers is probably the most appropriate. For large containers, such as the 3 litre PET bottle, where the purpose is to deliver "bulk" beverage, the volume delivered is probably the more appropriate.

Some care is therefore needed when comparing systems because the choice of an inappropriate normalising parameter can produce misleading results. Fig. 11, for example, shows the gross energy associated with the sale of fruit juice in two different sizes of aseptic brick.

Size (mL)	Energy per container (MJ)	Energy per litre (MJ)
125	2.19	17.52
2000	7.95	3.98

Figure 11. Gross energy required to sell fruit juice in aseptic bricks.

Measured as energy per container, the 125 ml brick appears to be the more energy efficient but measured as energy per unit volume, the 2 litre container is the more efficient. The 125 ml container is intended as a single serving container and is therefore the more energy efficient for this purpose; the 2 litre container is intended as a multiple serving container and is therefore the more energy efficient when viewed in this way. Fig. 11 is not therefore passing any judgement on either size of container; it is underlining a commonly observed property of all container systems — that larger containers consume less resources per unit volume delivered.

Results

Ecobalance calculations produce information under 5 main headings as shown in Fig. 12. Within these headings, the results are further subdivided; Fig. 12 gives a typical breakdown. As can be seen, the total number of different parameters needed to describe fully the performance of a system is large and they cannot be aggregated in any meaningful way to produce a single parameter that could be used as a measure of the overall environmental effect. Because of this it is meaningless to say that one system is "better" than another.

Presentation of Data

A large number of parameters is needed to describe fully the energy and raw materials requirements of any industrial system, together with solid waste emissions and emissions into air and water. In all the total number of parameters is approaching 200. While these data form the basic information needed to provide a complete description, only a selection is needed to consider, in detail, any specific problem.

Primary Fuels	Raw Materials	Solid Waste	Air Emissions	Water Emissions
Coal	Barytes	Glass containers	Dust	COD
Oil	Bauxite	Paperboard packs	CO	BOD
Gas	Brine	PET bottles	CO_2	Pb
Hydro	$CaSO_4$	Tinplate cans	SO_X	Fe
Nuclear	Chalk	Aluminum cans	NO_X	Na^+
Primary Feedstock	Clay	Plastic containers	H_2S	NO^{3-}
Coal	CoO	Paper/board refuse	Mercaptan	Hg
Oil	Copper	Plastics refuse	NH_3	NH^{4+}
Gas	Feldspar	Metal refuse	Cl_2	Cl^-
Wood	Ferromanganese	Organic refuse	HCl	CN^-
Biomass	Flourspar	Other refuse	F	F^-
	Iron chromite	Mineral waste	HF	S
	Iron ore	Slags/ash	HC	Dissolved organics
	Lead	Industrial waste	CHO	Suspended solids
	Limestone		Organics	Detergent
	Magnesium		Pb	Oils
	Manganese		Hg	HC
	Metallurgical coal		Metals	Phenol
	NiO			Phosphate
	Phosphate			Other N
	Rutile			
	Sand			
	Selenium			
	$NaNO_3$			
	Tin			
	Water			
	Wood			
	Zinc			

Figure 12. Typical headings under which LCA primary data might be reported.

It is tempting to seek ways of combining the data to reduce the number of parameters, for example by combining all energy sources into a single gross energy requirement or by combining air and water emissions into critical volumes. Whilst these techniques reduce the number of parameters, it is important to remember that they hide much valuable information.

Probably the most satisfactory approach is to identify the environmental problems that need to be considered and then select from the raw results those parameters that are needed to examine it. The problem here is that few environmental problems have been specified properly in any detail.

References

National Academy of Sciences & National Research Council (1969). *Resources and man*. W.H. Freeman, San Francisco.

Smith, H. (1969). The cumulative energy requirements of some final products of the chemical industry. *Transactions of the World Energy Conference*, **18**, Section E.

Meadows, D.H., Meadows, D.L., Randers, J. & Behrens, W.B. (1972). *The limits to growth*. ISBN 0-330-241699. Pan Books.

Club of Rome (1972). A blueprint for survival. *The Ecologist*, **2**, 1.

Hannon, B.M. (1972). System Energy and Recycling: A study of the beverage container industry. ASME Paper No 72-WA/Ener-3.

Boustead, I. (1972). *The milk bottle*. Open University Press, Milton Keynes.

The competitiveness of LDPE, PP and PVC after the 1973 oil crisis (1973). ICI Ltd., Welwyn Garden City, UK.

Sundstrom, G. (1973). Investigation of the energy requirements from raw materials to garbage treatment for four Swedish beer packaging alternatives. Report for Rigello Pak AB.

Hunt, R.G. and Franklin, W.E. (1974). Resource and environmental profile analysis of nine beverage container alternatives. Report to the U.S. Environmental Protection Agency (Contract 68-01-1848).

Schaefer, H. (1974). Fundamentals and methodology of investigating specific energy consumption. EC Contract 145-74-ECIC.

Semples, D.K. (1974). Energy in the automobile. Seminar at the Institute of Sciences and Technology, University of Michigan, Traverse City, Michigan.

Maddox, K.P. (1975). Energy analysis. Mineral Industries Bulletin, Colorado School of Mines, **18**, 1-18.

Leach, G. (1975). *Energy and food production*. International Institute of Environment and Development, London.

Campbell, I.M. (1977). *Energy and the atmosphere; A physical-chemical approach*. ISBN 0-471-99482-0. John Wiley, London.

Boustead, I. and Hancock, G.F. (1979). *Handbook of industrial energy analysis*. ISBN 0-85312-064-1. Ellis Horwood, Chichester/John Wiley, New York.

Fink, P. (1979). EMPA und Forschungsinstitut fur Adsatz und Handel der HSG, St. Gallen. Analyse der Verpackung von Joghurt-wirtschaftliche, technische und oekologische Bewertung.

Slesser, M. and Lewis, C. (1979). *Biological energy sources*. ISBN 0-470-26729-1. Spon, London.

Kindler, H. and Nikles, A. (1979). Energiebedarf bei der Herstellung und Verarbeitung von Kundstoffen. *Chem Ing-Tech.*, **51**, 11, 1125-1127.

Brown, H.L., Hamel, B.B., Hedman, B.A., Koluch, M., Gajanana, B.C. and Troy, P. (1980). Energy analysis of 108 industrial processes. Report to the US Department of Energy.

Boustead, I. and Hancock, G.F. (1981). *Energy and packaging*. ISBN 0-85312-206-7. Ellis Horwood, Chichester/John Wiley, New York.

Shaw, R. (1982). *Wave energy; a design challenge*. ISBN 0-85312-382-9. Ellis Horwood, Chichester/John Wiley, New York.

Bridgwater, A.V. and Lidgren, K. (1983). *Energy in packaging and waste*. ISBN 0-442-30570-2. Van Nostrand Reinhold (UK).

Boustead, I. and Lidgren, K. (1984). *Problems in packaging*. ISBN 0-85312-721-2. Ellis Horwood, Chichester/John Wiley, New York.

Lundholm, M.P. and Sundstrom, G. (1985). Tetra Brik aseptic environmental profile. Sundstrom AB, Malmo, Sweden.

EC Directive 85/339 (1985). Containers of liquids for human consumption.

Boustead, I. (1986). Energy Utilization. In Encyclopedia of packaging technology. ISBN 0-471-80940-3. John Wiley, New York.

Lundholm, M.P. and Sundstrom, G. (1986). Tetra Brik environmental profile. Sundstrom AB, Malmo, Sweden.

Boustead, I. and Hancock, G.F. (1989). EEC Directive 85/339: UK Data 1986. A Report to INCPEN, London.

Friends of the Earth (1989). *The heat trap; the threat posed by the rising levels of greenhouse gases*.

Franklin Associates (1990). *Energy and environmental profile analysis of childrens disposable and cloth diapers*. Prairie Village, Kansas.

Franklin Associates (1990). Environmental, safety and health impacts of polystyrene. Prairie Village, Kansas.

BUWAL (1991). Ecobalance of packaging materials: situation in 1990. Document 132 (in German) prepared by K. Habersatter for the Swiss Ministry of the Environment.

A technical framework for life-cycle assessment (1991). Society for Environmental Toxicology and Chemistry, Washington D.C.

Durrant, H.E., Hemming, C.R., Lenel, U.R. and Moody, G.C. (1991). Environmental labelling of washing machines. PA Consulting Group.

Ecobilan (1991). Comparaison des impacts sur l'environnement du recyclage et de la valorisation thermique des filmes plastiques usages: une approche a travers le cas des films agricoles polyethylene basse densite. Paris.

van Eijk, E., Nieuwenhuis, J.W., Post, C.W. and de Zeeuw, J.H. (1992). Reusable versus disposable — A comparison of the environmental impact of polystyrene, paper/cardboard and porcelain crockery. Ministry of Housing, Planning and the Environment, The Netherlands.

National Society for Clean Air (NSCA). Pollution Handbook. Annual publication. NSCA, Brighton, UK.

APME (Association of Plastics Manufacturers in Europe) — Ecoprofiles of the European Plastics Industry.
Report 1: Methodology. December 1992.
Report 2: Olefin feedstock sources. May 1993.
Report 3: Polyethylene and polypropylene. May 1993.
Report 4: Polystyrene. May 1993.
Report 5: Co-product allocation in chlorine plants. April 1994.
Report 6: Polyvinyl-chloride. April 1994.
Report 7: Polyvinylidene-chloride. In press.

DESIGN FOR OPERATIONS AND CONTROL

Iori Hashimoto
Chemical Engineering Department
Kyoto University
Kyoto 606-01, Japan

Evanghelos Zafiriou
Chemical Engineering Department and Institute for Systems Research
University of Maryland
College Park, MD 20742

Keywords

Controllability, Operability, Flexibility, Process uncertainty.

Introduction

Integration of process design and control system design is a research subject that is drawing increased attention by process systems engineers. An indication of that is the fact that an IFAC Workshop on this theme was held in London in September 1992 with large academic and industrial attendance, and in less than two years later, two weeks prior to FOCAPD'94, a second IFAC Workshop on the Integration of Process Design and Control was held in Baltimore, with similar success.

Needless to say, the design of a good control system is by nature closely related to the design of the process itself. Putting aside rigorous definition, let us consider controllability, which is a dynamic characteristic of process systems. As Prof. Morari often stresses by referring to early work of Ziegler and Nichols, the process Gp and the controller Gc in Fig. 1, form a unit G. Neither Gc nor Gp as independent elements are alone responsible for the characteristics of the closed loop, but it is G as a unit that is responsible.

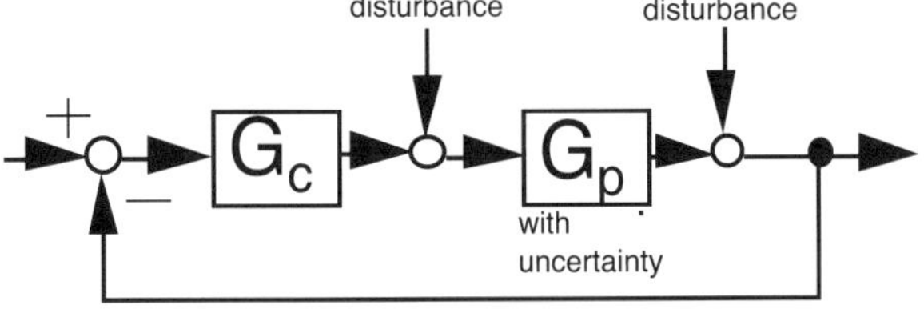

Figure 1.

It is often said that the cause of most control problems is poor design practice that is based only on steady state operation. Many of the attendees of the session know from experience that even the simplest controller can sometimes achieve good control performance for a properly designed process. Consider, e.g. the design of a continuous process, which starts with steady state operation. Traditionally, the design of this type of process is formulated as a static optimization problem with economics as the performance index. If the flow sheet has been decided, the process design can be accomplished by solving a parametric optimization problem. On the other hand, if the flow sheet has not been decided, that is, in the case of a synthesis problem, one must solve a combinatorial optimization problem. Only after this static design is accomplished, the control system that can ensure the steady state operation of the process can be designed, and finally the process flow diagram (PFD) be obtained.

At the next step, unsteady state operation, such as start-up and shut-down, is considered. Additional equipment such as pumps, tanks, pipes, control systems, which ensure unsteady state operation, are added to the PFD, and then, the piping and instrumentation diagram (PID) is obtained. As one proceeds from the steady state process design towards the design of the control systems, one needs more and more information on dynamical features of the process.

The system thus designed, must be able to cope with various uncertainties. One group of uncertainties consists of those emerging from within the process, such as uncertainties in reaction rates, catalytic activity, heat transfer coefficients, equilibrium relationships and so on. The other group consists of those uncertainties imposed from outside the process, such as product specifications, feed stocks, etc. In addition, various disturbances enter the process during the operation period. The designed process/control system must be capable of overcoming such uncertainties and disturbances and ensure sound operation of the process. In an attempt to realize such a system, the concepts of flexibility, operability and controllability have been introduced and rigorously studied. However, often even the definitions of these terms are not too clear. Hopefully, the papers in the session will help clarify them. In Fig. 2 we attempt to give a qualitative definition, which is not necessarily unique.

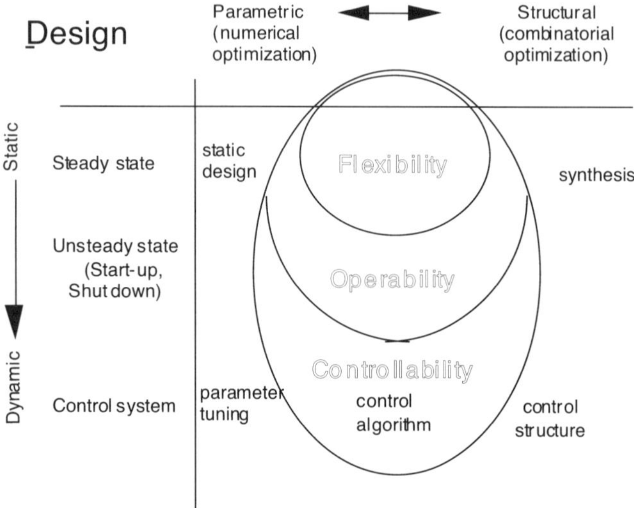

Figure 2.

Another question that should also be considered is whether controllability should be taken into account in all design problems. The answer is very likely "no." Assuming that one must consider controllability issues in designing a process, it may not be necessary to examine the whole plant all at once. It might be possible to decompose the plant into several subsystems from the viewpoint of controllability, and to enhance the controllability of the whole process by redesigning certain subsystems only.

Since the design of the process itself influences controllability, and since it is costly to redesign a plant already constructed, in an attempt to improve its controllability, one naturally wishes to take controllability into account at an early stage of the process design task. What is needed here is a good index or indices to help predict the controllability of the finished process at an early design stage. It is not clear yet, how to best define and quantify such indices, although several different approaches and methods have been presented in the literature over the years. However, an ideal method for designing a process with sufficient controllability is yet to be developed.

In the first paper of the session, Drs. Morari and Perkins give their careful review and assessment of the developments in the past decade and share their future perspective and prognostication. The review of different controllability indices clearly shows that the mathematical complexity increases with each further development. In an effort to develop quantitative indices that incorporate economic information, recent controllability measures maximimize performance indices over possible disturbances, thus resulting in an explicit dependence on the designer's disturbance estimates. Even indices that do not have such an explicit dependence, can often be interpreted only in relation to expected disturbances, e.g. indices that predict poor performance for certain disturbance directionalities. Experience has shown that for existing plants with vast amounts of past measurements available, the question of process monitoring to determine the nature and effect of disturbances on performance is still largely unanswered. How can one then expect good disturbance estimates for yet unbuilt plants? The ambitiousness of the controllability/operability measures has to be limited by our knowledge of what are likely disturbances. The discussion in the session will certainly help evaluate whether this leaves room for useful results.

The second paper in the session gives the perspective of the seasoned industrial practitioner. Drs. Downs and Ogunnaike share their experiences in order to help one see the issues from an industrial perspective. This paper should be of special interest as it covers not only technical aspects, but also includes cultural and managerial considerations. The concept of variability is discussed and its importance emphasized. There should be a clear relation between the process operability/variability concepts in the industrial paper, and the concept of process controllability in the academic paper in the session. But is there? The industrial notion of operability, as expressed in the second paper, seems broader but also more vague. How do we go about quantifying thc industrial concept both in systems theory and also economic terms? After all, how many dollars is a unit of dimensionless condition number worth anyway? The bottom line has to be part of the controllability indices if they are going to be accepted and used by the industry.

Bridging the gap between academia and industry is an urgent need. At the recent IFAC Workshop on Integration of Process Design and Control, in Baltimore, out of 40 papers presented, less than a third came from the USA, with most coming from Europe. For a meeting that was held in conjunction with the American Control Conference, where the US presence in the program is overwhelming, this may mean that US research in the area is drying up, possibly due to the lack of industrial support.

The two papers of the session provide some valuable insights, but also set the stage for an exchange of ideas that will start in the discussion section and which needs to be continued in the months to come.

DESIGN FOR OPERATIONS

Manfred Morari
California Institute of Technology
Pasadena, CA 91125

John Perkins
Imperial College of Science, Technology and Medicine
London SW7 2BY, UK

Abstract

Controllability is defined as the "best" dynamic performance (setpoint following and disturbance rejection) achievable for a system under closed loop control. By definition it does not depend on the controller but only on the plant itself. For at least 50 years it has been suggested to pay attention to controllability during the design phase of a plant and not to rely solely on the control system for achieving good dynamic performance. During the last decade valuable insights have been obtained on the effects of design on controllability. Significant progress has been made in the development of tools which allow the engineer to make controllability an integral part of the design objectives. The different approaches are reviewed and future needs are identified.

Keywords

Controllability, Disturbances, Uncertainty, Linear systems, Nonlinear systems, Synthesis.

Introduction

A processing system should be designed such that it has the ability to react to changes in demand, specifications, feedstock, and the environment in an economical and safe manner. During operation, the supervisory and regulatory layers of the control system are responsible for adapting the processing system to these changes. However, the performance of the control system carrying out this task does not only depend on the controller but also on the process itself. Indeed, the capabilities of a control system are actually quite limited. For example, the acceleration of a family sedan will never equal that of a Formula I racing car regardless of the quality of the fuel injection control system. It is evident that, in general, a system's ability to react to changes can only be guaranteed if the control requirements are considered at the design stage. At least conceptually, one can distinguish two types of requirement: flexibility, i.e. the ability of the system to handle a new situation at steady state, and controllability, i.e. the ability of the system to accomplish the dynamic transition between the operating states in an acceptable manner. Obviously, controllability requires flexibility. This paper is largely limited to a discussion of controllability.

The effect of design on controllability is certainly not a new discovery. The concept has been discussed in the literature for at least 50 years and reports on industrial problems which have surfaced over the last 30 years attest to its importance. In 1943, Ziegler and Nichols (1943) devote an entire paper to the topic of controllability.

> "In the application of automatic controllers, it is important to realize that controller and process form a unit; credit or discredit for results obtained are attributable to one as much as the other. A poor controller is often able to perform acceptably on a process which is easily controlled. The finest controller made, when applied to a miserably designed process, may not deliver the desired performance. True, on badly designed processes, advanced controllers are able to eke out better results than older models, but on these processes there is a definite end point which can be ap-

proached by instrumentation and it falls short of perfection."

Ziegler and Nichols proceed to define "controllability" as "the ability of the process to achieve and maintain the desired equilibrium value." They also introduce a "recovery factor" to classify processes in terms of controllability regardless of the controller used. Although their recovery factor does not have an entirely rigorous theoretical basis, it does express quite accurately the performance achievable in practice for single input-single output systems even when today's sophisticated control hardware is used.

One of the earliest detailed reports of industrial problems caused by a lack of controllability is due to Anderson (1966). A feed-effluent heat exchanger system was designed improperly. The energy recycle introduced positive feedback which de-stabilized the system for certain operating conditions (high throughput). Running the plant at its design capacity was only possible after a complete redesign of the feed-effluent heat exchange system.

The first ideas for systematically including controllability considerations into process design were introduced by Nishida and co-workers (1975, 1976). Some of the early work on controllability and flexibility is reviewed by Morari (1982, 1983a).

Has there been much progress made since that pioneering paper by Ziegler and Nichols and in particular in the last decade? The answer has to be a resounding yes when the work is judged against the fact that the problem is extremely difficult. It is broad and vague like most synthesis problems. In addition, the objective is not well-defined: apart from extreme cases, what is considered to be good or acceptably good control depends on personal preferences. Furthermore, contrary to economics, taking controllability into account in all design problems is neither necessary nor practical. Here the situation is very different from military aircraft. There, control is the dominant objective and not economics. Given the breadth of the field, it is not surprising that we find a diverse range of work which all in some sense contributes to the general goal. Eventually, it will lead to a set of design tools which will allow us to routinely assess the controllability of competing designs and make controllability an integral design objective.

Process Synthesis and Controllability

As defined in the review by Nishida et al. (1981), "process synthesis is an act of determining the optimal interconnection of processing units as well as the optimal type and design of the units within a process system." The structure of the system and the performance of the processing units are not determined uniquely by the performance specifications. "The task is (then) to select a particular system out of the large number of alternatives which meet the specified performance."

The important problems in process synthesis are defined to be the representation problem, (Can a representation be developed which is rich enough to allow all alternatives to be included without redundancy?), the evaluation problem (Can the alternatives be evaluated effectively so they may be compared?), and the strategy problem (Can a strategy be developed to locate quickly the better alternatives without totally enumerating all options?).

Most of the work on controllability has been focused on the evaluation problem. Ideally, one would like to assess the behaviour of a system with a controller of specified complexity which is tuned optimally. Even with current state-of-the-art optimal control theory and today's computing resources, this task would be enormous and not feasible for any but the simplest types of problems. Moreover, the formulation of a meaningful optimal control problem requires a wealth of parameters to be specified to define exactly the system model as well as the performance specifications. This is usually impossible until the completion of the design process. Thus, much effort has been expended toward the development of low-effort analysis tools which give a reasonable indication of the quality of closed-loop behaviour one can expect and which allow the designer at least to rank order alternatives according to controllability.

The following review organizes the various evaluation techniques according to the level of complexity of information required for the evaluation. Usually, the evaluation criteria are employed to guide the evolutionary development of the process. Design changes are postulated on physical grounds in an ad hoc manner. Only a few papers suggest algorithmic synthesis procedures which include some measure of controllability as part of the objective.

The review does not attempt to be complete in terms of the papers cited. It is our intention, however, to describe all the different schools of thought which have defined the work during the last decade. Strictly speaking, the selection of the control structure, i.e. the choice of the manipulated and measured variables, is a part of process design. It can dramatically affect control performance. For the assessment of different choices of manipulated variables, the same evaluation techniques as discussed in this paper can be utilized. As an example, consider the assessment of alternative control structures for distillation columns by Skogestad and co-workers (Skogestad and Morari, 1987a; Skogestad et al. 1990b).

Measurement selection is most important when the variables of interest cannot be measured directly. Alternative measurement sets can be interpreted as a very restrictive set of design changes and efficient strategies are available to assess their effect on performance. For example, techniques for selecting measurements for the control of distributed parameter systems have been studied by Kumar and Seinfeld (1978a, 1978b) and alternative measurement sets for inferential control have been investigated by Lee and Morari (1990, 1991).

In this paper the discussion of flexibility is limited to a few results with direct emphasis on control. For a review of the recent results on flexibility analysis see Grossmann and Straub (1991).

Controllability Evaluation Based on Steady-State Models

A series of papers by Douglas and co-workers (Fisher et al. 1988a; 1988b) describes a systematic procedure for assessing process controllability at the preliminary stages of a process design so that some of the economics associated with control can be used as an additional criterion for screening process alternatives. For improving controllability they consider (1) modifying the flowsheet to add more manipulated variables, (2) overdesigning certain pieces of equipment so that the process constraints never become active for the complete range of process disturbances, or (3) ignoring the optimization of the least important operating variables. The goal of their controllability analysis is to determine which of these alternatives has the best economic performance. In another related case study (Fisher et al., 1985) they demonstrate that for a distillation column overdesign in terms of the number of trays it is important to account for tray efficiencies, but such overdesign does not significantly increase the range of operability of a column. Overdesign of the column's vapour capacity, however, allows operation over a wide range of feed conditions.

The controllability problem for heat exchanger networks was defined first by Marselle et al. (1982). In this paper a procedure is developed to yield designs which handle stream flowrate and temperature fluctuations with maximum energy efficiency. The problem was studied further by Calandranis and Stephanopoulos (1986). Linnhoff and Kotjabasakis (1986) address the trade-offs between controllability and economics more effectively through a combination of "downstream paths" and "sensitivity tables." Georgiou and Floudas (1989) propose an automated synthesis procedure based on a superstructure approach which will either generate heat recovery networks with low capital costs and total disturbance rejection, or* if this is not possible, it will minimize disturbance propagation.

We suspect that at this stage of development the practical utility of all these approaches is somewhat limited. Marselle's approach leads to networks which are often not economical. Linnhoff's technique is hard to automate and becomes awkward for all but the simplest networks. Floudas' approach takes only structural effects into account though a quantitative evaluation of the trade-offs will usually be necessary in practice.

Shinnar (1981) introduces a new definition of the concept of controllability which takes into account model uncertainty and the fact that many important process variables may not be measurable in practice. The ideas are illustrated on a fluid cat cracker, a hydrocracker and a crystallizer.

Controllability Evaluation Based on Linear Dynamic Models

Given a linear dynamic model of the process, some performance objectives related to the desired closed-loop performance, and some restrictions on the controller structure (PI, decentralized, etc.), the objective is to find the optimal controller parameters and to compare the achieved optimal performance, a measure of controllability, among different designs. Several difficulties are associated with this approach. The traditional objective function in optimal control has been the integral square error (ISE). It was used to distinguish alternative designs, for example, by Lenhoff and Morari (1982). By itself the ISE is hardly a measure which is of direct interest in practice. Lee et al. (1972) used multiple performance criteria: integral square error, maximum deviation of exit response, maximum magnitude of control variable, and saturation magnitude. Any such assessment, however, would become prohibitively complex in the multi-variable context. A further difficulty with the constrained-structure optimal control approach is that the search for the control parameters is notorious for its multiple local optima.

Both of these drawbacks are alleviated to a large extent by adopting robust performance as a control objective (Morari and Zafiriou, 1989) and by removing any restriction on the controller structure. In the H_∞ context, the objective function is also scalar but it can be formulated to express frequency-dependent constraints on various outputs and manipulated variables as well as the effect of model uncertainty. Usually, the structured singular value μ is used as a normalized indicator, with values less than one signifying satisfaction of the performance specifications in the presence of uncertainty. Ever better methods for the computation of μ-optimal controllers (without complexity constraints) become available (Balas et al 1992). Nowadays, restricting control complexity is hardly an issue in any new installation. As predicted by Rinard (1975) complex multivariable controllers are widely accepted by industry. Applications of Model Predictive Control with as many as 20 manipulated and 40 controlled variables are being contemplated (Cutler and Yocum, 1991).

Skogestad et al. (1988) used μ-optimal controllers to distinguish various control structures for high-purity distillation columns. Jacobsen et al. (1991) employed μ-optimal PID controllers to evaluate various designs of a homogeneous azeotropic distillation column. Despite these successes, it must be stated that formulating a meaningful robust control problem and determining the optimal robust controller is far from trivial. Therefore, the search for alternative, simple criteria which would allow a rank ordering of design alternatives roughly consistent with that obtained from a μ analysis has continuing importance. Such criteria will be discussed next.

In 1983, Morari (1983a) identified the following process characteristics which limit the achievable control performance independent of controller design: non-minimum-phase behaviour (i.e. time delays and right-half-plane zeros), actuator limitations, and model uncertainty. Occasionally, measurement noise may have a dominant effect. Various researchers have proceeded to analyse the effect of time delays, right-half-plane zeros and model uncertainty individually on closed-loop performance.

Multivariable delays were analyzed by Holt and Morari (1985b) and by Perkins and Wong (1985). Exact bounds on achievable performance with and without decoupling constraints were identified through a mixed integer linear programming procedure by Psarris and Floudas (1990). The properties of multivariable right-half-plane transmission zeros were investigated by Holt and Morari (1985a) and Morari et al. (1987). Conditions under which decoupling is integral-square-error optimal were identified. This work was followed by a more complete treatment by Psarris and Floudas (1991a, 1991b) who deal with the case of an infinite number of zeros which usually arises in the presence of multivariable time delays.

The drawback of these analysis techniques is that in any but the simplest situations it is essentially impossible to rank order alternatives. What zero/delay structure is preferable depends in a complex manner on the performance specifications.

Morari (1982; 1983b) suggested the condition number of the plant transfer matrix as a function of frequency as an indicator of closed-loop sensitivity to model error. This criterion was first applied to a system of two CSTRs with heat integration (Morari et al. 1985). In the last ten years, we have gained some understanding of the role of the condition number (Skogestad and Morari, 1987b) but the basis for its use for controllability assessment is still somewhat tentative. The main problem is that all the conditions relating the condition number and closed-loop stability and performance (Morari and Zafiriou, 1989) are sufficient but not necessary. While we can say with certainty that the closed loop performance of low-condition-number plants tends to be insensitive to model error, we cannot reject high condition number plants with certainty though there are many indications that the performance for these types of plants will be bad. Over the years there also has been some discussion on the scalings of the plant inputs and outputs and their effect on the magnitude of the condition number. Our experience, supported by theory (Skogestad and Morari, 1987b), indicates that minimizing the condition number by input and output scaling tends to distort any conclusions on model error sensitivity.

Skogestad et al. (1991) make a strong case for using a combination of the frequency-dependent relative gain array (RGA) and the closed-loop disturbance gain (CLDG) to judge the relative controllability of alternative designs. These tools have been employed in many recent case studies. The conclusions tend to correlate well with the closed-loop performance obtained with m-optimal control systems. The steady-state RGA was introduced by Bristol (1966) and continues to be widely used in industry as a controllability indicator. While the steady-state RGA is indicative of the fault tolerance of multivariable control systems (Grosdidier et al., 1985), it can be very misleading as a controllability indicator (Skogestad et al., 1990a). The CLDG was introduced by Hovd and Skogestad (1992) based on the relative disturbance gain defined by McAvoy and co-workers (Stanley et al., 1985)[1]. Hovd and Skogestad found that the CLDG enters nicely into the relation between control error and disturbances, while the RGA enters in a similar way into the relation between control error and set point changes.

In the last few years, many case studies have appeared where these concepts were applied. An ethyl-benzene production facility and a two-column separator system in styrene manufacturing are analyzed at steady state in Mizsey and Fonyo (1991). The controllability of ordinary distillation columns is investigated by Jacobsen and Skogestad (1991). A fluid catalytic cracker is studied by Hovd and Skogestad (1991), heat exchanger networks by Wolff et al. (1991) and Mathisen et al. (1991), homogeneous azeotropic distillation columns by Jacobsen et al. (1991), and flotation circuits by Barton et al. (1991). It is often possible to interpret the behaviour of the frequency-dependent RGA and CLDG on physical grounds. This and their theoretical basis make the RGA and the CLDG highly appealing and easily applicable tools for controllability assessment.

Controllability assessment based on a linear analysis of the fundamental limitations to control performance identified by Morari (1983a) has reached a level of maturity to the point where the concepts are being applied in industry during process design (e.g. Bouwens and Kosters 1992). Nevertheless, these techniques are not ideally suited to exploring the trade-offs involved in design/control integration. While it is true that the results of these studies may be related to closed-loop performance, the impact on **economics**, the key driver of decision making during design, is less straightforward to determine. To overcome this difficulty, Narraway et al. (1991) propose a direct assessment of the impact of disturbances on system economics as a method of controllability analysis. The method is based on standard techniques of control benefits estimation (e.g. Marlin et al 1991), but can be used during design to analyse the relative benefits of various plant options.

The basic concept is illustrated in Fig. 1. The size of the "ball" around chosen steady state operating conditions for the plant depends on the disturbance regime, and on the control system implemented. The steady state operating point should be chosen so that all points in the ball are feasible with respect to operating constraints, and so that the predicted economic performance is as favourable as possible. The ideal situation is illustrated in Fig. 1, where the centre of the ball is as close to the steady-state optimum in the absence of disturbances as possible.

---

1. In the derivation of the RDG the "recovery factor" introduced by Ziegler and Nichols (1943) in 1943 was rediscovered.

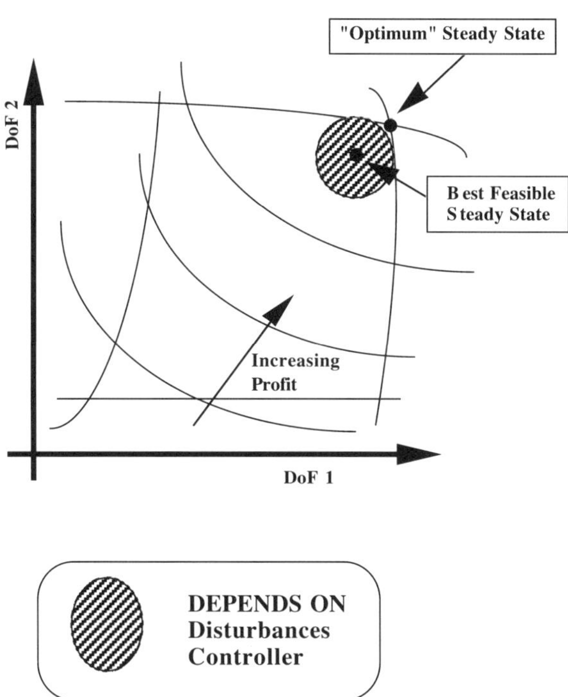

Figure 1. The economics of process control.

If the disturbances are characterised in terms of their frequency spectrum, the ball size for a given control structure may be calculated using standard frequency response techniques, assuming a linear plant model. The best steady-state operating point taking account of the ball size may then be determined by solving a linear programming problem (assuming a linear objective and linear constraints). Of course, the approach is also applicable to nonlinear systems, although the calculations can become significantly more involved when a general set of disturbances is considered.

Controllability Evaluation Based on Nonlinear Dynamic Models

In many cases, a controllability evaluation based on linear models as described above suffices, even when the system is strongly nonlinear and when a linear control system is inadequate. Often, it is reasonably easy to design simple static nonlinear compensators which remove most of the process nonlinearity. The compensated system can then be analyzed with the proposed linear techniques. As a typical example, high purity distillation columns can behave in quite a nonlinear fashion. However, when relative composition deviations are controlled or alternatively the logarithm of the compositions, the system is linearized sufficiently so that adequate performance is obtained with linear control systems. This was done, for example, by Skogestad and Morari (1988), where LV control for high purity distillation columns was investigated.

While it is possible to draw useful conclusions from the application of linear techniques to nonlinear systems before or after compensation, the designer may be left with a nagging doubt as to the validity of the results produced. Also, in rare but very important instances, the system can exhibit nonlinear behaviour which is not easily correctable with simple nonlinear transformations. For example, a reactor may display high parametric sensitivity and the temperature may rise to excessively high levels when small perturbations occur (Shinnar et al., 1992). Also, the system may exhibit multiple steady states, limit cycles, or even chaotic behaviour.

It was suggested recently that nonlinear characteristics should be examined at the design stage (Seider et al. 1990; 1991) and that nonlinear analysis techniques like bifurcation analysis and singularity theory should be used more routinely in process design. Indeed, this has been done in the area of reactor design for decades (see Uppal et al. (1974) for the first extensive application of bifurcation theory to a CSTR). The question is what to do with this type of analysis. Seider et al. (1990) suggest that modern nonlinear control algorithms allow us to deal with almost any difficult control situation, and consequently regions of unusual dynamic behaviour should not be avoided in process design. Carried to the extreme, one might speculate that nonlinear analysis is not needed at all at the design stage because any complex nonlinear behaviour can be fixed later on by the control algorithm.

It is possible that future developments may lead us somewhat in this direction (just as for linear systems, where the issue of control complexity became less and less important in the last decade), but we do not subscribe to this philosophy, at least not for now. First of all, nonlinear control theory is in its infancy. Even if the applicability of a particular algorithm (for example, nonlinear model predictive control) is established in principle, the control effort and the final control element required may be enormous (Shinskey, 1983), and totally uneconomical. Furthermore, in the case of control system failure, such a set-up may be disastrous.

It may be that increased quality standards, stricter environmental regulations, and economic pressures will push designs into regions which were previously avoided, and where unusual nonlinear behaviour occurs. Then nonlinear analysis techniques would increasingly be needed at the design stage. For chemical reactors, unusual dynamic behaviour is almost expected nowadays, but recently similar phenomena were discovered for other systems as well. The results obtained by Seader and coworkers (Chavez et al., 1986; Lin et al., 1987) suggest that multiplicities may be one reason why the so-called Petlyuk distillation configuration is not widely used in industry despite established energy advantages.

It is well known that heterogeneous azeotropic distillation columns can exhibit multiple steady states (Magnussen et al., 1979; Kovach and Seider, 1987; Rovaglio and Doherty, 1990). Rovaglio et al. (1991) have studied the control problems associated with these multiplicities and the parametric sensitivity also found in these columns. They discuss how to select the best operating decanter tie-line based on operability considerations. Recently Laroche et al. (1992a, 1992b) discovered multiple steady-states in homogeneous azeotropic distillation columns where such phenomena were believed not to exist. Fortunately, continuously improving software (e.g. Doedel 1986) allows today's designer to carry out bifurcation analyses on large systems (of the order of hundreds of differential equations in the case of the homogeneous azeotropic distillation columns) which were essentially infeasible a decade ago.

Often the bifurcation diagrams can be used to redesign the process to be more attractive from an economic and environmental point of view while at the same time avoiding complex dynamics. A nice example is presented by Ray (Ray 1992; Teymour and Ray 1989). At typical operating conditions solution polymerization of vinyl acetate carried out in a CSTR exhibits limit cycle behaviour. The bifurcation analysis shows that limit cycles can be avoided either by increasing the solvent fraction (a traditional technique which leads to costly recycle problems) or by decreasing the solvent volume fraction and increasing the initiator feed concentration.

Another example is the novel feed system for a CSTR for continuous emulsion polymerization, which was suggested by Penlidis et al. (1989). This new feed system was shown in experiments to remove the highly undesirable oscillatory (limit cycle) behaviour of conventional CSTRs. More examples are discussed in the review paper by Seider et al. (1991).

Though a particular system may not exhibit multiple steady states, limit cycles and other exotic dynamic behaviour, it may be extremely sensitive to disturbances and small changes in its operating parameters. This has been observed in catalytic reactors, where runaway has been the subject of numerous studies, starting with Wilson (1946) and Barkelew (1984) up to the recent work by Balakotaiah and Luss (1991). Shinnar et al. (1992) derived some simple design rules to avoid runaway. The conditions involve only parameters which can be easily obtained from pilot plant experiments.

To alleviate the uncertainties involved with the use of linear techniques to analyse the controllability characteristics of nonlinear systems, tests have recently been proposed which may be applied to the nonlinear model directly. The use of dynamic optimization as a tool for controllability analysis has been facilitated by the emergence of effective algorithms for the solution of optimal control problems (Perkins and Walsh 1994). The effects of time delays and bounded parametric uncertainty on disturbance rejection capability may be assessed by analysing the performance of an idealised controller under worst-case conditions (Walsh and Perkins 1992). The controller is idealised in that it is assumed to act perfectly as soon as knowledge of any disturbance becomes available. (Of course, such knowledge is delayed by the time delays in the system). Any real controller will deliver performance worse than this idealised case, so inadequate performance identified by the test implies inherent limitations in the plant. Walsh and Perkins (1992) present an analysis of a waste treatment plant design where many process and control options could be rapidly screened by the application of this test.

Continuing increases in computing power are facilitating computationally intensive approaches to controllability analysis which would have been impracticable a few years ago. The use of worst-case analysis techniques to evaluate the effectiveness of a candidate design in handling a given disturbance regime and range of uncertainty, and to select optimal design parameters for a range of scenarios, has been demonstrated to be a powerful tool in design/control integration (Walsh and Perkins 1994). Such an analysis may be applied both to steady-state and to dynamic models. In the former case, the analysis corresponds to the flexibility analysis of Grossmann et al (1983). The latter case represents an extension of these ideas to time-varying systems. There, care must be taken to avoid making unrealistic assumptions about the information available to any controller on the plant (Perkins and Walsh 1994).

Brengel and Seider (1992) proposed an approach to the co-ordination of design optimization with controllability analysis, the latter being measured through the performance of a Model Predictive Control algorithm on a range of disturbance scenarios.

Synthesis

Very little has been published on algorithmic synthesis techniques for processes which are both economical and controllable or where economics and controllability are traded off automatically in some intelligent manner. Apart from the early work by Ichikawa (Nishida and Ichikawa, 1975; Nishida et al., 1976), there is the more recent work by Floudas (Georgiou and Floudas, 1989), where the power of mixed integer nonlinear programming techniques is exploited for the synthesis of heat exchanger networks which exhibit minimal disturbance propagation. The same problem was studied by Huang and Fan (1992), where a knowledge engineering approach is proposed and mass exchanger networks are also considered. Luyben and Floudas (1992, 1994) used a multiobjective optimization technique to study the trade-off between various measures of steady state controllability and economics in the design of binary distillation columns.

The need to consider steady-state economics and controllability as separate objectives in a multiobjective formulation may be obviated if economic assessment of the effect of disturbances is employed (Narraway et al. 1991). This approach has been used by Narraway and Perkins (1993a, 1993b) for the synthesis of control systems for a given process design. A set of candidate manipulated variables and candidate measurements is given, together with costs of the instrumentation that would be necessary were any variable from these sets to be selected for the control structure. A disturbance regime is defined, and an optimal control structure is sought which minimises the sum of the economic giveaway implied by the ball-size analysis (Fig. 1) and the costs of instrumentation associated with the control system.

Narraway and Perkins present two special cases within this general framework. The first is based on a linear analysis, and approximates the performance of the controllers using the perfect control hypothesis. With this approach, whatever measurements are selected are assumed to be perfectly regulated; disturbance effects do not appear in these variables. The problem is then almost a mixed-integer linear programming problem. All nonlinearity is localised in the calculation of the ball-size for a given configuration. However, this nonlinearity includes discontinuities in first derivative due to the calculation of moduli of complex variables. A special algorithm, taking account of these properties is proposed by Narraway and Perkins (1993a) and tested on a number of case studies. Not surprisingly, it is found that the control structure chosen is critically dependent on the disturbance regime assumed. However, the relationship between disturbances and optimal control structure is not simple, with increasing levels of disturbance sometimes inducing more control (Heath and Perkins 1994) sometimes less (Narraway 1992)!

An attempt to address nonlinear systems directly, and to include realistic controller models, has been presented by Narraway and Perkins (1993b). A multiloop proportional plus integral control structure was assumed, which increases the combinatorics of the problem since pairings between measurements and manipulated variables must be selected in addition to the variables themselves. A highly simplified disturbance regime, consisting of a single sinusoidal disturbance was assumed. The formulation results in a mixed-integer optimal control problem, tackled by a version of the outer-approximation augmented penalty algorithm (Viswanathan and Grossmann 1990). An MILP MASTER problem is solved at each iteration to provide a new candidate control structure, whose performance is evaluated by solving a nonlinear optimal control problem to determine optimal controller tunings and the resulting economics. The performance of the method on case studies indicates significant nonconvexities in the formulation. However, the algorithm consistently suggested good structures for the cases considered.

It is not clear how useful the automatic synthesis techniques will be in the near future. Usually, they require a scalar performance index to be specified and they cannot exercise judgement on something like the behaviour of the frequency-dependent RGA and CLDG. The multiobjective approach introduced recently by Luyben and Floudas (1992) looks very promising although the controllability assessment is done at steady state and any implications for the dynamic behaviour are tenuous at best. The alternative approach of attempting to measure control performance directly in economic terms may turn out to be more effective in the long run. However, the strong dependence of the synthesised structures on the specified disturbances is a worrying feature of the current techniques. It is probably unrealistic to expect too much knowledge of likely disturbances at the process design stage. As the underlying analysis techniques improve, there is no doubt that the need for the automated methods which have been prototyped here will greatly increase.

Conclusions

While the definition of an economic objective for a design problem is relatively easy and unambiguous, there are many different ways to define the controllability of a system. What is ultimately of interest is the closed loop behaviour of the system once it is operating and subjected to disturbances, set point changes and other changes in its environment. Various indicators requiring more or less modelling information and computational effort have been developed to evaluate and predict the closed loop behaviour which can eventually be expected. Unfortunately, we often do not quite understand yet when and how these indicators should be used. Nevertheless they are applied widely and indiscriminately and sometimes lead to erroneous conclusions about controllability.

Steady state models can be used to assess various aspects of steady state controllability, for example, the ability to reject disturbances in steady state and the associated economic penalty. An analysis of steady-state multiplicities may also provide valuable design insights. Unfortunately, there is the unqualified use of steady state criteria (RGA, condition number, minimum singular value, etc.) to analyse the dynamic behaviour of processes. Although for specific physical systems there may be a correlation between steady state and dynamics, this is the exception rather than the rule.

A further pitfall is that frequently predictions are "verified" through simulations with controllers which have been tuned in an ad hoc manner or based on some rules of thumb. As a result, design modifications are identified as attractive which upon closer examination (Jacobsen and Skogestad, 1991) turn out to be irrelevant. The proper design of a multivariable controller according to a robust performance criterion which takes into account performance specifications on the controlled variables, manipulated variable constraints and model uncertainty takes skill and experience and can rarely be avoided for a final evaluation.

More research effort has to be devoted to the development of simple criteria for controllability evaluation and to clearly understand their limitations. Only then is it meaningful to formulate an algorithmic synthesis technique to trade off controllability and economics. Obviously, the objective of such a technique should not be "automatic synthesis," i.e. to remove the engineer from the loop, but it should constitute one tool in a big design tool kit, which includes the various other linear and nonlinear analysis tools. The algorithmic synthesis will allow the engineer to assess the trade-off between controllability and economics under precisely defined conditions. Such a tool kit has been proposed in Seider et al. (1990) and the general architecture of a comprehensive design package was discussed by Winter (1992). Usually, the final evaluation of the controllability of a system has to occur through simulation, in particular when nonlinear characteristics are important (Roat et al., 1986). Thus, an effective nonlinear dynamic simulator has to be an integral part of this tool kit.

Acknowledgments

Partial support for this work from the National Science Foundation, the US Department of Energy, the UK Research Councils (SERC, AFRC) and the Centre for Process Systems Engineering Industrial Research Consortium is gratefully acknowledged.

References

Anderson, J.S. (1966). A practical problem in dynamic heat transfer. *Chem. Engineer*, CE97-CE103.

Balakotaiah,V. and D. Luss (1991). Explicit runaway criterion for catalytic reactors with transport limitations. *AIChE Journal*, **37**, 1780-1788.

Balas, G., J.C. Doyle, K. Glover, A. Packard and R. Smith (1992). The μ-Analysis and Synthesis Toolbox, MATHWORKS INC.

Barkelew, C.H. (1984). Stability of Adiabatic Reactors, volume 237 of *ACS Symposium Series*, page 337. American Chemical Society.

Barton, G.W., W.K. Chan and J.D. Perkins (1991). Interaction between process design and process control: the role of open-loop indicators. *J. Proc. Control*, **1**, 161-170.

Brengel, D.D. and W.D. Seider (1992). Coordinated design and control optimization of nonlinear processes. *Comp. Chem. Eng.*, **16**, 861-886.

Bouwens, S.M.A.M. and P.H. Kosters (1992). Simultaneous process and system control design: an actual industrial case. In J.D. Perkins (ed.) *Interactions between process design and process control*. Pergamon, Oxford. 75-86.

Bristol, E.H (1966). On a new measure of interactions for multivariable process control. *IEEE Trans. Automat. Control*, AC-11, 133-134.

Calandranis, J. and G. Stephanopoulos (1986). Structural operability analysis of heat exchanger networks. *Chem. Eng. Res. Des.*, **64**, 347-364.

Chavez, C.R., J.D. Seader, and T.L. Wayburn (1986). Multiple steady-state solutions for interlinked separation systems. *Ind. Eng. Chem. Fundam.*, **25**, 566-576.

Cutler, C.R. and F.H. Yocum (1991). Experience with the DMC inverse for identification. In Y. Arkun and W.H. Ray (eds.), *Proceedings of Fourth International Conference on Chemical Process Control CPC IV*, Elsevier, Amsterdam, 297-318.

Doedel, E.J. (1986). *AUTO: Software for Continuation and Bifurcation Problems in Ordinary Differential Equations*. Concordia University, Montreal, Canada.

Fisher, W.R., M.F. Doherty, and J.M. Douglas (1985). Effect of overdesign on the operability of distillation columns. *Ind. Eng. Chem. Process Des. Dev.*, **24**, 593-598.

Fisher, W.R., M.F. Doherty, and J.M. Douglas (1988a). The interface between design and control. 1. process controllability. *Ind. Eng. Chem. Res.*, **27**, 597-605.

Fisher, W.R., M.F. Doherty, and J.M. Douglas (1988b). The interface between design and control. 2. process operability. *Ind. Eng. Chem. Res.*, **27**, 606-611.

Georgiou, G.A. and C.A. Floudas (1989). Simultaneous process synthesis and control: Minimization of disturbance propagation in heat recovery systems. In J.J. Siirola, I.E. Grossmann, and G. Stephanopoulos (eds), *Proc. of 3rd Conf. on Foundations of Computer Aided Process Design*, CACHE Publications, 435-450.

Grosdidier, P., M. Morari, and B.R. Holt (1985). Closed loop properties from steady state gain information. *Ind. Eng. Chem. Fundam.*, **24**, 221-235.

Grossmann, I.E., K.P. Haleman and R.E. Swaney (1983). Optimization strategies for flexible chemical processes. *Comp. Chem. Eng.*, **7**, 439-462.

Grossmann, I.E. and D.A. Straub (1991). Recent developments in the evaluation and optimization of flexible chemical processes. In *COPE-91*, Barcelona, Spain.

Heath, J.A. and J.D. Perkins (1994). Control structure selection based on economics. In *The 1994 IChemE Research Event*, Vol II, 841-843. IChemE Publications, Rugby.

Holt, B.R. and M. Morari (1985a). Design of resilient processing plants — VI. the effect of right-half-plane zeros on dynamic resilience. *Chem. Eng. Sci.*, **40**, 59-74.

Holt, B.R. and M. Morari (1985b). Design of resilient processing plants — V. the effect of dead time on dynamic resilience. *Chem. Eng. Sci.*, **40**, 1229-1237.

Hovd, M. and S. Skogestad (1991a). Controllability analysis for the fluid catalytic cracking process. In *AIChE Annual Meeting*, Los Angeles, California.

Hovd, M. and S. Skogestad (1992). Simple frequency-dependent tools for control systems analysis, structure selection and design. *Automatica*, **28**, 989-996.

Huang, Y.L. and L.T. Fan (1992). Distributed strategy for integration of process design and control. *Comp.Chem. Eng.*, **16**, 497-522.

Jacobsen, E.W., L. Laroche, M. Morari, S. Skogestad, and H.W. Anderson (1991). Robust control of homogeneous azeotropic distillation columns. *AIChE Journal*, **37**, 1810-1824.

Jacobsen, E.W. and S. Skogestad (1991). Design modifications for improved controllability of distillation columns. In *COPE-91*, Barcelona, Spain.

Kovach III, J.W., and W.D. Seider (1987). Heterogeneous azeotropic distillation: Homotopy-Continuation Methods. *Comp. Chem. Eng.*, **11**, 593-605.

Kumar, S. and J.H. Seinfeld (1978a). Optimal location of measurements for distributed parameter estimation. *IEEE Transactions on Automatic Control*, **AC-23**, 690-698.

Kumar, S. and J.H. Seinfeld (1978b). Optimal location of measurements in tubular reactors. *Chem. Eng. Sci.*, **33**, 1507-1516

Laroche, L., H.W. Anderson, and M. Morari (1992a). Homogeneous azeotropic distillation: Separability and flowsheet synthesis. *Ind. Eng. Chem. Research*, **31**, 2190-2209.

Laroche, L., N. Bekiaris, H.W. Anderson, and M. Morari (1992b). The curious behaviour of homogeneous azeotropic distillation-implications for entrainer selection. *AIChE Journal*, **38**, 1308-1328.

Lee, H.H., L.B. Koppel, and H.C. Lim (1972). Integrated approach to design and control of a class of counter current processes. *Ind. Eng. Chem. Process Des. Dev.*, **11**, 376-382.

Lee, J.H. and M. Morari (1990). Robust control structure selection and control system design methods applied to distillation column control. *In Proc. of IEEE Conf. on Decision and Control*, 2041-2046.

Lee, J.H. and M. Morari (1991). Robust measurement selection. *Automatica*, **27**, 519-527.

Lenhoff, A.M. and M. Morari (1982). Design of resilient processing plants. I. process design under consideration of dynamic aspects. *Chem. Eng. Sci.*, **37**, 245-258.

Lin, W.J., J.D. Seader, and T.L. Wayburn (1987). Computing multiple solutions to systems of interlinked separation columns. *AIChE Journal,* **33**, 886-897.

Linnhoff, B. and E. Kotjabasakis (1986). Downstream paths for operable process design. *Chem. Eng. Progress*, **82**, 23-28.

Luyben, M.L. and C.A. Floudas (1992). A multi objective optimization approach for analysing the interaction of design and control, part 1: Theoretical framework. In J.D. Perkins (ed.) *Interactions between Process Design and Process Control*, Pergamon, Oxford, 101-106.

Luyben, M.L. and C.A. Floudas (1994). Interaction between design and control in heat-integrated distillation synthesis. In E. Zafiriou (ed.) Integration of Process Design & Control, Pergamon, Oxford, 76-81.

Magnussen, T., M.L. Michelsen, and A. Fredenslund (1979). Azeotropic distillation using unifac. *In Third International Symposium on Distillation, I. Chem. E. Symposium Series No. 56*, 1-19. The Institution of Chemical Engineers.

Marlin, T.E., J.D. Perkins, G.W. Barton, and M.L. Brisk (1991). Benefits from process control: results of a joint industry-university study. *J. Proc. Control*, **1**, 68-83.

Marselle, D.F., M. Morari, and D.F. Rudd (1982). Design of resilient processing plants. II. design and control of energy management systems. *Chem. Eng. Sci.*, **37**, 259-270.

Mathisen, K.W., S. Skogestad, and E.A. Wolff (1991). Controllability of heat exchanger networks. In *AIChE Annual Meeting*, Los Angeles, California.

Mizsey, P. and Z. Fonyo (1991). Assessing plant operability during process design. In *COPE-91*, Barcelona, Spain.

Morari, M. (1982). Operability measures for process design. In *Understanding Process Integration, I. Chem. E. Sympo-sium Series No. 74*, pages 131-140. The Institution of Chemical Engineers.

Morari, M. (1983a). Design of resilient processing plants-III. a general framework for the assessment of dynamic resilience. *Chem. Eng. Sci.*, **38**, 1881-1891.

Morari, M. (1983b). Flexibility and resiliency of process systems. *Comp. Chem. Eng.*, **7**, 423-437.

Morari, M., W. Grimm, M.J. Ogelsby, and I.D. Prosser (1985). Design of resilient processing plants — VII. design of energy management system for unstable reactors — new insights. *Chem. Eng. Sci.*, **40**, 187-198.

Morari, M. and J.H. Lee (1990). Control structure selection: Issues and a new methodology. In F.A. Grunbaum, J.W. Helton, and P. Khargonekar, editors, *Signal Processing, Part II: Control Theory and Applications*, 195-219. Springer Verlag.

Morari, M. and E. Zafiriou (1989). *Robust Process Control*. Prentice Hall, Englewood Cliffs.

Morari, M., E. Zafiriou, and B.R. Holt (1987). Design of resilient processing plants-X. characterization of the effect of rhp zeros. *Chem. Eng. Sci.*, **42**, 2425-2428.

Narraway, L.T. (1992). Selection of process control structure based on economics. *Ph.D. thesis*. University of London.

Narraway, L.T., J.D. Perkins and G.W. Barton (1991). Interaction between process design and process control: economic analysis of proces dynamics. *J. Proc. Control*, **1**, 243-250.

Narraway, L.T. and J.D. Perkins (1993a). Selection of process control structure based on linear dynamic economics. *Ind. Eng. Chem. Research*, **32**, 2681-2692.

Narraway, L.T. and J.D. Perkins (1993b). Selection of control structure based on economics. *Comp. Chem. Eng.*, **18**, S511-5.

Nishida, N. and A. Ichikawa (1975). *Ind. Eng. Chem. Proc. Des. Dev.*, **14**, 236-242.

Nishida, N., Y.A. Liu, and A. Ichikawa (1976). *AIChE Journal,* **22**, 539-549.

Nishida, N., G. Stephanopoulos, and A.W. Westerberg (1981). A review of process synthesis. *AIChE Journal*, **27**, 321-351.

Penlidis, A., J.F. MacGregor, and A.E. Hamielec (1989). Continuous emulsion polymerization: Design and control of CSTR trains. *Chem. Eng. Sci.*, **44**, 273-281.

Perkins, J.D. and M.P.F. Wong (1985). Assessing controllability of chemical plants. *Chem. Eng. Res. Des.*, **63**, 358-362.

Perkins, J.D. and S.P.K. Walsh (1994). Optimization as a tool in design/control integration. In E. Zafiriou (ed) *Integration of process design and control*. Pergamon, Oxford, 1-10.

Psarris, P. and C.A. Floudas (1990). Improving dynamic operability in MIMO systems with time delays. *Chem. Eng. Sci.*, **45**, 3505-3524.

Psarris, P. and C.A. Floudas (1991a). Dynamic operability of MIMO systems with time delays and transmission zeroes. I: Assessment. *Chem. Eng. Sci.*, **46**, 2691-2707.

Psarris, P. and C.A. Floudas (1991b). Dynamic operability of MIMO systems with time delays and transmission zeroes. II: Enhancement. *Chem. Eng. Sci.*, **46**, 2709-2728.

Ray, W.H. (1992). Progress in pollution prevention through precision process modelling, design, and operation. In *Proceedings of EFChE Conference on Precision Process Technology — Perspectives for Pollution Prevention*, Delft, Netherlands.

Rinard, I.H. (1975). Process control of highly integrated processes. In *National Science Foundation Conference on Innovative Design Techniques for Energy Efficient Processes*, Northwestern University, Evanston, Illinois.

Roat, S.D., J.J. Downs, E.F. Vogel, and J.E. Doss (1986). The integration of rigorous dynamic modelling and control system synthesis for distillation columns: An industrial approach. In M. Morari and T.J. McAvoy, (eds), *Proceedings of the Third Int. Conf. on Chemical Process Control — CPC III*, Asilomar, California. CACHE-Elsevier.

Rovaglio, M. and M.F. Doherty (1990). Dynamics of heterogeneous azeotropic distillation columns. *AIChE Journal*, **36**, 39-52.

Rovaglio, M., T. Faravelli, and M.F. Doherty (1991). Operability and control of azeotropic heterogeneous distillation sequences. In *PSE'91 4th International Symposium on Process Systems Engineering*, Montebello, Quebec, Canada, II.17.1-II.17.15.

Seider, W.D., D.D. Brengel, A.M. Provost, and S. Widagdo (1990). Nonlinear analysis in process design. Why overdesign to avoid complex nonlinearities? *Ind. Eng. Chem. Res.*, **29**, 805-818.

Seider, W.D., D.D. Brengel, and S. Widagdo (1991). Nonlinear analysis in process design. *AIChE Journal*, **37**, 1-38.

Shinnar, R. (1981). Chemical reactor modelling for purposes of controller design. *Chem. Eng. Comm.*, **9**, 73-99.

Shinnar, R., F.J. Doyle III, H.M. Budman, and M. Morari (1992). Design considerations for tubular reactors with highly exothermic reactions. *AIChE Journal*, **38**, 1729-1743.

Shinskey, F.G. (1983). Uncontrollable processes and what to do about them. *Hydrocarbon Process*, 62.

Skogestad, S., M. Hovd, and P. Lundstrom (1991). Towards integrating design and control: Use of frequency-dependent tools for controllability analysis. In *PSE '91 4th International Symposium on Process Systems Engineering*,

Montebello, Quebec, Canada, pages III.3.1-III.3.15.

Skogestad, S., E.W. Jacobsen, and M. Morari (1990a). Inadequacy of steady-state analysis for feedback control: Distillate bottom control of distillation columns. *Ind. Eng. Chem. Research*, **29**, 2339-2346.

Skogestad, S., P. Lundstrom, and E.W. Jacobsen (1990b). Selecting the best distillation control configuration. *AIChE Journal*, **36**, 753-764.

Skogestad, S. and M. Morari (1987a). Control configurations for distillation columns. *AIChE Journal*, **33**, 1620-1635.

Skogestad, S. and M. Morari (1987b). Design of resilient processing systems — IX. effect of model uncertainty on dynamic resilience. *Chem. Eng. Sci.*, **42**, 1765-1780.

Skogestad, S. and M. Morari (1988). LV-control of a high-purity distillation column. *Chem. Eng. Sci.*, **43**, 33-48.

Skogestad, S., M. Morari, and J.C. Doyle (1988). Robust control of ill-conditioned plants: High purity distillation. *IEEE Trans. Auto. Control*, **33**, 1092-1105.

Stanley, G., M. Marino-Galarraga, and T.J. McAvoy (1985). Shortcut operability analysis. 1. the relative disturbance gain. *Ind. Eng. Chem. Process Des. Dev.*, **24**, 1181-1188.

Teymour, F. and W.H. Ray (1989). The dynamic behaviour of continuous solution polymerization reactors — IV. dynamic stability and bifurcation analysis of an experimental reactor. *Chem. Eng. Sci.*, **44**, 1967-1982.

Uppal, A., W.H. Ray, and A.B. Poore (1974). On the dynamic behaviour of continuous stirred tanks. *Chem. Eng. Sci.*, **29**, 967-985.

Viswanathan, J. and I.E. Grossmann (1990). A combined penalty function and outer approximation method for MINLP optimization. *Comp. Chem. Eng.*, **14**, 769-782.

Walsh, S.P.K. and J.D. Perkins (1992). Integrated design of effluent treatment systems. In J.D. Perkins (ed) *Interactions between process design and process control*. Pergamon, Oxford, 107-112.

Walsh, S.P.K. and J.D. Perkins (1994). Integrated design of waste water neutralization systems. In *ESCAPE 4 — 4th European Symposium on Computer-aided Process Engineering*. IChemE Publications, Rugby, 135-141.

Wilson, K.B. (1946). Calculation and analysis of longitudinal temperature gradients in tubular reactors. *Trans. Inst. Chem. Engrs.*, **24**, 77-83.

Winter, P. (1992). Computer-aided process engineering: The evolution continues. *Chem. Eng. Progress*, **88**, 76-83.

Wolff, E.A., K.W. Mathisen, and S. Skogestad (1991). Dynamics and controllability of heat exchanger networks. In *COPE-91*, Barcelona, Spain.

Ziegler, J.G. and N.B. Nichols (1943). Process lags in automatic-control circuits. *Transactions of the ASME*, **65**, 433-444.

DESIGN FOR CONTROL AND OPERABILITY: AN INDUSTRIAL PERSPECTIVE

James J. Downs
Eastman Chemical Company
Kingsport, TN 37662

Babatunde Ogunnaike
E. I. DuPont de Nemours & Co., Inc.
Experimental Station
Wilmington, DE 19880

Abstract

Over the last ten years issues centered on product quality and its relationship to customer demand have become increasingly important. It is no longer sufficient to focus only on plant designs that maximize raw material yield and minimize utility usage. Product quality is being measured as product variability and is quickly becoming a discriminator among chemical suppliers. Much of the current improvement work on existing processes is to reduce process and product variability. Achievements in this area have tied directly to expanded customer bases, higher plant utilization due to higher sales, and more consistent operation at optimum process conditions. Plant designs that take into account the issues of process and product variability are capable of yielding products that appear identical from shipment to shipment. This new focus on product variability has intensified the need to design processes that have good operability characteristics, are easy to operate, and result in low product variability. Current industrial practice to arrive at plant designs that include operability and variability notions is described from several viewpoints. These include (1) a political/cultural viewpoint, (2) recurring technical issues, (3) tools and techniques in use, and (4) future needs. There are many issues that are in need of work and innovation to achieve plant designs that weave in operability considerations. Those manufacturers building plants that can provide products with low variability will continue to capture an increasing share of the market and will be the preferred suppliers.

Keywords

Process operability, Process control, Process variability, Control strategy design, Industrial design practice, Chemical plant operation.

Introduction

The quality revolution that started in the consumer products industries has now reached several layers into the supply chain. The demand for increased consistency in the properties of products supplied by the chemical industry has led to decreased sales and lower prices for those companies not conforming to customer demands. In the past, companies could use additional energy to over-purify or rework out-of-specification material. Now companies can no longer rely on making products better than specification in purity alone to achieve sales volumes. The market now demands that products meet criteria related to product variability in addition to purity specifications. Today low quality is synonymous with high variability and the demand for product consistency is only going to become greater. One of the central issues in the manufacture of chemical products is that they conform to customer variability expectations.

As a result of the increasing importance of product variability, the ability of plants to exhibit stable, low variability operation will become a major factor in discriminating between competing designs. The design of the process control strategy, which in turn is dependent upon the process design itself, can modify the propagation of variability through a process. Often the design of the control strategy based on attenuating variability a certain way or redirecting variability may require that certain variables are available for manipulation. As pointed out by Fonyo (1994), controllability is enhanced — often without additional control structure complexity — by design modifications that create new design variables (i.e. those that increase the degrees of freedom). By including such flexibility at the design phase one can often avoid costly changes needed later to reduce product variability.

The design of plants with good variability propagation properties involves not only the equipment design and the control strategy design but also the integration of these two functions. This has caused the relationship between process design and control strategy design to change as the need for better integration has increased. There is movement among some companies to move their process design and control functions closer together both organizationally and geographically. (Sheffield, 1992; Downs and Doss, 1991).

Process operability is closely tied to the process control system not only through the variability issue but also through ease of operation of a process. Companies are beginning to see the necessity of designing process operability and ease of operation into a process instead of adding it later via an automatic control system. Designing for operability requires an operations perspective and a dynamic perspective to anticipate the disturbances, upsets, rate changes, etc., that the plant will experience in its day to day operation.

Process operability involves the coordinated effort of both the operator and the automatic control system. Plant operation has become less dependent on the operator and more dependent on the automatic systems in place. Plant operators are becoming more skilled and are used increasingly to handle abnormal situations as opposed to routine operation. There is significant effort in the chemical industry to standardize process operation to implement quality systems (ISO 9000 registration) and to address legal requirements of OSHA 1910. As a result, there are fewer operators dealing with the minute-to-minute operation and the need for processes to be easy to operate and to require minimal operator intervention is becoming greater. The requirement for human intervention usually results in different operators correcting problems in different ways, leading to different product attributes and higher product variability. Training and well documented procedures can mitigate the situation somewhat, but a plant requiring continual operator intervention usually has high process variability and is described as having poor operability characteristics. The product can be made — just not the same from day to day. Plants with good operability characteristics require little operator intervention.

Prior Work

The effect of process design on operability and controllability has always concerned industrial practitioners, perhaps more than is realized outside industrial circles. What is generally agreed on as the earliest published account of an actual industrial process controllability and operability problem arising from poor design is contained in Anderson (1966). This paper discusses how a poor design for the feed-effluent heat exchanger system prevented the process from being operable at high throughput rates. Regardless of control system design, stable process operation could not be achieved until the process was redesigned.

General issues concerning the implications of design on control and operability appeared even earlier than the Anderson paper, in Ziegler and Nichols (1943), a paper that is remarkable for its timeliness 50 years after its publication. Many of the issues that are now starting to gain currency were anticipated and presented in this 1943 paper. Morari (1992) provides a readable, succinct, but panoramic overview, which includes Anderson's pioneering application, Ziegler and Nichols' visionary concepts, as well as what has been done since. The conclusion in this survey paper is that while the open literature attests to the fact that much progress has been made, it is also clear that more work needs to be done. There is yet no systematic procedure (or theory) for assessing controllability objectively, or for assessing the tradeoffs between controllability and economics.

With the possible exception of Shinskey (1983) and Downs (1991, 1994) for example, the open literature contains little regarding industrial practice of process design for control and operability. In what follows, we shall attempt to discuss, and put in perspective, some of the important factors that currently influence industrial practice in this area.

Political and Cultural Issues

Operability Perspective

Most chemical engineering curricula focus the majority of time for the chemical engineering courses on design issues. The basics of how to apply heat and material balances to design problems, followed by the development of design equations, are typical fare. The design procedures are then enhanced to cover various equipment configurations and design objectives. The economic tradeoffs between operating and capital costs are then introduced. A process design course typically ties together the unit operation design and economics issues. Unit operations laboratories can range from bench scale operation of an experiment to computer simulation of a unit operation to the operation of pilot scale equipment. These unit operations courses are typically a small part of the entire curriculum. Process control courses typically focus on the dynamics and design of single-input single-output control loops.

As a result of the formal training that engineers receive, many enter the industrial workforce with strong backgrounds in design and evaluation and relatively weak backgrounds in plant operation. The perspectives of an engineer who is responsible for the safe, legal and productive operation of a chemical process are quite different from those of a design engineer. The non-technical issues that must be dealt with during the operation of a plant often dwarf the technical nuances that may consume hours of debate and study at the design phase. Often the results of hours of simulation and discussion needed to optimize a particular piece of a process are overridden once it is realized that an operations issue precludes the need for close design.

The design of chemical plants that combines the technical ingenuity needed to achieve minimum costs with the practical savvy that makes a plant pleasant to operate is the challenge. Process optimization that leads to increased process integration can lead to processes that are difficult to operate. Bringing to bear design and optimization technology on the big picture where large dollars are at stake is certainly paramount. The key is knowing where to draw the line between the benefits arising from process integration and the costs of diminished process operability that may result. Plants that are difficult to operate have long term costs that usually far exceed the marginal capital that may be saved on the front end. Conversely, processes that are designed to be easy to operate show not only expected profits and earnings but also are much more amenable to process improvement later on. Unless an operational perspective is applied early in the life of a project, plant designs may overlook important operational needs. Issues such as tank location and size, captive component inventory control strategies, reactor cooling strategies for hazardous situations, etc., can become established and be very difficult to alter later in the project. Usually a small amount of operational perspective can avert expensive revisions down the road.

The movement of the chemical industry toward globalization has prompted the design community to ask where to draw the line on process and operational complexity versus simpler and perhaps costlier designs. The trend toward globalization is expected to continue. The assessment of plant operability and plant maintainability in third world countries is critical to their success. The process design team needs to be aware of the expected qualifications of the operators, the capabilities of the control system hardware, and the propensity for future improvements by plant engineers. Knowing where a plant will reside and who will be operating it should influence design decisions.

Currently many methods are used to include the operability perspective into process designs. The inclusion of this perspective is certainly dependent on organizational structure, experience level of the design group, size of the project, and resources available. However, the important point is that someone who has an operational viewpoint needs to be involved early in the project. In a world where designers may never have to live with their design decisions day to day and may only be rewarded based on the minimization of initial capital outlay, there is great need to instill an operational focus in the design process.

Control Perspective

In the past, it was common to assume that suitable operation was attainable using control strategies developed after a process was designed. Chin (1979) provides a simple illustration of the shortcoming of this approach. He gives a description of 21 different equipment designs for distillation pressure control alone, all of which have significantly different operational characteristics. Such examples of the effect that design has on the operation and control provide incentive for careful consideration of control strategies during the design phase instead of relying on control strategies to handle whatever the design demands.

Despite the ever-increasing incentive to integrate the design and control functions, segregation of the process design and control tasks is still common. Two factors contributing to this segregation are: (1) the difficulty of changing from the historical approach of completing the process design before the control engineer becomes involved, and (2) the difference in the thought pattern of design and control engineers. The common notion is that the steady-state process design alone determines the process economics. While the nominal steady-state design point is very important, it loses its distinction if one is unable to maintain plant operation at the design point. Design decisions are often based on steady-state analysis without consideration of controllability, process and product variability, or plant-wide control issues. The basic thought pattern in the design stage usually follows the form, "Given these conditions, create a design to perform this function," (design question) as opposed to, "Given this design, how well will it perform its intended function?" (rating question).

The costs of installing additional equipment and altering process designs in the name of process operability or controllability improvement often come under great scrutiny. The forces present in the capital arena today push to minimize the up-front capital cost. The ability to assess the risk of design decisions is very important. One may envision a plant design with no intra-process inventory and no equipment overdesign over that implied by standard factors for uncertainty in physical properties and design correlations. Decisions based on steady-state considerations and standard overdesign factors may be misleading, because operability may be contingent upon additional overdesign. For example, even though the expected steady-state flow to a distillation column may not exceed 120% of design conditions, expected transients may require reboiler and condenser capacities that are 140% of design conditions. The tradeoffs between capital cost and operability considerations such as overdesign requirements must be rooted in a working knowledge of process dynamics.

As discussed, maintaining acceptable variability is an operability issue of increasing prominence. The management of process variability may require the addition of degrees of freedom or additional instrumentation and increased control structure complexity; therefore, there can be design implications and an initial capital cost associated with variability reduction. This cost must be weighed against the penalty for high variability and the potential savings afforded by built-in flexibility. The role of the automatic control system and its ability to diminish variability problems needs to be a part of the cost/benefit analysis.

The control perspective on process variability becomes more important when considering global plant sites. Customers demand products that are indistinguishable from shipment to shipment and from plant site to plant site. Such product consistency allows flexibility in controlling worldwide inventory and maximizing production at the efficient plant sites. This capability requires the plant designs to be resilient to the disturbances of the locality and feed-stock. Plant designs must also allow control strategies to be effective in maintaining consistent operation. The additional capital needed to insure smooth, consistent operation must be balanced by the marketing and management issues that will surely arise with customer complaints and lost sales.

Technical Issues

There are several recurring themes that arise during the process design phase that involve the tradeoff between cost of design and the cost of operability and control. Because the tools needed to quantify the benefits of improved operability are still under development, many technical issues are decided based on qualitative reasons. While these issues are inclined toward a technical solution, current general practice has not embraced any one procedure to address them.

Self-regulation vs. Active Control

One of the issues that most needs collaboration between design and control engineers is the decision to design for self-regulating operation. Often designs that self-regulate are easy to operate but they can limit the range of operation and the potential for optimizing operations. The capital costs of self-regulating designs are usually higher as well. Self-regulating designs affect the extent to which active control is necessary, which is tightly linked to plant reliability. Designs that rely on active control systems also rely on the underlying measurements and final control elements. Depending on the criticality of the automatic control system, plants can become almost inoperable if a measurement or final control element is unreliable.

Take for example a series of four continuous stirred tank reactors. If level measurements are easily obtained then the level control strategy is straightforward and many options are available. However, if the level measurements are anticipated to be difficult or unreliable, then the idea of a self-regulating control strategy is attractive. The design of a self-regulating control strategy may take the form of simply putting a weir into each reactor and allowing the contents to overflow to the next reactor. This option certainly has an impact on plant layout and structural requirements needed to arrange the equipment suitably. It may also limit the flexibility of changing reaction volume. The designer needs to study the consequences of limiting process conditions such as reaction volume to determine if the limitation is too restrictive based on design uncertainty. The process control engineer needs to assess the restrictions an overflow design will place on the overall plant control strategy. The need for this type of issue to be addressed early in a project in a collaborative way is essential because attempting to rework layout or design later may be too costly.

Process Measurements

As mentioned above, designing for self-regulation is influenced by the availability of measurements to establish reliably the state of the system. The number and types of measurements needed to know the state of a process is certainly a question that needs early attention. This is particularly important when the measurements anticipated are composition analyzers or product property measurements that are novel or under development. Plant designs that rely on the availability of such measurements may become operation nightmares when the measurements prove to be unreliable. Process designs that can tolerate the absence of information or the delays that occur when off-line sampling is necessary may justify the additional up front capital costs.

The anticipation of disturbances that the plant will experience is another key issue in the selection of measurement options. The measurement system must be able to detect the effect of disturbances entering the process to handle them effectively. Often the design of the process can make detection of specific disturbances easy or difficult. If the disturbance is one that can severely compromise the operation or the quality of the product, it may be justified to alter the process design and incur additional costs to make detection and control easier. Simulation can be used to evaluate the effectiveness of proposed measurements to detect disturbances in a timely fashion.

A common example is the use of temperature measurements for distillation control. When analytical composition measurement is difficult or impossible, it is necessary to infer product compositions using column temperature information. A tray temperature may be a sufficiently accurate indicator of product composition under normal circumstances. However, when a disturbance in the amount of a third component enters the column, the temperature used for control may no longer be appropriate. Strategies such as temperature differencing to detect feed composition changes may be needed. Adding temperature nozzles to the column specification at the design stage to handle this case is trivial — adding them to a fabricated and installed pressure vessel that had to have each weld x-rayed in the shop is another matter.

Manipulated Variables

Manipulated variables available for control can be classified into four main types: (1) the variable used to set the process throughput, (2) those variables used for inventory control, (3) those variables that affect product properties, and (4) the remaining variables that affect process costs. Each of the manipulated variable types has a relationship to the plant design; however, the decision of how and where to set the process throughput is usually the most important. This one decision determines much of the remaining inventory control decisions, which in turn determine many of the decisions concerning variables available for product quality and cost control.

Consider a reactor/separator/recycle process. Changing the process production rate requires changing the reaction rate. Changing the reaction rate can be done normally by changing the reactor temperature, reaction volume or reactant concentration. The choice of what to manipulate for control of the process production rate will have a profound effect on the remainder of the plant-wide control strategy. This in turn affects the variability propagation paths through the process as well as equipment decisions such as surge vessel location and size.

Another example that has significant impact on manipulated variables is combining unit operations into one vessel or using common utility equipment. This can result in measurements and capabilities for manipulations of intermediate streams being unavailable. A careful analysis of the cost savings is justified when the implications on the operation and control of the process are considered.

Advanced Control

The advent of multivariable and advanced control algorithms has provided the capability to retrofit and upgrade existing control strategies to squeeze out additional throughput and reduce costs. For the design of new processes the issue of when to base new designs on the capabilities of advanced control and the establishment of how much risk to take has not yet come to the forefront. It seems plausible that the design process could evolve to the point that cost-saving design decisions result in plants that are nearly uncontrollable by conventional means.

A familiar example is the naturally unstable, exothermic reactor. There is significant latitude in the reactor design. The design can be such that dynamics are slowed to the point that conventional or even manual control by an operator is easy and relatively straightforward. On the other hand, lower cost designs may require significant sophistication of the control system for the process to work.

As the success of advanced control applications become more widespread it is likely that the desire to reduce capital cost based on advanced control capabilities will increase. Understanding what advanced control can and cannot do is necessary at the design stage to incorporate this technology effectively.

Tools and Techniques Currently Used

Eastman Practice

When new processes are developed at Eastman, they may originate from a variety of places within the company. However, for all major new plants the Process Engineering Department of the Engineering and Construction Division generates a process design package. This package consists of process flow diagrams, the heat and material balance used as the design basis, process control strategy, safety interlock strategy, major equipment specifications, equipment layout, environmental and regulatory specifications, etc. This package is then used as the basis for the detailed design and construction of the plant. The process concepts for the design package may originate from our research organization, a process operating division, or may be licensed from another company. A team is formed to take the basic process concepts and to develop them into a process design package. This team consists of several engineering disciplines, including process design engineers and process control engineers. The team works with the originator of the process concepts to develop the process and complete the design package deliverables.

The organizational and geographical structure of the process design and the process control functions are arranged to promote a high degree of interaction. This interaction is not only formally set up on projects but is also nurtured and encouraged on other types of work. This arrangement has helped to generate a cooperative and synergistic atmosphere among the process design and process control functions. The process control engineers are chemical engineers, most of whom have a process control educational background.

The process development function is highly dependent on the interaction and discussion that takes place among members of the design team. As a flowsheet begins to coalesce, discussions among team members elicit strengths and weaknesses of particular designs. One of many perspectives that influences the direction a design can take is operability and control. The process control engineers study the design from an operating and process control viewpoint. Issues such as operating versus capital costs, safety considerations, environmental costs, etc., are studied from a steady state point of view.

Control strategy development requires an understanding of the primary control objectives and the primary process disturbances. As flowsheet options are considered, control strategies for overall inventory control and for setting of process production rate are developed. Discussion among process and control designers about the merits, liabilities, and risks associated with various design options characterizes this phase of work. Out of these discussions an understanding is gained of which parts of the process are routine and which parts deserve more detailed analysis and innovation. The key element is the discussion of how a certain design will perform. Once the pertinent issues are discovered, the design modifications needed to address them are usually straightforward. For those parts of the process that merit further design work due to operability concerns, the new design options and control strategies are developed together.

As the need to incorporate design and control has grown, two concepts have emerged as recurring themes: (1) understanding the overall component inventory control issues and (2) design of plants so that simple controls (e.g. SISO PID) work well and hold the process at desired conditions.

Component Inventory Control. Understanding how the inventories of each component in a process are controlled is a key to overall plant-wide control. The concept of self-regulating and integrating component inventories has proven useful in the analysis of plant-wide control strategies. The unit operations that separate components, such as distillation columns, play a central role in the control of component inventories. The choice of how to control each unit operation in a process associates closely with the overall control of component inventories. The control strategy for each unit operation must be developed within the framework of the overall component inventory control strategy. Each component, whether important or insignificant, must have a mechanism by which its inventory is controlled within each unit operation and within the process as a whole.

The control of component inventories may occur by process self-regulation or by manipulating process flows or conditions. For example, many processes have innumerable trace components that "recycle to extinction." That is, they are ever present in the process but have no obvious source or sink. Mechanisms such as reaction equilibrium or concentration-dependent irreversible reactions set the rate of formation and of consumption of these components equal. For components that occur in significant quantities, the control strategy for the unit operations must take into account how the inventory of each of them will be maintained.

The conscious effort to define the inventory control strategy of each component has led to significant process design innovations. Process designs that inadvertently trap components in a process section or that accumulate components where their inventory measurement would be difficult are sources of operability problems. The design for self-regulation of component inventories may prove to be the best option in some cases, whereas automatic control may be the best option in other cases. Downs (1992) gives several examples of using component inventory control to analyze process control strategies.

Design for Control Performance. Designing a process so that it requires only simple PI single-input single-output controllers has provided another avenue to create innovative process designs that have good operability characteristics. The notion is to conceive process designs that do not need any control algorithm beyond PI, that do not need measurements that are difficult or impossible to obtain, and that are inherently safe. This viewpoint promotes designs that incorporate self-regulation for process variables that would be difficult to measure or difficult to control using conventional hardware. It also suggests eliminating obvious control problems such as deadtime and inverse response. Plants that have a good first level control strategy are amenable to the application of higher level optimizing control systems once the plant is operating.

An example of this approach would be equipment design for linearity between the manipulated and controlled variable. This could be as simple as designing an overflow weir that has a linear increase in flow for a linear increase in level, or the orientation of a reflux drum vertically rather than horizontally to get a linear inventory measurement. The use of slave control loops to linearize the process response for a master loop is another common procedure. If particular disturbances have the effect of altering the process gain or process dynamics that a control loop will encounter, then the search for designs for which that is not the case may lead to superior design alternatives.

This concept is used routinely when designing for inherent safety. One can ask, "Is there a design that results in a safe plant when a given set of events occur?" Certainly for processes that contain significant safety hazards, the concept of designing so that it is impossible to reach those hazardous conditions is often preferable to the automated safety systems to handle the event. An example might be the use of a self-limiting heat source for a process that becomes unstable above a given temperature. By designing so that the heat source will pinch off as the temperature increases and in fact will become zero if the temperature goes above the supply temperature, self-regulation can eliminate the reliance on more exotic and possibly less reliable means.

DuPont Practice

The basic framework used for evolving new process designs at DuPont is essentially similar to those enumerated above for Eastman, even though specific implementation details may differ. For example, the spirit of Page Buckley still presides over the development of plant-wide control strategies at DuPont so that current practice necessarily reflects this particular "cultural" perspective. The principles employed have not deviated much from those first enunciated by Buckley (1964) and later refined over his many years of successful practice. In the same vein, regarding the use of simulation, the tendency at DuPont is to employ a specific home-grown, object-oriented, deliberately uncomplicated, modeling package for complete dynamic flowsheet simulation; the more intricate SPEEDUP package is employed for dynamic simulation of smaller subsystems. Each simulation tool is used to answer different questions at different stages of the design.

One concept that has recently been gaining currency at DuPont is the Taguchi technique for "off-line quality control". It is based on the concept that a process is easier to maintain "on-aim" if it is designed from the beginning to be intrinsically robust to the effect of disturbances. The technique, which employs statistical tools, is called "off-line quality control" in direct contrast to the "on-line quality control" techniques: the latter are concerned with holding the process "on-aim" during on-line operation while Taguchi's technique is concerned with making sure, off-line, that the variability to be dealt with once the process begins to operate on-line has been reduced, *by design*, as much as possible.

As discussed more fully in Ross (1988), there are three distinct phases in implementing the Taguchi technique:

1. System design: where new concepts and ideas are used to provide new and/or improved products or processes; the result is an overall process with specific parameters yet to be fully specified;
2. Parameter design: where the "free" process parameters are chosen to minimize the variation in the product quality variables; the objective function of this optimization problem is therefore specified explicitly in terms of the "cost of variation in product quality;"
3. Tolerance design: where the range of controlling factors are narrowed in order to achieve higher levels of quality than achievable by the "passive" parameter design.

In standard process control/process design parlance, "System design" is essentially equivalent to the selection of good plant-wide control structures, while "Tolerance design" is related to the selection of sensors and actuators as well as special high performance control algorithms to improve the overall performance of the control system. "Parameter design" on the other hand involves the selection of "free" design variables (for example, a reactor temperature or a reactor level) which may not be used directly for control but whose operating level may have significant impact on the sensitivity of the product quality to disturbances.

Identifying which process variables qualify for consideration during the "parameter design" stage, and the statistical procedure for determining their optimum settings is central to the Taguchi technique. More details are available, for example in Ross (1988); and a discussion of some applications are given in Wong and Newman (1991).

Future Needs

Education

To establish the knowledge and skill needed to understand and foresee operability problems, it is important for our control groups to gain experience operating process equipment, working with production departments to understand and appreciate non-technical production issues, and to gain the confidence of production personnel. It is also important for the control designer to have had experience designing process equipment. The knowledge regarding the latitude a designer has is important in understanding the possibilities available to avoid operability problems and their costs.

Because of the importance of the non-technical issues of design decisions, it is important for the process designer to be trained to understand the process life after the design. Once plants are started up there is a period of operational maturity. During this time knowledge is gained about the operating characteristics and improvement opportunities for the process. Processes that are too tightly integrated may not allow for the improvements to be made. The process designer also needs to understand the variety of long term expectations for new plant designs. For example, while some plants are designed for a specific product and are optimized for a particular operating point, others may need to be designed for a particular product with the expectation that many similar products will be campaigned on the equipment. Similarly, some processes will be staffed with technical crews that can see to the care and feeding of complex control or operating systems while others may be expected to work with a skeleton technical staff. Without the understanding of these issues, the best technical design may be misapplied.

There is a need for process designers and process control engineers to have a better understanding of process operability and how to operate process equipment. This understanding allows the answering of questions such as:

- What is meant by operability?
- What characteristics make process equipment difficult to operate?
- How does one startup process equipment such as a distillation column, an exothermic reactor, or a continuous crystallizer?
- If a process is not producing what it should, what action needs to be taken?

The skill in taking process knowledge, simulation tools and situational awareness and then developing the best design requires an operability perspective along with the technical skill.

Propagation of Variation Analysis

As stated previously, the design of processes that manage variability so that product streams are as immune to process upsets as possible is becoming more of a design criteria. Customers are beginning to request data on process capability indices (Oakland, 1986) to assess a supplier's ability to deliver consistent products. Process capability indices are a measure of the variability of a process relative to the product specification limits.

The cases where product consistency is critical have traditionally been addressed by blending techniques. In the future when product inventory and working capital need to be reduced to remain competitive, other, more cost-effective means will become necessary. Furthermore, the product attributes that define product quality are frequently non-additive and are not handled effectively by simple blending operations. Often one supplier is desired because of an impurity that is contained in the product. The effect of the impurity or even what the component is may not be understood. There is simply the request to make "exactly what you made yesterday" down to the last trace component.

These product characteristics demand that the process operation remain at nominal operating conditions throughout the process. It is beneficial to distribute the process variability among the tolerant unit operations. There are two variability issues that need to be addressed. The first issue is the distribution of variability among unit operations so that process indicators appear as steady state plus noise type signals. The second issue is the reduction of the noise component. This second issue influences process capability and is often used as a measure of process improvement efforts. As noted above, improvements in a processes capability index can have marketing impact.

Analysis tools that can capture some of these issues are principal component analysis (PCA) and partial least squares (PLS). Over the past 15 years, PCA has been adopted and applied to process data to perform multivariate process analyses. PCA was first popularized in the field of psychometry, due to the need to analyze data sets with many correlated variables and large amounts of noise and uncertainty. The development of PCA is one activity in the growing field of chemometrics, the science of relating measurements made on a chemical system to the state of the system via application of mathematical or statistical methods.

PCA takes a single set of data and characterizes it in terms of its "principal components" — the axes in the data space along which the data vary most widely. This decomposition, closely related to singular value decomposition, has been used to indicate variability in the process that is consistent with a base data set which is representative of normal operation. Indicators from the decomposition then can be monitored and used to make efficient use of the whole data set for diagnosis of process conditions and fault detection.

PLS is used to describe relationships when two sets of process data are available: a set **x**, whose values are routine process measurements, and a set **y**, whose value we wish to infer or predict using **x**. PLS takes into account variability in the **x** data and the **y** data simultaneously. This feature has led to its use in multivariate statistical process control and monitoring. MacGregor, et al. (1991) discuss PCA and PLS and describe their use for process monitoring and analysis.

The need in this area is to combine the proven data analysis techniques of PCA and PLS with the process design environment to analyze variability propagation behavior and monitoring at the design stage. These tools may provide an understanding of process movements by reducing the dimensionality of the problem and more clearly delineating the directionality of the problem. As the use of process monitoring tools becomes more widespread, the need grows to link the design of the process to the design of an on-line tool that indicates whether or not the process is performing as the designer intended.

Conclusions

The need for closer ties among the people who design processes, who design control systems for those processes, and who operate them is obvious. There is increasing importance of product quality as measured by product variability. The influence of product quality and customer expectations on the perspective that the control and operating personnel bring to the process design has been presented. Several of the more important technical issues that couple design, control, and operability have illustrated the overlap among these areas. A description of current practice at Eastman and DuPont illustrates the methods, thought patterns, and technologies used to develop process designs. The lack of quantitative measures to describe operability and controllability has resulted in an experienced-based approach to include these factors in the design function. Finally, two areas are discussed where it is believed significant benefits are realizable. One is the education of graduating engineers and experienced design engineers on the importance of control and operability on the design function. The other is the linkage of technology that can be to used to manage process variability with the process design function. The design of plants that are both economical and that produce products that meet customer's quality demands is critical to a company's long term success.

References

Anderson, J.S. (1966). A practical problem in dynamic heat transfer. *Chemical Engineer*, May, CE97-CE103.

Buckley, P.S. (1964). *Techniques of Process Control*, John Wiley and Sons, New York, NY.

Chin, T.G. (1979). Guide to distillation pressure control methods. *Hydrocarbon Processing*, **58**, 145-153.

Downs, J.J. (1992). Distillation control in a plantwide control environment. In W.L. Luyben (Ed.), *Practical Distillation Control*, Van Nostrand Reinhold, New York, NY, 413-439.

Downs, J.J. and J.E. Doss (1991). Present status and future needs — a view from north American industry. In Y. Arkun and W. H. Ray (Ed.), *Chemical Process Control CPC IV*. CACHE, Austin, TX and AIChE, New York, NY,

53-77.

Downs, J.J., A. C. Hiester, S.M. Miller, and K.B. Yount (1994). Industrial Viewpoint on Design/Control Tradeoffs. Presented at IFAC Workshop on Integration of Design and Control, Baltimore, MD, June 27-28.

Fonyo, Z. (1994). Design modifications and proper plantwide control. *Computers Chem. Engr.*, **18**, Suppl., S483-S492.

MacGregor, J.F., T.E. Marlin, J. Kresta, and B. Skagerberg (1991). Multivariate statistical methods in process analysis and control. In Y. Arkun and W. H. Ray (Ed.), *Chemical Process Control CPC IV*. CACHE, Austin, TX and AIChE, New York, NY, 79-99.

Morari, M. (1992). Effect of Design on the Controllability of Chemical Plants. Presented at IFAC Workshop on Interactions between Process Design and Process Control, London, September 6-8.

Oakland, J.S. (1986). *Statistical Process Control*, John Wiley and Sons, New York, NY.

Ross, P.J. (1988). *Taguchi Method for Quality Engineering*, McGraw-Hill, New York, NY.

Sheffield, R.E. (1992). An integrated approach to process and control system design. Presented at AIChE Spring National Meeting, New Orleans, LA.

Shinskey, F.G. (1983). Uncontrollable processes and what to do about them. *Hydrocarbon Processing*, **62**, 179-182.

Wong, H.H. and S. Newman (1991). Quality control off-line. *ChemTech*, July.

Ziegler, J.G. and N.B. Nichols (1943). Process lags in automatic-control circuits. *Transactions of the ASME*, **65**, 433-444.

PROCESS ENGINEERING SOFTWARE TOOLS AND ENVIRONMENTS

Jack W. Ponton
Department of Chemical Engineering
University of Edinburgh
The King's Buildings
Edinburgh EH9 3JL, Scotland

Introduction

In most of the sessions in this conference our concerns have been with the development of techniques which exploit the science of process engineering. For example, computer and mathematical techniques which help us to understand and control azeotropic distillation systems (Malone and Doherty 1994; Siirola 1994). Some of these *techniques* have been developed into *tools*, but for most of those people involved in the associated research, the "turning it into a tool" issues and activities have been regarded as a formally straightforward, and even boring, mechanical task.

Experience has shown that this last assumption is quite incorrect; issues of implementation, robustness and "usability" for sophisticated techniques turned into complex software are nontrivial. One of the topics in this session is therefore *Software Tools*. The other topic is *Software Environments*. What do we mean by an "environment" in this context, and how does it differ from a tool? Let me define these terms as I shall use them here:

- A **Tool** is a piece of software that does something useful.
- The **environment** is all the other software.

Hitherto, process engineers working in this general area, i.e. the turning of concepts into software, have mostly been concerned with developing tools. The environment in which these tools were used, we have in most cases accepted as provided by the computer vendor. It was usually the operating system of the computer. Why should we be concerned with our software environment? An obvious technical reason is that its what we see as users, and what the tools see, as shown in the figure.

The Tools here are shown as hammers, the trees are the Environment, and the User is looking confused! More generally, the reason for our concern for our software environment is the same as that for our concern about our "real" environment; because we have to live and work in it. A process engineer, confused by a multiplicity of tools in his or her environment.

Problems that have arisen with our software environment are similar to those of our other environment. We have too many tools, and they are all pouring out information. There are some particular problems which software developers thus have to address. The first problem is the problem of *information overload*. It is not just that the computer can generate information faster than it can be torn up and thrown away. This has been the case for a long time. However, designers have now such a wide range of tools available, all quick and fairly easy to use, that they *do* use them.

It is not the fault of the computer, nor the fault of the tools themselves, some of which produce quite concise and intelligible results, in the end it is *our* fault, we will keep using these tools to generate information. And this information is real and useful, it should not be thrown away just because there seems to be too much of it. It is important, because it is knowledge about the process that is being designed. The first problem is thus: how do we manage this information in order to

- Organize it, to get at it conveniently,
- Summarize it, so that we can understand it,
- Transfer it, to where we want to use it,
- Re-use it, the next time or place it is needed,
- Learn from it, to do better next time, and so on?

A second problem is also to do with this information explosion, and is also one of information management. The process engineering software tools we have today work, by and large, quite well. Once upon a time, process engineers were very pleased to have a program which would calculate a mass balance on a process. Now users of such programs expect them to model plant startup and probably tune the controls as well. Most particularly, users get annoyed when a programming system will not model a particular kind of reaction, type of thermodynamics, distillation column with a reactor in the reboiler etc., that happens to be required. And such a user response is quite reasonable. Given the hundreds of man years that have gone into building flowsheet modelling tools and models, the incremental effort required to add, say, a reaction to a distillation tray, should be extremely small; in practice, as those who have had this experience will know, it is not. The task should in principle be easy, because most of the information required to perform it already exists. A flowsheet model contains a huge amount of information and knowledge. However, most of this knowledge is not accessible, and is thus effectively lost to anyone who wants to build on it. This problem is a microcosm of the designer's problem; if the model had been designed correctly, the relevant information properly organized, all this knowledge could have been transferred and reused.

These two aspects of knowledge and information management, are addressed, in somewhat different ways, in all three papers. A third issue, which I believe is important, although not explicitly raised in any of the papers, is that of the aesthetics of our environment. Would we rather work here in Snowmass, Colorado or in the sort of places where most process engineers find themselves working? This is important in more than a cosmetic sense, since a good working environment can made the difference between being able to work well or not at all. The need for a good and appropriate environment provides justification for the involvement of the process systems community in this research. The reason why we as process engineers need to involve ourselves in the design of process software environments, is that if we leave this task to the suppliers of general purpose software, we are likely to end up working in the equivalent of one of the older and nastier chemical sites. To avoid this, we need to design our own environment, and make it look more like Snowmass.

The Papers

The three papers form a hierarchy. At the top is a literally global concept, the World Wide Web, which the authors, Robertson, Subramanian, Thomas and Westerberg, describe as a world wide information management system. On an only slightly smaller side they discuss the issues of computer information modelling and information management for the whole design process. The paper summarizes the state of the *n*-dim project which has been developing at Carnegie-Mellon over more than five years.

At the other end the hierarchy, Pantelides and Britt discuss design and simulation tools. Underlying most of what they say is the issue of the information and knowledge used in the model. For example, how the designer should express it, and how the same information can be used in different types of model. They address the specific question of how knowledge of the model can be incorporated in the model itself.

In the middle, the paper by Motard, Blaha, Book and Fielding is, appropriately, concerned with data interchange. Software environments have moved some way from requiring users to retype data from one program to another.

However, the generalization of information transfer over all process engineering activities is a complex and intractable one, but of crucial importance.

Further Questions

There area number of questions raised by these papers which I do not believe are fully answered. I would like to list and comment on these, to help concentrate the reader's mind on future issues, and to carry forward to the next FOCAPD meeting.

1. Can today's *tools* become tomorrow's *environments*?

Pantelides and Britt clearly believe that the tool which they describe is also an environment. This is not axiomatic, but if not, how do we maximize the reuse of the large investment in current tools?.

2. What is the significance of *history* in information management? We are concerned with the history:

- of the design,
- of the data and
- of the modelling process.

Only Robertson et al. address this problem explicitly, and they leave many questions unanswered. The state of any part of a design or model now is a function of its history. And how do we capture this history and exploit it?

- Should we represent the design or modelling *goals* as well as the design or model itself? And if so, how?

The need for this is, I believe, implied by Pantelides and Britt, although not discussed. The question might be rephrased as how to model the "why" as well as the "what" of design? Although not addressed directly by these papers, this issue is discussed elsewhere in the proceedings (Rene Banares-Alcantara et al. 1994). Are *standards*:

- desirable,
- possible,

and if so:

- who should define them, and
- how can they be changed?

Motard et al. imply positive answers to the first questions, and set about defining a standard. However, if a standard is not a good one, it might have been better if it had never been created. In any case, even the best standard will ultimately become out of date, so how might it be allowed to evolve gracefully?

References

Malone, M.F., and M.F. Doherty (1994). Separation system synthesis for nonideal liquid mixtures.

Rene Banares-Alcantara, R., J.M.P. King and G.H. Ballinger (1994). Extending a process design support system to record design rationale. In Doherty and Biegler (Eds.), *Foundations of Computer Aided Process Engineering*.

Siirola, J.J., (1994). An industrial perspective on process synthesis.

MULTIPURPOSE PROCESS MODELLING ENVIRONMENTS

Constantinos C. Pantelides
Centre for Process Systems Engineering
Imperial College of Science, Technology and Medicine
London SW7 2BY, UK

Herbert I. Britt
Aspen Technology Inc.
Cambridge, MA 02139

Abstract

Process modelling environments are software systems supporting the high-level definition of process models. They are multipurpose in nature, with the same model potentially being used for a variety of model-based applications. The scope of process modelling covers both the physical behaviour of the plant and the procedures used for its operation. This paper reviews recent developments in both of these areas, as well as in the key model-based applications of process simulation and optimisation, and identifies a number of promising directions for future work. The paper also advocates the adoption of open architectures for process modelling environments as a means for facilitating the development of model-based applications, and as a potential avenue for standardisation in modelling technology.

Keywords

Process modelling, Operating procedure modelling, Physical properties, Process simulation, Process optimisation, Open software architectures.

Introduction

Mathematical process models are currently employed, directly or indirectly, for almost all aspects of plant design and operation. Model usage covers the entire process lifecycle, from designing the basic process itself to designing the plant and its control system, training the personnel who are responsible for operating it and detecting and diagnosing faults in its operation.

Most early implementations of model-based techniques provided their own mechanisms for describing the underlying process models. Indeed, in some cases, the technique was inextricably intertwined with the model. A typical example is provided by steady-state flowsheeting packages based on the sequential modular approach, in which a model of each unit operation was coupled with mathematical solution methods for calculating the output streams of the unit given its inputs.

Several factors have increasingly been providing strong incentives for moving away from this strong coupling between applications on one hand and process models on the other. It is well-known that the cost of developing and validating any non-trivial process model can be quite substantial, and this naturally leads to the desire to make the most of any such model by reusing it in as many applications as possible. Another factor is the difficulty and cost of developing sophisticated model building tools: it is far too wasteful to develop a credible modelling tool simply to use it for supporting a single application.

We are therefore led to consider *multipurpose process modelling environments*, that is software tools which support the construction and maintenance of models irrespective of the application for which the latter are used. Of course, models on their own are of limited use, and practical implementations of modelling environments also support a number of applications such as various types of simulation and optimisation.

This paper presents a review of recent progress in process modelling environments, attempting to assess the current state of this field. It identifies what the authors consider to be the key outstanding issues and points out some potential directions for addressing them.

Scope of Process Modelling

It is worth considering first some issues pertaining to what a "process" really encompasses, and consequently the scope of problems that process modelling tools should be able to cover.

Traditionally process modelling has concentrated on describing the behaviour of process plants in terms of the physical, chemical and biological phenomena that take place in them. It should, however, be recognised that the behaviour and the performance of any process under both normal and abnormal conditions are determined not only by the physical characteristics of the plant, but also by the operating procedures and control mechanisms employed for its operation.

The primary purpose of modelling tools is to facilitate the construction of mathematical models of the processes under consideration. The mathematical description of the physical behaviour of process plant can be quite varied, its complexity depending on factors such as the nature of the applications for which the model will be employed, the degree of simplification that might be acceptable, and the capabilities of the available solution techniques and computer hardware. Thus the study of the transient behaviour of most processes requires the introduction of differential equations describing the variation of system properties with time. If, in addition to temporal variations, the spatial variation of properties within the plant equipment is also considered to be important, then the model will have to introduce partial differential equations used to describe such distributed systems. On the other hand, if spatial variations are negligible, then a lumped system approximation based on ordinary differential equations may be sufficient. The need for partial differential equations also arises in modelling processes in which some system properties are characterised in terms of distributions (e.g. chain length distributions in polymerisation, or crystal size distributions in crystallisation). In all cases, it is quite likely that the model will also involve a number of algebraic constraints; these are often used to describe phenomena that operate on (relatively) much shorter time scales.

Another important aspect of process design and operation that has been receiving increasing attention in recent years is that of the study of the effects of process uncertainty, and methods for dealing with it. It is therefore highly desirable that process modelling tools be capable of describing uncertainty both in the underlying physical plant behaviour and in the operating procedures applied to the plant.

To summarise, we have argued that

a. the scope of process modelling tools includes both the intrinsic physical behaviour of the plant and the procedures used for its operation under both deterministic and stochastic conditions; and
b. the wide range of processes and applications of interest implies a correspondingly wide range of mathematical descriptions that need to be supported.

Process Modelling Tools

It could well be argued that any software tool which covers, at least partially, the scope of problems described in the previous section is a process modelling tool. This would allow the term to be applied, for instance, to conventional steady-state flowsheeting packages in which plant models are built from a library of models of unit operations. It would also encompass general-purpose algebraic modelling languages (e.g. GAMS), continuous system simulation languages (e.g. ACSL) and many rule-based systems — as well as the derivatives of all of these adapted specifically for process applications.

A formal definition for a process modelling tool is rather difficult to produce, and, as might be evident from the above discussion, one based merely on strict *functional* capabilities would probably be too wide to be useful. Instead, we have to consider the *extent* to which different software tools support the process modelling activity. This includes the ease of use of the tools and the complexity of process that they can handle with reasonable effort, rather than merely what functions they might, in principle, be capable of. We will therefore restrict our attention to software that supports the *high-level declarative* definition of *mathematical* models of *complex* processes, as well as the construction of models of novel unit operations from first principles.

Physical System Modelling

We have already argued that process modelling comprises both the behaviour of the physical plant and the operating procedures applied to it. In this section, we concentrate on the former aspect attempting a review of the major trends that have been emerging in recent years. For presentation purposes, we sub-divide these in three parts.

First we consider work that aims to enhance the practical feasibility of handling increasingly complex processes without necessarily affecting the nature of the underlying mathematical descriptions. We then proceed to review developments aimed at widening the scope of mathematical models that can be described by process modelling tools. Finally, we focus on the special, but very important, issue of physical property calculations.

Managing the Complexity of Physical System Modelling

At the time of the last FOCAPD conference in 1989, the package that perhaps best fitted the description of a process modelling environment as outlined above was SpeedUp (Pantelides, 1988a). SpeedUp provides a high-level declarative language for describing mathematical models in terms of sets of variables and the ordinary differential and algebraic equations that relate them. Models of more complex unit operations (such as distillation columns), called "macros," may be formed from instances of the basic models. Finally, a model of the entire plant may be formed by combining instances of both models and macros into a flowsheet. SpeedUp supports directly a wide variety of model-based applications including steady-state and dynamic simulation, steady-state optimisation, parameter estimation and data reconciliation. It is now a commercial product (Aspen Technology, 1994) used widely in industry while undergoing continuing development to embody many of the concepts discussed in this paper.

A number of limitations can be identified with the modelling facilities provided by the early process modelling environments tools. Some of them are related to the mechanisms available for handling modelling complexity: a hierarchy with a fixed number of levels (3 in SpeedUp's case) may be ineffective for modelling really complex plants, forcing the user to confront a degree of complexity at the top-most level that is difficult to manage. Secondly, no mechanism is provided for re-using information employed in constructing a model of one unit operation in building the model of another. Thus, for instance, models of different types of heat exchangers have to be developed and maintained independently despite the fact that they may share many common characteristics.

Much of the research work since 1989 has aimed at addressing the above limitations. The use of object-oriented concepts for the organisation of modelling data was considered in the Ph.D. theses of Piela (1989), Nilsson (1989) and Andersson (1990), and was incorporated in the design of the ASCEND (Piela et al., 1991) and OMOLA (Mattsson et al., 1993) packages. By permitting models to contain sub-models which are instances of other lower-level models, these packages allow the construction of model hierarchies of arbitrary depth, thus supporting a top-down view of complex processes. The concept of inheritance is also employed to allow models to be organised into hierarchies of increasing refinement.

Similarly to SpeedUp, the lowest-level models in both ASCEND and OMOLA are mathematical in nature, expressed in terms of variables and the equations that relate them. The types of mathematical model description supported by these packages are sufficient for modelling directly a wide range of lumped parameter systems. These software tools do therefore make it possible to model large and complex processes, and substantial evidence for this capability being used in industrial practice has already been accumulated.

However, building correct and consistent models, especially for non-standard unit operations, still remains a difficult activity requiring both good physical understanding and special mathematical skills, a combination of requirements that have often been identified as a key factor preventing the even wider use of modelling in industry. It is therefore reasonable to seek alternatives that will allow many more users other than "modelling experts" to benefit from this technology. One possibility is to build large libraries of basic unit operation models which can then be used for building higher-level models, an approach employed very successfully by the traditional steady-state flowsheeting packages. However, for several reasons (Pantelides and Barton, 1993), it is not clear whether the compilation of comprehensive libraries of standard dynamic models is feasible or even useful, given the fact that dynamic process modelling is very often used precisely for studying the behaviour of non-standard processes, or that of standard processes under unusual conditions.

We are therefore faced with the problem of devising process model representations that lie somewhere between context-free mathematical descriptions and complete unit operation models, hopefully combining some of the flexibility of the former with the ease of use of the latter. One such approach is that proposed by Stephanopoulos et al. (1990) in their MODEL.LA package. In common with ASCEND and OMOLA, this system also adopts an object-oriented approach to data organisation. However, it goes beyond the domain of a generic modelling tool to define a set of modelling elements, such as units, ports and streams, that are specific to process engineering. More importantly for the purposes of the present discussion, the scope of the model, i.e. the assumptions and simplifications on which it is based, and the process constraints and process variables are also recognised explicitly as modelling elements. A comprehensive taxonomy of the different types of equation that are encountered in process engineering models is introduced as a sub-class of process constraints. Similarly, the various terms that may appear in these equations form a sub-class of the variables, and again a taxonomy of such terms is defined. A key part of the definition of a new model is the specification of the assumptions on which it is based, from which the necessary equations are generated automatically, each containing the appropriate set of terms.

More recent research on model construction has been aiming at establishing decompositions of process models in terms of entities that are simpler than the conventional unit operations, thus allowing the description of complex models in a structured fashion in terms of a relatively small number of concepts. Marquardt (1992, 1994) views processes as sets of *devices* linked via *connections* that carry fluxes of extensive quantities. Both devices and connections are hierarchical entities, thereby facilitating the description of complex systems in the manner described earlier. One important class of elementary devices is that of *generalised phases* which comprise one or more thermodynamic phases contained within a clearly delineated spatial region (e.g. a vessel). The structural description of the process is complemented by a behavioural description that characterises mathematically the behaviour of phases and connections in terms of process quantities and the relations between them.

An alternative structural decomposition of process models can be based on the simpler and well-established concept of *thermodynamic phases*. This approach views process systems as sets of thermodynamic phases exchanging mass, energy and momentum. The basic ideas can be traced to the work of Preisig et al. (1989, 1990) and more recently they have been elaborated by Vázquez-Román (1992) and Perkins et al. (1994). Multiple interacting phases may co-exist in a shared *region* (e.g. a vessel) and this may introduce additional constraints (e.g. total volume restrictions). This approach is conceptually simpler as thermodynamic phases can be characterised completely in terms of a small, standard set of quantities (e.g. component holdups, internal energy, etc.) and the corresponding standard conservation equations while the vast diversity of transport phenomena occurring in processes of practical interest is associated only with the connections between the phases.

To summarise,

- The work described here aims at facilitating the construction of process models by providing effective mechanisms for handling complexity. The latter may arise either from the sheer size of some processes (as evidenced, for instance, by large numbers of interacting units and streams), or from the intrinsic complexity of the interacting physical, chemical and biological phenomena taking place in certain unit operations.
- Some systems, such as ASCEND and OMOLA, are largely domain-independent mathematical modelling tools. Others incorporate a significant degree of understanding of the concepts of process modelling, their ultimate aim being the replacement of purely mathematical model descriptions by descriptions based on physical concepts.
- It is clear that there is substantial commonality between the various tools although this is sometimes disguised by differences in terminology, emphasis and implementation. Significantly different ideas *have* emerged in some specific areas, such as the definition of models of different complexity describing the same physical system. For instance, in MODEL.LA the explicit representation of modelling assumptions allows different forms of modelling equations to be generated depending on the simplifications specified by the user; alternatively, as proposed by Gerstlauer et al. (1993), equations representing various degrees of aggregation can be generated automatically by formally integrating general 3-dimensional conservation equations over one or more spatial domains; and in the approach described by Perkins et al. (1994), different assumptions are expressed through the specification of different transfer laws for the connections between phases (e.g. equilibrium vs. non-equilibrium).

To date, experience with the application of most of the above tools, some of which are still at the implementation stage, to realistic problems is rather limited. It is perhaps worth emphasising that implementation and practical application of this software should be viewed as an essential part of this work and not an undesirable diversion from higher academic pursuits. Indeed, we argue that convenience and ease of use are the key issues in the design of such tools, and it is from this angle that current and future progress in this area should be judged.

A common implementation issue for all systems that employ physical descriptions of plant behaviour is the automatic generation of the corresponding mathematical formulations. Recent work has resulted in much improved understanding regarding the way in which correct and, especially, well-behaved mathematical models can be written. For instance, the modelling of unit operations involving phase equilibria has been considered by Ponton and Gawthrop (1991), while systems with multiple reaction and phase equilibria have recently been studied by Pantelides (1994). However, it is not yet clear whether any of the modelling approaches described above has an advantage over the rest in terms of the ease with which they can make use of these results.

An aspect of process modelling which has received very little attention to date is that of accurately taking account of the geometry of process equipment. It is often assumed, either explicitly or implicitly, that, at least for lumped parameter systems, the effects of equipment geometry on the process model are confined to the introduction of simple volume and surface constraints. This is not generally true: for instance, the positioning of an outlet port on a process vessel along with the time-varying position of the boundaries between segregated phases in it may determine the type and/or flowrate of material that leaves through the corresponding connection. Some work on geometric modelling of process equipment has recently been reported by Telnes (1992), but much remains to be done in this area.

Extending the Scope of Physical System Modelling

Fundamentally the scope of the processes that can, in principle, be modelled by a certain tool largely depends on the underlying mathematical description employed by it, rather than the precise way in which this description is constructed. For instance, a system which supports only purely algebraic models cannot describe transient process behaviour irrespective of the sophistication and ease of use of its model building capabilities.

It has long been recognised (Pantelides et al., 1988) that a natural mathematical description of the transient behaviour of lumped parameter processes is in terms of mixed systems of ordinary differential and algebraic equations (DAEs) and that, in some cases, these systems may have to be discontinuous in order to represent physical discontinuities such as those caused by phase and flow regime transitions. These considerations were taken into account in the design of the SpeedUp system which provides a high-level language for the description of DAE systems. SpeedUp allows one or more of the equations to be subject to symmetric and reversible discontinuities, i.e. to exist in two distinct states (e.g. laminar and turbulent flow), with the logical condition that triggers the transition from one state to the other (e.g. the Reynolds number exceeding 2100) being the exact opposite of the condition triggering the reverse transition.

Much work in recent years has stemmed from the realisation that the above mathematical formalism is not sufficiently general for describing many process applications. One reason for this is that, as has already been mentioned, the spatial variation of system properties in many processes is far from negligible, and consequently their behaviour is naturally described mathematically not by DAEs, but in terms of mixed sets of partial differential and algebraic equations (PDAEs), or even integral, partial differential and algebraic equations (IPDAEs). Of course, in principle, all of these can be reduced to DAEs by appropriate discretisation of the spatial dimensions, but in practice the correct application of such manipulations requires a high degree of mathematical skill.

A second area of concern is the fact that many intrinsic discontinuities in physical systems are not symmetric or even reversible. An example of a reversible but asymmetric discontinuity is that of a safety valve exhibiting hysteresis because of different opening and closing pressures. On the other hand, the bursting of a safety disk once the pressure in a vessel reaches a certain level is an example of an irreversible discontinuity.

The *gPROMS* (general PROcess Modelling System) package (Barton, 1992; Pantelides and Barton, 1993; Barton and Pantelides, 1994) represents an attempt to address these issues. It employs a mathematical description of the physical process behaviour coupled with mechanisms for constructing model hierarchies of arbitrary depth. An important feature in the context of the present discussion is its support for the definition of general discontinuities as finite-state machines. This formalism provides a natural mechanism for the incorporation of reversible and symmetric, reversible and asymmetric, and irreversible discontinuities in process models.

gPROMS also supports the direct modelling of distributed parameter systems (Oh and Pantelides, 1994). This is achieved by allowing the process variables and equations to be distributed over one or more distribution domains. The latter may refer to variation with respect to spatial position or some other system property (such as polymer chain length). The hierarchical model description in *gPROMS* allows the modelling of complex processes involving a combination of lumped and distributed parameter units. However, even within a certain unit operation model, the degrees of distribution of different variables (or equations) need not be the same.

At present, the distributed system modelling capabilities in *gPROMS* are limited to regular geometries but can be applied, in principle, to an arbitrary number of dimensions. So far the package has been used successfully for modelling a variety of distributed processes with up to 3 distribution dimensions.

The incorporation of distributed system modelling capabilities in general-purpose process modelling systems represents a step change in the complexity and sophistication of unit operation models that can be defined. In due course, this will certainly be reflected by the wider use of more detailed models, removing the need for many common simplifying assumptions such as homogeneity and perfect mixing. It may also allow these modelling tools to transcend their traditional user base in the process systems community and start being used by a much wider audience engaged in detailed modelling in other areas or disciplines.

It is worth noting that packages for the construction of spatially distributed models of individual unit operations have existed for a long time, having been developed in the field of computational fluid dynamics (CFD) quite independently from process modelling tools. Until recently, the distinction between the two classes of software was reasonably clear, and indeed these were often perceived as complementary: typically, a process modelling tool could be used to develop a simple lumped-parameter model of a unit operation (e.g. a reactor) for the purposes of establishing overall material and energy flows in the process; a CFD package could then be employed for studying the detailed behaviour of the individual unit. However, the recent developments outlined above are inevitably beginning to blur such distinctions. Of course, process modelling tools have a long way to go to match the capabilities of CFD packages in describing and handling complex geometries, applying the most appropriate numerical method to each problem (see, for instance, Pfeiffer and Marquardt, 1993), and visualising the results. On the other hand, they offer much more flexibility both in constructing models of individual unit operations (as they are not limited to a fixed set of equations and boundary conditions) and in incorporating these within models of larger entities.

Physical Property Calculations

The fundamental data and calculation procedures used for the evaluation of thermodynamic and transport properties within a process model have a profound effect on the quality of the results obtained. It is essential in multipurpose modelling applications that accurate properties be used and that the same, or a consistent, set be employed for all applications. Furthermore, it is important that the properties be accurate over the entire range of possible application of the model. This can be a rather stringent requirement: for example, during plant startup and shutdown, or in batch processes, composition and condition can vary widely even during the same operation step.

As reported in Boston et. al. (1993), it is possible today to model most systems of industrial interest, including highly nonideal systems and electrolyte and polymer systems, with a high degree of accuracy over wide ranges of composition, temperature, and pressure. This capability is embedded in software packages involving hundreds of years of scientific and engineering development effort, and incorporating databanks of thousands of components, estimation systems, data regression systems, and interactive tools for data retrieval, reporting, and model discrimination, often backed by expert system guidance. These systems will continue to evolve as research in the areas of molecular modelling for estimation (Gubbins, 1993), equations of state for nonpolar systems (Wong and Sandler, 1992), and ever larger data compilations (Onken et al., 1989; Selover, 1993) continues.

It could reasonably be argued that physical properties are part of the physical description of the plant behaviour, and should therefore be treated using the already established mechanisms, for instance, by being incorporated as low-level models within a hierarchical description of the process. However, this is not practicable due to a variety of technical, organisational and commercial reasons. These include the difficulty of ensuring that a general-purpose algorithm for solving sets of nonlinear algebraic equations will always converge to the desired root of a cubic (or higher-order) equation of state; the very large number of possible calculation routes and methods that must be represented; the already very substantial investment in the development of physical property software systems; and the highly specialised nature of the development of physical property calculation methods and the proprietary nature of much of the available expertise in this area.

One solution to this problem, adopted for instance by the SpeedUp package (Aspen Technology, 1994), is the incorporation of physical properties within declarative models as *procedures* which calculate one subset of the model variables in terms of another. Separate mechanisms are provided for the specification of physical property related information (such as the sets of components and the calculation options) on a plant-wide basis. Procedures are in fact treated as black-box equations, differing from the rest of the model equations only in that their precise form is not visible to the modelling system. This may reduce the latter's ability to carry out symbolic manipulations on these equations (e.g. for generating partial derivatives), but modern physical property packages compensate for this by returning the sensitivities of the calculated properties with respect to temperature, pressure and composition.

An alternative to the above approach is provided by local property models (LPM) (Barrett and Walsh, 1979). These are relatively simple and mathematically well-behaved expressions for the explicit computation of individual physical properties. They contain adjustable parameters that can be used to fit the computed value to the rigorous properties over limited domains.

In principle, the use of LPMs is attractive as it permits the explicit incorporation of physical properties within the plant model, which may have important implications regarding the reliability and efficiency of solution. This is particularly valuable for optimisation or dynamic simulation of large, complex flowsheets where computing time might otherwise be prohibitive (Sørlie et al., 1992; Ledent and Heyen, 1994; Støren and Hertzberg, 1994; Anderko et al., 1994). It is also important for real-time plant modelling and operator training applications where fast solution times are critical.

In practice, the effectiveness of LPMs is determined by a number of factors such as the ability to select automatically a set of expressions that are appropriate for the system under consideration, the efficient estimation of the values of the adjustable parameters, the monitoring of the error of approximation during the course of a computation (Hillestad et al., 1989), and the efficient updating of the parameter values whenever this error exceeds certain tolerances. The latter feature, which is particularly important for enabling LPMs to represent properties over a wide range of conditions, inevitably introduces discontinuities which must also be handled efficiently by the numerical solution methods. A successful implementation of LPMs has been carried out in SpeedUp, and is largely transparent to the user. It employs LPMs in a manner similar to the inside-out concept proposed for the flash problem by Boston and Britt (1978).

Finally, an interesting question concerns the precise scope of the services provided by physical property packages within process modelling environments. More specifically, in addition to the computation of point values of thermodynamic and transport properties, many such packages offer capabilities for phase and chemical equilibria calculations using a variety of sophisticated iterative techniques. On the other hand, equilibria may also be represented within the process model itself in terms of the appropriate mathematical constraints, and this may improve the efficiency of the computation by removing the need for nested iterations. However, determining the correct number of phases present in a system is in itself a difficult problem (see, for instance, the recent work by McDonald and Floudas, 1994), and one which cannot be represented in a purely declarative model. Declarative modelling of systems with variable numbers of phases is also a difficult undertaking, but this is becoming less so with the advent of more general descriptions of intrinsically discontinuous systems. For instance, Barton (1992) presents a declarative dynamic flash model that represents both the two-phase and single-phase regimes, and the transitions between them. Also in a recent proprietary development of the SpeedUp package by Aspen Technology, the general problem of time-varying numbers of phases has been effectively addressed by an approach that does not involve discontinuities or phase transitions.

Overall the distinction between what is required of the physical property package and what is represented explicitly in the process modelling environment is not a clear one. A hybrid of the two extremes would be to use the physical property package to perform phase stability tests (e.g. Michelsen, 1982) that cannot easily be described explicitly; these would then effectively provide the conditions that trigger the appropriate changes in a discontinuous model defined in a declarative manner.

Operating Procedure Modelling

The procedures employed for the operation of process plants have traditionally been considered to be outside the scope of process modelling tools, being perhaps more relevant to real-time control systems which provided facilities for expressing and implementing them. Process modelling tools have instead concentrated on modelling the physical behaviour of the plant subjected to relatively simple manipulations, such as steady-state operation or dynamic operation around the steady-state.

This is clearly an unsatisfactory situation. From a philosophical point of view, it can be argued that it is the plant operation (rather than the plant equipment on its own) that achieves the desired process objectives, and therefore plant and operating strategy must be considered as two equally important facets of the process and its modelling. More pragmatically, studying the effects of operating strategies (comprising, for instance, aspects such as plant start-up, shut-down and control) is one of the main motivations for carrying out process modelling in the first place.

One way of supporting the modelling of plant operating procedures is by building interfaces between real-time control systems and process modelling tools, as proposed by Herman et al., 1985. However, one may then be restricted by the peculiarities of specific control systems, the difficulties of interfacing them to other software, and the degree to which they support the definition and execution of complex operating procedures.

An alternative which is preferable for both theoretical and practical reasons is to make operating procedure modelling an integral part of the process modelling tool itself. However, this is by no means a straightforward task:

- In contrast to the predominantly continuous nature of the physical plant behaviour, operating procedures often have a significant discrete component arising from discrete actions imposed on the system.
- While the physical plant behaviour can be described in a purely declarative manner, operating procedures are (by definition!) procedural. Consequently issues relating to the relative timing of different actions and the conditions under which they are triggered become important.
- The complexity of the operating procedures for realistic plants can be as large or even larger than the complexity of their physical behaviour. For instance, the physical behaviour of a simple batch reactor can often adequately be described in terms of a small number of differential and algebraic equations. However, the procedure for carrying out a single reaction batch (including loading the reactor, heating it up to start the reaction, controlling the temperature to follow a desired profile, quenching the reaction, emptying the reactor contents and carrying out a cleaning cycle) may easily involve hundreds of elementary actions.

One can attempt to deal with the above difficulties by adding some discrete event capabilities to existing modelling tools (see, for instance, Andersson, 1992a, b). Alternatively, one may design a tool specifically for processes with combined discrete and continuous characteristics. This is the approach adopted in the development of the *gPROMS* package (Barton, 1992; Barton and Pantelides, 1994). *gPROMS* explicitly recognises the dual nature of process modelling, defining two distinct types of fundamental entities, namely MODELs which describe physical behaviour of process plant, and TASKs describing the external actions and disturbances applied to the plant. MODELs correspond closely to similar entities in earlier process modelling packages, such as SpeedUp, ASCEND and OMOLA and can therefore utilise well-established ideas for handling complexity and other problems. On the other hand, TASKs are new concepts, and one is faced with addressing some basic issues such as, for instance, identifying a complete set of elementary building blocks for these entities, organising these elements into TASKs, devising ways of dealing with the complexity of realistic applications, and ensuring maximum TASK re-usability.

In designing *gPROMS*, the mathematical nature of combined discrete/continuous systems was analysed in order to derive a set of elementary external manipulations that can be applied to the mathematical description of the plant behaviour. Complexity is handled effectively by allowing higher level tasks to be built by arranging lower level ones in sequence, in parallel, or within conditional or iterative constructs. This establishes task hierarchies of arbitrary depth, with the lowest level corresponding to the elementary manipulations. Finally, task generality and re-usability is promoted by introducing concepts of parametrisation and inheritance. Applications of *gPROMS* to a wide variety of different processes, ranging from start-up and shut-down of continuous plants to modelling and simulation of entire batch processes have been described by Barton et al. (1991), Barton (1992), and von Watzdorf et al. (1994).

Stochastic Process Modelling

Process uncertainty may occur in the various parameters (e.g. thermodynamic, transport or kinetic constants) that appear in the model of the physical plant behaviour or in the operating conditions (e.g. ambient temperature, feed stream properties). It may also arise in the operating procedures reflecting, for instance, the probability of failure of plant control and management systems, or the lack of precise reproducibility of operator actions.

The effects of uncertainty on process design and operation have been receiving substantial attention in recent years, and effective techniques for managing it are beginning to emerge (see, for instance, Grossmann and Straub, 1991). However, most existing process modelling tools are entirely deterministic in nature. This is in sharp contrast to tools in other related areas, such as discrete event simulation, that have traditionally provided very strong support for the definition of uncertainty and the statistical analysis of the results obtained (see Kreutzer, 1986; Pritsker, 1986).

Nevertheless, a couple of attempts in incorporating uncertainty considerations within process modelling and simulation tools have been reported in recent years. Diwekar and Rubin (1991) have introduced parametric uncertainty in the public-domain version of the ASPEN steady-state simulator and used it for carrying out Monte-Carlo simulations using a variety of sampling techniques. Some modest stochastic modelling capabilities have also been introduced in *gPROMS* (Näf, 1994) and have been used for a variety of applications, including modelling of batch processes (von Watzdorf et al., 1994).

Model-based Applications

The discussion of modelling environments in this paper has so far been largely independent of the eventual uses to which the process models are to be put. We now turn our attention to model-based applications, and in particular, process simulation and process optimisation. These may be viewed as "core" applications as they often act as building blocks for higher-level model-based applications. For instance, operator training software utilises dynamic simulation to emulate the response of the plant to operator actions; and supervisory control systems may use process optimisation to calculate optimal settings for the regulatory plant control system.

As observed by Barton (1992), from the theoretical point of view, the idea of applications constructed around a process model fits the concept of an experiment (Oren and Zeigler, 1979; Zeigler, 1976) comprising the object being investigated (i.e. the process model), the experimental frame (i.e. the conditions under which the investigation is to take place), and the results generated by the experiment execution. It is interesting to note that the operating procedures can be viewed either as part of the experimental frame (Marquardt, 1994) or as part of the process model. The latter view may be preferable as the different types of experimental frame (e.g. simulation, optimisation etc.) are, in reality, applied not to the plant but to the process comprising both the plant and the procedures adopted for its operation. This is the approach adopted by *gPROMS*, where an experimental entity, the PROCESS, is applied to a MODEL of the physical plant and a TASK describing its operating procedures.

In the rest of this section, we attempt to highlight the key areas of progress in recent years, and to identify possible directions for extending the scope of these applications.

Process Simulation

The benefits of steady-state and dynamic process simulation are now well established and there is no need for repeating them here. Most of the earlier software packages offering simulation capabilities were specially designed and implemented for this purpose, providing large libraries of unit operation models but only limited facilities for the development of novel models. This category includes the steady-state sequential-modular flowsheeting packages (Westerberg et al., 1979) and several dynamic simulation packages (see, for instance the review by Marquardt, 1991). For the reasons discussed earlier in this paper, the more recent trend (although not without some exceptions!) is for these applications to be built around general process modelling tools.

Undoubtedly, much of the recent progress in this area can be attributed to advances in solution techniques which have resulted both in improving the reliability of the tools and in extending the classes of problems that can be handled. This has been particularly true for the solution of large sets of differential-algebraic equations where several important advances have been achieved. These include the detection and handling of discontinuities (Pantelides, 1988a; Smith and Morton, 1988; Preston and Berzins, 1991; Park and Barton, 1993), the detection, analysis and manipulation of high-index systems (Pantelides, 1988b; Chung and Westerberg, 1990; Mattsson and Söderlind, 1993; Cellier and Elmqvist, 1993) and the treatment of systems with fast transients (Jarvis and Pantelides, 1991; Jarvis, 1994). A more detailed review of this area has been given by Pantelides and Barton (1993). Here it suffices to stress that much of the progress in solution techniques involves the combined application of symbolic, structural and numerical techniques, which has been greatly facilitated by the high-level model definition supported by process modelling environments.

A second area of progress in dynamic simulation is closely related to the ability to represent complex operating procedures and/or external disturbances affecting the plant (Barton, 1992).

In view of these impressive advances, it is worth noting that the major problem affecting process simulation *still* remains the less than perfect robustness of methods for the solution of sets of nonlinear algebraic equations.

This is the mathematical problem underlying both steady-state simulations and the initialisation of dynamic simulations. While some progress in numerical algorithms for this problem has been reported in recent years[1], it is not clear to what extent these have been applied successfully in the context of general-purpose process simulation tools to date. In any case, the completely reliable, globally convergent, general-purpose nonlinear equation solver has so far proved a truly elusive goal!

An alternative way of improving the reliability of nonlinear equation solving is the provision or automatic generation of good initial estimates for the variables, subsequently relying on the local convergence properties of the numerical algorithms to locate the solution from these initial values. This is an area to which traditional steady-state flowsheeting packages have directed substantial attention. Thus, in addition to the basic modelling equations and the method for their solution, each unit operation module encapsulates an initialisation procedure. The latter is typically based on a detailed engineering understanding of the operation and may be quite sophisticated, for instance taking different paths depending on the specifications imposed on the unit operation and/or the thermodynamic properties of the system under consideration. In fact, it can be argued that it is precisely in these procedures that a significant part of the modelling expertise — and, consequently, the competitive advantage — of these packages lies.

The initialisation facilities provided by general-purpose process modelling environments are more limited. Several packages embody the concept of a "variable type" used to impart default initial values to wide classes of variables, and also provide facilities for overriding these values for specific variables. SpeedUp (Aspen Technology, 1994) also provides the user with a set of tools (SAVE/USE) for initialising appropriate subsets of the model variables using the results of one or more earlier simulations. This encourages evolutionary model development as a simpler version of model may be used to initialise a more complex one. Also ASCEND (Piela et al., 1991) allows models to contain sequences of initialisation statements assigning the model variables to expressions involving other variables and/or constants. At the start of a simulation, these are executed in a sequential manner to provide initial estimates for the variables.

It is obvious from the above discussion that a wide gap exists between, on one hand, the powerful but special-purpose initialisation procedures implemented in steady-state flowsheeting package and, on the other, the general but rather less effective facilities afforded by current process modelling environments. It is also almost certain that truly effective procedures will have to make use of understanding of both the specific process under consideration and the mathematical properties of its model — and that such understanding emerges gradually during the development of the model and the experience of its subsequent use. The interesting question which then arises in this context is: how can one formally incorporate this important knowledge within the model itself? The work in ASCEND is clearly aimed in this direction, but there are limits to what can be achieved by a sequence of explicit assignment statements. On the other hand, the SAVE/USE tools in SpeedUp are more powerful, but it is the responsibility of model users to formulate and implement the sequence of actions (e.g. simulations) that lead to the correct initialisation of their model.

We note that initialisation techniques are intrinsically procedural operations, which themselves can be viewed as simulation experiments in the sense introduced earlier in this paper. One can therefore envisage the definition of dynamic links between two or more simulation experiments so that the execution of one (the "primary" experiment) automatically initiates the execution of the others (the "secondary" experiments), the latter providing the initial estimates required for the execution of the former[2]. As an example, the execution of a dynamic simulation involving a rigorous distillation column could automatically trigger the execution of a steady-state experiment based on a constant relative volatility column model, or even a simple flash. In any case, a formal definition of experiments, such as that embodied in the *gPROMS* PROCESS entities (Pantelides and Barton, 1993), appears to be necessary for the practical implementation of these ideas. A degree of experiment parametrisation permitting the precise nature of the secondary experiment(s) (e.g. in the above example, the feed and operating conditions for the simple column model) to be determined dynamically by the primary experiment is also required.

Process Optimisation

Several general-purpose process modelling tools already have some form of process optimisation capability. Thus, both SpeedUp and ASCEND provide facilities for parametric steady-state optimisation using a variety of numerical algorithms and approaches (see Aspen Technology, 1994; Westerberg et al., 1991).

Parametric steady-state optimisation is useful for the off-line determination of optimal process designs and steady-state operating conditions. Recent years have also witnessed a rapidly growing interest in online optimisation for plant operations. One characteristic of such online applications is that the process model is typically used on a continual basis over long periods of time, over which it is often modified repeatedly to reflect changes in the plant or simply improved process understanding. The contribu-

1. Notably through the use of homotopy-continuation methods (see, for instance, Seader 1990).

2. This description naturally lends itself to the definition of trees of experiments of arbitrary depth, with each level of the tree initiating the execution of the lower level.

tion of process modelling environments in developing and subsequently maintaining a high-level model containing an up-to-date statement of what is known about the process is particularly important.

Despite its clear significance, it is important to recognise that parametric steady-state optimisation covers only a small subset of important process optimisation problems. It does not include, for instance:

- the determination of optimal operating strategies, and possibly plant equipment designs, for batch or continuous processes operating under transient conditions; and
- the determination of the optimal structure of both continuous and batch processes.

Mathematically, these respectively lead to dynamic optimisation and mixed integer programming problems. Significantly, there has been considerable progress in both of these areas over the past few years, in terms of both numerical algorithms and their practical implementation into reasonably reliable and efficient codes.

In dynamic optimisation, two main classes of techniques have emerged, differing primarily with respect to the degree to which the model equations are satisfied during the course of the optimisation. A review of much of the state-of-the-art in this area at the time of the previous FOCAPD meeting was presented by Biegler (1990). More recent work has aimed at improving the accuracy of the techniques (Vasantharajan and Biegler, 1990), dealing with larger systems (Logsdon and Biegler, 1992), and widening the class of mathematical problems that may be solved (Vassiliadis et al., 1994a, b). A general code for the solution of multistage optimal control problems described by differential-algebraic equations has been implemented by Vassiliadis (1993).

The current state of the art in mixed integer programming and, in particular, its application to process synthesis problems have recently been the subject of comprehensive reviews by Grossmann and Daichendt (1994) and Floudas and Grossmann (1994).

To date, most of the practical implementations of this new optimisation technology have been carried out outside the framework of general-purpose process modelling environments. For instance, the PROSYN package for process synthesis (Kravanja and Grossmann, 1990, 1993) is based on the GAMS system. However, the incorporation of these applications within process modelling environments could result in significant benefits, including facilitating the use of more complex process models and ensuring consistency between different but closely related applications such as dynamic simulation and dynamic optimisation, or process design and process synthesis.

The establishment of general and natural representations of process optimisation problems is a fundamental pre-requisite for significant progress in this direction. In this context, the modelling of operating procedures seems to be a necessary step for formalising the description of dynamic process optimisation problems. Consider, for instance, the operation of a batch distillation column. This typically involves a number of steps carried out in sequence, including the initiation and termination of the collection of the various cuts, and the variation of control variables such as the reflux ratio over time. It is clear that the optimisation of the column operation can be defined only in the context of a given operating procedure, for example by allowing some degree of freedom regarding the timing of the cuts and the precise reflux policy to be adopted.

In the area of structural process optimisation, a number of general representations already exist and could form the basis for the formal description of problems in this category. These include the superstructure and hyperstructure representations of continuous processes, and the state-task network representation for batch processes; and at a lower level, the mathematical mixed-integer programming description that relies on identifying certain variables as being restricted to integer or binary values. An important complication is that, as discussed by Floudas and Grossmann (1994), many of the currently available techniques for solving such optimisation problems rely heavily on the exploitation of problem structure. The real challenge for process modelling environments is to automate the analysis of this structure as far as possible, while providing convenient mechanisms for the user to specify any relevant information that cannot be deduced automatically.

Towards Open Architectures for Process Modelling Environments

There is an ever growing diversity of model-based applications, covering the entire life-cycle of the process. It is very unlikely therefore that, in practice, any single developer or vendor of process modelling environments has the skills, the resources and the commercial incentive to address all of these potential opportunities. Inevitably, this has led to:

- the developers of modelling tools concentrating on a few "core" applications, of the type discussed in the previous section; and
- the developers of model-based applications being forced either to rely on less sophisticated modelling facilities (e.g. procedural programming languages, such as FORTRAN, or general-purpose packages such as GAMS), or to collaborate with modelling tool developers to interface their software to the modelling environments.

We believe that this situation is unsatisfactory, indeed presenting a major obstacle to the effective use of the expertise and resources available within the process systems community.

The adoption of open architectures by process modelling environments would be a significant first step towards redressing this problem. One way of achieving this goal would be for the process modelling environment to act as a model server, i.e. a software package responding to requests issued by one or more applications concerning process models defined within the environment (see Fig. 1).

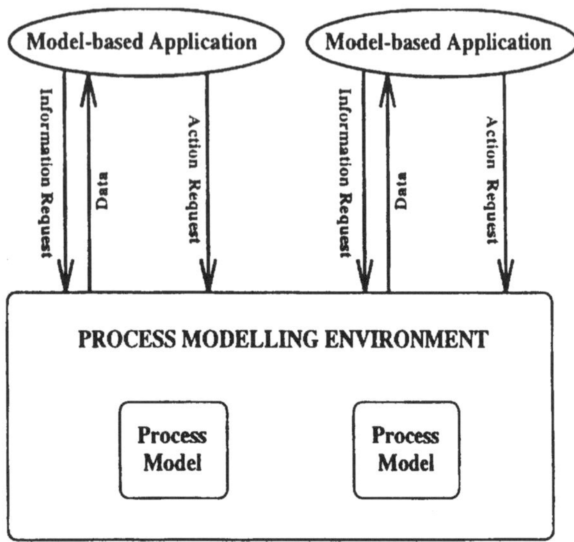

Figure 1. Process modelling environments as model servers.

Two types of request can be distinguished:

- *Information requests*
  These may concern the mathematical model description (e.g. its size and structure, the values of variables, equation residuals and their partial derivatives) as well as more general model properties. Clearly, some of this information (e.g. the residuals) may have to be computed on receipt of the request.
- *Action requests*
  These cause modifications to the internal state of the model, and are normally followed by one or more information requests. The actions involved could be simple (such as changing the current value of a variable), or more complex (such as carrying out an initialisation, or executing a dynamic simulation experiment). In the latter case, the modelling environment may need to make use of its core applications.

It is worth mentioning that a lower degree of openness is already provided by some of the current generation of process modelling environments in the form of direct communication between their core applications and foreign software. An example of this is SpeedUp's External Data Interface (Aspen Technology, 1994) which, for instance, permits the linking of dynamic simulation runs (a core application in SpeedUp) to external software such as real-time plant control systems.

The communication between model-based applications and modelling environments requires appropriate protocols, and this in turn raises the interesting proposition of establishing standards to which all process modelling environments should conform. An obvious benefit of this is that model-based applications would then be able to draw interchangeably, or even simultaneously upon a variety of sources for their process models. In fact, if the same communication protocols were to be adopted more widely, one could envisage situations in which some (or all) of these models do not actually originate from process modelling environments. Instead, they could be, for example, programs written in standard procedural languages, or unit operation modules from traditional flowsheeting packages. This may have important practical implications in terms of benefitting from the very substantial engineering know-how embodied in existing software (e.g. regarding initialisation procedures), and more generally maximising the utilisation of the associated investment.

The form of standardisation described above is aimed at facilitating the use of process modelling environments. Inevitably, before it can seriously be contemplated, some experience with open-architecture concepts of the type outlined earlier will have to be accumulated, concerning in particular the type and format of information and action requests that need to be accommodated in order to support a wide range of model-based applications.

Another type of standardisation, aimed at facilitating the *development* of process modelling environments, could be directed at some of the building blocks used for this development. One important area in this context is that of physical property calculations, where there has been massive investment in both in-house and commercial property calculation systems and data compilations. Standard interfaces to point property evaluations and property databanks would be particularly useful in this context.

A major problem with attempting to enforce standards in rapidly evolving fields, such as process modelling and model-based applications, is the difficulty of anticipating future technological developments and requirements. Two examples may serve to illustrate this point. Much of the recent progress in handling discontinuities in dynamic simulation stems from having direct access to the values of the logical clauses determining the status of the discontinuous equations; this goes far beyond the type of information required by standard numerical algorithms. Also in the area of physical properties, it is now accepted that it is highly advantageous for physical property packages to return the sensitivities of the properties being calculated in addition to their point values. Yet, in both cases it is only during the past decade or so that the benefit of the additional information was recognised. One could therefore plausibly argue that a set of rigidly enforced standards which had failed to anticipate these demands, far from being beneficial, would, in fact, have rendered a serious disservice.

Conclusions

The concept of multipurpose process modelling environments has gained wide acceptance over the past five years. Undoubtedly this is due to a large extent to their use in industrial practice, which has demonstrated the advantages of high-level declarative definition of process models, the truly multipurpose nature of the tools, and the ability of solution techniques to cope with many of the complications associated with the generality and complexity of the problems. Industrial usage has also indicated some of the current limitations of this technology, and much of the recent research effort has been directed towards addressing them.

The scope of process modelling encompasses both the physical behaviour of the plant and the procedures used for its operation. Recent developments in the former area have aimed at facilitating model definition by basing it on physical rather than mathematical descriptions; at widening the range of problems that can be represented through more general mathematical descriptions, covering for instance, systems with spatial property variation and general discontinuities; and at improving modelling accuracy and efficiency through developments in physical property calculations and their interfacing to modelling environments. Progress has also been achieved in the much newer area of operating procedure modelling, identifying the necessary building blocks, exploring the mechanisms necessary for handling the potentially immense complexity of this task, and demonstrating the feasibility and desirability of these ideas through practical implementations.

Process simulation and optimisation are prime examples of model-based applications. Much progress has been achieved in both of these, due to better solution methods that benefit from the increased numerical, structural and symbolic information made available to them by process modelling environments.

This paper has also argued for the adoption of open architectures for process modelling environments to allow them to form the basis for the wider development of model-based applications. In the longer term, open architectures raise the possibility of standardisation of various degrees and forms, with potentially significant benefits in terms of making the best use of past and future investment in modelling technology.

Of course, much work remains to be done in exploring and demonstrating the true potential of all these recent developments, as well as in pursuing the many important directions identified in this paper.

References

Anderko, A., J.E. Coon, and S.M. Goldfarb (1994). Local thermodynamic models for dynamic process simulation. *Proc. IFAC Symp. on Advanced Control of Chemical Processes* (ADCHEM'94), 270-275.

Andersson, M. (1990). OMOLA — *An object-oriented language for model representation*. Licentiate thesis, Dept. of Automatic Control, Lund Institute of Technology, Sweden.

Andersson, M. (1992a). Discrete event modelling and simulation in OMOLA. Paper presented at the 1992 IEEE Symposium on Computer-Aided Control System Design (CACSD'92), Napa, California, 17-19 March 1992.

Andersson, M. (1992b). Combined object-oriented modelling in OMOLA. Paper presented at the European Simulation Multiconference (ESM'92), York, England, 1-3 June 1992.

Aspen Technology, Inc. (1994). *SpeedUp User Manual*. Cambridge, Massachusetts.

Barrett, A., and J.J. Walsh (1979). Improved chemical process simulation using local thermodynamic approximations. *Comput. Chem. Engng.*, **3**, 397-402.

Barton, P.I. (1992). *The modelling and simulation of combined discrete/continuous processes*. PhD thesis, University of London.

Barton, P.I. and C.C. Pantelides (1994). Modeling of combined discrete/continuous processes. *AIChE J.*, **40**, 966-979.

Barton, P.I., E.M.B. Smith, and C.C. Pantelides (1991). Combined discrete-continuous simulation using gPROMS. Paper #152c, AIChE 1991 Annual Meeting, Los Angeles, California.

Biegler, L.T. (1990). Strategies for simultaneous solution and optimization of differential-algebraic systems. In J.J. Siirola, I.E. Grossmann, and G. Stephanopoulos (Eds.), *Foundations of Computer-Aided Process Design*, Elsevier, New York, 155-179.

Boston, J.F., and H.I. Britt (1978). A radically different formulation and solution of the single stage flash problem. *Comput. Chem. Engng.*, **2**, 109-122.

Boston, J.F., H.I. Britt, and M.T. Tayyabkhan (1993). Software: tackling tougher tasks. *Chem. Engng. Progr.*, November, 38-49.

Cellier, F.E., and H. Elmqvist (1993). Automated formula manipulation supports object-oriented continuous-system modeling. *IEEE Control Systems*, April, 28-38.

Chung Y., and A.W. Westerberg (1990). A proposed numerical algorithm for solving nonlinear index problems. *Ind. Eng. Chem. Res.*, **29**, 1234-1239.

Diwekar, U.M., and E.S. Rubin (1991). Stochastic modeling of chemical processes. *Comput. Chem. Engng.*, **15**, 105-114.

Floudas, C.A., and I.E. Grossmann (1994). Algorithmic approaches to process synthesis: Logic and global optimization. Paper presented at Foundations of Computer-Aided Process Design (FOCAPD'94) conference, Snowmass, Colorado.

Gerstlauer, A., M. Hierlemann, and W. Marquardt (1993). On the representation of balance equations in a knowledge-based process modeling tool. Paper presented at CHISA'93 conference, Prague, Czech Republic.

Grossmann, I.E., and M.M. Daichendt (1994). New trends in optimization-based approaches to process synthesis. In E.S. Yoon (Ed.), *Proc. 5th Intl. Symp. Proc. Syst. Engng.*, Korean Institute of Chemical Engineers, 95-109.

Grossmann, I.E., and D.A. Straub (1991). Recent developments in the evaluation and optimization of flexible chemical processes. In L. Puigjaner and A. Espuña (Eds.), *Computer-Oriented Process Engineering*, Elsevier, Amsterdam, 49-59.

Gubbins, K.E. (1993). Applications of molecular simulation. Fluid *Phase Equilibria*, **83**, 1-14.

Herman, D.J., G.R. Sullivan, and S. Thomas (1985). Integration of process design, simulation and control systems. *IChemE Symp. Ser.*, **92**, 505-511.

Hillestad, M., C. Sørlie, T.F. Anderson, I. Olsen, and T. Herzberg (1989). On estimating the error of local thermodynamic models — a general approach. *Comput. Chem. Engng.*, **13**, 789-796.

Jarvis, R.B. (1994). *Robust dynamic simulation of chemical engineering processes*. PhD thesis, University of London.

Jarvis, R.B., and C.C. Pantelides (1991). Numerical methods for differential-algebraic equations — new problems and new solutions. Paper #158c, AIChE 1991 Annual Meeting, Los Angeles, California.

Kravanja, Z. and I.E. Grossmann (1990). PROSYN — an MINLP process synthesizer. *Comput. Chem. Engng.*, **12**, 1363-1378.

Kravanja, Z. and I.E. Grossmann (1993). PROSYN — an automated topology and parameter process synthesizer. *Comput. Chem. Engng.*, **17S**, 87-94.

Kreutzer, W. (1986). *System simulation: programming styles and languages*. Addison-Wesley, Reading, Massachusetts.

Ledent T., and G. Heyen (1994). Dynamic approximations of thermodynamic properties by means of local models. *Comput. Chem. Engng.*, **18S**, 87-91.

Logsdon, J.S., and L.T. Biegler (1992). Decomposition strategies for large-scale dynamic optimization problems. *Chem. Eng. Sci.*, **47**, 851-864.

Marquardt, W. (1991). Dynamic process simulation — recent progress and future challenges. In Y. Arkun and W.H. Ray (Eds.), *Chemical Process Control*, CACHE-AIChE Publications, 131-180.

Marquardt, W. (1992). An object-oriented representation of structured process models. *Comput. Chem. Engng.*, **17S**, 329-336.

Marquardt, W. (1994). Trends in computer-aided process modeling. In E.S. Yoon (Ed.), *Proc. 5th Intl. Symp. Proc. Syst. Engng.*, Korean Institute of Chemical Engineers, 1-24.

Mattsson, S.E., M. Andersson, and K.J.Åström (1993). Omola — an object-oriented modeling language. In D.A. Linkens (Ed.), *CAD for Control Systems*, M. Dekker, New York, 31-69.

Mattsson, S.E., and G. Söderlind (1993). Index reduction in differential-algebraic equations using dummy derivatives. *SIAM J. Sci. Stat. Comput.*, **14**, 677-692.

McDonald, C.M., and C.A. Floudas (1994). Global optimization for the phase equilibrium problem using the NRTL equation. *Proc. 4th European Symp. on Computer Aided Process Engineering* (ESCAPE'4), 273-280.

Michelsen, M.L. (1982). The isothermal flash problem — Part I: Stability. *Fluid Phase Equilibria*, **9**, 1-19.

Näf, U.G. (1994). Stochastic simulation using gPROMS. *Comput. Chem. Engng.*, **18S**, 743-747.

Nilsson, B. (1989). *Structured modelling of chemical processes — an object-oriented approach*. Licentiate thesis, Dept. of Automatic Control, Lund Institute of Technology, Sweden.

Oh, M. and C.C. Pantelides (1994). A modelling and simulation language for combined lumped and distributed parameter systems. In E.S. Yoon (Ed.), *Proc. 5th Intl. Symp. Proc. Syst. Engng.*, Korean Institute of Chemical Engineers, 37-44.

Onken, U., J. Rareynies, and J. Gmehling (1989). The Dortmund data bank — a computerized system for retrieval, correlation and prediction of thermodynamic properties of mixtures. *Intl. J. Thermophys.*, **10**, 739-747.

Oren, T.I., and B.P. Zeigler (1979). Concepts for advanced simulation methodologies. *Simulation*, **32**, 69-82.

Pantelides, C.C. (1988a). SpeedUp — recent advances in process simulation. *Comput. Chem. Engng.*, **12**, 745-755.

Pantelides, C.C. (1988b). The consistent initialization of differential-algebraic systems. *SIAM J. Sci. Stat. Comput.*, **9**, 213-231.

Pantelides, C.C., D. Gritsis, K.R. Morison, and R.W.H. Sargent (1988). The mathematical modelling of transient systems using differential-algebraic equations. *Comput. Chem. Engng.*, **12**, 449-454.

Pantelides, C.C., and P.I. Barton (1993). Equation-oriented dynamic simulation: current status and future perspectives. *Comput. Chem. Engng.*, **17S**, 263-285.

Pantelides, C.C. (1994). *Dynamic modelling of systems with multiple reaction and phase equilibria*. Report C94-09, Centre for Process Systems Engineering, Imperial College, London.

Park, T., and P.I. Barton (1993). A new algorithm for the accurate and efficient location of state events. Paper #3b, AIChE 1993 Annual Meeting, St Louis, Missouri.

Perkins, J.D., R.W.H. Sargent and R. Vázquez-Román (1994). Computer generation of process models. In E.S. Yoon (Ed.), *Proc. 5th Intl. Symp. Proc. Syst. Engng.*, Korean Institute of Chemical Engineers, 123-125.

Pfeiffer, B-M., and W. Marquardt (1993). Symbolic semi-discretization of partial differential equation systems. *Proc. Intl. IMACS Symp. on Symbolic Computation* (SC'93), Villeneuve d'Ascq, France.

Piela, P. (1989). *ASCEND: an object-oriented computer environment for modeling and analysis*. PhD thesis, Carnegie Mellon University.

Piela, P., T.G. Epperly, K.M. Westerberg, and A.W. Westerberg (1991). ASCEND: an object-oriented computer environment for modeling and analysis: the modeling language. *Comput. Chem. Engng.*, **15**, 53-72.

Ponton, J.W., and P.J. Gawthrop (1991). Systematic construction of dynamic models for phase equilibrium processes. *Comput. Chem. Engng.*, **15**, 803-808.

Preisig, H.A., T.Y. Lee, F. Little, and B. Wright (1989). On the representation of life-support system models. Paper 891479, Society of Automotive Engineers, Warrendale, Pennsylvania.

Preisig, H.A., T.Y. Lee, and F. Little (1990). A prototype computer-aided modelling tool for life-support system models. Paper 901269, Society of Automotive Engineers, Warrendale, Pennsylvania.

Preston, A.J., and M. Berzins (1991). Algorithms for the location of discontinuities in dynamic simulation problems. *Comput. Chem. Engng.*, **15**, 701-713.

Pritsker, A.A.B. (1986). *Introduction to simulation and SLAM II* (3rd edition). John Wiley, New York.

Seader, J.D. (1990). Recent developments in methods for finding all solutions to general systems of nonlinear equations. In J.J. Siirola, I.E. Grossmann, and G. Stephanopoulos (Eds.), *Foundations of Computer-Aided Process Design*, Elsevier, New York, 181-207.

Selover, T.B., Jr. (1993). A status report on DIPPR in its fourteenth year. Paper presented at AIChE 1993 Spring Meeting, Houston, Texas.

Smith, G.J., and W. Morton (1988). Dynamic simulation using an equation-oriented flowsheeting package. *Comput. Chem. Engng.*, **12**, 469-473.

Sørlie, C., M. Hillestad, K. Strand, and P. Flatby (1992). Experiences of applying local thermodynamic models in dynamic process simulation. Paper presented at AIChE 1992 Spring Meeting, New Orleans, Louisiana.

Stephanopoulos, G., G. Henning and H. Leone (1990). MODEL.LA. A Modeling Language for Process Engineering. Part I: The Formal Framework. *Comput. Chem. Engng.*, **14**, 813-869.

Støren, S., and T. Hertzberg (1994). Local thermodynamic models applied in dynamic process simulation: a simplified approach. *Proc. 4th European Symp. on Computer Aided Process Engineering* (ESCAPE'4), 143-150.

Telnes, K. (1992). *Computer aided modeling of dynamic processes based on elementary physics*. PhD thesis, Norges Tekniske Høgskole, Trondheim, Norway.

Vasantharajan, S., and L.T. Biegler (1990). Simultaneous strategies for the optimization of differential-algebraic systems with enforcement of error criteria. *Comput. Chem. Engng.*, **14**, 1083-1100.

Vassiliadis, V.S. (1993). DAEOPT — A differential-algebraic dynamic optimization code, vs. 2.0. Technical Report, Centre for Process Systems Engineering, Imperial College, London.

Vassiliadis, V.S., R.W.H. Sargent, and C.C. Pantelides (1994a). Solution of a class of multistage dynamic optimization problems. Part I — Problems without path constraints. *Ind. Eng. Chem. Res.*, **33**, 2111-2122.

Vassiliadis, V.S., R.W.H. Sargent, and C.C. Pantelides (1994b). Solution of a class of multistage dynamic optimization problems. Part II — Problems with path constraints. *Ind. Eng. Chem. Res.*, **33**, 2123-2133.

Vázquez-Román, R. (1992). *Computer aids for process model building*. PhD thesis, University of London.

von Watzdorf, R., U.G. Näf, P.I. Barton, and C.C. Pantelides (1994). Deterministic and stochastic simulation of batch/semicontinuous processes. *Comput. Chem. Engng.*, **18S**, 343-347.

Westerberg, A.W., H.P. Hutchison, R.L. Motard, and P. Winter (1979). *Process flowsheeting*. Cambridge University Press, Cambridge, England.

Westerberg, A.W. P. Piela, R. McKelvey, and T. Epperly (1991). The ASCEND modeling environment and its implications. Report EDRC 06-110-91, Engineering Design Research Center, Carnegie Mellon University, Pittsburg, Pennsylvania.

Wong, D.S.H., and S.I. Sandler (1992). A theoretically correct mixing rule for cubic equations of state. *AIChE J.*, **38**, 671-680.

Zeigler, B.P. (1976). *The theory of modeling and simulation*. John Wiley, New York.

PROCESS ENGINEERING DATABASES – FROM THE PDXI PERSPECTIVE

R. L. Motard[1], M. R. Blaha[2], N. L. Book and J. J. Fielding
Department of Chemical Engineering
University of Missouri
Rolla, MS 65401

Abstract

The process industries are responding to driving forces that make advanced manufacturing concepts priority business issues. Data sharing (within an organization) and data exchange (between organizations) are critical to the successful implementation of these concepts. The Process Data Exchange Institute (PDXI) was constituted to promote electronic data sharing and exchange in the process industries.

PDXI has developed activity and data models for process engineering. The focus of the activity and data models has been the body of process engineering data that is shared and/or exchanged with high frequency. PDXI is also producing an application programming interface (API) based on the data models. The API enables the data captured by the data models to be easily shared and/or exchanged via a neutral file. The activity model, the data models, and the API are extensible to a larger body of data and are easily maintained. The implications of the data models for process engineering databases are discussed.

The OMT models have been translated into the ISO-STEP EXPRESS informational modeling language, but these results are not detailed here. The relation between OMT graphical constructs and EXPRESS are not one-for-one since EXPRESS is a data description language, not an object modeling language. Finally, the ISO-STEP developments for the actual implementation of data exchange programs and interfaces and the PDXI specifications are briefly mentioned. Full reports of the PDXI products are available from the American Institute of Chemical Engineers in New York.

Keywords

Design databases, Data models, Metadata, Data sharing, Data exchange, Process engineering.

Introduction

The current business climate for the process industries includes rapidly changing markets, the demand for an ever greater variety of products, more rigorous quality expectations, the emergence of global competition, concern for the natural environment, increased regulatory intervention and heightened awareness for the safety and welfare of employees and the public. One critical aspect of the new expectations is the capture of data and information from the entire life-cycle of a chemical processing venture, from conception, through the design, construction and operating phases, to decommissioning. Data exchange must therefore include database support for archival purposes, at least in addition to if not in parallel with sharing and exchange. State-of-the-art information management technology is a necessity for achieving world class manufacturing.

Data, Metadata and Information

Information has a broader context than does data. Data has syntax. Information is data with semantics. For example, an engineer could send the floating point number, 15.8, across a computer network to another engineer. If the second engineer received the value, 15.8, then data

1. Washington University, St. Louis

2. Consultant, St. Louis; formerly General Electric Co.

was successfully exchanged. However, if the value of 15.8 was used by the receiving engineer as a mass flow when, in fact, it was a mole flow, then the information exchange was unsuccessful. This demonstrates that one key to information exchange is metadata. Metadata is data about data. Metadata can be useful for expressing semantics and can transform data into information. Metadata can take on a variety of forms from the units of the data to a measure of the quality of the data to the organization of the data.

Motivation for PDXI

PDXI was established to study humanless and paperless data exchange, and a life-cycle database for a plant. Engineers in the process industries use numerous software packages, some commercial and some proprietary.

Process and equipment designs are passed from operating companies to engineering and construction (E&C) firms on paper, only to be converted back into electronic form. In turn, the E&C firm will pass paper specifications to fabricators and contractors.

Regulatory agencies require voluminous quantities of data for process safety management and permits. The data must be produced in a timely fashion, be accurate, and up-to-date, even after the plant is decommissioned. A plant database is clearly needed that contains the life-cycle data for the plant from cradle to grave. There needs to be a single copy of each data instance in the plant database. When there are multiple copies (probably in different databases), it is common for some, but not all copies, to get changed. This is an invitation to rework.

Data Exchange Methodologies

There are five approaches commonly used to exchange data between software systems (Doty, 1992). They are: 1) manual transcription of the data, 2) standardization on a single system, 3) direct or point-to-point translators, 4) translators to a common neutral exchange format, and 5) standardization on a common data format. The software can be data management systems or application software.

Manual transcription is slow and error prone. Standardization on a common system is not possible across organizations. Maintaining point-to-point translation is overwhelmingly costly and standardization on a common data format is too cumbersome to be dynamic.

Overview of Translators

Translators typically possess complicated logic to convert the data format of one system into that of another. The complicated logic makes each translator difficult and expensive to create and maintain. It is never safe to assume that any translator is perfect. Regardless of the deficiencies of translators, they are the normal exchange mechanism for legacy systems, between the internal and the common external format.

Translators to a Common or Neutral Exchange Format

A common or neutral exchange format can be defined to which each system can exchange data. The neutral exchange format must accommodate all systems and, therefore, must be a standard. PDXI was established for the purpose of defining a common neutral exchange format for process data.

Along with the neutral exchange format, PDXI was commissioned to develop an application programming interface (API) designed to simplify creation and maintenance of translators to a neutral exchange format and/or to simplify implementation of a standard internal data format. An API is essentially a standard translator. The API serves as an interface between an application program and a data repository (flat file or database) holding data in the standard format. Application programs can store, edit, and/or retrieve data in the repository through the API. The structured query language (SQL) for a database is an example of an application programming interface. The SQL provides a means for storing, editing, and/or retrieving data in the database.

PDXI has accepted the STEP physical file format as its neutral format (See the Appendix A for a discussion of STEP). PDXI has also undertaken the development of an API that will allow FORTRAN and C language programs to directly read and write STEP physical files. The mapping to the STEP physical file and the API are automated so that data that is compliant to an EXPRESS data model can be read or written to a STEP physical file. This automation will greatly simplify maintenance of the API, allow the data model to evolve with technology, and provide a means to compare data models for performance.

Data Models

The structure of the data exchange methodologies for translators to a neutral exchange format requires a standard format for storing the data in the repository. Methods are needed to define the protocols (a schema) for storing data in a repository. Data (or information) models express schema definitions. Good data models are independent of the type of repository (file or database).

Data models are graphical or lexical representations that describe how a body of data has been organized into a schema. The graphical representations are popular because can be easily and rapidly learned, large quantities of data can be modeled in rather small graphs, and the data models are highly "readable." NIAM, IDEF1X, EXPRESS-G, and the object modeling technique (OMT) are popular graphical syntaxes.

The object modeling technique (OMT) (Rumbaugh et al., 1991) allows for organization of the data using object oriented concepts. It has been found to be rapidly learned by application domain experts (Blaha et al., 1989). IDEF1X (Mayer, 1992) is based on relational concepts and NIAM (Nijssen and Halpin, 1989) is based on binary concepts. EXPRESS (Spiby, 1991) is a lexical information modeling syntax that is human readable and machine sensible. EXPRESS is based on object-oriented concepts and, though human readable, is not nearly as readable as a graphical data model for the same data. EXPRESS-G is a graphical syntax derived from EXPRESS.

The syntax of OMT data models is adequately illustrated in the figures contained in this article. Boxes on the diagram (Fig. 1) represent classes of objects into which the body of data has been organized. The bold text in the top section of each box is the name assigned to the object class. The bottom section of each box contains the names for the attributes used to characterize objects in that object class. Lines connecting boxes represent relationships that may exist between objects in those object classes (associations in OMT). Symbols on the lines indicate the type of association that exists. The OMT notation supports generalization or inheritance (a triangle as in Fig. 2), aggregation or is-composed-of (a diamond as in Fig. 8), multiplicity (one by no symbol, many by a filled-in circle, or zero-or-one by an open circle as in Fig. 3), qualifiers or identifiers (both used in Fig. 4), and link attributes or attributes of associations (as shown in Fig.5). Text near the lines indicates the roles that objects play in the association, as in Fig. 1.

Data models describe the schema for the data and not its instances (specific values) and are in a neutral form that is independent of the repository used to store the data. If a mapping (a set of rules to translate the schema defined by a data model onto a repository) can be devised, then the data model and mapping can be used to format the repository to accept instances of data. For example, the OMT methodology is supported by a softwarc product, OMTool, that allows graphical data models to be created. A second software product, SCHEMER, maps data models from OMTool into a relational database schema. The output from SCHEMER is data definition language (DDL) for relational databases. Other mappings could be developed to object-oriented databases or, for that matter, to a flat file. Once the DDL has been used to format a relational database, SQL can be used to populate, query or edit data in the database.

The first step in creating a data model is to define the body of data for the application domain. Documents containing application domain data and application programs are excellent sources of data entities to aid in specifying the initial scope of the data to be modeled. A business activity model that describes the flow of information between the activities in the application domain is very helpful in defining the scope of the body of data. PDXI has produced a business activity model modified and extended from an activity model for process plants developed by Pat Harrow in the UK. It uses the IDEF0 (United States Air Force, 1981) syntax that describes the activities and the information flows in process engineering. The PDXI activity model is recognized by the International Standards Organization (ISO) STEP community as the premier activity model for process plants.

The raw data entities must be scrutinized to identify and remove synonyms that frequently exist. It is critical that data models be created and carefully reviewed by a team of experts from the application domain. It is then required to identify the subtle differences between data entities and to ensure that all the important associations are captured. The names of object classes, attributes, and roles should be carefully chosen and an unambiguous definition created for each. The resulting compilation of names and definitions for the data entities in the data model is a data glossary, thesaurus or data dictionary (the term data dictionary is used in other contexts in information science and should be avoided).

Data modeling methodologies are still emerging. They do not support a full set of modeling constructs, although most data can be adequately modeled with a few basic constructs. The current mappings to data exchange technologies are not very sophisticated and, as a result, do not often produce efficient schema. Unsophisticated mappings producing workable schema may be an advantage at this stage of the technology

The PDXI Data Models

There are ten data models that have been produced for PDXI. They are:

- Planning Level
- Physical Property
- Process Materials
- Unit Operations
- Heat Transfer Equipment
- Material Transfer Equipment
- Separation Tower Equipment
- Process Vessel
- Simplified Geometry
- Metadata

There are well over 100 pages of graphical description of the models in the OMT notation.

This paper further explores key abstractions and design decisions for the current PDXI models. We present fragments of the OMT models, explore alternative modeling approaches and explain the rationale for our ultimate decisions. We attempt to provide introspection for some of our modeling decisions.

Some of the models have reached into equipment design. For core process functionality such as distillation, flash, compressors there is a sharp line between the functional abstraction of a unit operation and the physical reality of equipment. For specialized equipment such as crystallizers, filters and conveyors, this distinction is less clear.

Planning Level

This model describes the associations among process data at the highest level. In particular, geographical parameters such as site coordinates, area, elevation and meteorological information would be included here. The other models emanate from and are related through the planning level model.

Ports

A ***port*** is a portion of the boundary for a process plant through which material, energy or signals can flow. A port establishes connectivity between process equipment. Flow can occur from one piece of equipment to another only if there are ports which couple the equipment. The notion of a port is consistent with the fundamental chemical engineering concept of a boundary envelope.

An excerpt of our current PDXI model for ports is shown in Figure 1.

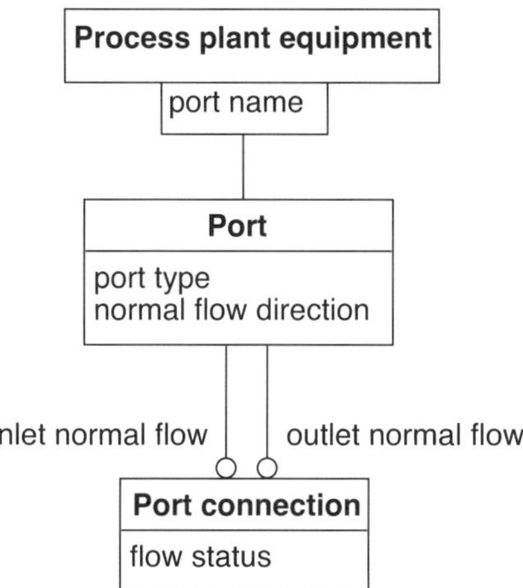

Figure 1. Current PDXI model of a port.

1. Process plant equipment can have multiple ports; each port is distinguished by a name.
2. There are three port types: material ports, energy ports and signal ports. A *material* port transfers chemical substances. An *energy* port transfers energy in the absence of any material flow. A *signal* port conveys information. A signal may be associated with a material flow or energy flow, however, the principal purpose of a signal port is to convey information.
3. A port has a normal flow direction that indicates the intended direction of flow. A port with an *inlet* flow direction normally fills its associated equipment; a port with an *outlet* flow empties its equipment. A port with a *static* flow does not ordinarily have flow except for special circumstances (such as a relief valve).
4. A ***port connection*** is a connection between exactly two ports of the same port type. One port has a normal flow direction of inlet and the other has a normal flow direction of outlet. The flow status through a port connection may be *normal* (occurring in the intended direction), *reverse* (occurring opposite of the intended direction) or *static* (not occurring). A collection of equipment is connected if there exists some path (material, energy and/or signals) through port connections that reaches all the equipment. This flow path constitutes the process topology.

Figure 2 presents an alternate model of a port from one of our earlier papers [1]. In this figure a piping component, such as an elbow, pipe or tee, has pipe nodes. Thus a pipe has two nodes and a tee has three nodes. Similarly a piece of equipment has nozzles for connection. Plant topology is captured by the connectivity of pipe nodes to other pipe nodes and equipment nozzles.

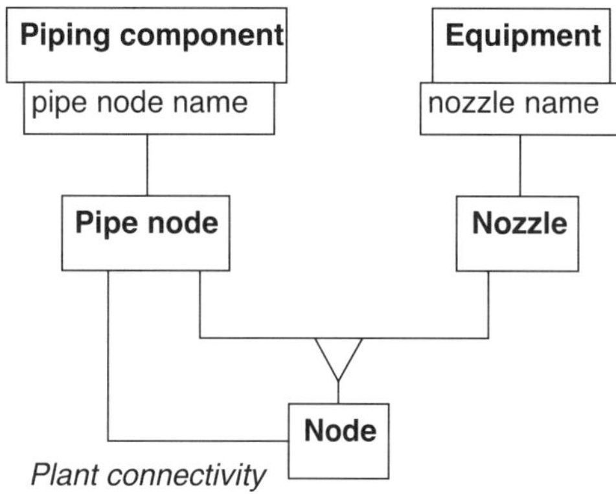

Figure 2. Past model for equipment connectivity.

Both Figure 1 and Figure 2 can represent 3D piping and physical equipment connectivity. Figure 1 can also represent heat transfer interfaces, mechanical or electrical connections and control information.

The model in Figure 2 is less abstract and less powerful than that in Figure 1. However the model in Figure 2 is probably easier to understand. With modeling there is often a trade-off between power and simplicity.

We believe that Figure 1 is a more versatile model than Figure 2. However, Figure 1 is still not entirely satisfactory.

1. The port paradigm breaks down somewhat for permeable membranes and heat transfer. The surface along which exchange occurs is diffuse. Ports only approximate the physical reality.
2. There is still some discomfort with the notion of a port among the PDXI team. The notion of a port may be more abstract than what is needed. It is not sufficient that a model be

correct. A model must to some extent also be minimal and avoid introducing superfluous concepts.

Simulator Topology

There are different ways of modeling the connectivity between unit operations and streams which affect understandability and enforcement of constraints.

Fig. 3 shows our current model for simulator topology. A process simulation has many streams and unit operations. Streams internal to the simulation must have a source and sink. Net input streams lack a source; net output streams lack a sink.

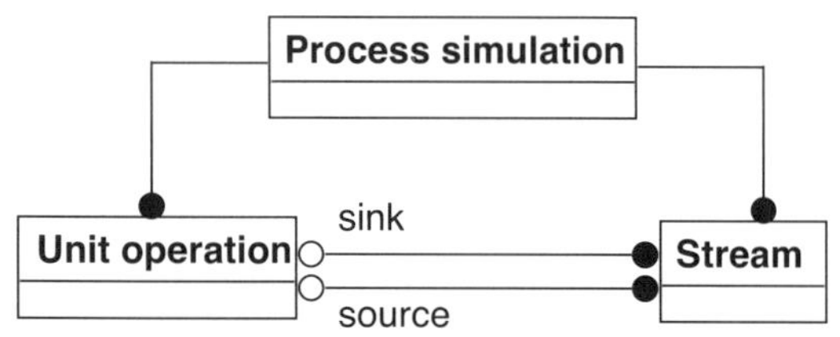

Figure 3. Current PDXI model for simulator topology.

Fig. 4 shows two alternate models for representing simulator topology. One can just relate streams and unit operations with a many-to-many association and note for each association link whether the stream connects with the unit operation as a source or sink. Or one may qualify the many-to-many association to distinguish between source and sink.

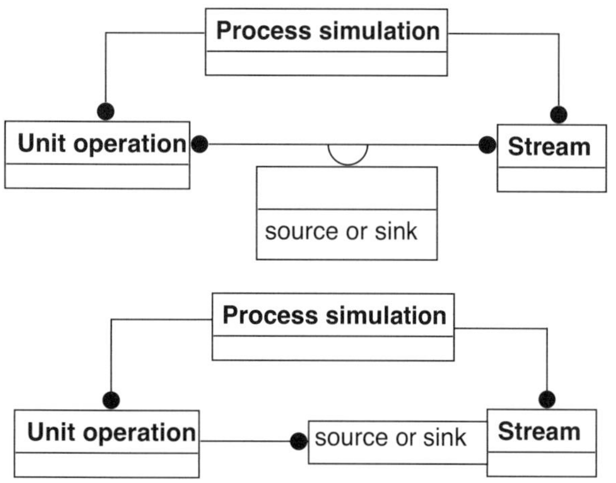

Figure 4. Alternate models for simulator topology.

The representation of simulation topology reduces to the general problem of representing a directed graph.

Sample Tables for Simulator Topology

In general many possible relational database implementations of an object model are possible. Similarly, a relational database implementation may correspond to multiple object models. There is a many-to-many relationship between object models and relational database implementation.

The model in Fig. 3, "Current PDXI model for simulator topology," could be implemented with tables 1 and 2 by burying the associations in the table for the many side. The underline denotes the primary key.

Table 1. "Unit operation."

unit operation ID	process simulator ID

Table 2. "Stream."

stream ID	process simulator ID	sink unit op	source unit op

Alternatively the model in Fig. 3 could be implemented with tables 1, 3, 4 and 5 by implementing the associations as dedicated tables.

Table 3. "Stream."

stream ID	process simulator ID

Table 4. "Sink unit op."

stream ID	sink unit op

Table 5. "Source unit op."

stream ID	source unit op

The top model in Fig. 4 would most naturally be implemented with tables 1, 3, and 6.

Table 6. "Unit op—Stream."

stream ID	unit operation ID	source or sink

The bottom model in Fig. 4 would also most naturally be implemented with tables 1, 3 and 6. However this object model specifies an additional candidate key of stream + sourceOrSink. The qualified association specifies that stream + the qualifier yields one unit operation.

Physical Property

Substances can be chemical species, pseudochemical species, or mixtures. Physical properties can be expressed as fundamental physical constants (for example, critical temperature, normal boiling point, or open cup flash temperature), as equations with coefficient values (for example, the Antoine vapor pressure equation or the Peng-Robinson equation of state), or as tables of point values generated by experiment, correlation or retrieval.

Recursive Composition of Substances

Fig. 5 shows an excerpt from the current PDXI physical properties model. A ***substance*** is any chemical entity which may be of interest to chemists or chemical engineers. A substance may be a component, mixture, or an amorphous solid. An amorphous solid is a substance that cannot be characterized with a molecular weight. An amorphous solid is characterized by various solid analyses, such as ultimate, proximate, maceral, and others.

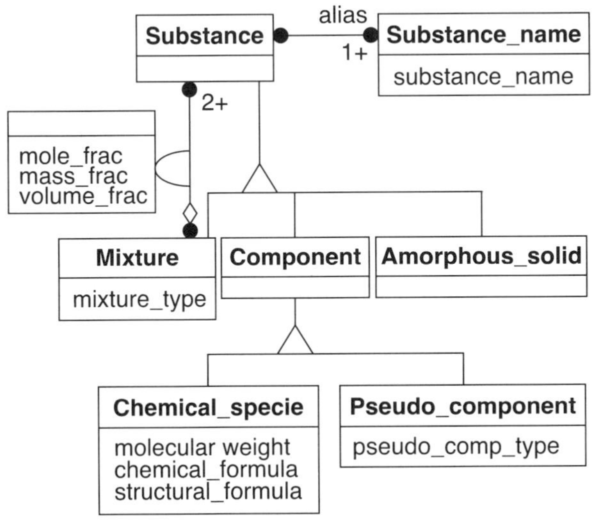

Figure 5. Recursive composition of substances.

A component encompasses both pure chemical species and pseudo components such as petroleum fractions and coal fractions. A ***pseudo component*** is a substance that can be adequately characterized by an average molecular weight that is measured or estimated. A ***chemical specie*** is a substance that has a single known chemical formula.

In reality, a pseudo component consists of a combination of chemical species but in practice it can be difficult to identify all of them. Often for simulators and other software it is adequate to pretend that a pseudo component is like a pure specie that is empirically characterized.

We recursively define a mixture as a combination of substances; each substance within a mixture has a corresponding mole, mass and volume fraction. A substance can then be a component, amorphous solid or a mixture with further substructure. Petroleum (which consists of multiple petroleum fractions), coal (which includes multiple coal fractions), biomass, and tar sands are examples of mixtures.

Such a recursive structure often occurs with models. Recursion provides a natural way to represent a hierarchy of arbitrary depth.

For example, tar sands could be treated as a mixture of sand and tar. One could regard the sand in tar sands as an amorphous solid; the sand is empirically characterized and is of little interest aside from the difficulty of separating the sand from the latent tar. The tar could be considered a mixture of pseudo components.

Each substance may be referenced by multiple aliases. For example, propylene may be referred to as *propylene* and *C3H6*. Various mixtures of ethylene glycol and automotive additives may have the alias of *antifreeze*.

One of our examples was a crude oil boiling point assay from a major oil company. A subset of the crude properties to be modeled are shown in Table 7.

Table 7. Alaska North Slope Crude Oil.

Fraction	Mass fraction	Volume fraction
60-400F	0.195	0.233
60-200F	0.064	0.084
200-400F		0.149
200-300F	0.069	0.080
300-400F	0.062	0.069
400-500F	0.106	0.112
500-650F	0.164	0.166
650+ F	0.536	0.490
650+ F, 0.33 mm vac. flash overhead	0.679	0.700
650+ F, 0.33 mm vac. flash bottoms	0.321	0.300
650+ F, 16.1 mm vac. flash overhead	0.473	0.496
650+ F, 16.1 mm vac. flash bottoms	0.527	0.504
900+ F		0.247

Sample Tables for Crude Oil Example

Relational database tables for Table 7 in the context of the data model of Fig. 5 are shown in Appendix B.

Point Properties vs. Parametrized Properties

The Physical Properties Model distinguishes between a point property and a parameterized property. Fig. 6 and Fig. 7 summarize the current PDXI model for point properties and parameterized properties.

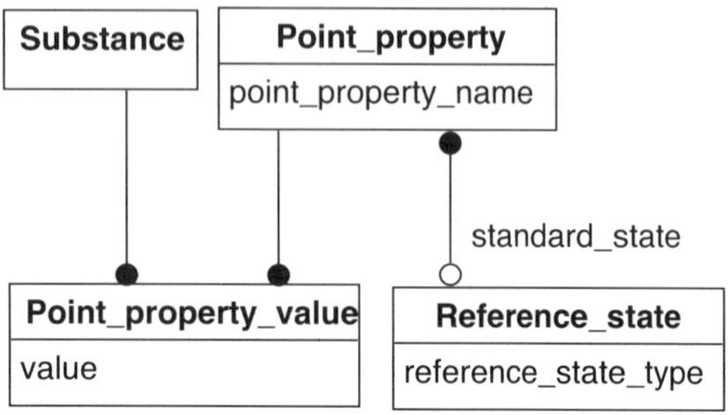

Figure 6. Point properties.

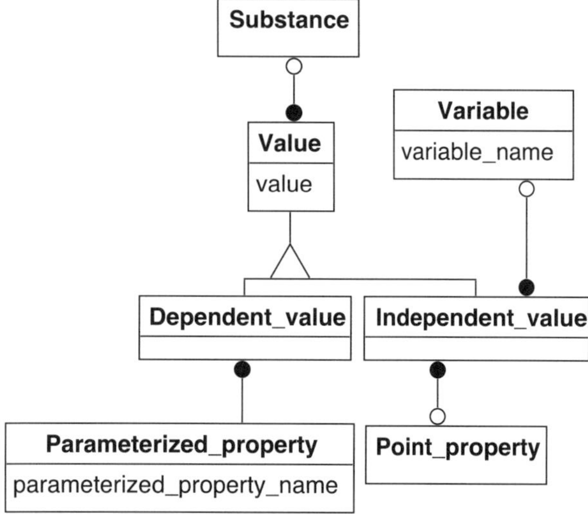

Figure 7. Parameterized properties.

A ***point property*** is a property that has a specific value for a substance and does not depend on temperature, pressure or other experimental and simulation conditions.

1. Some point properties, such as the ideal gas enthalpy of formation, are not absolute values and must be expressed with respect to some reference state.
2. Point properties for mixtures are often computed with mixing rules. Fig. 6 accommodates measurements of point properties for mixtures as well as components. Some point properties are intended for chemical species and others are intended for pseudo components; we have not attempted to enforce this dichotomy with the structure of the model.
3. Our organization for point properties is 'flat'; we have not imposed a taxonomy. We feel that there is no obvious, best canonical structure.

A ***parameterized property*** is a property that is a function of one or more variables.

Be careful not to confuse a parameterized property with a point property. A point property has a specific value for a substance; the only variation in the value of a point property is due to experimental or computational error. For example, critical temperature is a point property for a substance.

In contrast a parameterized property depends on one or more other properties. Sometimes a property such as Tc is used as a "pseudo property" in a semi-theoretical equation. If Tc can be adjusted to improve the fit of the equation, then Tc is a coefficient. If the value of Tc is the best estimate for a substance then Tc is a point property. A point property is an intrinsic value not subject to manipulation.

Process Materials

The quantities and the state of materials as they exist in the process such as, the flow rate of material in a process stream or the amount of material in a process vessel and its state (temperature, pressure, density, enthalpy, etc.) are captured in this data model.

We found it awkward to directly refer to substances in the simulation and equipment models because the physical properties model only describes intensive properties of substances. We invented the notions of ***material flow*** and ***material amount*** to standardize our treatment of the extensive property of quantity.

Unit Operations

The data associated with the engineering models of process equipment as used in chemical process simulators.

Heat Transfer Equipment

The design specification data for process equipment used to effect heat transfer. The model is finely detailed only for shell and tube heat exchangers.

Material Transfer Equipment

The design specification data for process equipment used to effect material transfer. The model is detailed for centrifugal pumps and compressors.

An abstract of the Centrifugal Compressor Model is shown in Fig. 8. What is evident are the many levels of inheritance, i.e. generalization in the PDXI data models. Aggregations do not imply inheritance so that inheritance, if it exists, must be represented in parallel with aggregation or composition, e.g. compressor is composed of stages but both the machine and each stage inherit attributes from gas transfer machine. A triangle in an association merely acts as a place marker for other classes that generalize with the abstracted class.

Some components of fluid transfer machines can be linked to the generic class of fluid transfer equipment, namely liquid and vapor machines, and are not exclusive to gas compressors.

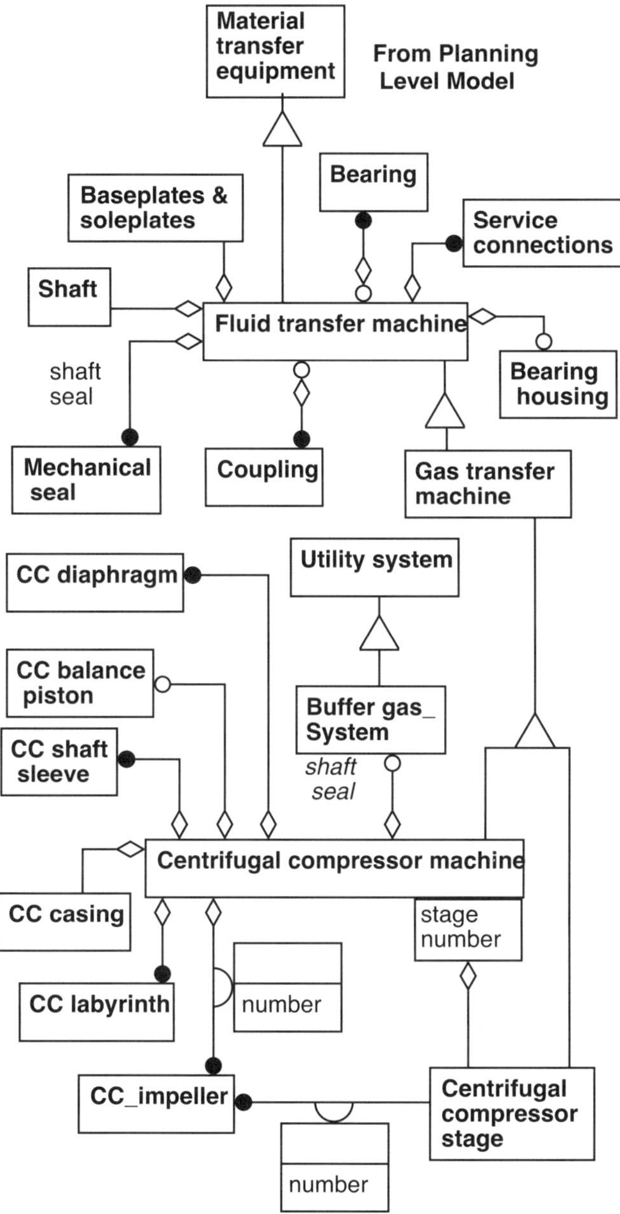

Figure 8. Centrifugal Compressor Model.

Separation Tower Equipment

The design specification data for process equipment used to effect chemical separations. The model is detailed for distillation columns fitted with sieve trays.

Process Vessels

The design specification data for process vessels. This model includes pressure vessels, tanks, accumulators and towers.

Simplified Geometry

Several of the PDXI models (machinery, heat exchanger, process vessel, distillation tray) must deal with the geometry of physical equipment. We have developed a simple standard representation for three dimensional geometry which is used throughout PDXI. It is consistent with the STEP geometry model.

Metadata

The metadata associated with the data in any of the other models. Units of measure, dimensions, data source, data accuracy, etc., are all captured in this model.

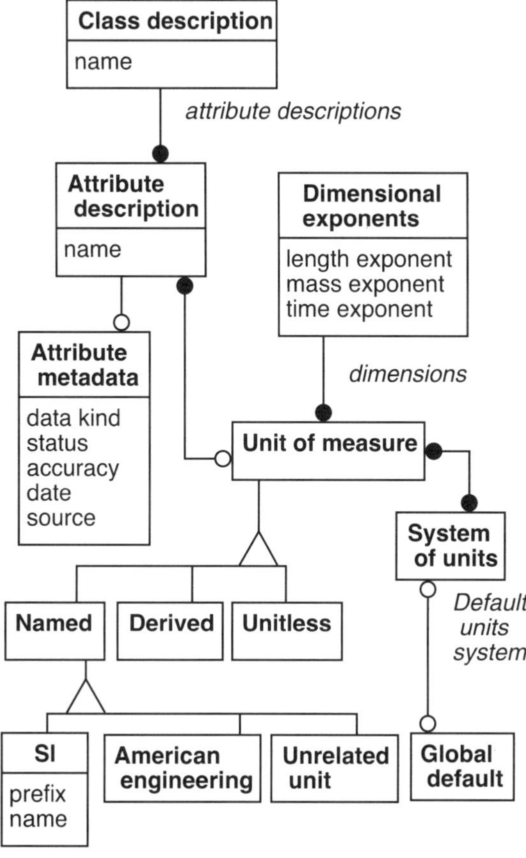

Figure 9. Metadata Model.

Fig. 9 is an inexact abstract of the actual metadata model for the sake of simplicity. Each class can be addressed by name in order to retrieve information about its attributes. One can retrieve the kind of data an attribute represents such as, estimated, calculated, literature, experimental, etc. The status of the data is retained such as preliminary, approved or frozen. The accuracy, date and source are also included. There follows the encoding of units of measure, their dimensional exponents, and the type of unit system used. SI units provide for a prefix such as kilo- or milli-.

Packaging the Data Models

So far, we have described some examples of OMT graphical notation in several data models. Each data model is then packaged into a module of five parts: 1) a textual description of the scope of the data modeled, 2) a graphical data model using the Object Modeling Technique (OMT) of Rumbaugh et al. (1991), 3) a translation of the OMT model to the EXPRESS information modeling language, 4) a complete data glossary/dictionary that defines each object class, attribute and enumerated variable in the model and indicates its variable type (real, integer, string, etc.) and dimension (length, time, velocity, etc.), and 5) the definition of ANSI C language elements such as,

- data structures for storing instances of each object class in the model and
- function calls that allow navigation through the models so that the desired instances of any object class can be identified.

Referring to these modules simply as models in the larger sense, we find that the ten models are highly interrelated. They have been tested for interoperability by populating a STEP physical file (manually instantiated using the NIST Data Probe software tool) with the data describing the simulation and design specification of a large, three-stage, centrifugal compressor with intercoolers.

Despite their interrelation, a number of models have been developed so that they can be used in a stand-alone fashion. For example, the Physical Property model can be used independently to exchange physical property data.

Data Model Based Information Management Systems

The data model is metadata for the body of data modeled. The schema defined by the data model can be mapped onto a repository for storing instances of data. The data model can also be used by an API to identify an instance of data, its syntax and to indicate how and where to store, retrieve or edit it. Thus, an information management system can be created in which the definition of each data item, its realization and the means to access it is self-contained. There is sufficient metadata to make information management systems from data management technologies.

Extensibility and Maintenance

Extensibility is, in principle, inherent in that data models from differing application domains can be integrated into a spanning model provided they are properly harmonized. Maintenance implies changing the data model. Changing the data model is trivial and the existence of automated mappings to repositories and application programming interfaces means that the information management system will be instantly prepared to accept data under the new schema. Tools for migrating data stored under one schema to storage under a new schema do not currently exist, although they will, eventually. The schema can evolve with the applications if the data model itself is modeled and packaged with the application. This requires another kind of metamodel (Blaha, 1992) and the applications then become oblivious to the scheme under which the data is stored.

Standards

Standards are not a panacea and tend to stifle innovation. Standards, therefore, must be dynamic in order to avoid discouraging innovation. The STEP standards have attempted to provide a trade-off between standardization and innovation by ruling that the parts of the STEP standard will be static for three years. This period of time is hoped to be sufficiently long to encourage developers to produce compliant technologies and short enough to allow innovation to proceed.

Conclusions

Data models are compact, efficient methods for defining the protocols for storing data. Coupled with mappings to repositories and application programming interfaces, data models provide a means for managing data that makes applications independent of the data repository. Applications can be integrated and data can be exchanged or shared through such a repository. Information management systems based on data models are readily extensible and easily maintained. They provide the means to eliminate large quantities of manual, repetitive, error-prone activities that are, by and large, no-value-added.

Data models allow for the development of a plant database or even a corporation database (actually multiple, distributed, and likely heterogeneous databases with a single manager). Each data instance is stored exactly once. There is no need to have multiple copies of the same data instance stored in several different databases. Everyone that has the need to access the data instance can do so through the API. If a change is made to the data instance, everyone that has an interest in the change can be automatically notified. Data can be maintained throughout the lifecycle of a plant to simplify compliance with the regulatory agencies.

The STEP standards for exchanging product data are based on data models. Many of the current standards for product data were developed by groups of application domain experts. There was little, if any, harmonization between the groups, therefore, the standards were limited to data exchange within the application domain. The ease of extending and maintaining data model based systems like STEP allows for the possibility of a single, consistent standard that covers all application domains.

References

Blaha, M.R., N.L. Eastman and M.M. Hall (1989). An Extensible AE&C Database Model. In *Computers and Chemical Engineering*, **13**, 753.

Blaha, M.R. (1992). Models of Models. In *Journal of Object Ori-*

ented Programming, September, 1992.

Blaha, M.R. and W Premerlani (1995) *Object-Oriented Modeling and Design for Database Applications*. To be published by Prentice Hall.

Clark, S.N. (1990). *An Introduction to the NIST PDES Toolkit*. NISTIR 4336, National Institute for Standards and Technology, Gaithersburg, MD.

Doty, R. (1992). *An Introduction to STEP*. Digital Equipment Company, Chelmsford, MA.

Mayer, R.J., ed. (1992) *IDEF1X Data Modeling: A Reconstruction of the Original Air Force Wright Aeronautical Laboratory Technical Report Developed Under Air Force Contract #F33615-80-C-5155*.Knowledge Based Systems, Inc., College Station, TX.

Morris, K.C. (1991). *Architecture for the Validation Testing System Software*. NISTIR 4742, National Institute for Standards and Technology, Gaithersburg, MD.

Nijssen, G.M., and T.A. Halpin (1989). *Conceptual Schema and Relational Database Design*. Prentice-Hall, Englewood Cliffs, NJ.

Rumbaugh, J., M. Blaha, W. Premerlani, F. Eddy and W. Lorensen (1991). *Object-Oriented Modeling and Design*. Prentice-Hall, Englewood Cliffs, NJ.

Spiby, P. (1991). *EXPRESS Language Reference Manual*. ISO Document TC184/SC4/WG5/N9.

United States Air Force (1981). ICAM Architecture Part II, Volume IV — Function Modeling Manual (IDEF0). Report number AFAWL-TR-81-4023, United States Air Force Wright Aeronautical Laboratories, Wright-Patterson Air Force Base, OH.

Wilson, P.R., ed. (1993). *Processing Tools for EXPRESS*. National Institute for Standards and Technology.

Appendix A

The STEP Project

The STandard for the Exchange of Product model data (STEP) is an evolving international standard for the exchange of a broad spectrum of product data throughout a lifecycle. STEP is data model based. It is being developed within the International Standards Organization (ISO), Industrial Automation Systems and Integration Technical Committee (TC 184), Industrial Data and Global Manufacturing Programming Languages Subcommittee (SC 4). Various parts of the STEP specification are currently being balloted as ISO standard 10303 by the 25 member countries. Product areas currently being addressed by STEP include electrical products, mechanical products, 3-D drafting, kinematics, and architecture, engineering and construction (AEC). Product Data Exchange using STEP (PDES), formerly the Product Data Exchange Specification, is the US contribution to the STEP specification.

STEP development methodology has become a technology of its own. The STEP standard is developed and balloted in parts, allowing experts to devote attention to their application domain and allowing each part to be standardized as it matures. The parts are organized into five basic categories:

- descriptive methods, such as the EXPRESS information modeling language used for describing product data,
- implementation methods, including the specification of the STEP physical file format,
- general resource and application resource information models, used to support and integrate the STEP parts,
- application protocols (AP), which define the standard in terms specific to the application domain and explain how the standard should be implemented to ensure unambiguous exchange, and
- conformance and tools, specifying testing methodology and the requirements for conformance testing.

The STEP physical file (STEP exchange file or STEP level 1 file) is a human readable and machine sensible ASCII flat file exchange format. The STEP specification includes the format of the STEP physical file and a defined mapping from EXPRESS to the physical file. The definition of the EXPRESS language is also a part of the STEP specification. Product data whose definition is captured by an EXPRESS data model can be exchanged via a STEP physical file. ASCII files can be exchanged with virtually all systems, thus, the STEP physical file provides the means to exchange data between numerous, diverse systems. The STEP physical file is intended for data exchange and is not intended to be an efficient method for storing or randomly accessing data.

The specification for an application programming interface, the STEP Data Access Interface (SDAI), is also a part of the standard. The SDAI specification defines the functionality required of application programming interfaces between application domain software and product data in STEP compliant data repositories.

Application protocols contain the data definitions for a specific application domain. Harmonization with other application protocols is required and general resource and application resource information models support subdomain areas that are common to a number of application domains.

Levels of STEP Implementation

Four levels of target implementation for the STEP specification have been informally defined. The first implementation level is passive file exchange involving data translation via the STEP physical file. The second level of implementation is active file exchange in which in-memory product data definitions (working form) are exchanged. The third implementation level is shared database, which implies that several applications share a common STEP compliant database. The fourth level is implementation via a shared knowledge base and is currently not well defined.

Tools for STEP Implementation

A rich set of tools is under development that support the STEP standard. The tools are being developed in commercial and open system forums (Wilson, 1993). The National Institute of Standards and Technology (NIST) is developing a public domain toolkit (FEDEX+) to support the construction of the first three levels of STEP implementations (Clark, 1990 and Morris, 1991). Commercial tools will soon be available to translate IDEF1X, NIAM, and EXPRESS-G data models to EXPRESS. Several compilers and parsers are available to manipulate EXPRESS models and several automated mappings for commercial relational and object oriented databases exist or soon will. Prototypes for the SDAI and the STEP data probe (a tool for interactively creating and editing STEP physical files) activities are under development as well.

Appendix B

Relational database tables for crude oil example in Table 7.

Table 8. Substance Name.

substance name ID	substance name
sn-1	Alaska north slope crude oil
sn-2	Alaska north slope crude — 60-200F fraction
sn-3	Alaska north slope crude — 200-400F fraction
sn-4	Alaska north slope crude — 400-500F fraction
sn-5	Alaska north slope crude — 500-650F fraction
sn-6	Alaska north slope crude — 650+ fraction
sn-7	Alaska north slope crude — 900+ F fraction
sn-8	Alaska north slope crude — 60-400F fraction
sn-9	Alaska north slope crude — 200-300F fraction
sn-10	Alaska north slope crude — 300-400F fraction
sn-11	Alaska north slope crude — 650+ 0.33 mm overhead
sn-12	Alaska north slope crude — 650+ 16.1 mm overhead
sn-13	Alaska north slope crude — 650+ 0.33 mm bottoms
sn-14	Alaska north slope crude — 650+ 16.1 mm bottoms

Table 9. Substance — Substance Name.

substance ID	substance name ID
sub-1	sn-1
sub-2	sn-2
sub-3	sn-3
sub-4	sn-4
sub-5	sn-5
sub-6	sn-6
sub-7	sn-7
sub-8	sn-8
sub-9	sn-9
sub-10	sn-10
sub-11	sn-11
sub-12	sn-12
sub-13	sn-13
sub-14	sn-14

Table 10. Mixture — Substance.

mixture ID	substance ID	mole fraction	mass fraction	volume fraction
sub-1	sub-2		.064	.084
sub-1	sub-3			.149
sub-1	sub-4		.106	.112
sub-1	sub-5		.164	.166
sub-1	sub-6		.536	.490
sub-1	sub-7			.247
sub-1	sub-8		.195	.233
sub-1	sub-9		.069	.080
sub-1	sub-10		.062	.069
sub-6	sub-11		.679	.700
sub-6	sub-12		.473	.496
sub-6	sub-13		.321	.300
sub-6	sub-14		.527	.504
sub-1	sub-11		.364	.343
sub-1	sub-12		.254	.243
sub-1	sub-13		.172	.147
sub-1	sub-14		.282	.247

MANAGEMENT OF THE DESIGN PROCESS: THE IMPACT OF INFORMATION MODELING

Jerry L. Robertson
Consultant
20 Westwind, Sand Springs, OK 74063

Eswaran Subrahmanian
Engineering Design Research Center
Carnegie Mellon University
Pittsburgh, PA 15213

Mark E. Thomas and Arthur W. Westerberg
Dept. of Chemical Engineering and Engineering Design Research Center
Carnegie Mellon University
Pittsburgh, PA 15213

Abstract

An information revolution is impacting how companies will operate in the future. Companies are looking for models that they can use to improve organizational efficiency. Early models decomposed organizations into workers and management and ascribed roles to each; newer models such as that by Deming concern themselves with integration of the parts into the context of the outside world. We suggest an analogy to decomposing processes into unit operations which must then be integrated into total processes within the world market, suggesting that, while the words are different, the process design activities may be very similar. Other important chemical engineering concepts such as process control and information transfer appear to be applicable to organizational management as well.

After examining management processes and how information flows in a company, we explore different technologies for information management/consolidation, noting how well each permits the gathering, structuring and using of the data both to operate a company and to understand and improve its performance, particularly in its ability to carry out a complex process such as design. Particular attention is paid to databases, the World Wide Web, and a new information management system under development at Carnegie Mellon University called *n*-dim.

The paper ends with a brief view into Etopia, an information society some time far into the future.

Keywords

Corporate organization, Information flow, Information modeling, Process synthesis, Retrofit design, World Wide Web.

Introduction

This paper speculates about how changes in work processes and the way information flows in companies along with new developments in information management techniques will create opportunities for improving the design process. It describes how models of the design and operation of organizations seem to parallel models of chemical engineering processes and suggests that similar information management techniques will apply to both. It describes some of the developments in information technology that will apply to both.

Trends in Corporate Organization

Major transformations are underway in management literature and practice. Almost all institutions are reshaping their patterns for organizing work and the work place. They are scrutinizing traditional folklore about management truths and testing new management ideas, while also examining concepts about managing for and achieving quality products. We speculate that the activities of institutions, in their current search for improved work processes and organizational design, parallel our activities aimed at improved chemical and petroleum process design and that computer-based information technology will have major impact on both.

Breaking Down Work Efforts — Decomposition

Taylor's (1911) classic of management science literature has strongly influenced the configuration and activities of the workplace. Taylor proposed to improve the efficiency of the organization by substitution of science for individual judgment. There was only "one best method" which guaranteed maximum efficiency. The worker was considered to be part of the machine. It was, therefore, unnecessary for a worker to understand why or exactly how the system performed. Explicit job descriptions, time and motion studies, and work method standards were expected to assure work efficiency and free the worker from the drudgery of work practices. Management's functions in his model are to enforce work standards and provide the best tools and working conditions. It provides the understanding of the overall goal and the means to achieve it. As organizations grew in size and complexity in the early 20th century, this model enabled great increases in efficiency. It remains central to many organizations today.

Drucker's (1973) view of management is broader. "Management has to give direction to the institution that it manages." This includes establishing mission, setting objectives, and providing resources for the institution. Management skills include: "communications within organizations; making decisions under conditions of uncertainty; and strategic planning." The importance of the customer is emphasized: "There is only one valid definition of business purpose: *to create a customer*."

We suggest that Taylor's view of organizational effectiveness and Drucker's view of management tasks might parallel chemical engineers' view of their domain in terms of unit operations and unit processes. Decomposing the process systems into their units and understanding the detailed functioning of these units enables the understanding of the overall process just as decomposing organizational work functions enabled improving organizational efficiency.

Analysis of Work Activities

In the early 1980s many observers viewed US industry as being in organizational crisis. Peters and Waterman (1982) searched for excellence in America's best run corporations and found, among other things, three major characteristics: 1) that customers reign supreme, 2) that a high level of employee dedication and enthusiasm was essential and 3) that trial and error was not only accepted but encouraged. The late W. Edwards Deming taught his 14 Essential Points for Managers (Walton, 1986; Scherkenbach, 1986). The Deming method emphasis is also on the customer as well as improving consistency, instituting leadership, eliminating employee fear, and training and retraining. Some essential elements of Total Quality Management (TQM) include: 1) an accurate and complete understanding and view of the customer; 2) a need to understand the entire work process; 3) systematic measurement of performance to assure minimum variation; and 4) the need to continuously improve the work performing system.

Several paths are being used to achieve TQM including assimilating the Deming steps, utilizing the Baldridge Award criteria as measurement techniques, and undertaking ISO 9000 certification. In many ways these paths are similar. An advantage of the Baldridge Award criteria is that it provides metrics for measurement of the state of the organization and, therefore, also can be used to indicate improvement quantitatively. While it can be argued that these metrics are arbitrary, they nonetheless provide valuable indicators to management.

The ISO 9000 series is a set of five quality system standards which include documentation of activities and procedures, guidelines for management reviews both in frequency and content, provisions for organizational self review and training. Rather than identifying and distinguishing measurable characteristics as found in most technical standards, the ISO 9000 series provides management guidelines and models that are aimed at assuring customers that their expectations will be met. They are characterized by "considerations" rather than "directives."

ISO 9000 certification is becoming essential for business, particularly in the European Community (Corrigan, 1994). Customer requirements were cited as the leading reason for registration by 26% of 110 chemical companies in a recent survey. ISO 9000 registration reached about 3,600 in the first quarter of 1994 compared to 1,500 a year ago and 400 two years ago (Thayer, 1994).

We suggest that if the Taylor/Drucker methods for analysis of management is parallel to chemical engineering unit operations, the Deming/TQM approach is analogous to overall process descriptions — the integration of parts and their context with the outside world. Analysis of the overall process as an entity and examination of the interactions of the components to determine how to improve the process appear to be characteristics of both. Using techniques that prove successful in management may be beneficial for improving process design methods.

Synthesis of New Work Processes

More recently, the impact of the information age and the necessity of examining the details and objectives of the entire work process is being described. Naisbitt (1985) described the transformation of organizations and specific work activities that were beginning to emerge as a result of the information age. Flatter organizations with less bureaucracy, networks where everyone learns from everyone, where intuition becomes valuable because there is so much data, where much of the work is performed by contract rather than hired staff are part of the "Re-invented Corporation." Middle management is replaced by the computer and people managers by independent, competent and self confident, self managers.

"The *fundamental* rethinking and *radical* redesign of *business processes* to achieve *dramatic* improvements in critical contemporary measures of performance ..." is how Hammer (1993) describes the activity of *Re-engineering the Corporation*. He emphasizes the difference between incremental continuous improvement and fundamental change to the work processes, organization and culture. The business process is the complete action of creating and delivering the desired products to the customer and must be understood in its entirety. In contrast to Taylor's work breakdown, Hammer's emphasis is on work integration. Activities of organizational entities must bring added value to the core business as a whole rather than be justified on an internal unit goal. Even inter-company organizational boundaries are crossed. Information technology is a key enabler for the re-engineered corporation. It does this by creating opportunities to change the work process dramatically and not just doing the same thing faster. Radical redesign of the complete organization from top to bottom and a dramatic change in the way of doing business is also cited by the National Research Council (1991) as an ingredient for Designing for Competitive Advantage.

Work and Chemical Processes — The Future

We suggest that the most recent approach to organizational management analysis strongly parallels the state of chemical engineering process design analysis when process synthesis was embryonic. We see parallels between Hammer's work and retrofit design. Institutional management may benefit from the systematic techniques developed and proven for process synthesis. Furthermore, the needs of the process design community and the general business community in terms of information consolidation and management appear to be strongly congruent. The challenge will be to integrate developments in the common elements of these communities which have diverse histories, cultures and even language. It is apparent that one of the certainties of the future is uncertainty and that changes will occur not only in information technology, but in all aspects of business. The Learning Organization as advocated by Senge (1990) will be better able to respond to these changes. "Total Systems Thinking," seeing the whole ("The Fifth Discipline"), is a major attribute of this organization. It should have the same characteristics as a self-adaptive organism, adjusting to changes in the business environment and technology. System archetypes (Senge, 1990) which help understanding of business detail and dynamic complexity and which can be connected to form different business structures might be a useful starting point for business process synthesis. Furthermore, concepts in chemical engineering process control might be useful in the analysis of business environment and organizational dynamics.

The preceding review focused on elements common to chemical engineering and business processes. A basic difference is the human element in business processes. Morale, participation with high performance teams, innovation, motivation to learn and just plain thinking hard are human characteristics for which information technology has little to offer at present. Both business and process design success relies heavily on these characteristics.

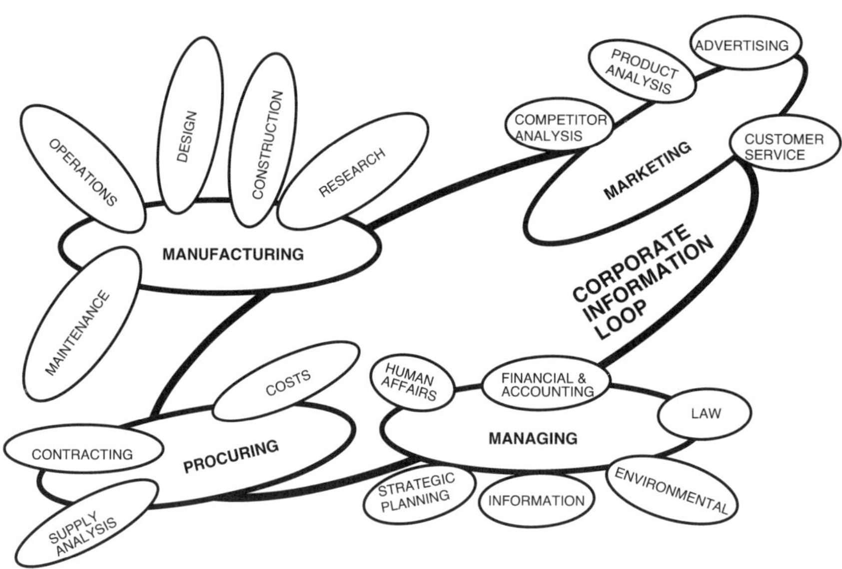

Figure 1. Total corporate information flow loop for a typical chemical or petroleum industry institution.

The Corporate Information Loop

As we discuss later, information consolidation technology is evolving rapidly. While the design activity represents only a small fraction of total resources involved or total information flow in any corporation, it is strongly influenced by changes in information flow. Fig. 1 shows one perception of a total corporate information flow loop for a typical chemical or petroleum industry institution. All commercial businesses utilize work processes to convert some input to some output. Each of the nodes represents a subdivision of the overall organization. In this simple example, the Procuring node represents obtaining raw material or other inputs; Manufacturing represents the transformation process and Marketing represents dialogue with the customer. Corporate Management makes up the final of the four major nodes. Further sub-nodes depict major activities within each of the major divisions. It is likely that there are further sub divisions in most organizations. (For example, offsite design as a sub division of design and cooling water system design as a sub division of offsite design).

There are three points to make with this diagram: 1) almost everything done within the design sub-node is influenced by some information from one or more of the other sub-nodes, 2) organizational control, as usually practiced, requires that information based on current data be approved by the management of each of the sub-nodes before it is shared with other nodes and 3) all information is not shared among all nodes. As information becomes more broadly shared and organizational culture changes, it is likely that the information available for design activities will also change. It appears likely that decisions that were made by parts of the organization and documented by providing "functional or job specifications" are more likely to be made by the design team. For example, data about costs of equipment, catalysts, utilities, etc. will be available in "real time." Product quality requirements and projected sales data would also be current as would knowledge about the availability financial resources. This activity would have such broad corporate ramifications that the team might be referred to as the "future team" rather than the design team.

Information

We are in the age of Information Technology. But what is information technology? Let us first define information. One often hears discussions about the differences among data, information and knowledge. Typically this discussion proposes that *data* is unprocessed, plentiful and likely to contain both meaningful and irrelevant things. An example is all the readings from a set of sensors for a plant. When one takes lots of data, removes the irrelevant parts, discovers the hidden messages in it and organizes it to expose patterns, one is transforming data into *information*. When one converts information into insights that aid in one's work, one is transforming it into *knowledge*. These steps are abstraction processes in that lots of data are summarized to form a more useful and much more compact way to describe it. An example of abstraction is to take a parts list and abstract it by saying this list describes a distillation column, an interpretation that immediately gives considerable understanding of it to a chemical engineer. It becomes evident that one person's knowledge can be someone else's data. These definitions are, therefore, relative to one's perspective. In what follows we shall be concerned with aiding this abstraction process.

Information Management/Consolidation Technology

Information management/consolidation involves the discovering and/or creating, the abstracting, the interpreting, the organizing, the using, and the sharing of information. The computer, and very specifically the networking of computers, is at the heart and will continue to be at the heart of all new information management/consolidation technology.

Consolidation technology involves the use of analysis tools to abstract information content out of data. An example would be to input the data from a set of process sensors into a computer model which then determines the material and heat flows for that process. Another would be to take noisy data and smooth it using digital filtering techniques.

While such analysis tools are extremely important and worthy of a lengthy discussion, we shall concentrate in this section what we call information management (IM). We shall, in particular, emphasize ways to discover information located elsewhere, to organize information for the purpose of exposing relationships among it and for sharing it both directly and by making it available for others to discover.

A simple form of managing information is to organize files containing it under a directory/subdirectory tree as is done in all computer operating systems. One typically organizes such a directory system to put things which are related into the same subdirectory. Through the use of "links" (a special pointer to a file), a file can appear to be in more than a single subdirectory. If one browses a subdirectory, the operating system lists a link exactly as it lists a file name. Clicking on a link will move one to the subdirectory containing the file and will open the file. The operating system will now be pointing in the subdirectory where the file is located with the disadvantage that it (and often the user) will not remember how it got there. Files can contain any of a variety of things: text, drawings, audio recordings, film strips and so forth.

A second way to organize information is to place it in a database, and a related third way is a document management system. We shall assume the reader is familiar with the concepts related to both of these technologies.

A fourth approach is to organize information into a hypertext. A *hypertext* is a text or graphics document with buttons attached to it. Clicking a mouse on one of these buttons causes a hypertext system to open up another page of text that the author has *linked* with the current document by associating it with the button. On-line help systems today often are hypertexts. The author places buttons, which are often invisible, under a piece of text (or within a drawing) so one gets the impression that one is clicking on the text. Doing so will bring up a page relevant to that piece of text. A hypertext is, therefore, a complex interlinked document which one can read by following several different paths through it. The hypertext concept is an important ingredient of the world wide web (WWW).

Placing meaningful labels on the links which connect two documents in a form the IM system recognizes extends the capability to relate information ever further and is part of the semantic network modeling one often uses to create data models for databases. Such labeling is also an important ingredient of our information modeling environment *n*-dim (***n*-d**imensional **i**nformation **m**odeling).

There are many properties we hypothesize we want to have in IM systems. We want to keep and link information on people, on their activities, on the formal and informal organizational structures relating those people, on the numerical and symbolic data describing physical artifacts such as the data for a flowsheet computation, on the group development of arguments such as described by gIBIS (Conklin and Begman, 1988), on drawings, on correspondence and e-mail files, and so forth. We want to be able to store tremendous amount of information. We want to be able to find things in this information by browsing, by formal searching and by being pointed to it because someone else or the IM system concludes it might be interesting to us. We usually want very fast responses. We will often want the IM system to record enough history of the data creation process that we can study the process and/or reproduce the arguments behind the *why* of the data and not just values of data. The argument that the annotated blueprint for a design is not enough supports such capture. Often, especially for a new activity, deciding what information to gather, how to structure it and how to share it is a major part of the effort one makes in managing it. We want these systems to support exploration among alternatives for organizing and interrelating it.

IM cannot be solved through technology alone. With more information kept, a company may expose itself to litigation where the evidence needed by the other party will likely be sitting there in the IM system. There are also significant "people" issues one must understand also when developing and using this type of technology. For example the maintenance of a more complete history may make a participant very nervous about how his/her boss will use it when the annual review comes around.

We shall now examine the World Wide Web (WWW) and *n*-dim in more detail. We shall argue then how databases, the WWW and *n*-dim are each useful for information modeling, assessing both advantages and shortcomings.

The World Wide Web

The World Wide Web is a system that is supporting the creation and sharing of a global hypertext web of information. It physically exists now and is growing very rapidly, both in the information it contains and in the facilities available over it.

Two tools provide access to the WWW. The first displays any document written in the HyperText Markup Language, HTML. An HTML file is an ASCII text file with instructions mixed among the text telling how to display it. The approach is very similar to having a text document in Latex (Lamport, 1986) or Scribe (Unilogic, 1985). The second tool is the locator tool and is one that, when given the address of an object in the form of a Uniform Resource Locator (URL), will automatically retrieve it from across the Internet using its own HyperText Transport Protocols, HTTP. (These protocols are similar to the commonly used File Transfer Protocols (FTP) and to those underlying X-windows.)

A very common program that people use to interface these WWW access tools is Mosaic. One can also interface them using the Emacs editor, for example. Anyone with Internet connections can access the World Wide Web by installing Mosaic. To retrieve the Mosaic interface software:

> Anyone on the Internet can obtain an executable and/or source code needed to run Mosaic. Getting the code is via anonymous FTP. At present support is available for several workstations but not all. FTP to the Internet address *ftp.ncsa.uiuc.edu*. Use *anonymous* as the response to *name* and follow instructions. The binaries are in the subdirectory Mosaic/Mosaic-binaries. They have been compressed using the *gzip* program which is available by anonymous FTP to *prep.ai.mit.edu*. *gzip* is in the subdirectory */pub/gnu*.

When one starts the Mosaic interface, the locator tool retrieves and displays, using the display tool, a designated HTML file called a "home page." This display will contain text intermixed with graphics. Within the text are highlighted phrases (underlined and blue). Each indicates the existence of a link to another computer file elsewhere in the World Wide Web. As one places the mouse arrow (without clicking) over a highlighted phrase, the Mosaic interface displays the associated file address (its URL) in a small window at the bottom of the main Mosaic window. Clicking on the phrase invokes the locator tool which then retrieves the object and opens it, or, if it is a program, launches it (i.e. causes it to start executing).

This object may be another page, a graphics object such as the latest US infrared weather map or an executable such as the program on the library computer at Carnegie Mellon University that allows one to search CMU's database of article titles, authors, keywords and abstracts. The locator tool is capable of deciding the kind of file it has retrieved and launching it if it is an executable or starting a local software package that can display it, if that software exists; if not available, one gets a simple apology and no display.

A person can write his/her own pages in HTML using a text editor and add him/herself to the WWW at any time.

n-dim

n-dim is an information modeling environment on which the three authors of the this article from Carnegie Mellon University, along with others in the *n*-dim group, have been working for the last four years. From observing different design projects (new control system for power generation with Westinghouse, a connector design with three other Engineering Research Centers and a Database design with Alcoa), we noticed the importance of managing the information that the different groups created and the difficulty all the project participants had in sharing it and establishing a shared understanding of it. Two subsequent projects, one with ABB and one with Union Switch and Signal, have provided us with more empirical evidence of the importance of handling information within a design activity. There are also several other projects reported in the literature that come to the same conclusions (Schmidt, 1993, Grønbæk, et al., 1993).

To aid the handling of information, we hypothesized that a design team was always *modeling* things. Team members are continually sketching and revising models of the organization, of the process they will undertake — perhaps as an activity diagram, as an equational model of the physical artifact they wish to design, etc. We proposed a general purpose *modeling* environment where one can interrelate all types of models. We also wanted to support activities where the design team needs to learn what information to gather, how to structure it and how to share it. Finally, like the Internet, we argued that *n*-dim has to support users at their desks while they are geographically apart.

The environment has to record the process occurring more precisely than anyone can do at present, annotating the mistakes or the reasoning used as well as the final results. It should allow a future team to use this recording as a starting point for a similar design. It should also allow the company to have the information needed to improve the work process itself (i.e. the work process is also a designable artifact).

Recording also suggests big brother sitting there watching (including governmental agencies who demand information). Recording is needed. However, one cannot ignore the social concerns discussed here; we must consider them along with the technical solutions.

n-dim has progressed through three generations of development from a Macintosh hypertext version to a demonstration multi-machine, multi-user IM system. The current version is a complete rewrite, comprising the BOS (Basic Object System) in C and supported by an interactive language called STITCH. We can implement new ideas very rapidly and model hundreds of thousands of objects.

Information Modeling. *n*-dim uses nodes and links models. Fig. 2 illustrates. Consider first only the files and the links; ignore the grouping. Labeled links as we show here are not possible in organizing a typical file system. The labels can tell us how we think the files are interrelated. This type of structuring must be supplemented with grouping so we can appreciate which "Data file" leads, for example, to which "Printout." The same file — for example, "My code for X" — should exist in several groups. Fig. 3 shows actual groups typical of models in *n*-dim. They appear as boxes within boxes for the has-part link (the upper box has two files which are part of it: "Object 1" and "My code for X") and with explicit display of any labeled links, here the "annotates" link.

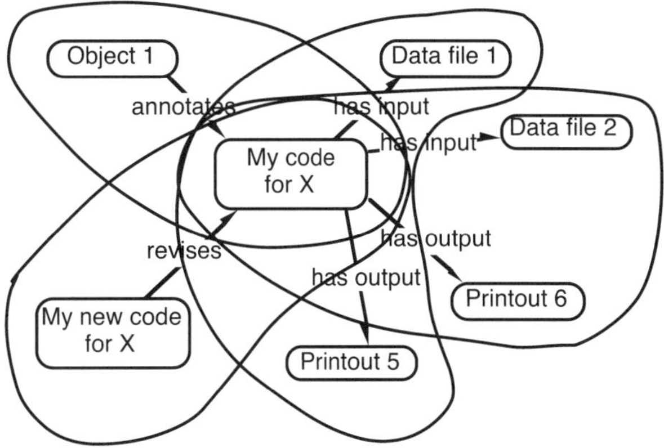

Figure 2. Grouping files into different views.

Atomic model object types in *n*-dim include integers, reals, complex numbers, frames, file pointers and the like. From these and other previously generated models we construct new *n*-dim models, each of which is a list of links — pure and simple. Links are objects that point from one object to another. In *n*-dim they always have an associated label. *n*-dim contains special "has-part" links. Each points from the identifier of the model itself to the identifier of an object that model "contains." It is the display of these objects that gives one the impression that the parts are "in" the model. For convenience, we will often speak of an object being in an *n*-dim model when it fact the *n*-dim model only contains a has-part link from the model identifier to the object identifier. Semantically, has-part links are the links one creates when organizing a set of files into a directory/subdirectory tree or in creating a hypertext. The parts of a directory are its contained files and subdirectories. The parts for a hypertext are those objects associated with its buttons.

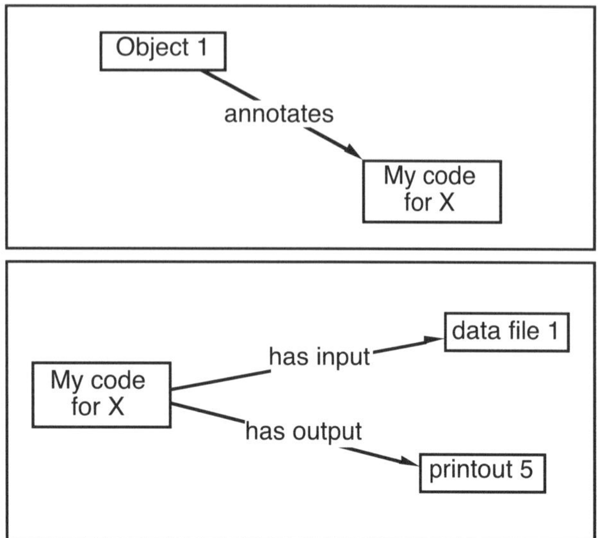

Figure 3. Nodes and links models typical of those created in n-dim. *These models relate to two of the views shown in Fig. 2.*

A user can also include any number of user-defined, labeled links to express relationships between pairs of model parts (including self links). For example, one could use a link labeled "is-preceded-by" to relate two activities which are part of an activities model. In usual directory/subdirectory file management schemes, such links are not available nor are they explicitly available in a hypertext.

Conceptually objects in *n*-dim exist in a flat space. One imposes an organizing structure over these objects when one relates them by placing them into an *n*-dim model. Each object needs to exist only one time but can be a part of any number of models.

Publishing, Revision Management, Prescriptions and Access Control. Publishing is the mechanism n-dim uses to make an object persist. n-dim uses publishing to maintain history and to facilitate communication. A proper metaphor is the library. Once one publishes an article, it will always exist and cannot be altered. Any two people claiming to read the same article actually have read different copies which are guaranteed to be the same. We can partition the objects to which n-dim can point into two types: those which the user has given to n-dim to store and manage and those which the user manages outside n-dim, such as files in his/her file system or external programs belonging to the system as a whole.

A user creating an object in *n*-dim will do so by an editing process. To share a model with others, its creator must *publish* it. *n*-dim will not allow a published object to be modified by anyone, not even its creator/owner. The user must first publish all the parts in a model that *n*-dim manages. If *n*-dim manages everything the object knows about either directly or indirectly through its parts, then a published object is guaranteed to be immutable.

n-dim must also support revision management since a published object may not be considered finally correct. The original object will continue to exist along with the revision. Anyone can relate a revised object to an original by creating a separate *revision model* containing the two objects with a revision link connecting them. By treating revision links as special, *n*-dim can locate revisions rapidly.

Often a person wants to open the latest revision for an object rather than the original. To accommodate this need, we created the notion of a *prescription* model. It has two parts: the original object and an *n*-dim model representing a person or group of persons. If a user clicks on a prescription model, *n*-dim does not open it. Rather it locates the latest revision of the original object made by anyone in the group and opens that.

Two benefits accrue. First, one gets the latest revision even for published parts and, second, *n*-dim keeps all revisions as a history of the object. Two people revising a published document can only copy it and edit the copies, sidestepping the so called *long term check-out problem* with databases and document management systems. They will have to reconcile any differences later.

A user can control who has *access* to any of the objects s/he publishes. A person can only see an object created by someone else if it is published and if the creator gives that person access to see it. A conflict occurs between access and publishing. Access is an attribute of an object. How can one change the access for a published object? First of all we add the attribute of access to an *n*-dim model by adding a special part to each model that points at an *n*-dim model of a person or a group of persons. Then we resolve the publishing issue by requiring the access model to be a prescription model so the pointing is indirect. With this approach *n*-dim knows who currently has access and everyone who has ever had access and when.

Rules. Users want *n*-dim to tell them if certain events occur. For example, if someone in a group creates a revision link to an agenda model the user published last week. We introduced *rules* that events known to *n*-dim trigger. Publishing a rule (an *n*-dim model) turns it on; revising a rule allows one to alter it or turn it off. *n*-dim remembers a history of all rules and when they were active because they are *n*-dim models.

Launching External Programs. A user can structure data in an *n*-dim object type we call a *frame*. These data can become the input for an external program such as a simulator. For each type of external program, *n*-dim must have a special purpose *wrapper* program to translate data from an *n*-dim frame into a suitable input file for the program. A completed wrapper will also translate output from the program into an *n*-dim frame. The external program is an object at which *n*-dim can point. Thus an icon for it can appear as a part of an *n*-dim model. Double clicking with the mouse on this icon *opens* the external program which in this case launches it by first translating the data and then triggering its execution.

Storage of Objects. Each *n*-dim object has several attributes such as a unique *n*-dim identification string, the name of an owner, a time of creation, a time of publishing, and a language used to create it. It also has its contents which, for a model, is a list of has-part and labeled links. Graphical objects contain graphics, etc. Text (for example, a whole article), graphical, audio and animated objects can be very large.

n-dim stores all attributes except the contents for large objects in any of a number of databases throughout the web. Large objects are stored as files or, if the database system will allow it, inside the database. We use the ISIS system (Birman et al., 1992) to connect workstations over a network. Available both as "freeware" until recently and in a commercial version, ISIS takes care of all the messy coordination problems that occur when sending messages back and forth among networked computers. Each participating workstation has a piece of ISIS code on it that sends and receives messages from the other participating workstations. To find an object, one can ask the local ISIS code to broadcast the request to all the other workstations. It passes responses back to the requester.

By storing all the model attributes, one can construct a query to find a model which may be stored anywhere in the web by its attributes. For example one can ask for all activity models which Richard Smith owns and which he created after January 15, 1994. ISIS broadcasts such a query to databases throughout the *n*-dim web.

Languages: A Mechanism to Support Inductive Learning. A language is a special *n*-dim model that any user can construct. Using a model as a language when creating a new model limits the new model to having only the object types and link types listed in the language model. A universal language allows one to construct a model using any type of part and any type of labeled link. Using the universal language allows users to try out new types of models which they may then share with others. For example, suppose a member of a design team sees the need for an activity model to organize his activities. He will create instances of such a model allowing it to contain only atomic activities (as frames containing activity names, start times, etc.), other instances of activity models and precedence links. He may share this with others in the group who adopt it. One of them may write a method to find critical paths. Finally, to pass model type and its supporting method to others, someone in the group creates a language for constructing a legal activity model. Until the language exists, a model author enforced these restrictions manually. We see the capabilities evolving with time in this manner. The generalization process needed to hypothesize the usefulness of a type of model and then to create a language to pass it to others is an *induction* process. *n*-dim does not do induction, but it supports its users in this process.

Using Information Management Technology

We shall discuss three key uses for information management technology: to generate standards, to support collaboration, and to permit an organization to reuse and to improve its work processes.

We have argued above that we can use databases, document management systems, the World Wide Web and *n*-dim for information management. Since *n*-dim uses databases to store the things it controls, we view it as the software one will add to a database system to effect the type of information management we have described above. We, therefore, shall generally limit our comparisons here to using *n*-dim vs. using the WWW. We believe the following to be the key comparison issues.

Structuring Information

The WWW is a hypertext and, as such, supports the has-part link but not the notion of a labeled link directly. A user can group items by referring to them from a single page. The text on the page can serve as the annotation for the grouping, describing why each item is there and how all the items relate to each other. Thus one can informally set up the equivalent of semantic nets that underlie the design of data structuring in databases. This capability makes the WWW very useful for structuring information. *n*-dim, with its labeled links, allows the equivalent structuring to occur.

Allowing Ad Hoc Views

There are two extremes for allowing users to choose how to structure information: complete anarchy or complete control. Anarchy allows users to discover new ways to structure the information — i.e. it supports induction, but it precludes users easily attaching analysis codes and using the data as input. Complete control supports attaching of analysis codes, but it restricts use to routine, predefined structuring of information.

The WWW is complete anarchy. Users are free to structure information anyway they wish. At the other extreme is the complete control database managers often impose on users of corporate-wide databases. Typically, experts have predefined all the schemata allowable for users to enter data. For some database managers, users can invent their own schemata for their own data in their own space but cannot readily share it in that form with others. *n*-dim supports both approaches. Languages other than the Universal Language support predefined structuring while the Universal Language allows complete anarchy. Users can share all models. Schemata, like other *n*-dim models, can be redefined.

Finding Information

A major activity in an IM system is locating things which are useful. Searching for interesting things in the World Wide Web has generally been by browsing (called *surfing* by some) around the Web. One can place the names of interesting pages in a "Hotlist" that the interface program Mosaic maintains for each user, making it easier to relocate interesting objects.

With the development of a new program called the WWW Worm (WWWW) (McBryan, 1994), one can now also locate objects using text strings that occur in their title, their URL, or in any button text and URL they contain. The WWWW runs in off-peak hours[1]. It discovers more and more HTML files by recursively scanning through all HTML files of which it is aware. For each HTML document, the WWWW tabulates its title, its URL and each URL the document mentions together with the text annotating that URL.

n-dim stores several attributes in searchable databases for all objects it maintains. In particular, it keeps all the links contained in a model as part of its searchable attributes. Thus one can search over all has-part and labeled links, owner's name, modeling language, date of creation/publishing and so forth, to find things of interest. Also one can simply browse.

Maintaining History

The WWW has no specific mechanisms for maintaining a history of the objects in it. Participants can edit any object they own, and the new version automatically replaces the old provided the author alters neither its name nor its location. Thus one will always access the latest version of an object. A group of collaborating users of the WWW can agree informally to maintain a history by the following mechanism. Before editing a document, copy it into a file with a name such as fileX.revision12. Then point at this revision in the new version. This form of revision management is limited to a chain of revisions that the owner maintains.

Many databases also do not have a history mechanism, though some newer ones now keep everything that has ever been put in them, and users can ask for the answer that would have been given at noon three days ago.

A main feature of *n*-dim is that published objects are both immutable and permanent. In principle, no one can change them nor remove them (of course, a superuser or loss by failing to backup information can defeat any such mechanism). *n*-dim users must publish objects to share them so they will publish lots of things. *n*-dim has a complete revision management/recording capability. Anyone can revise any other model, even those they do not own. Prescription models allow one to open only the latest revision for an object, allowing one to find only those revisions made by an identified group of people. Access control uses prescriptions so *n*-dim records who has ever had access and when.

[1] To find the WWWW, open the URL http://www.cs.colorado.edu/homes/mcbryan/public_html/bb/summary.html which is the home page for the Mother of all Bulletin Boards.

Speed of Access

The WWW and most databases have each object stored in one location. Thus speed of access depends on how rapidly one can copy the object from that location. The publishing paradigm of *n*-dim guarantees the immutability of objects. Thus it can keep multiple copies dispersed strategically around the web, offering the opportunity to speed up searches. For example, *n*-dim can place copies of objects one accesses frequently in one's local database.

Controlling Access

The WWW defers access control to that of the operating system for the computers on which the information resides. *n*-dim maintains its own access management system on top of that for the operating system. We described it previously.

Controlling Who Can Join

One of the beautiful characteristics of the World Wide Web is that no one is in control of it nor who can join it. Anyone can join at any time by fetching and installing the right software. New users must make their files accessible and both accessible and known if they want to be a source of information for the community, something which happens when others point at their objects.

Through a policy decision, *n*-dim, like a database community, controls who is in and who is not.

An Application: Generating Standards

Standards generation is a process by which the practice and experience with a product designed and used over time is codified into a set of rules, representations and methods. Information management technology provides a means to aid this process by allowing the creation and coexistence of alternate models of standards in the same computational environment. In a product design organization, this technology creates a new possibility where standards creation is not an ad hoc periodic process but a continuous process where the practitioners can incorporate changes in the social and technical requirements as clearly and as fast as possible to minimize variations in the designs produced.

The WWW is not intended to support multiple classifications and the history that are critical to the standards generation process. However, a standards group could get input from a large external audience by making the standards documents available for comment as we do today on paper.

An Application: Supporting Collaboration

Imagine a group of designers working together. They generally work at their own desks, meeting together every two to three weeks. The design is something new for them. Their first tasks are to gather information, establish a working vocabulary, and define and create a common understanding of the problem and possible approaches to solving it. They agree to scan images into the computer of all hand written documents and to save electronic objects such as e-mail messages, drawings, and so forth. Databases, documentation management systems, the WWW and *n*-dim offer very useful ways to keep and structure this information. By using the protection schemes on their computers they can protect their files from others.

n-dim is our hypothesis for the type of support needed for collaboration. Our next step is to prove this hypothesis by actual case studies. Our beliefs are that maintaining history is important (it is frustrating to have an object you count on disappear from sight, even one that has errors in it), ad hoc models are a necessity as no one yet knows what information they will collect nor how they want to structure it, and being able to find things by attribute is crucial — browsing is not sufficient.

An Application: Reusing and Improving Work Processes

To reuse and/or improve a work process, we believe one must maintain a high fidelity history of instances of that process. The company can then use the history as a starting point for reuse. It can also study these recorded instances to discover what really occurs and to look for ways to make improvements. Interviewing people after they complete an activity does not give an accurate view of what they have done as they will rationalize their actions. One can demonstrate this assertion by video taping a problem solving session and comparing it with a follow-up interview.

Conclusions

Greatly expanded information management is now possible. It will take time to discover all the attributes support systems must have for information management as no one has yet really tested these systems. But we shall. This is an exciting but in some sense troubling time as shown in the next section.

A Small Glimpse of Etopia — an Information Society Some Time Far Into the Future

Past attempts to characterize an ideal world include Sir Thomas More's *Utopia* written in the early 16th century, Edward Bellamy's *Looking Backward* and William Morris's *News From Nowhere* in the late 19th century. They offered solutions to perceived abuses in the cultural and social systems of the times. The value of these solutions in these works continue to be debated but many have been adopted; the credit card from Bellamy and self education from Morris are examples. We thought that it might enliven discussion to speculate about the future of process design in an hypothetical future electronic information age; we call it *Etopia*.

In Etopia work falls into three categories of activities: 1) Education, 2) Business and 3) Consulting. Business activities include the transformation of material or knowledge inputs into outputs. Businesses are privately held reporting to stockholders or publicly held reporting to Oracles. Educators consolidate what is known about reality into what is necessary to teach students about how to apprentice in specific aspects of business or consulting. Consultants provide the service of conjecturing about what is presently unknown to businesses. Businesses provide material needs, social welfare and public infrastructure to all inhabitants. Representatives from each of the three work categories are elected by their underwriters to manage businesses. An agent, called Network, provides all tools and data about all work activities for all institutions in Etopia. Each institution can maintain some data and a number of methods for confidential use (similar to other privately held property). Businesses and individuals pay taxes based on both 1) the amount of data, methods and other capital assets which they hold privately and 2) on the value their business adds. The taxes support Network and the social and infrastructure activities.

There is a capability for all to share information from publicly funded research. However, the Oracles have fixed delays for access according to type of information (such as medical or defense types) as well as to which operational organization the agent seeking the information belongs. At present there is a maximum 18 month time delay. Initially it was four years, but, as the evaluation systems improved to recognize ideas that truly benefited society, the Oracles decided to shorten the delay period.

Network automatically translates all information into different languages including nuances and syntax of meaning. Where meanings are ambiguous, as was often the case in previous technical papers, Network asks the author for clarification.

The Etopian society fosters teamwork by rewarding and recognizing outstanding team performance. All members of the team receive the same level of reward. Contributors learn to work together without concern about individual glory. Dissension among team members is accepted, but Network evaluation processes aid the rapid formation of a consensus.

The society encourages personal mastery of special disciplines. It also promotes Network dialogue among people who are working on similar ideas. Network records who invented what first. The Oracles establish criteria for novelty. Oracles are elected from three highest levels of apprentice and full consultants. Network makes the election based on the record of the value it gives to the consultants' judgments. An Oracle's term of office ranges from three to twelve years, based on random assignment at election.

Etopians have solved the problem of ownership of ideas. Network monitors the misuse of or taking credit for others ideas, and Oracles punish offenders by ostracizing them from the information community. Network, with oversight by the Oracles, patents and values ideas. Institutions and individuals, as users of ideas, remunerate idea owners and their organizations. The society encourages the use of others ideas.

Education in Etopia is truly a pleasure. For education through the BS level, students and mentors view the same virtual screen. However, students communicate verbally through Network by asking questions about the subject being discussed. Instructors have provided several levels of help in the form of a hypertext. Network allows users to select a depth or a breadth view through the hypertext information tree. As the deeper levels of information become more esoteric, Network automatically supplies required underlying knowledge and theory needed to interpret the given information. As the level deepens further, Network opens actual experimental databases. If the user selects breadth, Network provides associations with other similar branches of knowledge. If depth in a hypertext is analogous to a dictionary, breadth is analogous to a thesaurus. Educators learn because Network gives them instantaneous feedback on the questions students ask, the paths they take to solve problems and on whatever Network perceives is missing. They can interrupt student learning sessions to provide guidance. Educators continually refine the information trees. The breadth dimension of the trees has grown so that all former educational disciplines now overlap. The core curriculum, always the center of much controversy, now consists of a path connecting the nodes of a few fundamental subjects. Students no longer attend lectures but are connected to a virtual interactive discussion session when they complete the exploration of a node.

Society encourages all citizens of Etopia to take risks with new processes and products without undue concern for liability. Initially, all citizens were concerned about the ability of Network to record timing and ownership of all activities. This lead to a reluctance for users to take advantage of the capabilities of the system. Decision makers feared that knowledge about who made which decision would lead to legal liability far into the future. Someone wrote a novel about the family of an engineer who lost everything in a class action suit because Network had documented that this engineer made an incorrect judgment. He had selected a new ceramic material for a bearing based on superior performance in the laboratory over a three year period. The ceramic actually underwent a very slow denitration reaction in the presence of some newly introduced lubricant additives. As a consequence of this slow reaction, the wheel bearings on several very high speed passenger trains failed over a period of a week. There was significant damage and loss of life. It turned out that the reaction in question had been mentioned in passing in one of the Network documented courses that the engineer had taken in industrial chemistry seven years before. In the novel the Oracles agreed that the engineer should have known about the reaction. Because of public outcry about the novel, and the recognition that these concerns were real, legal rules were rewritten to limit both the scope and the time that historical data from Network could be used for evidence in litigation as well as the evaluation of personal performance. Furthermore, the capability of Network to associate and evaluate marginal information was enhanced so that it was less likely for information to be obscure. Only rarely now are legal questions about liability tried before the Oracles.

Etopians appreciate change. For eons humans and their evolutionary predecessors were conditioned not to accept new things into either their everyday life or their thinking patterns. A random selection of inhabitants now test new products and ideas. It has become part of the cultural entertainment to be a participant in any testing program.

Etopians understand well the importance of complete system modeling. A Model Building Tool (MBT) reached a high state of development as a result of great attention by many consultants. MBT suggests all potential constructs for the system being modeled. As the user pares the structure, MBT automatically generates a list of assumptions. Thus, the documentation for any model contains the list of underlying assumptions. A surprising result from using MBT is that in some cases model building is an initial valued problem. Some Consultants have found that slightly different models result depending on the urgency of the situation for which the model is needed. This significant finding is the center of much consternation in the Consultant community and an active area of investigation.

Professional societies have melded together to form a consortium of specialized technical consultants. Multiple hierarchies of consultants each have their own bulletin boards for communication. Consultants are trained by taking virtual courses in their specialty. The experiences in these courses include emergency situations from actual equipment and system operating history. Societies use licensing exams to qualify consultants, extending the BS education system. They score apprentice consultants based on the speed, originality and soundness of their responses and actions. Consultants' rankings are based on the same system as "Go" players. There are 30 levels of entry apprentice, six levels of apprentice and nine levels of full consultants. There are only three or four consultants at level 9 in any field at any one time. Project Managers bid for consultants at any point in time with higher level consultants being more expensive.

Network provides immediate costs for equipment, labor, modeling, design, laboratory work, etc. Network establishes consulting and business fees and adjust them according to demand, utility and uniqueness of products. Inter-business supplier--purchaser relationships have been firmly established on the basis of contractual auditing procedures. These procedures allow the purchaser to receive certain information about the cost of manufacturing or otherwise creating products. This allows purchasers to make better decisions. One result is that now each vendor supplies to the purchaser the best estimate about cost of equipment items based on the probability that equipment will actually be purchased, the delivery schedule needed, the costs of labor and raw material and the process and mechanical specifications. Vendors also provide information about costs and characteristics of alternative equipment which they have under development, thus assuring a plant will be built or revamped with state of the art equipment.

Network has been given the capability to act as an independent "third party" to search certain aspects of all institution information bases. Thus Network can be used to automatically "benchmark" the performance of similar institutions, point out the differences, and recommend improvements. Where it finds similar data (for example corrosion data for various materials in various chemical atmospheres), data bases are adjoined, greatly improving the accuracy of models used for equipment design.

These are only a few of the fascinating aspects of Etopia that relate to process design. Indeed, Network is in the process of writing a virtual book about Etopia.

References

Bellamy, Edward (1982). *Looking Backward*, Ed. Cecelia Tichi, Viking Penguin Inc. New York.

Birman, K., et al. (1992). *The ISIS Distributed Systems Systems Toolkit Programmers Manual*, ISIS Distributed Systems, Inc., Ithaca, New York.

Conklin, J. and Begman M. L. (1988). gIBIS: A Hypertext Tool for Exploratory Policy Discussion. *ACM Transactions on Office Information Systems*, **6**(4), 303-331.

Corrigan, James P. (May 1994). Is ISO 9000 the Path to TQM? *Quality Progress*.

Drucker, Peter F. (1973). *Management — Tasks, Responsibilities, Practices*, Harper and Row, New York.

Grønbæk, K., Kyng, M., and Mogensen, P. (1993). CSCW challenges: Cooperative design in engineering Projects. *Communications of the ACM* 36, **6**, 67-77.

Hammer, M, J. Champy (1993). *Reengineering the Corporation*, Harper Collins, New York.

Lamport, L. *Latex: Users Guide and Reference Manual*, Addison Wesley, Reading, Mass, 1986.

McBryan, O.A. (May 1994). GENVL and WWWW: Tools for Taming the Web, in Nierstrasz, O. (ed), *Proc. First Intn'l WWW Conference*, CERN, Geneva.

More, Sir Thomas (1975). *Utopia*, Translated and Edited by Roger M. Adams, W.W. Norton & Co. New York.

Morris, William (1962). *News from Nowhere and Selected Writings and Designs*, Edited by Asa Briggs, Penguin Books Ltd. London.

Naisbitt, J., P. Aburdene (1985). Re-*inventing the Corporation*, Warner Books, New York.

National Research Council (1991). *Improving Engineering Design — Designing for Competitive Advantage*, National Academy Press, Washington, D.C.

Peters, Thomas J., Robert H. Waterman Jr. (1982). *In Search of Excellence*, HarperCollins, New York.

Scherkenbach, William W. (1986). *The Deming Route to Quality and Productivity — Road Maps and Roadblocks*, CEE-Press Books, Washington, D.C.

Schmidt, K (1993). Modes and Mechanisms of Interaction in Cooperative Work. Eds. Carla Simone and Kjeld Scmidt, *COMIC Project Deliverable* 3.1, 21-104, Computing Department, Lancaster University, Lancaster, U.K.

Senge, Peter M. (1990). *The Fifth Discipline: The Art and Practice of the Learning Organization*, Doubleday, New York.

Taylor, Frederick Winslow (1911). *The Principles of Scientific Management*, Harper, New York.

Thayer, Ann M. (April 1985). Chemical Companies See Beneficial Results from ISO 9000 Registration. *C&EN*.

Unilogic (1985). *Scribe Document Production Software: User Manual*, Unilogic Ltd., Pa.

Walton, Mary (1986). *The Deming Management Method*, Dodd, Mead & Co., New York.

METHODS DRIVEN BY ADVANCED COMPUTING ENVIRONMENTS

J. F. Pekny
School of Chemical Engineering
Purdue University
West Lafayette, IN 47907-1283

Introduction

As expected, computer technology has continued its explosive growth since FOCAPD89. From a broad perspective, this growth is evolving into three major trends, (i) a dramatic improvement in price/performance/capability of available desktop systems, (ii) extensive integration of computers through networking, and (iii) a trend towards pervasive embedding of sophisticated digital technology that encompasses portability of computers and increased accuracy/volume of information available about processes. As an example of this growth tied to the FOCAPD conference series, consider comparing advertisements for desktop PCs in a major computer magazine, such as BYTE, between 1989 and 1994. A typical $2,500 system circa 1989 included an Intel 80386 CPU at 20 MHz, 1 MByte RAM, and 80 MByte hard disk drive while a $2,500 system circa 1994 includes an Intel Pentium at 60 MHz, 8 MByte RAM, and 540 MByte hard disk drive.

Harder to quantify, but more important to engineering productivity, are the types of algorithms, software environments, and tools that are evolving as a natural consequence of hardware advances. Indeed the work underlying both presentations in this session emphasized the fact that advances in computation are making what passed for large scale problems only a few years ago routinely solvable on modestly priced hardware. Both speakers discussed the use of their methodology in contexts that made a quantifiable impact on the time and money to complete design tasks. Dr. John Betts of Boeing Computing Company described the new Boeing 777 airplane as a milestone from the standpoint of computer use in automating design including aerodynamics studies, project documentation, visualization, and performance optimization. Dr. Scott Kirkpatrick of IBM discussed applications within electronics design and manufacture which would be impossible without the use of computers. In fact, Dr. Kirkpatrick's discussion pointed out that the current rapid advances in computer technology would be impossible without automated support driven by existing computing environments.

Optimization methods were the centerpiece of both presentations. Dr. Betts described state-of-the art methods in nonlinear programming (NLP) with specific examples of application in optimal control and data fitting. After briefly reviewing the strategy underlying Sequential Quadratic Programming (SQP), he focussed on the need to exploit sparsity and structure in solving Quadratic Programming (QP) subproblems. In fact, the exploitation of sparsity in practical applications has been the major factor underlying performance gains in SQP methods, dominating those even attributable to hardware improvements. In addition to exploiting sparsity in linear algebra operations, performance gains also result from using a scheme that parametrically updates an initial factorization of the Kuhn-Tucker system. Dr. Betts illustrated how SQP methodology can be used to exploit problem structure by showing how it is tailored to the Nonlinear Least Squares Problem. An example of the use of SQP methodology to determine the climb trajectory for an aircraft to reach a given altitude in minimum time yielded a counterintuitive result that has the aircraft dive during a portion of the trajectory. This result was explained to be a consequence of the relationship of thrust to altitude and speed and how the aerodynamics changes during the climb.

Dr. Kirkpatrick reviewed the principles behind the simulated annealing optimization paradigm which he introduced in the early 1980s along with coworkers at IBM. Two critical features are the annealing (or cooling) strategy and the moves used to generate changes in an incumbent solution. He pointed out that annealing strategies were an initial focus of research but that over the last decade a generally applicable cooling strategy has been developed based on an analysis of the variance in the objective function over the space of solutions. Dr. Kirkpatrick illustrated through layout problems drawn from the electronics industry how simulated annealing moves must be tailored to problem structure if the algorithm is to obtain high quality answers in reasonable time. Thus the success of simulated annealing hinges on the implementation details for individual problems. He referred to several simulated annealing codes at IBM that were a product of significant engineering effort. Much of the effort has been directed at novel data structures that support analysis of moves and reduce solution time so that algorithms are useful for screening purposes in design applications. Dr. Kirkpatrick pointed out that simulated annealing works best on problems in which feasible solutions are easily obtained so that effective moves may be constructed from one solution to another. Problems in which feasible solutions are difficult to obtain are more problematic for the simulated annealing methodology.

THE APPLICATION OF OPTIMIZATION TECHNIQUES TO AEROSPACE SYSTEMS

John T. Betts
Boeing Computer Services
Seattle, WA 98124-0346

Abstract

The development of complex systems encountered in the aerospace industry often leads to the formulation of a constrained optimization problem. The equations of motion describing the flight dynamics of a vehicle can be converted to a nonlinear programming problem involving many variables and constraints. Similarly the mathematical description of the vehicle dynamics often requires representation of physical phenomena described by experimental test data. This paper will describe a technique for solving large sparse optimization problems, including least squares applications. The techniques will be illustrated on three separate problems which are derived from the same engineering example involving the trajectory of a high performance fighter aircraft.

Keywords

Sparse optimization, Nonlinear programming, Optimal control, Data fitting.

Introduction

The aerospace industry builds very complex mechanical systems. The modern airplane or space vehicle must be efficiently produced, and be of very high quality. Across the spectrum of disciplines within the industry it is possible to encounter a very broad range of applications which naturally can be formulated as optimization problems. Manufacturing and production models often require solution of integer and linear programming problems. Computational solution of systems described by partial differential equations, such as aerodynamic fluid flow, and electromagnetic wave propagation, typically lead to very large systems of nonlinear equations which are solved using iterative methods. This paper will focus on the solution of problems which can be posed as large sparse nonlinear programs amenable to solution using direct factorization methods for the linear algebra. After a brief presentation of the optimization methods, we describe a linear least squares problem in constrained data fitting. Then a nonlinear programming problem derived from an optimal control problem is described, followed by a nonlinear least squares example.

Sparse Nonlinear Optimization

The NLP Problem

Let us begin the discussion with an overview of the general nonlinear programming problem (NLP). It is desired to find an n-vector $\mathbf{x}$ to minimize

$$f(\mathrm{x}) \tag{1}$$

subject to

$$\bar{\mathrm{b}}_{\ell} \leq \begin{bmatrix} \mathrm{c}(x) \\ x \end{bmatrix} \leq \bar{\mathrm{b}}_{u} \tag{2}$$

where $\mathrm{c}(x)$ is an m-vector, with $f(x)$ and $\mathrm{c}(x)$ having continuous second derivatives.

An NLP is considered *sparse* when most of the elements in the m x n Jacobian matrix

$$\mathrm{G} = \nabla \mathrm{c}(x) \tag{3}$$

and the Hessian of the Lagrangian

$$H_L(x,\lambda) = \nabla^2 f(x) - \sum_{i=1}^{m} \lambda_i \nabla^2 c_i \tag{4}$$

are zero. For a sparse application typically less than 1% of the elements are nonzero. In these expressions λ is an *m*-vector of Lagrange multipliers.

Optimization Strategy

A general purpose algorithm for solving large sparse nonlinear programming problems is described in [10], [6] and [7]. A fundamental feature of the approach is to compute a sequence of estimates for the variables according to the formula

$$\bar{x} = x + \alpha p \tag{5}$$

where p is referred to as the search direction, and the scalar step length α is adjusted at each iteration to produce a sufficient reduction in an augmented Lagrangian merit function suggested by Gill, et al. [13].

Each search direction solves a quadratic programming (QP) subproblem: i.e. minimize

$$g^T p + \frac{1}{2} p^T H p \tag{6}$$

subject to

$$\bar{b}_\ell \le \begin{bmatrix} Gp \\ p \end{bmatrix} \le \bar{b}_u \tag{7}$$

The QP subproblem is well posed provided the *projected Hessian* is positive definite for all active set changes. Unfortunately the projected Hessian $Z^T HZ$ is not readily available for large sparse problems and even if it could be formed in general it would be dense. Furthermore, for most iterations the correct active set is unknown. Consequently we correct the defective QP subproblem by modifying the *full Hessian*

$$H = H_L + \tau(|\sigma| + 1)I \tag{8}$$

where σ is the Gerschgorin bound for the most negative eigenvalue of H_L. The Levenberg parameter, $0 \le \tau \le 1$ is chosen such that projected Hessian is positive definite which can be inferred from calculations performed during the QP solution process. In addition, at each iteration the Levenberg parameter is modified using a trust region strategy such that $\tau \to 0$. Observe that the essential feature of the modification is to augment the Hessian of the Lagrangian with a multiple of the identity matrix. Since the size of the modification is determined by the scalar quantity $\tau(|\sigma| + 1)$ it is adequate to "scale" the correction using the somewhat conservative value provided by the Gerschgorin bound. Numerical experience suggests that efficient algorithm performance is affected by the strategy used to adjust the parameter τ, but is not significantly altered by the accuracy of the parameter σ.

Solving the Sparse QP Subproblem

The efficient solution of the sparse quadratic programming subproblem (6)-(7) is achieved using a technique based on the Schur-complement method of Gill, Murray, Saunders, and Wright [14]. The fundamental G^T idea of this method is to solve the large sparse symmetric indefinite Kuhn-Tucker (KT) system

$$\begin{bmatrix} H & G^T \\ G & 0 \end{bmatrix} \begin{bmatrix} p \\ \pi \end{bmatrix} \equiv K_0 \begin{bmatrix} p \\ \pi \end{bmatrix} = \begin{bmatrix} g \\ 0 \end{bmatrix}. \tag{9}$$

This linear system is factored only *once* using the *multifrontal* algorithm developed by Duff, Ashcraft, Grimes, Lewis, Peyton and Pierce [3], [4]. Subsequently changes to the QP active set can be computed using a *solve* with the previously factored KT matrix K_0 and a solve with a small dense "Schur-complement" matrix. As a part of the solution process it is also possible to compute the inertia of the matrix K_0 where the inertia $[k_1, k_2, k_3]$ is defined as the number of positive, negative, and zero eigenvalues respectively. It is then possible to infer that the projected Hessian is positive definite if the inertia of K is $[n, m, 0]$ If at some point during the QP solution process the inertia of K is incorrect it is necessary to terminate the QP solution process and modify the Hessian using the Levenberg parameter as in (8).

The Nonlinear Least Squares Problem

In contrast to the general nonlinear programming problem as defined in (1)-(2), the *Nonlinear Least Squares Problem* (NLS) is characterized by an objective function

$$f(x) = \frac{1}{2} r^T(x) r(x) = \frac{1}{2} \sum_{i=1}^{\ell} r_i^2 \tag{10}$$

where $r(x)$ is an *l*-vector of *residuals*. As for the general NLP it is necessary to find an *n*-vector *x* to minimize $f(x)$ while satisfying the constraints (2). The *l*- x *n Residual Jacobian* matrix R is defined by

$$R^T = [\nabla r_1, \ldots, \nabla r_l] \tag{11}$$

and the gradient vector is

$$g = R^T r = \sum_{i=1}^{\ell} r_i \nabla r_i \tag{12}$$

And finally the Hessian of the Lagrangian is given by

$$H_L(x,\lambda) = V + R^T R \tag{13}$$

where

$$V = \sum_{i=1}^{\ell} r_i \nabla^2 r_i - \sum_{i=1}^{m} \lambda_i \nabla^2 c_i \tag{14}$$

The matrix $R^T R$ is referred to as the *normal matrix* and we shall refer to the matrix V as the *Residual Hessian.*

Sparse Least Squares

Having derived the appropriate expressions for the gradient and Hessian of the least squares objective function one obvious approach is to simply use these quantities as required within the framework of a general NLP. Unfortunately, there are a number of well known difficulties which are related to the normal matrix $R^T R$. First, it is quite possible that the Hessian H_L may be dense even when the residual Jacobian R is sparse which can occur for example when there is one dense row in R. Even if the Hessian is not completely dense, in general formation of the normal matrix introduces "fill" into an otherwise sparse problem. Secondly, it is well known that the normal matrix approach is subject to ill-conditioning even for small dense problems. In particular, formation of $R^T R$ is prone to cancellation and the condition number of $R^T R$ is the square of the condition number of R. It is for this reason that utilization of the normal matrix is contraindicated. For small dense least squares problems the preferred solution technique is to introduce an orthogonal decomposition for the residual Jacobian without forming the normal matrix. Unfortunately these techniques do not readily generalize to large sparse systems when it is necessary to repeatedly modify the active set (and therefore the factorization). A survey of the various alternatives is found in [16]. In order to ameliorate the difficulties associated with the normal matrix, while maintaining the benefits of the general sparse NLP Schur-complement method [15] a "sparse tableau" approach has been adopted.

If we proceed formally to derive the quadratic objective function for the QP subproblem, combining (6) with (14) yields

$$\begin{aligned} & g^T p + \frac{1}{2} p^T H p \\ &= g^T p + \frac{1}{2} p^T \left[V + R^T R\right] p \\ &= g^T p + \frac{1}{2} p^T V P + \frac{1}{2} p^T R^T R p \end{aligned} \tag{16}$$

This form suggests introducing new variables w and constraints w = Rp. With this identification made we then solve an augmented QP subproblem for the variables $q^T = (p,w)^T$ i.e. minimize

$$\frac{1}{2}(p,w)^T \begin{pmatrix} \overline{V} & 0 \\ 0 & I \end{pmatrix} \begin{pmatrix} p \\ w \end{pmatrix} + g^T p \tag{17}$$

$$\begin{pmatrix} 0 \\ b_\ell \end{pmatrix} \le \begin{pmatrix} R & -I \\ G & 0 \\ I & 0 \end{pmatrix} \begin{pmatrix} p \\ w \end{pmatrix} \le \begin{pmatrix} 0 \\ b_u \end{pmatrix} \tag{18}$$

As in the general NLP problem it is necessary that the quadratic programming subproblem be well posed. Consequently to correct a defective QP subproblem the *residual Hessian* is modified

$$\overline{V} = V + \tau(|\sigma| + 1)I$$

where σ is the Gerschgorin bound for the most negative eigenvalue of V. Notice that it is not necessary to modify the entire Hessian matrix for the augmented problem as given in (16) but simply the portion which can contribute to directions of negative curvature. As for the general NLP, we choose the Levenberg parameter, $0 \le \tau \le 1$ such that the projected Hessian of the augmented problem is positive definite, and modify the Levenberg parameter, at each iteration such that ultimately $\tau \to 0$. Observe that for linear residuals and constraints the residual Hessian V = 0 and consequently $|\sigma| = 0$. Thus in the linear case no modification is necessary unless R is rank deficient. Furthermore, for linearly constrained problems with small residuals at the solution as $\|r\| \to 0$, $|\sigma| \to 0$ so the modification to the Hessian is "small" even when the Levenberg parameter $\tau \ne 0$. And finally for large residual problems, i.e. even when the $\|r^*\| \ne 0$ the accelerated trust region strategy used for the general NLP method adjusts the modification $\tau \to 0$ which ultimately leads to quadratic convergence. It is important to note that when $\tau = 0$ the exact Hessian information is used (both for least squares and general NLP problems), and when this occurs the method becomes a full Newton method.

Minimum Time to Climb Trajectory Example

In order to illustrate the behavior of the sparse optimization algorithm let us consider a series of problems derived from a well-known trajectory optimization problem. The original minimum time to climb problem was presented by Bryson, et al. [11] and has been the subject of many analyses over the past 25 years. The basic problem is to choose the optimal control function $\alpha(t)$ (the angle of attack) such that an airplane flies from a point on a runway to a specified final altitude, as quickly as possible. In its simplest form, the planar motion of the aircraft is described by the following set of ordinary differential equations:

$$\dot{h} = v \sin \gamma \tag{19}$$

$$\dot{v} = \frac{1}{m}\left[T\cos(\alpha) - D\right] - \frac{\mu}{R^2}\sin\gamma \tag{20}$$

$$\dot{\gamma} = \frac{1}{mv}\left[T\sin(\alpha) + L\right] + \cos\gamma\left[\frac{v}{R} - \frac{\mu}{vR^2}\right] \tag{21}$$

$$\dot{w} = \frac{-T}{I_{sp}} \tag{22}$$

where $R = R_e + h$ and h is the altitude, v the velocity, γ the flight path angle, w the weight, $m = w/g_0$ the mass, with $g_0 = 32.174$, $\mu = 0.14076539 \times 10^{17}$ the gravitational constant, and R_e the radius of the earth.

The aerodynamic forces on the vehicle are defined by the expressions

$$D = \frac{1}{2}C_D S\rho v^2 \tag{23}$$

$$L = \frac{1}{2}C_L S\rho v^2 \tag{24}$$

$$C_L = c_{L\alpha}(M)\alpha \tag{25}$$

$$C_D = c_{D0}(M) + \eta(M)c_{L\alpha}(M)\alpha^2 \tag{26}$$

where D is the drag, L is the lift, C_L and C_D are the aerodynamic lift and drag coefficients respectively, with S the cross reference area of the vehicle, and ρ the atmospheric density. Although the results presented utilized a cubic spline approximation to the 1962 Standard Atmosphere, qualitatively similar results can be achieved with a simple exponential approximation to $\rho(h)$.

The following constants complete the definition of the problem:

$h_0 = 0.$	$h_f = 65600.0$
$v_0 = 424.260$	$v_f = 968.148$
$\gamma_0 = 0.$	$\gamma_f = 0.$
$w_0 = 42000.0$	$S = 530$
$I_{sp} = 1600.0$	$R_e = 20902900.$

Tabular Data

As with most real aircraft, the aerodynamic and propulsive forces are specified in tabular form. For the sake of completeness the data as it appeared in the original reference is given in Tables 1 and 2. There are a number of significant points which characterize the tabular data representation. First both the thrust and aerodynamic table values are given to limited numeric precision (i.e. approximately two significant figures). Perhaps the most obvious explanation for the limited precision is that the data probably was originally obtained from experimental tests and truncated at the precision of the test equipment. Unfortunately the statistical analysis (if any) of the original test data has long since been forgotten. Consequently, it is common to assume that the table values are "exact" and correct to full machine precision. A second difficulty which is evident in the bivariate thrust data is that apparently data are missing from the corners of the table. Of course the data are "missing" because a real aircraft can never fly in these regimes (e.g. at mach = 0, and h = 70000 ft.). In fact for most experimentally obtained data it can be expected that data will be missing for unrealistic portions of the domain. In view of these realities it is common to linearly interpolate and never extrapolate the tabular data.

While a linear treatment of tabular function may be adequate for simply evaluating the functions it is totally inappropriate if the functions are to be used within a trajectory simulation and/or optimization.

The principle difficulty stems from the fact that most numerical optimization and integration algorithms assume that the functions are continuously differentiable to at least second order. Thus, just to propagate the trajectory using an integration algorithm such as Runge-Kutta, or Adams-Moulton, it is necessary that the right hand side of the differential equations (19)-(22) have the necessary continuity. Similar requirements are imposed when a numerical optimization algorithm is used to shape the trajectory, since the optimization uses second derivative (Hessian) information to construct estimates of the solution.

The most direct way to achieve the required continuity is to approximate the data by a tensor product cubic B-spline of the form

$$T(M,h) = \sum_{i=1}^{n_1}\sum_{j=1}^{n_2} c_{i,j}B_i(M)B_j(h)$$

In order to utilize this approximation it is necessary to compute the coefficients $c_{i,j}$. The simplest way to compute the spline coefficients is to force the approximating function to *interpolate* the data at all of the data points. However, for a unique interpolant, data are required at all points on a rectangular grid. Since data are missing in the corners of the domain typically this difficulty is resolved by adding "fake" data in the corners using an "eyeball" approach. Although this "eyeball" approach permits the use of standard fitting software, it can introduce unpredictable physical behavior that is not inherent in the fundamental data, i.e. the computer simulation of the trajectory may not match the true flight profile! It is common to ignore the limited precision of the data and simply treat data as though it were of full precision.

Table 1. Propulsion Data.

M	0	5	10	15	20	25	30	40	50	70
0.0	24.2									
0.2	28.0	24.6	21.1	18.1	15.2	12.8	10.7			
0.4	28.3	25.2	21.9	18.7	15.9	13.4	11.2	7.3	4.4	
0.6	30.8	27.2	23.8	20.5	17.3	14.7	12.3	8.1	4.9	
0.8	34.5	30.3	26.6	23.2	19.8	16.8	14.1	9.4	5.6	1.1
1.0	37.9	34.3	30.4	26.8	23.3	19.8	16.8	11.2	6.8	1.4
1.2	36.1	38.0	34.9	31.3	27.3	23.6	20.1	13.4	8.3	1.7
1.4		36.6	38.5	36.1	31.6	28.1	24.2	16.2	10.0	2.2
1.6				38.7	35.7	32.0	28.1	19.3	11.9	2.9
1.8						34.6	31.1	21.7	13.3	3.1

Thrust T(M,h) (thousands of pounds)
Altitude h (thousands of ft.)

Table 2. Aerodynamic Data.

M	0	.4	.8	.9	1.0	1.2	1.4	1.6	1.8
$c_{L\alpha}$	3.44	3.44	3.44	3.58	4.44	3.44	3.01	2.86	2.44
c_{D0}	.013	.013	.013	.014	.031	.041	.039	0.36	0.35
h	.54	.54	.54	.75	.79	.78	.89	.93	.93

Finally by using divided difference estimates of the derivatives at the boundary of the region, the spline coefficients are uniquely determined by solving a sparse system of linear equations.

The cubic spline interpolant does provide C^2 continuity as needed for proper behavior of integration and optimization algorithms. Unfortunately the approximation produced by simply interpolating the raw data does not necessarily reflect the qualitative aspects of the data. In particular it is common for the interpolant to introduce "wiggles" which are not actually present in the tabular data itself. This is clearly illustrated in the aerodynamic data for $M \le .8$ in Fig. 1. It is obvious that the interpolant does not reflect the fact that η is constant for low Mach numbers. A second drawback of the cubic interpolant is the need for data at all points on a rectangular grid when constructing an approximation in more than one dimension.

Minimum Curvature Spline

To resolve these deficiencies we introduce a rectangular grid with k_1 values in the first coordinate and k_2 values in the second coordinate. We require that all data points lie on the rectangular grid. However not all grid points need have data (i.e. data can be missing). Then let us consider introducing a quartic spline with; (a) double knots at the data points in order insure C^2 continuity and, (b) single knots at the midpoints of each interval. Then let us determine the spline coefficients $c_{i,j}$ which minimize the curvature

$$f\left(c_{i,j}\right) = \sum_{k=1}^{L} \left[\frac{\partial^2 T}{\partial M^2}(M_k, h_k)\right]^2 + \left[\frac{\partial^2 T}{\partial h^2}(M_k, h_k)\right]^2 \tag{28}$$

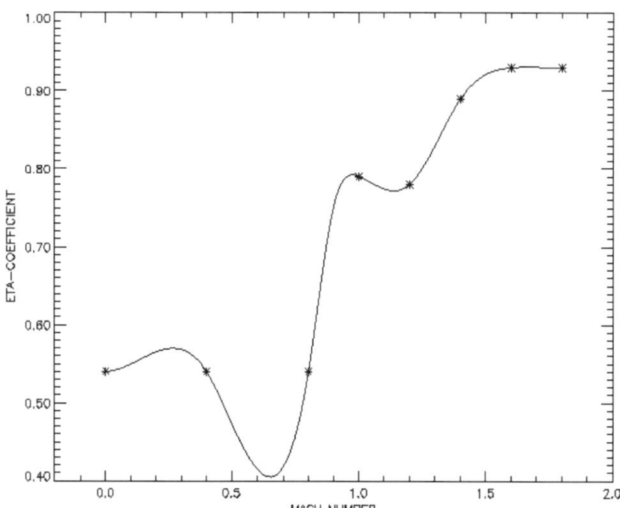

Figure 1. Cubic spline interpolation for aerodynamic data.

and satisfy the data approximation constraints

$$\hat{T}_k - \varepsilon \le T(M_k h_k) \le \hat{T}_k + \varepsilon \tag{29}$$

for all data points $k = 1, \ldots, l$, where ε is the data precision. In order to insure full rank in the Hessian of the objective we evaluate the curvature at points determined by the knot interlacing conditions. Furthermore, it is possible to introduce constraints on the slope of the spline approximation to reflect the monotonicity of the data; i.e.

$$\left[\frac{\partial T}{\partial M}\right] \ge 0 \quad \text{if } \left[\hat{T}_{k+1} - \hat{T}_k\right] \ge 0 \tag{30}$$

$$\left[\frac{\partial T}{\partial M}\right] \ge 0 \quad \text{if } \left[\hat{T}_{k+1} - \hat{T}_k\right] \le 0 \tag{31}$$

for $M \in [M_k, M_{k+1}]$. Similar constraints can be imposed to reflect monotonicity in the h-direction. The coefficients which satisfy these conditions can be determined by solving a sparse constrained linear least squares problem.

The resulting approximations are illustrated in Fig. 3 and Fig. 2. For this particular fit the least squares problem had 900 variables, with 988 constraints of which 77 were equalities. In general the number of variables n for a bivariate minimum curvature fit is $n = 9k_1k_2$. For the general case the minimum curvature formulation requires imposition of m constraints where $10k_1k_2 \le m \le 14k_1k_2$ and the exact number depends on the number of algebraic sign changes in the slope of the data. In addition, if interpolation is required the number of equality constraints m_e is $m_e \le k_1k_2$.

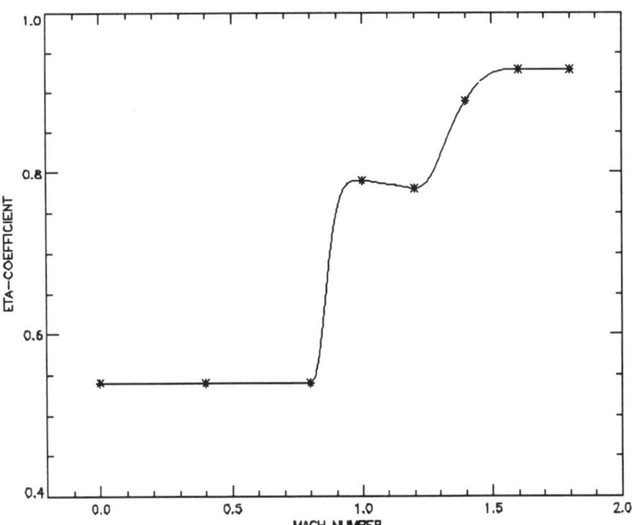

Figure 2. Minimum curvature spline for aerodynamic data.

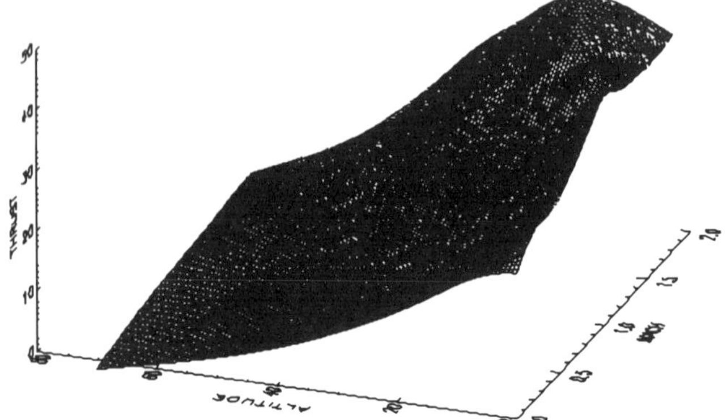

Figure 3. Minimum curvature spline for thrust data.

Transcription of the Optimal Control Problem

The original minimum time to climb trajectory problem is an example of an optimal control problem which can be solved using a *direct transcription method* [1], [2], [9], [5], [6], [7], [11], [14], [16], [17]. Essentially the method can be described as follows:

1. Transcribe the optimal control problem into a nonlinear programming (NLP) problem by discretization.
2. Solve the nonlinear program using the method previously described.
3. Check the accuracy of the discretized solution
   - If the accuracy is acceptable —stop;
   - otherwise refine the discretization and return to step 1.

In general the optimal control problem requires finding the n_u-dimensional control vector $u(t)$ to minimize the performance index $\phi[y(t_f),t_f]$ evaluated at the final time t_f. The dynamics of the system are defined by the state equations

$$\dot{y} = h[y(t), u(t), t] \tag{32}$$

where y is the n_e dimension state vector. Initial conditions at time t_0 are defined by

$$\psi[y(t_0), u(t_0), t_0] \equiv \psi_0 = 0, \tag{33}$$

and terminal conditions at the final time t_f are defined by

$$\psi[y(t_f), u(t_f), t_f] \equiv \psi_\phi = 0, \tag{34}$$

In addition, the solution must satisfy path constraints of the form

$$\Psi_L \le \Psi[y(t), u(t), t] \le \Psi_U, \tag{35}$$

where Ψ is a vector of size n_p, as well as simple bounds on the state variables

$$y_L \le y(t) \le y_U, \tag{36}$$

and control variables

$$u_L \le u(t) \le u_U. \tag{37}$$

All transcription approaches divide the time interval into n_s segments

$$t_0 < t_1 < t_2 < \ldots < t_f = t_{n_s}$$

where the time points are referred to as mesh points, grid points, or nodes. Let us introduce the notation $y_j \equiv y(t_j)$ to indicate the value of the state variable at a grid point. In like fashion denote the control at a grid point by $u_j \equiv u(t_j)$. For the trapezoidal discretization, the NLP variables are

$$x = [y_0, u_0, y_1, u_1, \ldots, y_f, u_f, t_f]^T. \tag{38}$$

The state equations are approximately satisfied by setting the *defects*

$$\zeta_j = y_j - y_{j-1} - \frac{1}{2}k_j\left[h_j + h_{j-1}\right] \tag{39}$$

to zero for $j = 1, \ldots, n_s$. The step size is denoted by $k_j \equiv t_j - t_{j-1}$, and the right hand side of the differential equations evaluated at grid point j is denoted by the vector h_j.

As a result of the transcription process the differential-algebraic system defining the optimal control problem is replaced by the NLP constraints

$$c_L \le c(x) \le c_U \tag{40}$$

where

$$c(x) = [\zeta_1, \zeta_2, \ldots, \zeta_f, \psi_0, \psi_f, \Psi_0, \Psi_1, \ldots, \Psi_f]^T \tag{41}$$

with

$$c_L = [0, \ldots, 0, \Psi_L, \ldots, \Psi_L]^T \tag{42}$$

and a corresponding definition of c_U. The first $n_e n_s$ equality constraints require that the defect vectors from each of the n_s segments be zero thereby approximately satisfying the differential equations. The boundary conditions denoted by ψ are enforced directly as equality constraints, and nonlinear path constraints denoted by Ψ are imposed at the grid points. Note that nonlinear equality path constraints are accommodated by setting Ψ_L, Ψ_U.

Ultimately the accuracy of the discrete solution as an approximation for the continuous one is dictated by a number of factors. In general the accuracy can be improved by increasing the order of the discretization, increasing the number of grid points, or improving the location of the grid points. Strategies for efficient *mesh refinement*, are discussed in [1], [9] and [19]. Although the results presented below were obtained with equidistributed mesh points, significant computational benefits can be realized for many aerospace applications by more sophisticated approaches.

The Optimal Trajectory

The previous sections have described two distinctly different sparse optimization problems. First a smooth B-spline approximation to tabular data was constructed by solving a sparse quadratic programming problem. Using this representation it was possible to define an optimal control problem which described the minimum time to climb trajectory. This optimal control problem was solved by transcription into a large sparse nonlinear programming problem. The solution of the trajectory problem is illustrated in Fig. 4.

One obvious attribute of the solution which has been the subject of considerable interest is the fact that there is a dive during part of the climb. Essentially the aircraft dives to trade potential for kinetic energy thereby reaching the final altitude sooner. Despite the counterintuitive nature of the solution, the validity has been demonstrated in practice, and the original paper [11] has lead to the "energy-state" treatment of trajectory problems. In order to illustrate a more benign climb trajectory let us suppose that the desired climb path is described by

$$\hat{h}(t) = c_0 + c_1 t + c_2 t^2 + c_3 t^3 \tag{43}$$

where the coefficients are chosen such that $h(0) = h0$ and $h(t_f) = h_f$, $\dot{h}(t_f) = 0$, for a specified final time $t_f = 350$ and free final velocity v_f. Let us now attempt to choose the control to minimize the deviation from the desired path, that is,

$$f = \int_0^{t_f}\left[h(t) - \hat{h}(t)\right]^2 dt \tag{44}$$

$$\approx \sum_{k=0}^{n_g} a_k\left[h(t_k) - \hat{h}(t_k)\right]^2 \tag{45}$$

$$= \sum_{k=0}^{n_g} r_k^2 \tag{46}$$

where the nonlinear *residuals* are defined by

$$r_k = \sqrt{a_k}\left[h(t_k) - \hat{h}(t_k)\right]. \tag{47}$$

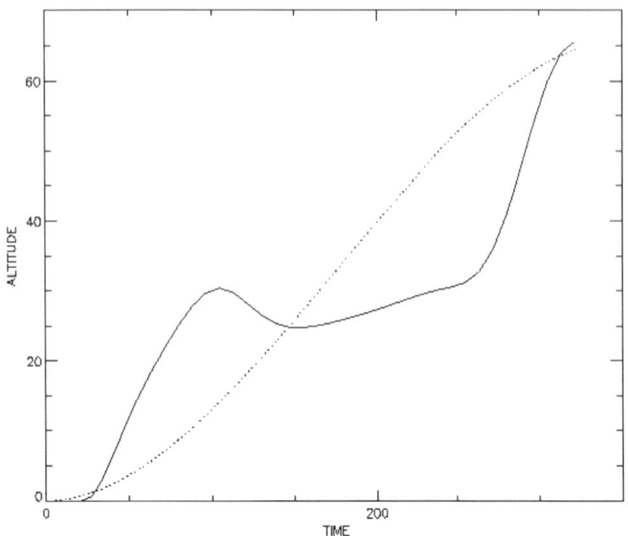

Figure 4. Minimum time to climb trajectory.

The actual choice for the coefficients in the integral approximation a_k depends on the quadrature rule being used. To illustrate the approach the transcription method in [8], [6] and [7], was used to solve the problem and the results are illustrated in Fig. 5. For this example a trapezoidal discretization with 100 equally spaced grid points was used, which results in a sparse nonlinear least squares problem with 500 variables, 396 nonlinear constraints and 86 degrees of freedom at the solution. In this instance the Jacobian matrix has 4460 nonzero elements (1.8\% sparse), and the Hessian matrix has 1500 nonzero elements (1.2\% sparse). Using a linear initial guess for the state and control variables the solution was obtained in 41.6 seconds on a SUN Sparc 20 workstation. During the iteration the trajectory functions were evaluated 281 times, and the Hessian matrix was computed 10 times.

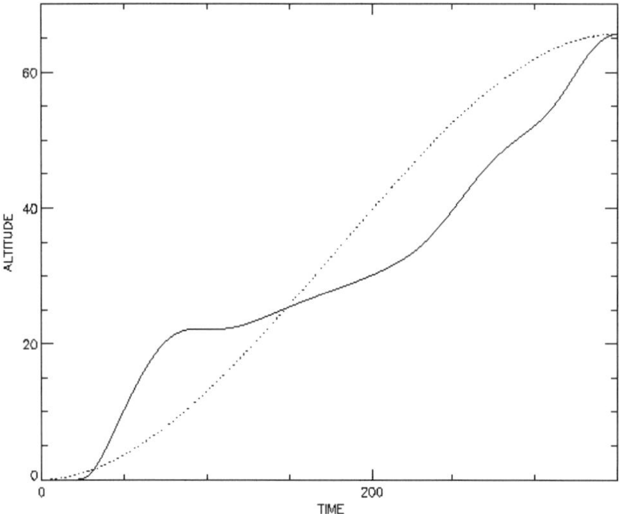

Figure 5. Minimum deviation climb.

Conclusions

This paper describes the application of sparse nonlinear programming methods to problems arising from the design of an aerospace vehicle trajectory. First we describe a method for solving a sparse nonlinear program using a technique based on a Schur-complement implementation of a sequential quadratic programming approach. An extension of the technique to sparse nonlinear least squares is then outlined. The detrimental affects of the normal matrix on matrix sparsity and conditioning are avoided by utilizing a sparse tableau formulation which results in an augmented quadratic programming subproblem. A fundamental feature of the method is to modify the residual Hessian matrix using a diagonal matrix whose size is determined by the Gerschgorin bound for the most negative eigenvalue and a Levenberg parameter which is adjusted utilizing an accelerated trust region strategy. Since projected and/or reduced Hessian approximations are not explicitly constructed the algorithm is efficient even when the number of degrees of freedom is large.

Three different cases are described to illustrate the technology. First, the linear least squares capability is illustrated on a data modeling example. In this case we introduce a technique for constructing a minimum curvature tensor product spline approximation to tabular data which requires solving a large sparse linear least squares problem. The data approximation exhibits the desired continuity and differentiability, without introducing "wiggles" that are not actually in the original data. The technique is applicable even when the data is not available on a uniform rectangular grid. The monotonicity and precision of the data are treated directly by formulating linear inequality constraints. Finally the knot locations for constructing the approximation are defined without manual interaction. As a second example the general nonlinear programming technique is illustrated on a sparse optimization problem derived from a discretization of a trajectory design application. Finally, the nonlinear least squares capability is illustrated by solving a minimum deviation trajectory optimization problem.

Acknowledgments

The author wishes to acknowledge the valuable contributions of Mr. Richard Mastro during the development and implementation of the minimum curvature spline approximation procedure. The author also gratefully acknowledges the valuable interaction with Dr. Paul Frank concerning implementation of the sparse least squares capability.

References

Ascher, U., Christiansen, J., and Russell, R. (1981). Collocation Software for Boundary-Value ODEs, *ACM Transactions on Mathematical Software*, **7**, 2, 209-222.

Ascher, U., Mattheij, R., and Russell, R.D. (1988). *Numerical Solution of Boundary Value Problems for Ordinary Differential Equations*, Prentice Hall, Englewood Cliffs, N.J.

Ashcraft, C.C. (1987). *A Vector Implementation of the Multifrontal Method for Large Sparse, Symmetric Positive Definite Linear Systems*, Technical Report ETA-TR-51, Boeing Computer Services.

Ashcraft, C.C., and Grimes, R.G. (1988). *The Influence of Relaxed Supernode Partitions on the Multifrontal Method*, Technical Report ETA-TR-60, Boeing Computer Services.

Betts, J.T., and W.P. Huffman (1991). Trajectory Optimization on a Parallel Processor, *Journal of Guidance, Control, and Dynamics*, **14**, 2.

Betts, J.T., and W.P. Huffman (1992). The Application of Sparse Nonlinear Programming to Trajectory Optimization, *Journal of Guidance, Control, and Dynamics*, **15**, 1.

Betts, J.T., and W.P. Huffman (1993a). Path Constrained Trajectory Optimization Using Sparse Sequential Quadratic Programming, *Journal of Guidance, Control, and Dynamics*, **16**, 1, 59-68.

Betts, J.T., and W.P. Huffman (1993b). *Accuracy Refinement in Transcription Methods for Optimal Control*, BCSTECH-93-006, Boeing Computer Services, February 10, 1993.

Betts, John.T. (1993c). Using Sparse Nonlinear Programming to Compute Low Thrust Orbit Transfers, *The Journal of the Astronautical Sciences*, **41**, 3, 349-371.

Betts, J.T., and P.D. Frank (1994). A Sparse Nonlinear Optimization Algorithm, *Journal of Optimization Theory and Applications*, **82**, 3.

Bryson, A.E., M.N. Desai, and W.C. Hoffman (1969). Energy-State Approximation in Performance Optimization of Supersonic Aircraft, *Journal of Aircraft*, **6**, 6.

Dickmanns, E.D. (1980). Efficient Convergence and Mesh Refinement Strategies for Solving General Ordinary Two-Point Boundary Value Problems by Collocated Hermite Approximation, 2nd IFAC Workshop on Optimisation, Oberpfaffenhofen.

Gill, P.E., W. Murray, M.A. Saunders, and M.H. Wright (1986). *Some Theoretical Properties of an Augmented Lagrangian Merit Function*, Report SOL 86-6, Department of Operations Research, Stanford University.

Gill, P.E., W. Murray, M.A. Saunders, and M.H. Wright (1987). *A Schur-Complement Method for Sparse Quadratic Programming*, Report SOL 87-12, Department of Operations Research, Stanford University.

Hargraves, C.R., and S.W. Paris (1987). Direct Trajectory Optimization Using Nonlinear Programming and Collocation, *J. of Guidance, Control, and Dynamics*, **10**, 4, 338.

Heath, M.T. (1984). Numerical Methods for Large Sparse Linear Least Squares Problems, *SIAM J. on Scientific and Statistical Computing*, **5**, 3, 497-513.

Kraft, D. (1985). On Converting Optimal Control Problems Into Nonlinear Programming Problems, NATO ASI Series, **F15**, *Computational Mathematical Programming*, Springer-Verlag.

Russell, R.D. and L.F. Shampine (1972). A Collocation Method for Boundary Value Problems, *Numer. Math.*, **19**, 1-28.

Vasantharajan, S. and L.T. Biegler (1990). Simultaneous Strategies for Optimization of Differential-Algebraic Systems with Enforcement of Error Criteria, *Computers Chem. Engng.*, **19**, 10, 1083-1100.

INTEGRATING DESIGN CRITERIA

William R. Johns
Intera Information Technologies Ltd.
Henley-on-Thames
Oxfordshire, RG9 1AT, United Kingdom

Abstract

Computer Aids to Process Design are reviewed in the context that design is a human activity. The aids have the objective of reducing human perspiration and stimulation human inspiration. There will, never be a design method that can provide a mathematical guarantee of optimality. The space of all possible designs is impossible to define in mathematical terms. Alongside computer aids there are other methodical procedures that aid the innovation process, and there is a place for integrating computer-oriented and human-oriented design approaches. The discussion focuses specifically on three methods, a (manual) means-ends analysis shown to be practically useful in industry, a mathematical programming method (Mixed-Integer Non-Linear Programming, MINLP) and an Artificial Intelligence (AI) implementation of an (originally) manual hierarchical approach. The AI framework aims to provide a link to enable possible MINLP refinement of the designs produced. It is concluded that, whilst integration between Design Synthesis methods is advancing, there is only limited progress in integrating design criteria.

Introduction

Design is an essentially human activity requiring both inspiration and perspiration. The objective of the conference is to review the extent to which available computer aids reduce the perspiration and stimulate the inspiration.

"Design" covers a wide range of activities from conceptual design (what chemistry should we use, what main process steps?) through to detailed mechanical design of individual components. At this conference, we concentrate on Process Design and are not concerned with mechanical design beyond a level needed to a achieve realistic cost estimates. The "Process Synthesis," section of this overview discusses the range of activities falling under design synthesis and the relevant tools available to the designer. "Integration of Criteria" discusses some of the criteria for judging a good design and how those criteria may be integrated into the design process. It is concluded that, although Process Synthesis is advancing rapidly, there are currently limited computer-aids available to the practising designer. Furthermore, little attention has yet been focussed on integrating alternative design criteria such as process operability.

Process Synthesis

Figure 1 illustrates a classification of levels of Process Synthesis. The design space is divided into four areas: "Overall Design," "Discrete Synthesis," "Superstructure Synthesis" and "Sub-system Synthesis," which tend to fall within one another as shown.

Overall Design Synthesis

There is no globally guaranteed solution because we can only guarantee optimal solutions when we fully know the objectives and the constraints. We never fully know the objectives or the constraints, because one of the synthesis objectives is to relax the constraints by discovering new catalysts, new reactants, new extractants, new pieces of equipment, and new products with the same or improved functionality. Such discoveries are dominated by human innovation and have the greatest potential for generating radically improved designs. Some synthesis methods aim to stimulate innovation but this is the area most difficult to automate. It is such new inventions, often of single key stages of the process, that constitute major advances. The means-ends approach of Siirola [1] aims both to stimulate such innovation and to ensure the best use of known technology. The method provides no guarantee of optimality and the only "computer aids" used are tools such as flowsheeting (simulation) systems that reduce the labour of evaluating the alternatives generated.

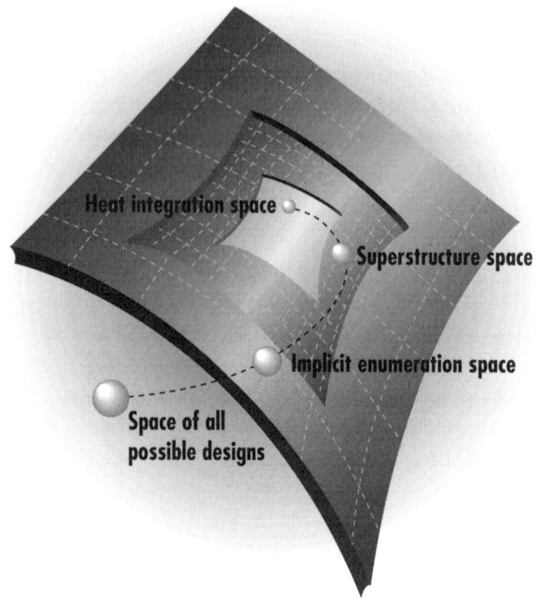

Figure 1. Synthesis within the design space.

Discrete Synthesis

Discrete synthesis covers the more tractable area of combining currently known chemistry and operations in the most effective manner. To generate such units as Eastman's Reactive Extractive Distillation, the synthesis must go beyond merely assembling a 'kit' of standard unit operations; it must look at the chemical and physical processes that go on within the units to generate new operations that combine the known processes in cleaner, more economic packages. The number of combinations, even of the known unit operations that can be presented to a synthesis program, is astronomical. It is beyond the bounds of any currently conceivable technique to guarantee generating an optimal combination which also incorporates fully optimized operating conditions. The computer aids being developed thus take some short cuts which allow a very wide area to be covered, and give a reasonable assurance that the result will be much better than is likely to be achieved by incremental evolution of the previous plant design. Methods applicable include implicit enumeration (Dynamic Programming, Branch and Bound etc.), pseudo-random methods (Genetic Algorithms, Simulated Annealing and mixed and related methods), Artificial Intelligence (AI), Fuzzy AI, and incremental AI.

For implicit enumeration, the problem is largely discretized with, for example, a finite number of possible component flowrates, equipment sizes and operating conditions. The recursive formulation leading to a Dynamic Programming approach, as described by Johns and Romero [2], is illustrated in fig.2. Each possible initial decision decomposes the overall process synthesis into smaller problems each of which is similar to the original in having defined inputs and outputs. With only a finite number of possible streams, solutions to these intermediate problems can be recorded to give an efficient overall optimization procedure. AI methods use heuristic (or provable) rules to enable design decisions to be made sequentially without recursion. Careful choice of the hierarchical sequence in which decisions are made minimizes the coupling between successive decisions thus minimizing the penalty of omitting the recursion. Such a system is described by Douglas and Stephanopoulos [3]. It will be noted that these authors also allow some recursion, i.e. decisions at a higher level may be reviewed in the light of optimization at a lower level, in this way providing a system combining features of AI and implicit enumeration. An alternative enhancement to AI is "fuzzy AI" as described by Meshalkin and co-workers [4]. In this approach the AI "rules" are ranked by degree of "belief" and all processes within a certain overall belief margin generated. The belief weightings are revised by matching against detailed evaluation of the processes synthesized and the overall synthesis repeated until convergence of ranking is achieved. Connolly and Prince [5] describe an evolutionary application of AI in which an initially simple process is augmented until the cost of further refinement gives no further benefit.

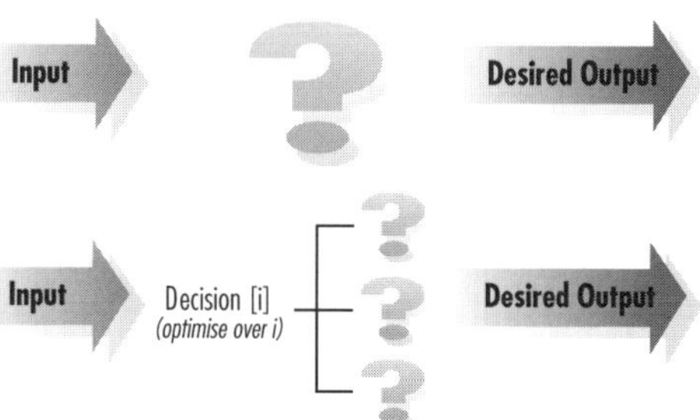

Figure 2. Dynamic programming.

Designs generated by Discrete Synthesis, although assembled from known steps, can give rise to radical departures from current practice, to the extent that experimental programmes may be necessary to validate the designs. As early as the 1960's the method of ref [2] was applied to a process for the hydrodimerization of acrylonitrile to adiponitrile. It produced a variant requiring less units, less capital and using 18% less energy. It was necessary to build a small-scale unit to verify that it worked before considering patent protection. The recent work by Connolly and Prince [5] on minerals processing also shows substantial savings through radical departure from current practices, and experimental verification of the synthesized processes is in hand.

Superstructure Synthesis

Where the number of discrete alternatives is small, all the alternatives can be included within a superstructure which can be optimized by rigorous methods, e.g. MINLP. The paper by Floudas and Grossman [6] takes MINLP a substantial step towards practicality by tackling two of the major aspects limiting its application, binary variables and non-convexity. Binary variables cover the selection/non-selection of units and the connections of streams between them; non-convexity gives rise to multiple local optima and arises, for example, as a consequence of the characteristic "economy of scale" cost relationships. The authors show that applying logical relationships can increase the efficiency of handling binary variables by up to two orders of magnitude whilst a sequence of lower-bounding convex functions can identify a global optimum amongst multiple local optima.

Subsystem Synthesis

Subsystems, such as heat exchanger networks, can be optimized when the major features of the process flowsheet are fixed. A fixed flowsheet limits the options available, but significant benefits can nevertheless be achieved. Thus, for heat exchanger networks, the "pinch" method is widely used and has resulted in substantial energy savings. Computerized enhancements of the method are available from a number of sources, e.g. "Advent" (from Aspen Technology Inc.) and "Supertarget" (from Linnhof March Ltd). Alternative approaches using Mathematical Programming are also under development [7].

The range of approaches described does not have fixed boundaries and the methods are not necessarily competitive or mutually exclusive. Thus Douglas and Stephanopoulos describe how their AI structure can incorporate MINLP to give mutually beneficial performance. These authors also provide a human interface to stimulate the introduction of human innovation into the AI synthesis. Similarly Evans and Johns [8] show how implicit enumeration can allow an "impossible" step (i.e. not foreseen in the data base) which, if it radically improves the overall process, can stimulate the designer to devise a practical implementation of the step. Note also that innovations developed by methods such as those described by Siirola, can be incorporated into any of the computer-based methods, to reduce manual effort and give a greater assurance that the innovations are being deployed most effectively. Efficient methods of predicting the optimal performance of subsystems (e.g. "pinch target") can also be employed to improve the efficiency of either computer-based or manual methods of whole-process synthesis.

Design Criteria

Current computer-aided process synthesis methods employ, as a criterion of optimality, combined capital and running costs for a process operating at nominal capacity. There are, however, a range of additional criteria, mostly related to process "operability." These criteria include: Safety, Environmental Impact, ability to start-up and shutdown, Availability, Maintainability, turn-down operability, Controllability, and Flexibility, and are also relevant to Process Synthesis. For example, in studying safety, HAZOP analysis [9] frequently stimulates the invention of intrinsically safer process variants. In general, however, these distinct criteria are not formally integrated; they are informally brought together by industrial design teams to give, what they hope are, economic and operable processes. Limited formal work in linking the criteria has been undertaken in the area of flexibility where Johns et al [10] and, more recently, Straub and Grossman [11] have modified the conventional design criteria on a statistical basis. Both approaches discretize the uncertain outcomes through quadrature; although neither allow the optimization to synthesize radically different flowsheets. (Straub and Grossman use Guassian quadrature whilst Johns et al use an n-dimensional Hermite quadrature which eliminates the necessity to truncate the distribution). Ponton and Laing [12] have also initiated work on establishing a consistent hierarchical approach for concurrently synthesizing process flowsheets and control systems. Their method is inspired by the Douglas [3] approach.

Conclusions

Considerable advances are being made in providing interfaces between manual and computer-based design synthesis methods. Computer aids to design, as currently used in industry, are mainly restricted to evaluation tools. Computer-aided Process Synthesis tools are, however, approaching the stage at which industrial use is becoming practicable. It must be anticipated that they will become increasingly used in practice and will be important in developing the next generation of more economic, cleaner processes. Considerable further work is, however, needed before the important design operability criteria are included in practical computer-aids to process synthesis; the crucial decisions in process design are expected to be made by human beings for the foreseeable future.

References

Connolly, A.F. and Prince, R.G.H. (1994). The Evolutionary Synthesis of Mineral Processing Flowsheets. 4th European Symposium on Computer-Aided Process Engineering (ESCAPE 4), *I. Chem. E.,* 359-365.

Douglas, J.M. and Stephanopoulos, G. (1994). Hierarchical Approaches in Conceptual Process Design: Framework and Computer-Aided Implementation. FOCAPD 94. CACHE Corporation 1994.

Evans, G.H. and Johns, W.R. (1982). Automatic Generation of Process Flowsheets with non-predefined Unit Operations. I. Chem. E Symposium Series No. 74, Understanding Process Integration.

Floudas, C.A. and Grossmann, I.E. (1994). Algorithmic Approaches to Process Synthesis: Logic and Global Optimization. FOCAPD 94. CACHE Corporation 1994.

Johns, W.R., Marketos, G. and Rippin, D.W.T. (1978). The Optimal Design of Chemical Plant to meet time-varying Demands in the Presence of Technical and Commercial Uncertainty. *Trans. I. Chem. E.*, **56**, 249-257.

Johns, W.R. and Romero, D. (1979). The Automated Generation

and Evaluation of Process Flowsheets. *Comp. and Chem. Engin.*, **3**, 251-260.

Johns, W.R. and Williams, M.J. (1994). Cost Optimal Heat Exchange Network Synthesis. 4th European Symposium on Computer-Aided Process Engineering. (ESCAPE 4), *I. Chem. E.*, 247-256.

Kletz, T. (1986). HAZOP and HAZAN. *I. Chem. E.*

Meshalkin, V. and co-workers: Moscow Mendeleev Institute; various papers in Russian.

Ponton, J.W. and Laing, D.M. (1993). A Hierarchical Approach to the Design of Process Control Systems. *Trans. I. Chem. E.*, **71**, 181-188.

Siirola, J.J. (1994). An Industrial Perspective on Process Synthesis. FOCAPD 94. CACHE Corporation 1994.

Straub, D.A. and Grossman, I.E. (1993). Design Optimization of Stoichastic Flexibility. *Comp. and Chem. Engin.*, **17**, 339-354.

HIERARCHICAL APPROACHES IN CONCEPTUAL PROCESS DESIGN: FRAMEWORK AND COMPUTER-AIDED IMPLEMENTATION

James M. Douglas
Chemical Engineering Dept.
University of Massachusetts
Amherst, MA 01003

George Stephanopoulos
Chemical Engineering Dept.
Massachusetts Institute of Technology
Cambridge, MA 02139

Abstract

Hierarchical refinement and decision-making is a commonly employed strategy in a variety of design problems. In this paper we examine its use on the synthesis of conceptual processing schemes. It is argued that a hierarchically evolving definition of the conceptual design problem is a necessary condition for the development of a computationally tractable design procedure, and that it can provide sufficient conditions for the generation of "optimal" processing schemes. Furthermore, the paper discusses the knowledge base that drives the formulation of a specific hierarchical decision-making procedure for conceptual design, and outlines the features of the approach. Finally, the paper outlines the essentials of a software system that automates large segments of the hierarchical conceptual design methodology, focusing on the following pivotal components; (i) a computational model of the design process, (ii) a high-level design-oriented language and (iii) a human-computer interface to integrate the human into the design process.

Keywords

Hierarchical design, Process synthesis, Automation in design.

Introduction

Increasing international competition in the processing industries and public awareness of their environmental impact has made it quite apparent that it is imperative to find ways of developing both more competitive and "cleaner" chemical processes, and bringing them on-line in shorter time-to-market periods. Finding simpler chemical routes, utilizing less expensive raw materials and smaller amounts of energy, producing fewer and in smaller amounts hazardous byproducts, inventing new processing technologies (e.g. new types of reactors, separation techniques) which reduce both costs and waste loads, shortening the time to commercialization, have become pivotal goals in many industrial research and development departments around the world.

It usually takes between eight to ten years from the time that a new chemical production route has been discovered until the commercial plant has been brought on-line. For specialty chemicals, pharmaceuticals, biochemical products it can take longer, given the stringent requirements imposed by the process of clinical testing and FDA approval, or the securing of the necessary permits of environmental compliance. In these cases, the processing scheme is "frozen" very early on, thus making it virtually impossible to contemplate later-stage process improvements. Bringing a product to market in very short periods of time provides a tremendous competitive advantage. Producing this product through a cost-competitive, clean and efficient process, guarantees sustenance of a market share over a long period of time. The new DuPont process for producing an alternative to Freon refrigerant was commercialized in four years from inception, and is a convincing illustration of the significant improvements that are possible.

Nevertheless, it should be born in mind that the success rate for developing and commercializing a new process is only about one percent. Furthermore, although a processing route may be overall very attractive for commercialization, it is possible that its *economic robustness* to market conditions is low, i.e. a different chemical route, using different raw materials, may offer a far more cost-effective and environmentally benign production alternative. Therefore, it is of paramount importance to be able to screen potentially new chemical routes and terminate poor ideas for new processes as efficiently as possible.

In view of the above, it is not surprising that the established practice approaches the design and evaluation of new processing schemes in a hierarchically evolutionary manner. At the initial level of the hierarchy, one considers the most abstract description of the processing scheme, suppressing as much detail as possible (only the dominant process units are considered and rough cost estimates are used). If the results of this first design indicate that additional effort can be justified, rigorous models are used to optimize the process flowsheet at the second level of the hierarchy. At subsequent levels, more detailed process descriptions are generated and the accuracy of the ensuing evaluations improves, as additional equipment are introduced, the control strategies are considered, and safety issues and potential hazards are examined. Pikulik and Diaz (1977) describe a possible set of levels in the hierarchical evolution of a process design and the amount of detail as well as the accuracy achieved at each level. However, different companies use a different number of levels and different error tolerances at each level.

If the systematic synthesis and evaluation of conceptual chemical processing schemes provides the competitive edge, it is the availability of a computer-based design environment, which supports the rapid generation and screening of many ideas, that enables the harvesting of the promised fruits. However, such a software system is quite different from the traditional computer-aided design environments. It is called upon to automate substantial segments of the design approach, and to become a natural extension of the human designer's decision-making process.

The present paper has been structured around the following two themes, which are pivotal in addressing the needs, described above:

1. The structure and characteristics of a hierarchical approach, as the backbone of the systematic procedure for the generation of attractive processing schemes.
2. The design of a software system that can provide extensive automation of the above design procedure.

Conceptual Design of Chemical Processing Schemes

The chemical processing industries employ chemists, microbiologists, molecular biologists and material scientists to discover new reactions, new or genetically modified microorganisms and new materials. Once a new production idea has been discovered, many companies attempt to develop a first design within a very short period of time (e.g. two weeks), in order to evaluate the idea's potential for a commercial process. The information available is scarce; desired product purity and flow, the set of reactions, catalysts or solvents for the reaction, and some data on product distribution. No data is yet available on the physical properties of new components, the catalyst deactivation rate, the vapor-liquid, liquid-liquid, or solid-liquid equilibrium, the reaction kinetics, crystallization kinetics, etc.

The First Design

The goal of the first design, which we call a *conceptual design*, is to determine if a process based on the discovery will be profitable, and if so, to determine the best few flowsheets. Only a small fraction of the total design effort (5 to 10%) is spent on the conceptual design, which, nevertheless, fixes about 85% of the total manufacturing cost, once the flowsheet has been fixed. The invention of a flowsheet requires (i) the specification of the process streams and units that are needed to transform the raw materials into the desired product and byproducts, (ii) the specification of a separation system that generates the desired exit and recycle streams from the reactor effluent and (iii) the specification of an energy management system. In addition, the need for reactor diluents, quench streams, reaction or separation system solvents must be recognized, these materials must be identified, and any waste treatment and disposal problems must be determined. Finally, a set of attractive alternative flowsheets must be generated. All of the above are important decision-making steps and lead to the structure of the first design(s) (and represent the synthesis portion of the design activity).

The means-ends procedure, used by Siirola and Rudd (1971) to develop the AIDES system, and the BALTAZAR system of Mahalec and Motard (1977a, b), which used mechanical theorem-proving techniques to resolve conflicts and synthesize process flowsheets, are among the earliest efforts in systematizing the synthesis of first designs.

Hierarchical Evolution as a Necessary Condition

The conceptual design is a very difficult problem because in its initial statement is grossly underspecified. There are two general approaches that one could take to complete the definition of the problem:

1. Define completely the problem at the outset by specifying the following: all possible types of processing units and all possible

connections among these units; all possible ancillary process materials (e.g. diluents, heat carriers, quench materials, reaction and product solvents, separation system solvents); all possible destinations of subsidiary streams (e.g. liquor streams leaving liquid-solid separators); all possible energy management systems; all possible ways of treating waste byproducts (remove as a solid, remove in solution, etc.).
2. Progressively evolve the definition of the design problem, through a series of partial solutions with increasing amount of detail and accuracy.

The first approach is a formidable task, if at all possible, and may bring into the problem's formulation elements that are quite irrelevant or redundant. Its logical consequence is to utilize an implicit enumeration algorithm in order to extract the most attractive solutions imbedded in the problem statement. The sheer size of the problem, as stated, makes this approach computationally intractable.

On the other hand, the evolutionary definition of the design problem can follow one of two distinct tracks:

(a) Through ad hoc decision-making, based on intuition and analogies to parts of other processes, an experienced designer may complete the problem definition, synthesize attractive alternatives and eliminate poor ones. In such cases, the risk of overlooking attractive alternatives is usually high.
(b) A structured hierarchical procedure can be stated in a concise manner, so that it can progressively evolve the problem definition, through a series of partial solutions and account for all process alternatives. The ensuing search for the identification of the "best" solution(s) may follow different strategies to accommodate the designer's needs. The approach fashioned by Douglas (1985, 1988) clearly follows the latter track.

Based on the above observations, we postulate that, the hierarchically evolving definition of a conceptual design problem, *is a necessary condition for the development of a computationally tractable procedure,* which leads to the generation of "optimal" conceptual designs. In subsequent sections we will discuss how this postulate shapes the form of the proposed hierarchical approach.

Sufficiency in the Optimality of Hierarchical Designs

As the definition of the conceptual design problem evolves through a series of hierarchical descriptions with an ever increasing detail, one may see this as a *constraint-posting* process, which defines a progressively evolving space of feasible flowsheets. In searching for the "optimal" flowsheets, a designer may adopt one of the following two generic strategies:

1. Maintain all the process alternatives implied by the set of constraints, which have been identified through the final level of the evolutionary problem definition, and search for the optimum through an implicit enumeration algorithm. Such an approach is based on mixed-integer nonlinear programming (MINLP) formulations, which are solved by various breeds of branch-and-bound algorithms.
2. Proceed from one level of the hierarchy to the next by making decisions over a select set of design variables. Such an approach can still guarantee the optimality of the generated processing schemes, provided that *the design decisions made at each level of the hierarchy bound the "optimal" solution(s).* As we will see in subsequent sections, the sufficiency in locating the optimum can be guaranteed by one or more of the following mechanisms: (i) progressive tightening of the bounds on the optimal value of the objective function (e.g. total annualized cost), (ii) progressive reduction in the space of feasible solutions, which include, nevertheless, the "optimal" one, or/and (iii) generation of a base-case design, a list of the alternatives and implicit enumeration of them in order to locate the optimum. Clearly, an optimum solution cannot be guaranteed, unless the method, used to generate all of the process alternatives, is in fact complete.

Although the hierarchical approach can be structured so as to guarantee the optimality of the generated solution(s), its versatility allows a multitude of search strategies to suit the needs of the designer. For example, Douglas (1985, 1988) has opted for a depth-first search leading to rapid generation of very good "base-case" designs, and the expeditious evaluation of new ideas. Sensitivity analysis of the base-case design identifies the dominant design decisions, which can then be adapted to improve the overall process economics.

Hierarchical Evolution of Conceptual Process Designs

In this section we will describe the generic features of a hierarchical approach, initially proposed by Douglas (1985, 1988), for the synthesis of conceptual process designs. It provides a well-defined structure of how the synthesis of conceptual processing schemes proceeds from abstract, vague and incomplete, to fairly detailed, exact and quite complete flowsheets. Despite minor variations, its essential structure has been used to synthesize conceptual designs for a variety of diverse processes; single-product, continuous processes with single-, or multi-step reactions (Douglas, 1985; 1988; 1990), processes with solids (Rossiter and Douglas, 1986, 1988; Rajagopal et al., 1992), polymer processes (Malone and McKenna, 1990), biochemical processes (Siletti and Stephanopoulos, 1990) and batch processes for pharmaceuticals and specialty chemicals (Stephanopoulos et al., 1994). In the next section we will describe the recent modifications of the methodology.

Hierarchical Planning of the Design Process

It is common to characterize a design history as a series of transformations from the initial specifications to the final design. However, every design methodology does not handle the gap between "specifications" and "final design(s)" in one sweeping structure of design steps. Also, the discussion in the previous section indicated that a hierarchically evolving definition of the conceptual design problem, is a necessary condition for the development of a computationally tractable procedure, which leads to the generation of "optimal" conceptual designs.

In recognition of these necessities, the first component of the approach proposed by Douglas is a *hierarchical planning of the intermediate design milestones*, organized as follows:

Level 0. Input Information
Collect data on chemistry to be used, feedstocks available, product specifications (amount, purity), economic constraints (e.g. maximum product unit cost, minimum ROI, maximum allocatable capital) and prevailing regulations (environmental, health, safety).

Level 1. Number of Plants for Multistep Reactions

Level 2. Input-Output Structure and Plant Connections
Specify the feed, exit and recycle streams for each "plant," specify the interconnections between plants for multistep reaction processes and estimate the annualized capital and operating costs associated with the feed and exit streams.

Level 3. Recycle Structure
For each "plant" specify the reactor recycle streams and flows, the type of reactor, as well as the reactor size and cost, the gas recycle compressor's and feed dissolver's (for solid raw materials) sizes and cost, if any.

Level 4. Separation System
Specify the separation system for each "plant" and estimate its annualized cost. Additional hierarchical refinement reveals seven sub-levels of intermediate design milestones at this level (see subsequent section).

Level 5. Energy Integration
Specify a heat exchanger network to accomplish heat and power integration.

Level 6. Create and Evaluate Alternative Designs.

The series of the above milestones does not describe but *prescribes* the state of the intermediate, partial conceptual designs with increasing amount of detail. The *top-down abstract refinement* of one intermediate design milestone to the next proceeds in two steps; *a decomposition* and *a coordination* step. Specific engineering knowledge guides the decomposition of e.g. an input-output structure into the abstract reaction and separation systems of the recycle structure, or of the abstract separation system into the network of generalized separation subsystems (e.g. vapor recovery subsystem, solid recovery subsystem and liquid separation subsystem). At the coordination step, the inputs and outputs of all resulting subsystems, as well as their interconnections are identified. Furthermore, at the coordination step *constraint propagation* mechanisms (see following paragraph) are used, to achieve consistency between the different subsystems which were produced by the decomposition. Through the top-down abstract refinement the number of decisions that is required to make at each level is of the same order, so that the combinatorial explosion is avoided.

Bridging the Gap Between Design Milestones

While the top-down abstract refinement is quite satisfactory for the hierarchical prescription of the intermediate conceptual designs, it is possibly incorrect as a model to describe the intricate web of design actions that lead from one milestone to the next. However, certain conceptual groupings of design tasks are common and repeated throughout every level of the design procedure, as follows:

Synthesis Tasks. They represent the design activities and associated decisions, which generate the alternative structures of the intermediate design milestones, through heuristic reasoning, or numerical procedures.

Analysis Tasks. They set up and solve the material and energy balances, and using short-cut design and cost models they generate cost estimates for the partial conceptual designs.

Evaluation Tasks. They evaluate a series of performance measures and compare them against predefined bounds. Thus, if the economic potential becomes negative the corresponding partial design is rejected. Similarly, an intermediate design is rejected if any of the cumulative capital cost, the unit-product cost, the ROI, the load of waste streams, etc. violates a predefined bound. On the other hand, any alternative which satisfies the performance constraints, or is within a certain percentage of the annualized cost of the best alternative, is retained. Consequently, one may employ a variety of search techniques for the identification of attractive alternatives, which can range from aggressive depth-first search to exhaustive enumeration of retained alternatives.

Constraint Propagation Tasks. The synthesis and analysis tasks at each level of the hierarchy impose constraints on the feasible or attractive intermediates at the next level. The following mechanisms for constraint propagation can be and have been used in hierarchical conceptual design:

1. *Propagation of constraint values*. For example, the feed and exit streams at the recycle structure should have the same values, or obey the same constraints as their counterparts in the input-output structure.
2. *Induction of new constraints*. The economic potential at the input-output structure level is a function of the reaction's selectivity (or equivalently, of the per pass conversion, x) and must be $EP = f(x) > 0$. At the recycle level where a compressor has been revealed and the reactor type has been specified, the following constraint is induced; $f(x)$ - $Annualized\_compressor\_cost(F_{RC})$ - $Annualized\_reactor\_cost(F_{RC}) > 0$, where F_{RC} is the flow of the gas recycle stream.
3. *Generation of context-dependent constraints*. The partial design created at any level of the hierarchical procedure, generates a context which gives rise to certain constraints as to what is feasible or attractive at the next level. For example, if the loss of valuable organics in a purge stream is smaller than the amount that would justify the addition of a vapor recovery system, then no attempt should be made to consider a vapor recovery system. Similarly, if a non-condensable reactant is present in a vapor phase reaction system at Level 0, then the use of gas recycle and purge stream must be considered at Level 2 and Level 3, normally only plug flow reactor combinations are considered at Level 3, a flush drum is needed at Level 4 and a vapor recovery system is a potential option at Level 4.

Thematic Variations: Taxonomy of Design Problems

The hierarchical decision procedure is applicable to (almost) all types of conceptual design problems, which can be organized in a concise taxonomy, according to the characteristics shown below, most of which have been considered in various implementations of the hierarchical procedure.

A. *Type of pivotal operations involved*
   1. Blending, separations (e.g. minerals), reactions
B. *Continuous or batch plants*
C. *Product slate and plants*
   1. Single product vs. multiple products
   2. Single plant vs. multiple plants
D. *Type of Product*
   1. Pure chemical
   2. Mixture, e.g. gasoline
   3. Distribution function, e.g. polymer, solids
E. *Phases*, present in the process
   1. Vapor, organic liquids, aqueous solutions, solids
F. *Type of industry*, based on size of molecules and types of unit operations
   1. Petrochemical, bulk commodity chemicals
   2. Polymers; requiring special unit operations
   3. Biochemicals; requiring special unit operations
   4. Specialties; e.g. monomers, pharmaceuticals, herbicides, pesticides, dyes
G. *Production volume and value-added*
   1. Therapeutic proteins
      1 kg/yr $\$1x10^6$/kg
   2. Specialties
      $1x10^6$ lb/yr $20/lb
   3. Commodities
      $100x10^6$lb/yr $0.5/lb
   4. Basic (e.g. ethylene)
      $1000x100^6$lb/yr $0.15/lb

Some of the specific knowledge, e.g. types of unit operations considered, design and cost models, is different for different classes of problems, but some remain the same for several classes of problems, e.g. the overall and recycle material balances, the reactor models, selection rules for various separation systems, etc. In addition, the hierarchical decision procedure provides the same framework for finding the best few flowsheets. The classification of new design problems according to the above taxonomy, allows the efficient pruning of the search space, because many processing alternatives do not need to be considered.

Summary of Features

Following the logical implications of the discussion in the preceding paragraphs of this section, we can summarize the important features of the hierarchical decision procedure as follows:

1. *Generality*. The basic procedure can be used for any type of processing system.
2. *Problem definition*. It provides a model for the design process with a systematically evolving definition of the design problem, through a series of partial solutions. The partial solution at each level provides a context for the redefinition of the system and a set of constraints for the design at the next level.
3. *Creation of feasible designs*. The hierarchically consistent formulation and propagation of the constraints guarantees the feasibility of the design, or provides an efficient way of terminating poor ideas for design.
4. *Tractability*. At the low levels of the hierarchy, the reaction chemistry and the ambient phases of the components usually provide strong constraints on the selection of the

feed, exit and recycle streams, so that only a few alternatives need to be considered. The constraints generated at the low levels often make it possible to eliminate complete sets of design alternatives at subsequent levels of the hierarchy. The number of decisions at any level of the hierarchy remains essentially of the same order, thus leading to computationally tractable implementations of the procedure.

5. *Flexibility in the search for the "best" designs*, including aggressive depth-first searches, extensive enumeration of alternatives, or a mixture of the two.

Decision-Making and the Knowledge Base

In this section we will discuss the recent modifications of the knowledge base used in the decision-making process for the hierarchical design of conceptual processing schemes, following the six-level hierarchy discussed in the previous section.

Level 0 — The Input Information

The input information that we would like to have available at the start of a design project is given by Douglas (1985, 1988). In addition to the chemical constraints listed for the reaction chemistry, environmental constraints, toxicity and safety constraints must also be included.

All of the input information is seldom available at the start of a project, and we need to make reasonable guesses to complete the first conceptual design. The results of the base-case design and sensitivity analysis studies around it, identify the most important components of the missing information and thus set the experimental program that will lead to their acquisition.

Level 1 — Number of Plants

The rules, proposed by Douglas (1990), for the determination of the number of plants are unchanged. Plants though, which are used to produce or degrade homogeneous catalysts, need to be distinguished from plants used to produce intermediates and the final product. Homogeneous catalysts should be produced immediately before they are needed and removed after the last plant where they are used. For single product plants, this part of the synthesis procedure identifies the forward connections between the plants to produce the intermediates and the final product. The rules for the connections between the plants have been moved to Level 2.

Level 2 — The Input/Output Structure of the Flowsheet

Once the individual plants have been identified, the feed and exit streams for each plant are specified, considering the whole plant as an abstract unit with no internal structure. The destinations of all chemical components, appearing in the reaction and the list of feed streams, are determined according to the following revised table:

Table 1. Component Classification and Destination

Component Classification	Destination
Reactant (liquid)	Recycle (exit)
Reactant (solid)	Recycle, or Waste
Reactant (gas)	Gas recycle & Purge, Vent
Byproduct (gas)	Fuel, or Flare
Byproduct (reversible reaction)	Recycle, or Exit
Reaction intermediate	Recycle (exit)
Product	Product tank
Valuable byproduct	Byproduct tank
Fuel byproduct	To fuel supply
Waste byproducts	(Pay for disposal)
Aqueous waste	Biological treatment
Incineration waste	Incinerator
Solid waste	Landfill
Feed impurity	Same as byproduct
Homogeneous catalyst	Recycle
Homogeneous catalyst activator	Recycle
Reactor or product solvent	Recycle (exit)

Note that differing destinations for waste byproducts imply that these components must be separated before they are treated or disposed. Also note that all treatment or disposal problems associated with the reaction chemistry are identified at the very beginning of a design. Obviously, treatment problems associated with reactor diluents, heat carriers, or reaction solvents, and the waste problems emanating from the separation system will have to be considered at a later stage of the design, when the reactor system (Level 3) and the separation system (Level 4) are designed.

As it was discussed in the previous section, the design activities at every level of the design hierarchy can be grouped into synthesis, analysis, evaluation and constraint propagation design tasks. Let us examine those for Level 2.

Synthesis. Most of the decisions at this level are associated with the specification of destinations for each component, in order to fix the input/output structure of the flowsheet for each "plant." The original list of decisions provided by Douglas (1985, 1988), needs to be supplemented by a decision that determines the removal of solids (intermediates and byproducts) from a plant as crystals, as a melt, as a slurry, or in solution. This decision affects the types of equipment that will appear in the flowsheet at subsequent levels and determines the extent of the environmental impact of the process at large.

It should be noted that the synthesis decisions made at Level 2 lead to different sets of component destinations, and these in turn lead to different process alternatives. A complete list of decisions leads to the generation of *all* alternative component destinations, and thus is essential for the successful deployment of the hierarchical conceptual process design.

After the feed and exit streams for each of the individual "plants" have been specified, the recycle connections between the plants must be specified. The general rule is to recover byproducts in plants where they are formed and send them to plants where they are reactants. However, new alternatives can be generated by adding fresh reactants upstream of the plants where they are needed for the reaction (so that they can act as heat carriers, solvents, etc.) and to send mixtures that require separation to downstream plants, where the separation might be easier than in the plant where they originated (although it will be necessary to oversize some equipment to accommodate the increased load).

Analysis and Evaluation. The analysis and evaluation procedures for Level 2 are essentially the same as described by Douglas (1985, 1988) with the following two exceptions:

a. The strategy for the solution of material balances has been modified so that it can account for multistep reaction processes.
b. Waste treatment costs are also included in the computation of the economic potential.

Since the separation system has not as yet been specified, the material balances have been drawn on the assumption that 100% of all materials are recovered in the separation system. These short-cut balances are used to prescribe the scope of the separation system and help identify its essential structure, and will be (or, can be) revised after the separation system has been specified.

The treatment costs at this stage reflect only the cost for treating the generated reaction byproducts. Additional treatment costs, associated with the reactor and separation systems, will be identified and computed at later stages of the design methodology, where these systems are considered.

The economic potential must be kept larger than a certain lower bound. If this constraint is violated, the alternatives are examined, and if no alternative is acceptable, the design project is terminated. This constraint generates in turn ranges of acceptable values for the design variables of the feasible structures.

Constraint Propagation. The structure of the feed and exit streams, their destinations, as well as the ranges of feasible values for the design variables are propagated into the next level of the hierarchy.

Level 3 — The Recycle Structure

At this level, each of the abstract units describing the corresponding whole "plant" is decomposed into two subsystems; a general reactor system and a general separation system. The design goals include; fixing the recycle streams between the two subsystems, the selection of the reactor type and its operating conditions for each "plant," the possible inclusion of a melter or a dissolver for the transfer of solids from one plant to an other, and the computation of the annualized capital and operating costs for the units associated with the recycle flows. The recycle connections between the various plants was discussed in Level 2.

Synthesis. Most of the decisions that are used to specify the recycle structure have been discussed by Douglas (1985, 1988). However, the use of a reactor-separator (which might need to be supplemented by additional separators) should be added to the list of potential alternatives. The use of a reactor-separator usually eliminates the need for some recycle streams. Similarly, operation at very high conversions might eliminate the need for the recycle of a reactant.

For feed components that are supplied as solids to liquid reaction system, either a melter or a dissolver need to be added to the flowsheet. If the melting point of the solid is less than the reactor temperature, a melter is selected, otherwise a dissolver is chosen. We prefer to use a component in the process as a solvent, and if the dissolution kinetics are faster than the reaction kinetics, it might be possible to dissolve the feed in the reactor vessel. In some cases a solvent might be added to keep an intermediate or a byproduct (which is solid) in solution, in order to facilitate its transfer from the reactor to a downstream plant (if it is an intermediate), or to facilitate its removal (if it is a byproduct).

Analysis and Evaluation. The analysis and evaluation procedures in Level 3 are the same as those described by Douglas (1985, 1988), with the exception that the melter or dissolver must be sized and the corresponding capital cost must be estimated. The base-case design is updated by specifying the values of all design decisions at Level 3. Furthermore, the list of feasible process alternatives, generated at Levels 1 and 2, is supplemented with the feasible alternatives introduced in Level 3. The constraints on the values of the design variables, generated at Level 2, are propagated and produce the following new information:

a. Constraints on the feasible values of the flows of the recycle streams.
b. Tighter bounds on the values of the design variables of Level 2.

Furthermore, the fact that the updated version of the economic potential must satisfy the same positive lower bound, generates new constraints associated with the values of the design variables introduced at this level.

Constraint Propagation. The decisions that determine the configuration of the alternative feasible recycle structures, along with the quantitative constraints on the values of the design variables are propagated to the next level of the hierarchy.

Level 4 — Specification of the Separation System

The synthesis of the separation system is usually the most challenging component in a conceptual design. However, the problem itself has an internal structure that can be exploited to simplify its complexity. Specifically, in the most general case the effluent of the reactor subsystem could be a multiphase stream (or, several distinct streams coming from several reactors) with gaseous, several liquid (e.g. organic and aqueous) and solid phases. Consequently, one immediately discerns a separation of phases, followed by the design of specific separation subsystems, in a loose hierarchy of design sub-levels (Douglas, 1992), as follows:

Level 4a	General separation structure; identification of specific separation subsystems.
Level 4b	Vapor recovery system; to recover valuable liquid components from a gas stream.
Level 4c	Solid recovery system; to recover valuable components held up in the pores of a filter cake.
Level 4d	Liquid separation systems(s); to separate mixtures of liquid components.
Level 4e	Gas separation system; to recover valuable gaseous component(s) from a gas stream.
Level 4f	Solid separation system; to separate a mixture of solids.
Level 4g	Combine the separation systems for multiple plants.

The above hierarchy allows a very effective decomposition of the overall separation system design, rendering problems with manageable complexity. Specifically, once the phase splits for the reactor exit stream have been considered (at Level 4a), we can establish the *general separation system superstructure* with all potential separation subsystems, such as the one shown if Fig. 1.

For each of the subsystems in Fig. 1 we can identify a set of separation techniques, which are appropriate, with the following being among the most common:

A. *Vapor recovery system*: chilled condenser, absorber, adsorber, membranes
B. *Gas separation system*: cryogenic distillation, pressure-swing adsorption, membranes
C. *Solid recovery system*: single wash, counter-current wash and re-slurry, re-crystallize
D. *Solid separation system*: flowsheets with dissolvers, crystallizers and distillations
E. *Liquid separation system*: simple, reactive, azeotropic, or extractive distillation, stripping, extraction, crystallization, adsorption

However, for certain types of processes specific unit operations must also be considered, e.g. devolatilizing extruders and wiped film evaporators (for polymer processes), affinity chromatography, electrophoresis and others (for protein recovery and purification), or micellar extraction and liquid membranes (for large pharmaceutical molecules).

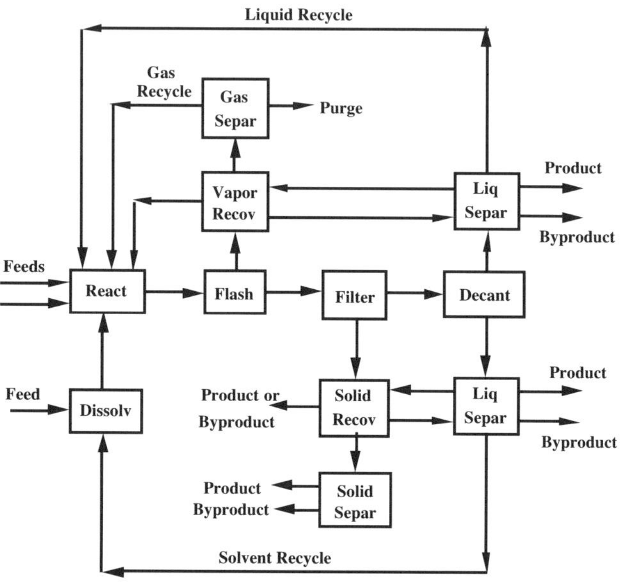

Figure 1. General structure of a separation system.

The separation system superstructure of Fig 1, along with the alternative separation methods for each subsystem (listed above), lead to a fairly complex *superstructure of separation units*, which can be represented by a mixed-integer nonlinear programming formulation. Solution of this problem through a branch-and-bound implicit enumeration of the alternatives can lead to the identification of the "optimal" separation system. Although such implicit enumeration strategies exist and can tackle fairly large problems (Grossmann, 1990), in this paper we will continue the emphasis on discussing the engineering knowledge-base, which can render simpler problem formulations.

Let us, therefore, discuss the specifics of the knowledge base and the decisions which must be made before the complete structure of the separation system has been specified. In doing so we will follow the six levels of hierarchical planning outlined earlier in this subsection.

Level 4a — Phase Separations

The first task is to identify the requisite phase separators for the reactor effluent(s), so that we can determine the resulting separation subsystems. The following design rules guide this selection (Douglas, 1992):

1. The rules for removing gases are the same as those described by Douglas (1985, 1988).
2. If the reactor exit stream (after any gases have been removed) contains a mixture of three phases, i.e. aqueous liquid, organic liquid, solid, we normally use a liquid-solid separator to remove the solids first, and then split the two liquid phases in a decanter. Normally, we filter before we decant to prevent solids settling in the decanter.
3. If the reactor contains a gas-solid mixture, we remove the solids using a cyclone.
4. The liquid-solid separator might require a thickener (a hydrocyclone, or a gravity settler), a rotary vacuum filter, a pusher centrifuge, etc. Selection rules for various liquid-solid separators have been published by Purchas (1981) and Ernst, et al. (1991).
5. If the reactor exit is all vapor, we cool the reactor exit stream to 100 °F (cooling water temperature) and then flash, filter and decant, depending on the phases obtained. Sometimes both a high and a low pressure flash are desirable.
6. If the reactor exit stream is a liquid, in some cases we cool the stream to crystallize a component. In other cases we can add water to an organic stream to cause a phase split and to recover a component. Similarly, we can add an organic to an aqueous exit stream to cause a phase split and recover another component. Still an other alternative is to add an anti-solvent to cause a component to crystallize.

The general rules, given above, of removing gases, then solids and splitting the liquid phases in a decanter, are applicable for reasonably well balanced flows, otherwise the following rules produce additional alternatives:

7. If a vapor-liquid mixture is mostly liquid, it is often better to send the stream to distillation column and to remove the gases overhead in the column, using a partial condenser. If the mixture is mostly a gas, send the stream directly to a gas absorber.
8. For liquid-liquid-solid streams containing only a small amount of solids, it is often better to decant before filtration.

Level 4b — Vapor Recovery System

The purpose of the vapor recovery system is to remove valuable liquid components from a gas stream, and the most common units used were listed earlier in this section. Douglas (1985, 1988) has listed several rules, which are used to place the vapor recovery system on the purge stream, the gas recycle stream, or the flash vapor stream. In addition he has also discussed the decisions associated with the selection of the type of the vapor recovery system, as well as the short-cut models which should be used for determining the size and cost of the vapor recovery system.

Level 4c — Solid Recovery System

Filter (centrifuge) cakes usually contain valuable liquid components held up in the pores of the cake. A single washing is usually adequate for a solid byproduct exit stream. Countercurrent washing with a reslurrying step is more common for a solid product stream. A single wash liquid or different wash liquids can be used for multiple washing stages. For very high purity products, a re crystallization step is often required after an initial washing. A decision must be made on whether to mix the wash liquid with the mother liquor leaving the filter (centrifuge), or recover and recycle the wash liquid separately. In many cases when water is used as a wash liquid, the aqueous stream is sent to a waste treatment facility after the valuable components have been recovered; a practice that may be precluded in the future by changing EPA requirements.

Level 4d — Liquid Separation System

Distillation is normally the preferred separation for the two liquid mixtures (organic and aqueous), produced from a decanter, or the single-phase, liquid stream exiting the flash unit. According to Barnicki and Fair (1990), simple distillation is almost always the least expensive if the relative volatility is $\alpha > 1.5$, while other separation techniques might be competitive for $1.1 < \alpha < 1.5$, and are usually cheaper if $\alpha < 1.1$. Separation techniques other than simple distillation should (or, must) be considered if the liquid mixture contains monomers, temperature sensitive materials, dissolved solids, azeotropic mixtures with separation boundaries, or if a dilute binary mixture is encountered.

For case where simple distillation can be used for all of the separations, the determination of the best distillation sequence provides a complete separation system. For cases where extraction, crystallization, extractive and azeotropic distillations, separation of mixtures of solids, separation system recycle loops are introduced and so the list processing procedures, used for selecting a separation technique on the basis of an ordered list of physical properties, cannot and do not identify a complete separation system flowsheet (Douglas, 1992).

Distillation Column Sequencing. There is a vast literature concerning the sequencing of simple distillation columns for single feed streams (see for example, Seader and Westerberg, 1977; Modi and Westerberg, 1992). Numerous heuristics, often contradictory, have been used, but even in cases where the heuristics agree, the resulting solution may not be the best (Glinos and Malone, 1984). Exhaustive enumeration of all alternatives, a feasible proposition for feeds with a small number of components (Glinos and Malone, 1985), or an implicit enumeration of alternatives through branch and bound strategies (for feeds with a large number of components), are more reliable strategies for identifying the best sequences. In the PIP program (Kirkwood, et al., 1988), exhaustive enumeration was used to select the best distillation sequence for every value of the design variables. At Level 4d the utilities costs were included, but the utilities costs were deleted and the best sequence was determined again at Level 5 in conjunction with heat integration. Unfortunately, so far there are no conclusive studies for the sequencing of distillation columns with multiple feeds; a case arising when, for example, a liquid stream leaving the reactor, a liquid stream leaving a gas absorber and a third stream used to wash a filter have similar components and could be fed to the same distillation sequence. Similarly, if different plants in a multistep reaction process have the same distillation separations, it is necessary to reconsider the column sequencing by accounting for multiple feeds. Further work is needed to provide more concrete design strategies for the sequencing of columns with multiple feeds.

Douglas (1988, 1990) has provided a set of rules for addressing the following issues, within the scope of distillation sequencing:

1. Remove light ends to avoid product contamination. The alternatives are shown in Douglas (1988).
2. Identify the scope of separation subproblems, which must be handled by other than simple distillation separations e.g. close boilers, temperature sensitive materials, monomers (prefer not to distill, but if distillation is selected, use vacuum column), dissolved solids (add a crystallizer-filter, or centrifuge-dryer subsystem to the flowsheet), and cases where distillation boundaries are encountered.
3. Separate mixtures with grossly imbalanced distribution of components, using complex columns (see Glinos and Malone, 1985; Tedder and Rudd, 1978).

Obviously, if a dissolved solid is present in the mixture, then a crystallizer-filter (or, centrifuge)-dryer configuration must be added to the flowsheet.

Distillation Boundaries — Special Distillation Subsystems. If azeotropes are present, then it is necessary to see if distillation boundaries are present and what distillation sequences are feasible (Foucher, et al., 1991). In some instances the boundaries are sufficiently curved so that boundary crossings are possible (Doherty and Caldarola, 1985). In other cases, two distillation columns operating at different pressures can sometimes be used to cross the limiting boundaries. In still other cases, the addition of an entrainer that causes a liquid-liquid phase split, where the exit streams are in two different distillation regions, can be used to accomplish a separation (Doherty and Caldarola, 1985; Wahnschaft, et al., 1992). These boundary crossing distillation problems identify the need for and the scope of special separation subsystems in the flowsheet, and normally give rise to a small number of distinct feasible configurations.

Extraction — Solvent Recycle Loops. Water soluble components can be removed from an organic mixture by extraction with water, which is then sent to a treatment facility after the valuable components have been reclaimed; a practice which may be precluded in the future due to changing EPA requirements. Similarly, organic components dissolved in an aqueous stream can be recovered by extraction with an organic solvent. An organic solvent must be recovered and recycled in a local column, or the mixture can be sent to a train of distillation columns used to separate the other components. Extraction can also be used to separate mixtures of organics that are in different chemical families. In general, solvent selection rules are not available and additional research is needed in this area.

Solvent selection has a significant effect on the structure of the extraction subsystem flowsheet. For the extraction of a binary mixture an extraction column and two distillation columns with a decanter, or three distillation columns are needed in the extraction subsystem. Different solvents have different residue curve maps thus leading to different "best" ways in which the extractor is connected to the auxiliary distillation columns. Siirola has shown how the structure of the "best" extraction subsystem changes as various solvents are selected to extract acetic acid from its mixture with water (see Figure 15 in Siirola, 1994). Consequently, solvent selection impacts on the structure of the subsystem flowsheet and involves more than just physical property considerations.

Crystallization Subsystems. A crystallization separation is usually required when a product or a byproduct that is solid at ambient conditions is dissolved in a liquid (in some cases byproducts can be removed in solution and products can be passed on to a downstream plant in solution). The list processing synthesis procedures that select crystallization as an alternative based on differences in freezing points between various components, do not develop a complete crystallization subsystem.

The selection of a crystallization subsystem requires the specification of the type of crystallizer, a liquid-solid separator, a solid recovery system, a dryer, the destination of the mother liquor leaving the filter (or, centrifuge), the destination of the wash liquid and a purge stream, if needed. The set of crystallizers that can be used includes the following types; (a) evaporative (to remove some solvent if the product *is not* temperature sensitive), (b) flash (to remove some solvent if the product *is* temperature sensitive), (c) direct cooling (adding a liquefied gas to obtain a low temperature), (d) indirect cooling (use of an external cooling loop) or (e) salting out crystallizer (change the solubility of the dissolved solid through the addition of an organic solvent or water).

The selection of the liquid-solid separator and a solid recovery system are based on the same rules that are used for liquid-solid phase splits and cake washing. Dryer alternatives include cocurrent rotary vacuum dryers (for large particles and temperature sensitive materials), countercurrent rotary vacuum dryers (for large particles and temperature insensitive materials), fluid bed dryers (for small particles), etc.

Crystallizations are usually carried out at only 3 to 5 percent levels of supersaturation to avoid the rapid formation of crystals with inclusions. Thus, a crystallizer may contain a number of stages, with lower pressures or temperatures between each stage. In some instances two or more units are placed in series with the removal of solids between stages. Solids are sometimes recycled from the second filter (or, centrifuge) back to the first crystallizer to provide seed crystals. Similarly, the mother liquor from the filters, along with wash liquid streams can be recycled to the first crystallizer, or somewhere upstream of the first crystallizer, to prevent the loss of some of the product in solution. Usually, it is necessary to add a purge stream to the mother liquor recycle loop to prevent the accumulation of trace components. If very high purity products are required it is often necessary to re dissolve and re crystallize the solid product.

If eutectic mixtures are encountered, separation boundaries are present and special flowsheet subsystems are needed to completely separate the components. One approach is to add a new component that is easy to distill from the original components. For a binary mixture, the addition of the entrainer often introduces two additional binary eutectics and a ternary eutectic. Complete separations can be accomplished using an extractive distillation subsystem (Rajagopal, et al., 1991), in which case only two types of subsystems need to be considered.

Level 4e — Gas Separation System

Gas separations are seldom used and are present when the flow of the gas mixture exceeds 10 to 20 tons/day (Barnicki and Fair, 1992) or a membrane is used to recover some raw material from the gas purge stream.

Level 4f — Solid Separation System

The separation of mixtures of solids can be accomplished using a special crystallizer-dissolver configuration (Rajagopal, et al., 1988), but normally an attempt is made to avoid co-crystallization in a process.

Level 5 — Energy Integration

The extensive literature on heat exchanger synthesis assumes that all of the process flows are fixed (Gunderson and Naess, 1988; Linnhoff, 1993). However, the optimum flows depend on the heat exchanger network selected. Terrill and Douglas (1987) indicated that if various heat exchanger networks are considered for a complete plant and the flows for each flowsheet are optimized, the optimum costs (with different flows) are about the same. In addition, the area estimation technique of Townsend and Linnhoff (1984) gives essentially the same result. Thus, for the initial conceptual design, where the objective is to identify the best few flowsheets, we can use the pinch approach to compute the minimum utilities requirements, and the area estimation approach of Townsend and Linnhoff to estimate the capital costs, every time we attempt to estimate the optimum cost for each alternative and the range of the design variables, where the optimum is flat.

Once a small number of flowsheets has been identified, and a computer-aided support system is in place for more detailed material balances and improved estimates of equipment sizes and costs, the optimal network can be developed through an enriched pinch technology, the solution of a MINLP problem with various implicit enumeration strategies, or simulated annealing Dolan, et al., 1987).

Computer-Based Automation of Conceptual Design

The discussion in the previous section revealed that the complete structure of tasks during the synthesis of conceptual designs and the knowledge base that supports them can be very large, detailed and complex for any human designer to document and mentally carry with him/her. To the extent that we can untangle and make explicit segments of the design procedure, the corresponding design tasks of the hierarchical synthesis of conceptual designs can be mechanized. The benefits are many and diverse (Han, Stephanopoulos and Douglas, 1994):

1. Rapid creation and evaluation of a base-case conceptual design to assess the attractiveness of an idea.
2. Explicit documentation of the conceptual designs, as well as easy verification and modification.
3. Explicit documentation of the design process itself: why certain goals were set during design and how they were achieved; how design decisions were made; what assumptions and simplifications were involved; what models were used at the various stages of the design; what alternatives were considered and why certain ones were selected over others.
4. Improvements in cost and reliability of the preliminary conceptual design work.
5. Creation of a depository for the organization of new empirical knowledge and the systematic incorporation of new theoretical results and analytic tools, all of which tend to expand the automation of the design process and improve the quality of new designs.

To construct a software system that can emulate the hierarchical conceptual process design, one needs (i) a model of the conceptual design process, (ii) a high-level design-oriented language for implementing (i) and (iii) a human-computer interface to support the integration of the human in the design process. In the following paragraphs we will outline the essentials of a software system that implements the hierarchical design of processing schemes. For more technical details see Han, Stephanopoulos and Douglas (1994), or Han (1994).

Model of the Conceptual Design Process

This is based on the hierarchical procedure, whose characteristics were discussed in the previous sections and whose software implementation has been built around the following entities:

Planner, which defines the hierarchical planning of design activities, by prescribing the characteristics of the intermediate design milestones that the design process must go through, i.e. the state of the design at each of the six levels of the hierarchical approach, discussed earlier.

Scheduler, which determines the sequence of the design steps to be taken as one attempts to advance the current state of a processing scheme to the next milestone, defined by the Planner, above.

Designer, which maintains the representation of the processing scheme being synthesized and other domain specific knowledge to do the following; acquire necessary data, reason with a specific set of rules, execute a design algorithm, carry out an optimization procedure, update the state of the evolving design, etc.

To represent the Planner we have used the *abstract refinement* model (Mitchell et al., 1981; Stephanopoulos, 1990), which is deployed through a "tree of non-interacting goals." Each goal prescribes the required conceptual design at each level of the abstraction. For example, the top goal is "design the conceptual processing scheme for a multistep reaction plant." Six subgoals emanate from the top goal, each corresponding to a level of the hierarchical procedure. Successive accomplishment of these goals leads from the specifications to the final design. In the next subsection we will how these "goals" are represented by specific computational elements.

The abstract refinement, represented by the Planner's "tree of non-interacting goals," *is not* a good model for representing the design tasks of the Scheduler. The *transformational* model (Balzer et al., 1976; Stephanopoulos, 1990), which converts specifications into implementations through a series of correctness-preserving transformations, is more appropriate for maintaining consistency between the successive intermediate designs of the hierarchical approach. For example, Fig. 2 shows the tree that captures the goals (tasks), which must be satisfied, as we try to achieve the design milestone "specify plant connectivity," starting with the available input data. The goals (tasks) of the transformational model are computational elements, which operate over several components of an evolving flowsheet, thus maintaining design integrity. A tree of tasks, such as that of Fig. 2, should be viewed as an executable program.

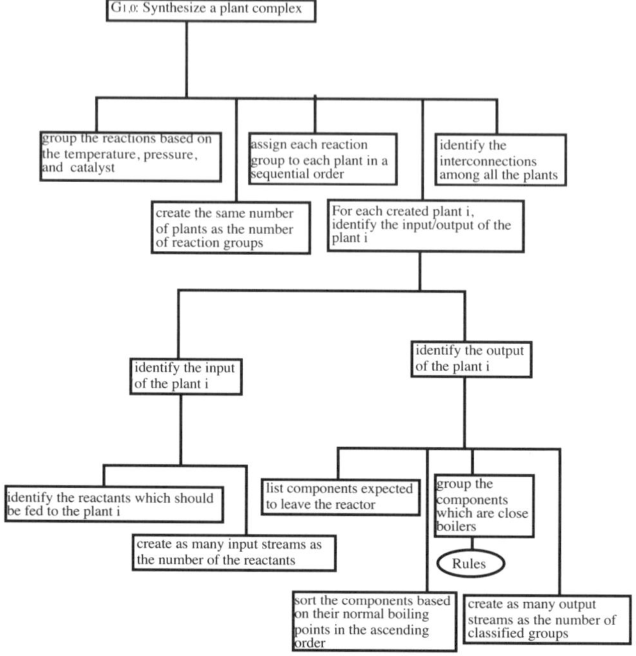

Figure 2. Goal structure for the goal $G_{1,0}$, "Synthesize the plant complex structure."

Each of the tasks in the tree is represented by specific computational entities, whose characteristics will be discussed in the next subsection. There exists an extensive "one-to-one" mapping between the engineering tasks and the computational entities in the trees of tasks. As a result, future modifications of the software system can remain closely coupled to the modifications of the engineering methodology itself. For the detailed description of the other trees of tasks in the overall system, see Han, Stephanopoulos and Douglas (1994).

The formalized computational model, outlined above, satisfies the following design principles, which were required for the engineering design methodology itself:

1. Hierarchical planning of the design methodology.
2. Successive refinement of specifications into implementations.
3. Propagation of constraints, using the mechanisms discussed in an earlier section.
4. Hierarchical decomposition of successive designs with ensuing coordination of the subsystems.
5. Context-based design.

A Hierarchical Design-Oriented Language

The value of the computational model of the conceptual design process, discussed above, rests with the effectiveness of the representation schemes that one employs to describe the "trees of tasks," the "tasks" themselves, and the declarative and procedural knowledge they contain. To achieve this, we need a well-formulated, high-level design language, which can be used to express the following:

a. The evolving structures of the conceptual processing schemes in sufficient detail, including all the semantic relationships between the various components (e.g. abstract or specific processing units, streams, materials, modeling relationships, variables, decision-making procedures, numerical algorithms, etc.).
b. The design methodology itself, i.e. the goals and the trees of goals used to model the activities of the Planner, Scheduler and Designer, as described above.

Let us provide a brief overview of the language's characteristics. For more details the reader is referred to Han, Stephanopoulos and Douglas (1994), or Han (1994).

Multi-Faceted Modeling of the Conceptual Design State

To describe the conceptual flowsheet of a processing scheme, as it evolves through the various intermediate milestones of the hierarchical procedure, the modeling language uses a series of modeling elements (shown in Courier font), which capture the following aspects, at any level of abstraction:

1. The structure of the process flowsheet, as a complete system (`Flowsheet`), or in terms of its units (`GenericUnit`), and their interconnections (`Port, Stream`).
2. The characteristics and behavior of units, streams, reactions and materials (`GenericVariable, Reaction, Compound`).
3. The maintenance of alternative process flowsheets, generated by the hierarchical decision procedure (`Project`).

The semantic relationships among the various modeling elements of the language, establish

a. refinement of an abstract system into a network of subsystems,
b. propagation of design values and constraints from one intermediate design to the next, more refined, one,
c. relationships among the hierarchical descriptions of the same process, e.g. the relationship between the input-output structure and the subsequent reaction and separation subsystems,
d. relationships among the various alternative designs of the same process.

Modeling of the Design Tasks

Any design task is a transformation process, and possesses its own goal(s) and a plan on how to accomplish the goal. Therefore, any of the goals (tasks) employed by the Planner, Scheduler, or Designer (see above), are represented by the following modeling elements of the language:

A. `ProjectManager`. This generic modeling element provides the information and the procedures, which are needed in order to define, plan and carry out the hierarchical planning of conceptual design. Specifically, the `ProjectManager` has access to the input information defining a project, and provides the information and schedules the execution of the `IOManager, RecManager, GSPManager,` etc., which have the requisite knowledge to carry out the specification of the input-output structure, recycle structure, general separation system structure, etc. respectively.
B. `DesignManager`. This modeling element contains all the necessary information for the creation of designer managers with design-specific plans. The `IOManager, RecManager, GSPManager`, etc., are specialized descendants of the `DesignManager` and possess the data and procedures to define, schedule and monitor the execution of the corresponding design tasks, in order to carry out the desired transformation. For example,

```
aRecManager(aIOStructure) ->
      aRecycleStructure
```

C. `DesignAgent`. This is also a generic element, which is used for the construction of specialized design agents. The specialized descendants of this element capture the common grouping of tasks, as these groups were identified in our earlier discussion. For example, the specialized element, `StructureAgent`, is used to capture the design tasks revolving around the specification of the flowsheet structure, the `CoordAgent` to model the tasks for the coordination of subsystems resulting from the refinement of an abstract unit, the `ModelAgent` to prepare the required models for a given flowsheet, the `EvalAgent` to organize the tasks for the simulation and evaluation of flowsheets, and the `DrawAgent` to develop iconic depictions of the process flowsheets. Further specialization of the design agents captures the corresponding activities at the various levels of the hierarchical procedure. For example, the `RecStructureAgent` contains the data, rules and procedures for the generation of the recycle structures.

A series of semantic relationships among the modeling elements, representing design tasks, allow

1. the refinement of a design task into simpler ones,
2. the transformation of a task into an other (very useful when the change of the design task creates new alternative designs),
3. the merging of tasks into a higher-abstraction task.

The design-oriented modeling language has been implemented in an object-oriented programming environment, with all the modeling elements representing distinct classes organized in hierarchical inheritance trees, and all their semantic relationships encoded by generic class methods. Figure 3 shows part of the object model for the design managers and design objects. Notice how the structuring of these elements reflects the engineering logic of the hierarchical synthesis of conceptual processing schemes.

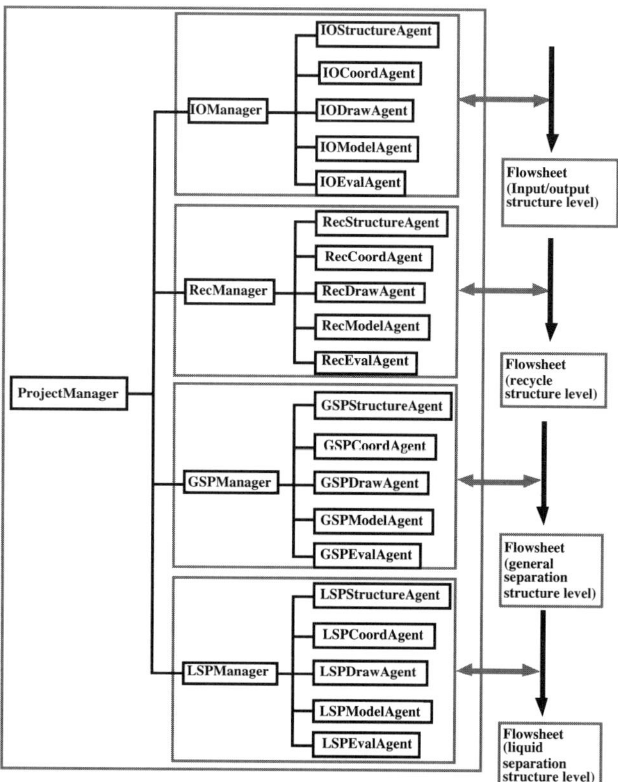

Figure 3. Object model for design managers and design agents.

Human-Computer Interface

The computer-based automation of the hierarchical conceptual design procedure integrates the human designer at every stage of the evolving design by allowing the user to define the scope of design, guide the direction of the design evolution, or/and select among competing designs in a rapid and efficient manner. This interaction is supported by a rich interface that allows the human user to respond to queries posed by the design methodology, provide data and make decisions. In addition, the software system should inform the user about the current state of the design, the rationale for the design decisions it has reached, and provide a recorded history of how a design has evolved to its current state. In view of these requirements, one should never underestimate the importance of an effective interface between the human and the computer. Han, Stephanopoulos and Douglas (1994) describe the characteristics of a graphic interface, which can support most of the above requirements.

References

Balzer, R., N. Goldman and D. Wile (1976). On the transformational implementation approach to programming. *Proceedings 2-nd Int'l Conf. Software Engng.*, 337.

Barnicki, S.D. and J.R. Fair (1990). Separation system synthesis: A knowledge-based approach. 1. Liquid mixture separations. *I&EC. Res.*, **29**, 421.

Barnicki, S.D. and J.R. Fair (1992). Separation system synthesis: A knowledge-based approach. 1. Gas/vapor mixtures. *I&EC. Res.*, **31**, 1679.

Doherty, M.F. and G.A. Caldarola (1985). Design and synthesis of homogeneous azeotropic distillations. 3. The sequencing of columns for azeotropic and extractive distillations. *I&EC. Res.*, **24**, 474.

Dolan, W.B., P.T. Cummings and M.D. LeVan (1987). Heat exchanger network design by simulated annealing. *Foundations of Computer-Aided Process Operations*, G.V. Reklaitis and H.D. Spriggs (eds.), CACHE.

Douglas, J.M. (1985). A hierarchical decision procedure for process synthesis. *AIChE J.*, **31**, 353.

Douglas, J.M. (1988). *Conceptual Design of Chemical Processes*, McGraw-Hill, NY.

Douglas, J.M. (1990). Synthesis of multistep reaction processes. *Foundations of Computer-Aided Process Design*, J.J. Siirola, I.E. Grossmann and G. Stephanopoulos (editors), Elsevier, 79.

Douglas, J.M. (1992). Synthesis of separation system flowsheets. Paper presented at the Annual AIChE Meeting, Miami Beach, Florida, November, 1992

Ernst, R.M., H.M. Talcott, H.C. Romans and G.R.S. Smith (1991). Tackle solid-liquid separation problems. *Chem. Eng. Prog.*, July, **22**.

Fair, J.R. (1969). Mixed solvent recovery and purification. *Washington Univ. Design Case Study No. 7,* B.D. Smith (ed.), Washington Univ., St.Louis, Mo.

Foucher, E.R., M.F. Doherty and M.F. Malone (1991). Automatic screening of entrainers for homogeneous azeotropic distillation. *I&EC Res.*, **30**, 760.

Glinos, K. and M.F. Malone (1984). Minimum reflux, product distribution and lumping rules for multicomponent distillation. *I&EC Process Des. Dev.*, **23**, 764.

Glinos, K. and M.F. Malone (1985). Minimum vapor flows in a distillation column with a side stream stripper. *I&EC Process Des. Dev.*, **24**, 1087.

Grossmann, I.E. (1990). MINLP optimization strategies and algorithms for process synthesis. *Foundations of Computer-Aided Process Design*, J.J. Siirola, I.E. Grossmann and G. Stephanopoulos (editors), Elsevier, 105.

Gunderson, T. and L. Naess (1988). The synthesis of cost optimal heat exchanger networks. An industrial review of the state of the art. *Comp. and Chem. Engng.*, **12**, 503.

Han, C. (1994). *Human-Aided, Computer-Based Design Paradigm: The Automation of Conceptual Process Design*. Ph.D. thesis, MIT.

Han, C., G. Stephanopoulos and J.M. Douglas (1994). Automation in design: The conceptual synthesis of chemical processing schemes. *Paradigms of Intelligent Systems in Process Engineering,* (chapter 2), G. Stephanopoulos and C. Han (eds.), Advances in Chemical Engineering Series, Academic Press

Kirkwood, R.L., M.H. Locke and J.M. Douglas (1988). A prototype expert system for synthesizing chemical process flowsheets. *Comput. Chem. Engng.,* **12**, 329

Linnhoff, B. (1993). Pinch analysis. A state of the art review. *Trans. IChemE,* **71**, 503.

Mahalec, V. and R.L. Motard (1977a). Procedures for the initial design of chemical processing systems. *Comput. Chem. Engng.,* **1**, 57

Mahalek, V. and R.L. Motard (1977b). Evolutionary search for an optimal limiting process flowsheet. *Comput. Chem. Engng.,* **1**, 149

Malone, M.F. and T.F. McKenna (1990). Process design for polymer production. *Foundations of Computer-Aided Process Design,* J.J. Siirola, I.E. Grossmann and G. Stephanopoulos (eds.), Elsevier, 469.

Mitchell, T., L. Steinberg, G. Reid, P. Schooley, H. Jacobs and V. Kelly (1981). Representations for reasoning about digital circuits. *Proceedings IJCAI-81*

Modi, A.K. and A.W. Westerberg (1992). Distillation column sequencing using marginal prices. *I&EC Res.,* **31**, 839.

Pikulik, A. and H.H. Diaz (1977). Cost estimating major process equipment. *Chem. Eng.,* **84**(2), 106.

Purchas, D.B. (1981). *Solid Liquid Separation Technology*. Uplands Press, Croyden.

Rajagopal, S., K.M. Ng and J.M. Douglas (1988). Design of solids processes. *I&EC RES.,* **27**, 2071.

Rajagopal, S., K.M. Ng and J.M. Douglas (1991). Design and economic trade-offs of extractive crystallization processes. *AIChE J.,* **37**, 437.

Rajagopal, S., K.M. Ng and J.M. Douglas (1992). A hierarchical decision procedure for the design of solids processes. *Comput.Chem.Eng.,* **16**, 675.

Rossiter, A.P. and J.M. Douglas (1986). Design and optimization of solids processes: 1. A hierarchical decision procedure for process synthesis of solids systems. *Chem. Eng. Proc. Des.,* **64**, 175.

Rossiter, A.P. and J.M. Douglas (1988). Trade-offs and optimization in continuous solids processes. *Filtration and Separation,* Sept/Oct., 348.

Seader, J.D. and A.W. Westerberg (1977). A combined heuristic and evolutionary strategy for synthesis of simple separation sequences. *AIChE J.,* **23**, 951.

Siirola, J.J. and D.F. Rudd (1971). Computer-aided synthesis of chemical process designs. *Ind. Eng. Chem. Fundamentals,* **10**, 353

Siirola, J.J. (1994). An industrial perspective on process synthesis. *Proceedings of FOCAPD '94*, Snowmass, Colorado, July 10-15, 1994.

Siletti, C.A. and G. Stephanopoulos (1992). BioSep Designer: A knowledge-based process synthesizer for bioseparations. *Artificial Intelligence Approaches in Engineering Design, Vol. I,* C. Tong and D. Sriram (eds.). Academic Press, 295.

Stephanopoulos, G. (1990). Artificial Intelligence and Symbolic Computing in Process Engineering Design. *Foundations of Computer-Aided Process Design*, J.J. Siirola, I.E. Grossmann and G. Stephanopoulos (eds.), Elsevier.

Stephanopoulos, G., C. Han, A. Linninger, S. Ali and E. Stephanopoulos (1994). The concept of ZAP in the synthesis and evaluation of batch pharmaceutical processes. Paper to be presented at the Annual AIChE Mtg., San Fransisco.

Tedder, D.W. and D.F. Rudd (1978). Parametric studies in industrial distillation. *AIChE J.,* **24**, 303, 316.

Terrill, D.L. and J.M. Douglas (1987). Heat exchanger network analysis: Part 1. *I&EC Process Des. Dev.,* **26**, 685.

Townsend, D.W. and B. Linnhoff (1984). Paper at the annual meeting of the Instit. Chem. Engrs, Bath, UK., April.

Wahnschaft, O.M., L.E. Rudulier, J.P. Blania and A.W. Westerberg (1992). SPLIT II: Automated synthesis of hybrid liquid separation systems. *Comput. Chem. Eng.,* **16**, 305 (suppl.).

ALGORITHMIC APPROACHES TO PROCESS SYNTHESIS: LOGIC AND GLOBAL OPTIMIZATION

Christodoulos A. Floudas
Department of Chemical Engineering
Princeton University
Princeton, NJ 08544

Ignacio E. Grossmann
Department of Chemical Engineering
Carnegie Mellon University
Pittsburgh, PA 15213

"In memory of Professor David W. T. Ripping whose work in Process Systems Engineering has been a source of inspiration for us and many other researchers."

Abstract

This paper presents an overview on two recent developments in optimization techniques that address previous limitations that have been experienced with algorithmic methods in process synthesis: combinatorics and local optima. The first part deals with the development of logic based models and techniques for discrete optimization which can facilitate the modelling of these problems as well as reducing the combinatorial search. It will be shown that various levels can be considered for the integration of logic in mixed-integer optimization techniques. The second part deals with the development of deterministic optimization methods that can rigorously determine the global optimum in nonconvex optimization models. It will be shown that this can be effectively accomplished with algorithms that exploit identifiable nonlincar structures. Examples are presented throughout the paper and future research directions are also briefly discussed.

Keywords

Process synthesis, Design optimization, Mixed-integer programming, Nonconvex optimization.

Introduction

Process synthesis continues to be a major area of research in process systems engineering. Significant advances have been achieved in terms of developing synthesis methods for subsystems (reactor networks, separation systems, heat exchanger networks) and for total flowsheets. Earlier reviews on general developments can be found in Hendry, Rudd and Seader (1973), Hlavacek (1978) and in Nishida, Stephanopoulos and Westerberg (1981). A review on algorithmic methods based on MINLP was given by Grossmann (1990a) at the previous FOCAPD meeting in Snowmass. A recent review and trends in MINLP based methods were recently presented by Grossmann and Daichendt (1994) at the PSE'94 meeting in Korea. As for the synthesis of subsystems, reviews have been given by Gundersen and Naess (1988) on heat exchanger networks, and by Westerberg (1985) and Floquet, Pibouleau and Domenech (1988) on separation systems. From these reviews it is apparent that some of the major trends in the synthesis area include an increasing emphasis on the use of algorithmic methods that are based on MINLP optimization and their combination and integration with other design methodologies.

It is important to note that from a practical point of view a major motivation behind algorithmic techniques is the development of automated tools that can help design engineers to systematically explore a large number of design alternatives. From a theoretical point of view a major motivation is to develop unified representations and solution methods. Given the clear progress that has been made in the last decade in algorithmic techniques, and given the advances that have taken place in optimization and computer technology, the debate of heuristics or physical insights vs. mathematical programming has become largely irrelevant. It has generally become clear that a comprehensive approach to process synthesis will require a combination or integration of the different types of approaches. It has also become clear that significant work and progress are still required in the underlying methods that support each of these approaches. It is precisely this issue that is considered in this paper in the context of algorithmic methods.

This paper concentrates in two fundamental areas of optimization techniques that are used to support algorithmic methods in process synthesis. Specifically, we present an overview of two major advances that have recently taken place: (a) the incorporation of logic in mixed-integer optimization methods to reduce the combinatorial search and to facilitate problem formulation; (b) the development of rigorous global optimization techniques that can handle nonconvexities in the model and avoid getting trapped in suboptimal solutions. These advances have been largely motivated by two major difficulties that have been encountered in the solution of MINLP models for process synthesis: combinatorics and local optima. The former are due to the potentially large number of structural alternatives that arise in process synthesis; the latter are due to the nonconvexities that arise in nonlinear process models. The negative implication in the former is often the impossibility of solving large synthesis models; the negative implication of the latter is generating poor suboptimal designs.

While new developments are still under way, a review of the progress achieved up to date in logic based methods and in global optimization would seem to be timely as this might hopefully promote further research work. These algorithmic techniques are also significant in that they can be applied to other areas such as process scheduling and process analysis. The paper is organized as follows. We first discuss general aspects of process synthesis to see how the work described in this paper fits in the overall scheme. We next present a motivation section to illustrate difficulties in existing algorithmic methods with combinatorics and nonconvexities. The remaining part of the paper then concentrates in providing the overview of the new developments in logic and global optimization. Finally, we present the conclusions where we indicate future directions for research.

General Components Of Process Synthesis

Algorithmic methods in process synthesis are rather general in scope and they involve the following four major components:

a. Representation of space of alternatives.
b. General solution strategy.
c. Formulation of optimization model.
d. Application of solution method.

The representations can range from rather high level abstractions such as is the case of targeting methods, to relatively detailed flowsheet descriptions such as is the case of superstructure representations. It is important to note that these representations are in fact commonly closely related as their difference lies in the level of abstraction that is used.

Having developed a representation, the next step to consider is the general solution strategy. The two common and extreme solution strategies are the simultaneous and the sequential approaches. The simultaneous strategies attempt to optimize simultaneously all the components in a synthesis problem in order to properly capture all the interactions and economic trade-offs. While conceptually superior, these strategies may give rise to larger problems. The sequential approach on the other hand has the advantage of dealing with smaller subproblems since they sequentially decompose the problem, although often at the expense of sacrificing optimality.

The nature of the optimization models is of course heavily dependent on the type of representation as well as on the general solution strategy being used. Target models often involve only continuous variables since they usually do not generate topologies nor do they consider capital cost as they deal with higher level objectives (minimize utility consumption, maximize yield). Therefore, these models commonly give rise to linear (LP) or nonlinear programming (NLP) problems. At the other extreme superstructure models determine topologies and operating conditions, and account for capital costs, often requiring 0-1 and continuous variables giving rise to mixed-integer linear (MILP) or mixed-integer nonlinear (MINLP) optimization models. Within each of the levels of representation the degree of rigorousness of the model can of course also range from the simpler short-cut models to detailed simulation models. As for the solution methods a global optimum solution can be guaranteed if the problem can be posed as an LP or MILP problem. Furthermore, in the case of LP models efficient solution times can be expected since these problems are theoretically solvable in polynomial time. This is however not the case of the MILP problems which generally are NP-complete, and therefore may have exponential time requirements, at least in the worst case. If the problem is posed as an NLP or MINLP the first drawback is that a unique global solution can only be guaranteed if the NLP or the continuous relaxation of the MINLP are convex. This is of course only a sufficient condition. But nevertheless, nonconvexities are prevalent in synthesis problems, often giving rise to multiple local solutions, or in fact even preventing convergence to feasible solutions with conventional NLP techniques. In addition to the numerical and theoretical difficulties of handling nonconvex models, there is the added difficulty of potential combinatorial explosion for the MINLP case. In the context of process synthesis a good example of the dilemma between the use of MILP and MINLP models are the approaches for superstructure optimization of flowsheets by Papoulias and Grossmann (1983) and by Kocis and Grossmann (1989). The advantage of the former is that the global optimum can be guaranteed but at the expense of using a discretized and approximate process model. The advantage of the latter is that nonlinear process models can be explicitly handled, but with the disadvantage that the global optimum cannot be guaranteed.

Based on the above discussion, it is clear that in order to properly support the development of algorithmic techniques, whether for targeting or superstructure models, or for simultaneous or sequential approaches, it is imperative that limitations due to combinatorics and nonconvexities be addressed. It is in this context that the two motivating examples below are presented.

Motivating Examples

MILP Model for Heat Integrated Distillation Sequences

In order to illustrate potential combinatorial difficulties with synthesis problems, consider the MILP model reported in Raman and Grossmann (1993a) in which heat integration is considered between different separation tasks in the synthesis of sharp distillation sequences (see also Andrecovich and Westerberg (1985) and Floudas and Paules (1988)). An example of a superstructure for 4 components is given in Fig. 1. For the heat integration part, it is assumed that the pressures of the columns can be adjusted in such a way that the condenser of every column can potentially supply heat to the reboilers of the other columns as shown in Fig. 2 (multieffect columns are not considered). The MILP model involves as 0-1 variables the potential existence of columns and the potential heat exchanges between columns and reboilers, and as continuous variables the flows, heat loads and temperatures of condensers and reboilers, with which pressure changes are accounted for. The objective function consists of the minimization of the investment cost of the columns and the operating cost for the utilities. The constraints involve mass and heat balances, and logical constraints that enforce feasible temperatures if heat exchange take place and zero flows and heat loads if the corresponding 0-1 variables are set to zero.

For a four component system such as the one in Fig. 1 the MILP model involves 100 0-1 variables, 191 continuous variables and 258 constraints. The 100 binary variables are split as follows: 10 to model the existence of the distillation columns and 90 to model the existence of heat exchange matches between the reboilers and condensers of the various columns. The computer codes ZOOM, OSL and SCICONIC were tried for solving this problem. The three of them were not able to even find a feasible solution after enumerating more than 100,000 nodes and after running more than 1 CPU hour on an IBM RISC/6000! A major reason for this performance was that the relaxation gap is very large in this problem; the LP relaxation in which the binary variables are treated as continuous the optimum is \$1,117,000/yr. while the optimal MILP solution is \$1,900,000/yr. As will be shown later in the paper, by using logic rigorous optimization of this problem can be achieved in only few seconds!

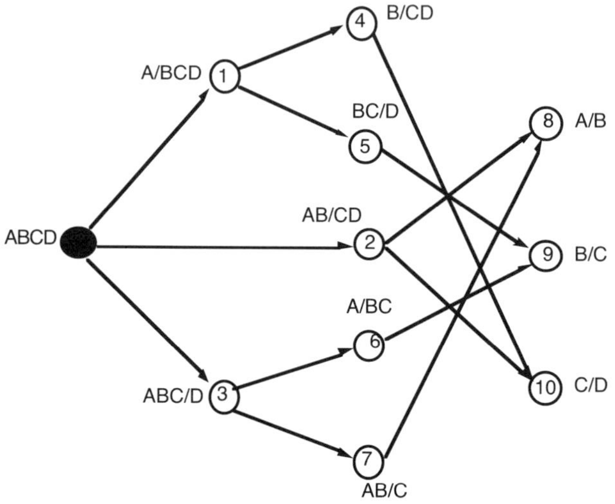

Figure 1. Superstructure for 4-component example.

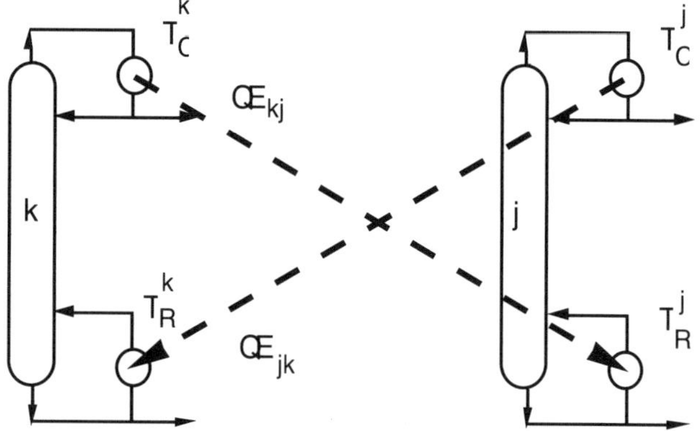

Figure 2. Heat integration between different separation tasks.

Nonconvex Model for Pooling/Blending Problems

To illustrate the potential difficulties associated with the existence of multiple solutions in nonlinear optimization NLP problems, we will consider as motivating example the pooling problem proposed by Haverly (1978) which is shown in Figure 3. Three crudes A, B and C with different sulfur contents are to be combined to form two products x and y which have specifications on the maximum sulfur content. Note that streams A and B are mixed in a pool and it is the existence of such a pool that introduces non-convexities in the mathematical model in the form of bilinear terms between the sulfur quality of the streams exiting the pool, denoted as p, and flowrates P_X, P_Y of the pool exiting streams. The objective in this pooling problem is to maximize the profit subject to (i) linear overall and component balances, (ii) bilinear pool quality and product quality constraints, and (iii) bounds on the products and sulfur quality. This problem has been studied using several local nonlinear optimization algorithms which have been reported to either obtain suboptimal solutions or fail to obtain even a feasible solution (see Floudas and Aggarwal, 1990 for a review of previous approaches and a decomposition strategy which alleviates but does not eliminate the multiplicity of local solutions problem). Table 1 presents results of local optimization algorithms (e.g. MINOS) for several starting points.

The non-convex nature of this pooling problem is better illustrated via Figure 4 where the optimal solution of the pooling model is shown for different values of the of the pool quality p. Note that the global optimum occurs at p = 1.5, while there exists a local optimum at p = 2.5 and between p = 1.5 and p = 2.2 (approximately) the optimal solutions are of the form of constant line. As a result, several starting points for p in the flat region or the region close to the local optimum terminate with the local solution or even fail to obtain a solution.

Table 1. Local Optimization for the Pooling Problem.

		Solution Found	
No.	Starting Quality	Objective Value	Quality P
1	1.00	-750.000	1.50
2	1.25	-750.000	1.50
3	1.50	-750.000	1.50
4	1.75	0.0000	1.75
5	2.00	0.0000	2.00
6	2.25	-125.0000	2.50
7	2.50	-125.0000	2.50
8	2.75	-125.0000	2.50
9	3.00	-125.0000	2.50

Floudas and Visweswaran (1990) applied the decomposition global optimization approach GOP, which is discussed in the global optimization section, to this pooling problem, as well as large instances of other pooling problems and multiperiod tankage problems (see also Visweswaran and Floudas, 1993) where the global optimum is obtained regardless of the starting point.

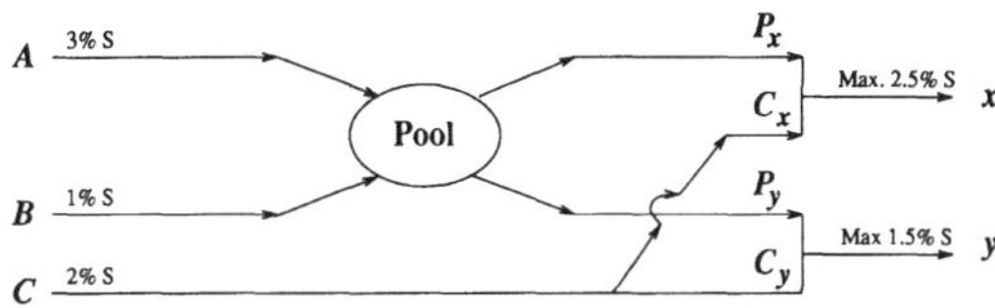

Formulation

$$\min \quad 6A + 13B + 10(C_x + C_y) - 9x - 15y$$

s.t.

$$P_x + P_y - A - B = 0 \quad \text{pool balance}$$

$$\left.\begin{array}{l} x - P_x - C_x = 0 \\ y - P_y - C_y = 0 \end{array}\right\} \quad \text{component balance}$$

$$p.(P_x + P_y) - 3A - B = 0 \quad \text{pool quality}$$

$$\left.\begin{array}{l} p.P_x + 2.C_x - 2.5x \le 0 \\ p.P_y + 2.C_y - 1.5y \le 0 \end{array}\right\} \quad \text{product quality constraints}$$

$$\left.\begin{array}{l} x \le 100 \\ y \le 200 \end{array}\right\} \quad \text{Upper bounds on products}$$

$$1 \le p \le 3 \quad \text{Bounds on sulfur quality}$$

Figure 3. Motivating example (pooling problem).

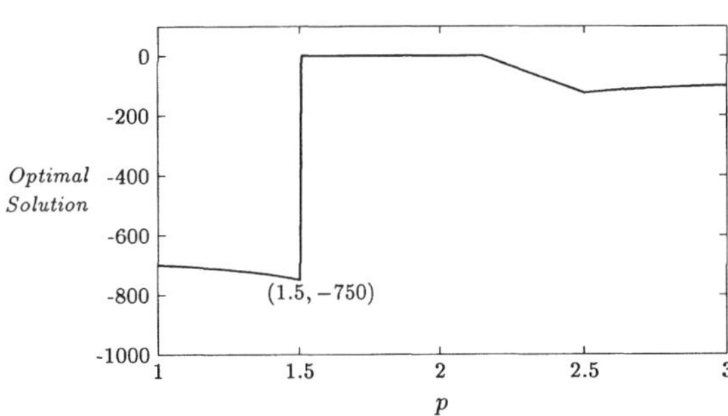

Figure 4. Optimal solution in projected space.

Integration Of Logic In Mixed-Integer Programming

In this section we present a brief review of previous work on the modelling and solution techniques of logic based discrete optimization. We also review basic concepts for the representation of logic and inference problems. We then describe our recent work at Carnegie Mellon on the integration of logic in mixed-integer optimization which has been primarily motivated by process synthesis problems.

Review of Previous Work

A major issue in the application of mixed-integer programming is the efficient modelling of discrete decisions. Different representations are often possible for the same model, each of which may be solvable with varying degrees of difficulty. In some cases it is possible to even formulate an MILP problem so that it is solvable as an LP, or else, so that its relaxation gap is greatly reduced. While some basic understanding has been achieved on how to properly formulate special classes of mixed-integer programs (see Rardin and Choe, 1979; Nemhauser and Wolsey, 1988), the modelling of general purpose problems is largely performed on an *ad hoc* basis. The use of propositional logic, however, offers an alternate framework for systematically developing mixed-integer optimization models as discussed by Jeroslow and Lowe (1984) and by Williams (1988).

The role of logic at the level of modelling of discrete optimization problems has also been studied by Balas (1974, 1985) who developed Disjunctive Programming (DP) as an alternate representation of mixed-integer programming problems. In this case, discrete optimization problems are formulated as linear programs in which a subset of constraints is expressed through disjunctions (sets of constraints of which at least one must be true). An interesting feature in the disjunctive formulation is that no 0-1 variables are explicitly included in the model, which is the more natural form to model some problems as, for instance, in the case of job-shop scheduling problems. Also, as noted by Balas (1985), every mixed-integer problem can be reformulated as a disjunctive program, and every bounded DP can be reformulated as a mixed-integer program. The reason the disjunctive programming formulation has not been used more extensively is that very few methods have been proposed to explicitly solve the problem in that form. Most of the research has focused on characterizing the convex hull of disjunctive constraints and on the generation of strong cutting planes which are included in the corresponding mixed-integer problem to strengthen the LP relaxation (Balas, 1985; Jeroslow and Lowe, 1984). The only reported method, to our knowledge, that explicitly solves problem is the algorithm by Beaumont (1991) for the case where the functions are linear and there is only one constraint in each term of every disjunction. The method is similar to a branch and bound search except that the branching is done directly on the disjunctions. This requires the addition and deletion of the corresponding disjunctive constraints in the LP subproblems. Although this may increase the overhead in the computations, Beaumont showed that the number of nodes required for the enumeration of the branch and bound tree can often be significantly reduced. In terms of integrating logic explicitly for improving the solution efficiency of mixed-integer programs, aside from our own work which will be described in the next section (Raman and Grossmann, 1991, 1992, 1993a, b, 1994), Lien and Whale (1991) considered the addition of a subset of unit resolution cuts for the branch and bound solution of MILP problems which produced large reductions of enumeration of nodes in the MILP formulation for heat integrated synthesis by Andrecovich and Westerberg (1985). It should also be mentioned that logic has been considered earlier in process synthesis with the purpose of performing high level decisions in the structuring of process flowsheets (Mahalec and Motard, 1977).

Representations of the Logic

Most of the work described above has been restricted to the form of logic called propositional logic for developing modelling and solution techniques for discrete optimization problems (see Mendelson, 1987, for general review on logic). The basic unit of a propositional logic expression, which can correspond to a state or to an action, is called a literal which is a single variable that can assume either of two values, true or false. Associated with each literal Y, its negation NOT Y ($\neg Y$) is such that [Y OR $\neg Y$] is always true. A disjunctive clause is a set of literals separated by OR operators [$\vee$], and is also called a disjunction. A proposition is any logical expression and consists of a set of clauses P_i, i = 1, ..., r that are related by the logical operators OR [$\vee$], AND [$\wedge$], IMPLICATION [$\Rightarrow$].

In synthesis logic propositions usually refer to relations of existence of units in a superstructure. These are commonly expressed by a set of conjunctions of clauses,

$$\Lambda = \{L_1 \wedge L_2 \wedge \ldots \wedge L_q\} \tag{1}$$

where L_i is a logical proposition expressed with boolean variables Y_i in terms of implications, OR, EXCLUSIVE OR and AND operators. In synthesis problems Y_i is a boolean variable representing the existence of unit i and $\neg Y_i$ its nonexistence. There are two ways of transforming the propositions in Λ. In the simplest case, the logic propositions are converted into the conjunctive normal form [CNF] by removing the implications through contrapositions in each of the clauses L_i in (1) and applying De Morgan's Theorem. In this way each clause in the CNF from consists of only OR operators with non-negated and negated boolean variables as follows:

$$\Omega_C = \left[\bigvee_{i \in P_1} (Y_i) \vee \bigvee_{i \in \overline{P_1}} (\neg Y_i)\right] \wedge \left[\bigvee_{i \in P_2} (Y_i) \vee \bigvee_{i \in \overline{P_2}} (\neg Y_i)\right] \wedge \ldots \wedge \left[\bigvee_{i \in P_s} (Y_i) \wedge \bigvee_{i \in \overline{P_s}} (\neg Y_i)\right] \tag{2}$$

where P_i and $\overline{P}_i$ are subsets of the boolean variables that correspond to some of the 0-1 variables, and s is the number of clauses.

In the second representation, the logic propositions in the CNF form are converted into the disjunctive normal form [DNF] (see Clocksin and Mellish, 1984) by moving the AND operators inwards and the OR operators outwards by applying elementary boolean operations. The DNF form is as follows:

$$\Omega_D = \left[\underset{i \in Q_1}{\wedge} (Y_i) \underset{i \in \overline{Q_1}}{\wedge} (\neg Y_i) \right] \vee \left[\underset{i \in Q_2}{\wedge} (Y_i) \underset{i \in \overline{Q_2}}{\wedge} (\neg Y_i) \right] \vee \left[\underset{i \in Q_r}{\wedge} (Y_i) \underset{i \in \overline{Q_r}}{\wedge} (\neg Y_i) \right] \quad (3)$$

where Q_j and $\overline{Q}_j$ are the index sets of the boolean variables which correspond to a partition of all the 0-1 variables Y_i, $i = 1, \ldots, p$ in non-negated and negated terms. Each clause separated by a disjunction represents the assignment of units in a feasible configuration, where it is assumed that each boolean variable has a one-to-one correspondence with the 0-1 binary variables of the MILP model. Therefore, r represents the number of alternatives in the superstructure. While the DNF form is more convenient to manipulate, the drawback is that the transformation from CNF to DNF has exponential complexity in the worst case.

To illustrate the CNF and DNF representations in (2) and (3), consider the small example problem shown in Fig. 5. The following propositional logic expressions apply:

$$L_1: Y_1 \vee Y_2 \Rightarrow Y_3$$

(process 1 or process 2 imply process 3)

$$L_2: Y_3 \Rightarrow Y_1 \vee Y_2$$

(process 3 implies process 1 or process 2)

$$L_3: \neg Y_1 \vee \neg Y_2$$

(do not select process 1 or do not select process 2)

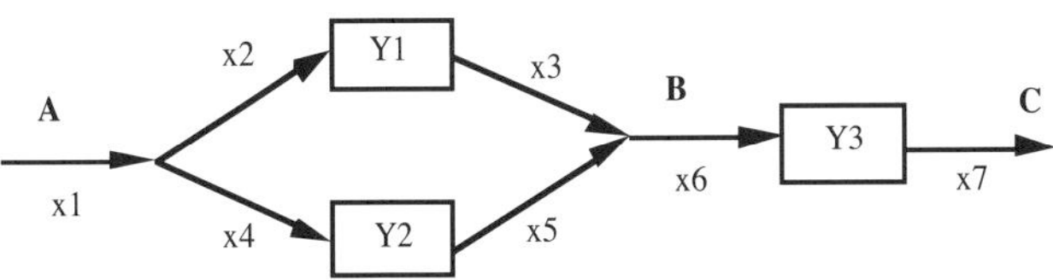

Figure 5. Superstructure for small example.

Applying the contrapositive to L_1 and L_2, and using De Morgan's theorem, the corresponding CNF representation for the logic is:

$$\Omega_C = (\neg Y_1 \vee Y_3) \wedge (\neg Y_2 \vee Y_3) \wedge (\neg Y_3 \vee Y_1 \vee Y_2) \wedge (\neg Y_1 \vee \neg Y_2) \quad (4)$$

Distributing the OR over the AND operators, the corresponding DNF representation is given by:

$$\Omega_D = (Y_1 \wedge \neg Y_2 \wedge Y_3) \vee (Y_2 \wedge \neg Y_1 \wedge Y_3) \vee (\neg Y_1 \wedge \neg Y_2 \wedge \neg Y_3) \quad (5)$$

Note that the disjunctions in (5) represent the three alternatives in Fig. 5.

In order to obtain an equivalent mathematical representation for any propositional logic expression, this can be easily performed using the CNF form as a basis. We must first consider basic logical operators to determine how each can be transformed into an equivalent representation in the form of an equation or inequality. These transformations are then used to convert general logical expressions into an equivalent mathematical representation (Cavalier and Soyster, 1987; Cavalier, et al., 1990).

To each literal P_i, a binary variable y_i is assigned. Then the negation or complement of P_i ($\neg P_i$) is given by $1 - y_i$. The logical value of true corresponds to the binary value of 1 and false corresponds to the binary value of 0. The basic operators used in propositional logic and the representation of their relationships are shown in Table 2. With the basic equivalent relations given in Table 2 (e.g. see William's, 1985), one can systematically model an arbitrary propositional logic expression that is given in terms of OR, AND, IMPLICATION operators, as a set of linear equality and inequality constraints. One approach is to systematically convert the logical expression into its equivalent *conjunctive normal form* representation which involves the application of pure logical operations. The conjunctive normal form is a conjunction of clauses, $Q_1 \wedge Q_2 \wedge \ldots \wedge Q_s$. Hence, for the conjunctive normal form to be true, each clause Q_i must be true independent of the others. Also since a clause Q_i is just a disjunction of literals, $P_1 \vee P_2 \vee \ldots \vee P_r$, it can be expressed in the linear mathematical form as the inequality.

$$y_1 + y_2 + \ldots + y_r \geq 1 \quad (6)$$

Symbolic and Mathematical Methods for Logic Inference

The most common logic inference problem is the satisfiability problem where, given the validity of a set of propositions, one has to prove the truth or validity of a conclusion which may be either a literal or a proposition. This inference problem is one of the basic issues in artificial intelligence and data bases. However, the general satisfiability problem for propositional logic is NP-complete (Cook, 1971; Karp, 1972). Therefore, research has focused on identifying classes of problems within the general satisfiability problem that can be solved efficiently. Knowledge based systems commonly require the use of Horn clauses systems which have at most one non-negated literal in each clause. The inference problem for this class of propositional logic problems can be solved in linear time using unit resolution (Dowling and Gallier, 1984). The unit resolution technique (e.g. see Clocksin and Mellish, 1981) is one of the most common inference techniques, and in simple terms, it consists of solving sequentially each logic clause one at a time. Chandru and Hooker (1988) have extended the class of problems that can be solved in linear time to include extended Horn clause systems. One of the most effective logic-based methods for solving the general satisfiability problem is the algorithm of Davis and Putnam (1960) as treated by Loveland (1978). This approach is closely related to the branch and bound method for mixed-integer programming. Jeroslow and Wang (1990) have developed branching heuristics to improve the performance of the Davis-Putnam procedure. It must be noted that although the previous work has been restricted to propositional logic, the techniques used for this class are essential to higher order representations like predicate logic which involve additional logic operators like for all [∀] and it exists [∃].

Table 2. Representation of Logical Relations with Linear Inequalities.

Logical Relation	Comments	Boolean Expression	Representation as Linear Inequalities
Logical OR		$P_1 \vee P_2 \vee \ldots \vee P_r$	$y_1 + y_2 + \ldots + y_r \geq 1$
Logical AND		$P_1 \wedge P_2 \wedge \ldots \wedge P_r$	$y_1 \geq 1, y_2 \geq 1, \ldots y_r \geq 1$
Implication	$P_1 \Rightarrow P_2$	$\neg P_1 \vee P_2$	$1 - y_1 + y_2 \geq 1$
Equivalence	P_1 if and only if P_2 $(P_1 \Rightarrow P_2) \wedge (P_2 \Rightarrow P_1)$	$(\neg P_1 \vee P_2) \wedge (\neg P_2 \vee P_1)$	$y_1 = y_2$
Exclusive OR	exactly one of the variables is true	$P_1 \underline{\vee} P_2 \underline{\vee} \ldots \underline{\vee} P_r$	$y_1 + y_2 + \ldots + y_r = 1$

Since the logical propositions can be systematically converted into a set of linear inequalities, instead of using symbolic inference techniques, the inference problem can be formulated as an integer linear programming problem. In particular, given a problem in which all the logical propositions have been converted to a set of linear inequalities, the inference problem that consists of proving a given clause,

$$\begin{aligned} &\text{Prove} \quad P_u \\ &\text{st} \quad B(P_i) \quad i = 1, 2, \ldots, q \end{aligned} \tag{LIP1}$$

can be formulated as the following MILP (Cavalier and Soyster, 1987):

$$\begin{aligned} &\text{Min } Z = \sum_{i \in I(u)} \alpha_i y_i \\ &\text{st} \quad A\, y \geq a \\ &\quad y \in \{0,1\}^n \end{aligned} \tag{LIP2}$$

where $A\, y \geq a$ is the set of inequalities obtained by translating $B\, (P_1, P_2, \ldots, P_q)$ into their linear mathematical form, and the objective function is obtained by also converting the clause P_u that is to be proved into its equivalent mathematical form. Here, $I(u)$ corresponds to the index set of the binary variables associated with the clause P_u. This clause is always true if $Z = 1$ on minimizing the objective function as an integer programming problem. If $Z = 0$ for the optimal integer solution, this establishes an instance where the clause is false. Therefore, in this case, the clause is not always true. In many instances, the optimal integer solution to problem (LIP2) will be obtained by solving its linear programming relaxation (Hooker, 1988). Even if no integer solution is obtained, it may be possible to reach conclusions from the relaxed LP problem (Cavalier and Soyster, 1987).

The qualitative knowledge available about the design of a system can be classified as one of the following two types — hard logical facts or uncertain heuristics. Hard, logical facts are never violated — for example, the reaction NaOH + HCl —> NaCl + H_2O holds from basic chemical principles. Qualitative knowledge in the form of heuristics on the other hand are just rules of thumb which may not always hold. Therefore all the knowledge for synthesizing a design may not be consistent since the heuristics may contradict one another; for example, a rule that suggests to use higher temperatures to increase yield may conflict with a rule that suggests to use lower temperature to increase selectivity. Resolution of conflicts is an important part of reasoning. In general, one must violate a weaker (more uncertain) set of rules in order to satisfy stronger ones. Therefore, it becomes necessary to model the violation of heuristics, which is done as follows (Post, 1987),

$$\textit{Clause} \text{ or } V \tag{7}$$

where either the clause is true or it is being violated (V). In order to discriminate between weak and strong rules, penalties are associated with the violation v_i of each heuristic rule, i = 1, ..., m. The penalty w_i is a non-negative number which reflects the uncertainty of the corresponding logical expression. The more uncertain the rule, the lower the penalty for its violation. In this way, the logical inference problem with uncertain knowledge can be formulated as an MILP problem where the objective is to obtain a solution that satisfies all the logical relationships (i.e. $Z = 0$), and if that is not possible, to obtain a solution with the least total penalty for violation of the heuristics:

$$\text{Min} \quad Z = w^T v \qquad \text{(LIP3)}$$

$$\text{st} \quad A\,y \geq a \quad : \quad \text{Logical facts}$$

$$B\,y + v \geq b \quad : \quad \text{Heuristics}$$

$$y \in \{0,1\}^n,\ v \geq 0$$

Note that no violations are assigned to the inequalities $Ay \geq a$ since these correspond to hard logical facts that always have to be satisfied. The solution to (LIP3) will then determine a design that best satisfies the possibly conflicting qualitative knowledge about the system.

Logic-based Formulations for Discrete Optimization

Given a superstructure of alternatives for a given design problem, the general form of the mixed-integer optimization model is (Grossmann, 1990a),

$$\begin{aligned} \text{Min} \quad & Z = c^T y + f(x) \\ \text{st} \quad & h(x) \leq 0 \\ & g(x) + My \leq 0 \qquad \text{(DP1)} \\ & x \in X,\ y \in Y \end{aligned}$$

where x is the vector of continuous variables involved in design like pressure, temperature and flow rates, while y is the vector of binary decision variables like existence of a particular stream or unit. Integer variables might also be involved but these are often expressed in terms of 0-1 variables. Also, model (DP1) may contain among the inequalities pure integer constraints for logical specifications (e.g. select only one reactor type). If all the functions and constraints are linear (P1) corresponds to an MILP problem; otherwise it is an MINLP. For the sake of simplicity, we assume that $f(x)$, $g(x)$ and $h(x)$ are convex, differentiable functions. The case of nonconvexities will be addressed later in the paper.

The mixed-integer program (DP1), is not the only way of modelling the discrete optimization problem in a superstructure. As has been shown by Raman and Grossmann (1994) that problem can be formulated as the generalized disjunctive program:

$$\begin{aligned} \text{Min} \quad & Z = \sum_i \sum_k c_{ik} + f(x) \\ \text{st} \quad & h(x) \leq 0 \qquad \text{(DP2)} \\ & \bigvee_{i \in D_k} \begin{bmatrix} Y_{ik} \\ g_{ik}(x) \leq 0 \\ c_{ik} = \gamma_{ik} \end{bmatrix} \qquad k \in SD \\ & \Omega(Y) = True \\ & x \in R^n,\ c \in R^m,\ Y \in \{true, false\}^m \end{aligned}$$

in which Y_{ik} are the boolean variables that establish whether a given term in a disjunction is true $[g_{ik}(x) \leq 0]$ or false $[g_{ik}(x) > 0]$, while $\Omega(Y)$ are logical relations assumed to be in the form of propositional logic involving only the boolean variables. Y_{ik} are auxiliary variables that control the part of the feasible space in which the continuous variables, x, lie, and he variables c_{ik} represent fixed charges which are activated to a value γ_{ik} if the corresponding term of the disjunction is true. Finally, the logical conditions, $\Omega(Y)$, express relationships between the disjunctive sets. In the context of synthesis problems the disjunctions in (DP2) typically arise for each unit i in the following form:

$$\begin{bmatrix} Y_i \\ g_i(x) \leq 0 \\ c_i = \gamma_i \end{bmatrix} \vee \begin{bmatrix} \neg Y_i \\ B^i x = 0 \\ c_i = 0 \end{bmatrix} \qquad (8)$$

in which the inequalities g_i apply and a fixed cost γ_i is incurred if the unit is selected (Y_i), otherwise ($\neg Y_i$) there is no fixed cost and a subset of the x variables is set to zero with the matrix B^i. An important advantage of the above modelling framework is that there is no need to introduce artificial parameters for the "big-M" constraints that are normally used to model disjunctions.

An interesting question that arises with problem (DP2) is whether it always pays to convert the general disjunctive program into mixed-integer form. To answer this question for the case of linear functions and constraints, Raman and Grossmann (1994) have developed the concept of w-MIP representability which is defined as follows:

Definition: The disjunction $\bigvee_{i \in D_k} [A_{ik} x \geq b_{ik}]$ is *w-MIP representable* if the following conditions hold:

i. There exists an $\hat{\imath} \in D_k$ for which the convex hull of the disjunction is reducible to the constraint:

$$A_{\hat{\imath}k}\, x \geq b_{\hat{\imath}k}\, y_{\hat{\imath}k} \qquad 0 \leq y_{\hat{\imath}k} \leq 1$$

ii. Every feasible solution

$$x' \in F = \{x \mid \bigvee_{i \in D_k} [A_{ik} x \geq b_{ik}]\}$$

for which $A_{\hat{\imath}k}\, x' \geq b_{\hat{\imath}k}$, $A_{ik}\, x' < b_{ik}$, $i \neq \hat{\imath}$ implies that $y_{\hat{\imath}k} = 1$ *and* $y_{ik} = 0\ \forall i \neq \hat{\imath}$.

Thus, in general, we can consider a partly transformed form of problem (DP2) where mixed-integer equations are used for the w-MIP constraints part of the problem, while the rest is kept in disjunctive form, as this part is "poorly-behaved" in equation form. In general, this partially reformulated problem has the form,

$$\text{Min} \quad Z = \sum_{k \in SD^1} \sum_{i \in D_k} \gamma_{ik} y_{ik} + \sum_{k \in SD^2} \sum_{i \in D_k} c_{ik} + f(x)$$

$$\text{st} \qquad h(x) \le 0 \qquad \text{(DP3)}$$

$$r(x) + By \le 0$$

$$Ay \ge a$$

$$\bigvee_{i \in D_k} \begin{bmatrix} Y_{ik} \\ s_{ik}(x) \le 0 \\ c_{ik} = \gamma_{ik} \end{bmatrix} \qquad k \in SD^2$$

$$\Lambda(Y) = True$$

$$x \in R^n,\ y \in \{0,1\}^q,\ Y \in \{true, false\}^m$$

in which the subset of disjunctions $SD^1 \subseteq SD$, which are w-MIP representable, have all been converted into mixed-integer form. The inequalities $r(x) + By \le 0$ correspond to these constraints and to subsets of the inequalities $g_{ik}(x, c_{ik}) \le 0, i \in D_k, k \in SD^2$, which have also been converted into mixed-integer form. Finally, $s_{ik}(x, c_{ik})$ are the remaining inequalities which appear explicitly in the disjunctions $k \in SD^2$.

Note also that a subset of the logical constraints in $\Omega(Y)$ = True, which are required for the definition of the discrete optimization problem, have been translated to the form of linear inequality constraints $Ay \ge a$. The simplest option is to convert the propositions into CNF which can then be translated readily into inequalities as was discussed in the previous section. In cases where the number of these constraints become large, the generation of a smaller number of tighter constraints through the application of cutting plane techniques may be useful. The rest of the logic constraints, $\Lambda(Y)$ = True, which are redundant and correspond to logic cuts that do not alter the optimal solution (Hooker et al, 1993), have been left in symbolic form in order to improve the enumeration in a branch and bound search.

It should be noted that a particular case of (DP3) of interest is when the entire problem is converted into mixed-integer form, but the logic cuts $\Lambda(Y) = True$ are included as part of the formulation:

$$Min\ Z = \sum_{i=1}^{m} \gamma_i y_i + f(x)$$

$$\text{st} \qquad h(x) \le 0 \qquad \text{(DP4)}$$

$$r(x) + By \le 0$$

$$Ay \ge a$$

$$\Lambda(Y) = True$$

$$x \in R^n,\ y \in \{0,1\}^m,\ Y \in \{\text{true, false}\}^m$$

Solution Methods

As was mentioned in the review section there are still few methods for solving mixed-integer optimization problems that incorporate propositional logic. As shown below, methods have been developed for addressing linear and nonlinear problems. Obviously some of the methods are equally applicable to both cases. However, for the sake of clarity, and to also emphasize the more useful methods in each case, we will distinguish between methods for linear and nonlinear problems.

For linear problems the simplest case is when logic cuts $\Lambda(Y)$ = True are added to an MILP problem as in (DP4). These cuts, which represent redundant constraints in high level form, can be systematically generated for process networks as discussed in Raman and Grossmann (1993a). As an example, the logic cuts for the network in Fig. 1 in terms of the potential existence of the 10 columns are given by the propositions:

$Y1 \Rightarrow Y4 \vee Y5$	$Y6 \Rightarrow Y3 \wedge Y9$
$Y2 \Rightarrow Y8 \wedge Y10$	$Y7 \Rightarrow Y3 \wedge Y8$
$Y3 \Rightarrow Y6 \vee Y7$	$Y8 \Rightarrow Y2 \vee Y7$
$Y4 \Rightarrow Y1 \wedge Y10$	$Y9 \Rightarrow Y5 \vee Y6$
$Y5 \Rightarrow Y1 \wedge Y9$	$Y10 \Rightarrow Y2 \vee Y4$

There are two basic ways of handling these cuts. One is to convert them into inequalities and add them to the MILP (Raman and Grossmann, 1992). While this will increase the number of constraints, it generally reduces the relaxation gap. The other extreme is to process the logic symbolically as part of the branch and bound search for the MILP. In this case the logic is used to select the branching variables and to determine by inference whether additional binary variables can be fixed at each node (Raman and Grossmann, 1993a,b). This can be accomplished by treating the logic either in CNF form as in (2) or in DNF form as in (3). The former requires unit resolution for the inference, while the latter involves the solution of Boolean equations. Although the DNF form is generally more expensive to obtain, its nice theoretical property is that *one can guarantee that in the worst case the number of enumerated nodes does not exceed twice the number of clauses in (3) minus one* (see Raman and Grossmann (1993a) for proof). A third alternative is to use a hybrid approach in which only violated inequalities at the root node are included to strengthen the LP relaxation, but the remaining enumeration is performed by solving the logic symbolically. For the case that the discrete optimization problem is formulated as in (DP3) by involving both disjunctions and mixed-integer constraints, Raman and Grossmann(1994) proposed an extension of the hybrid branch and bound method for (DP4) in which the disjunctions are converted for convenience into mixed-integer form, but the branching rule is altered to recognize the fact that no branching be performed on disjunctions that are logically satisfied, even if the corresponding 0-1 variables are non-integer. Note that such an algorithm can also be applied to problem (DP2). Finally, it is worth to mention that Beaumont (1991) has proposed an algorithm that applies to (DP2) in the case that only one equation is involved in each disjunction. In this algorithm constraints are successively added or deleted as needed in the branch and bound search.

Similarly as in the linear case, the simplest way to integrate logic in nonlinear discrete models is to add the logic cuts to an MINLP as in problem (DP4) (see Raman and Grossmann, 1992). If these are converted to inequalities this has the effect of reducing the relaxation gap. This has the important effect of strengthening the lower bound that is predicted by the master problem in the Generalized Benders decomposition method by Geoffrion (1972). As has been shown by Sahinidis and Grossmann (1991) the "optimal" formulation for the GBD method is when there is no gap between the relaxed and the integer optimum solution. In the case of the outer-approximation method by Duran and Grossmann (1986) the quantitative or symbolic integration has the effect of potentially reducing the branch and bound enumeration at the level of the MILP master problem. An interesting variation of the above idea is to integrate the logic inference problem with heuristics (LIP3) in the MILP master problem as was proposed by Raman and Grossmann (1992). First assume that given the solution of K NLP subproblems the MILP master problem is represented by:

$$\begin{aligned} \text{Min} \quad & \alpha \\ \text{st} \quad & \alpha \geq \phi_k(x,y) \\ & x,y \in \Omega_k \qquad k = 1,\ldots,K \\ & y \in INT_k \\ & x \in X,\ y \in Y \end{aligned} \qquad \text{(M1)}$$

in which $\phi_k(x, y)$ represents either the Lagrangian in the GBD method or an objective linearization in the OA method, Ω_k is the linear approximation to the continuous feasible space and INT_k represents integer cuts to exclude configurations that were previously analyzed. The integer programming model (LIP3) can be integrated in the above master problem(M1) by minimizing the weighted violation (plus an extra term to reflect the cost) and subject to constraining the lower bound to the current upper bound; that is,

$$\text{Min } [\, w^T v + \bar{w}(\alpha - LB)/(UB^K - LB) \,]$$

$$\begin{aligned} \text{st } \alpha &\geq \phi_k(x,y) \\ x,y &\in \Omega_k \\ y &\in INT_k \end{aligned} \left.\vphantom{\begin{aligned}a\\a\\a\end{aligned}}\right\} k = 1,\ldots,K$$

$$\begin{aligned} & y \in INT_k \\ & Ay \geq a \\ & By + v \geq b \\ & \alpha \leq UB^k \\ & x \in X,\ y \in Y \\ & \alpha \in R^1,\ v \in \{0,1\} \end{aligned} \qquad \text{(M2)}$$

in which $\bar{w}$ is a penalty chosen such that $\bar{w} = \min_i \{w_i\}$, LB is a valid lower bound to the solution of the MINLP (e.g. solution to the relaxed NLP problem or some reasonable but valid bound) and UB^K is the current upper bound of the objective at iteration K. The interesting feature with the master problem (M2) is that optimality can still be guaranteed (within convexity assumptions) even though heuristics are used as part of the search. The master problem (M2) is especially appropriate for the GBD method because of the loose approximation that is obtained with that method. It is also important to note that the master problem (M2) can be used when applying Benders decomposition (Benders, 1962) in the solution of MILP problems.

For the case that the nonlinear discrete optimization problem is formulated as the generalized disjunctive program in (DP2) one can develop corresponding logic-based OA and GBD algorithms as described in Turkay and Grossmann (1994). First, for fixed values of the boolean variables, $Y_{\hat{i}k}$ = true and Y_{ik} = false, the corresponding NLP subproblem is as follows:

$$\begin{aligned} \text{Min} \quad & Z = \sum_{i=1}^{m} c_{ik} + f(x) \\ \text{st} \quad & h(x) \leq 0 \end{aligned} \qquad \text{(SP)}$$

$$\left\{\begin{array}{c} c_{\hat{i}k}=\gamma_{\hat{i}k} \\ g_{\hat{i}k}(x) \le 0 \end{array}\right\} \text{for } Y_{\hat{i}k}=true \qquad k \in SD$$

$$c \;=0 \quad for\, Y \;=false \quad i \ne \hat{i}$$

$$x \in R^{n}, \; c_{ik} \in R^{m},$$

Note that only constraints corresponding to true boolean variables are imposed. Also fixed charges γ_{ik} are only applied to these terms. Assuming that K subproblems (SP) are solved in which sets of linearizations l=1, ..., K are generated for subsets of disjunction terms $L(ik) = \{l \mid Y^l_{ik} = true\}$, one can define the following disjunctive OA master problem:

$$\text{Min} \quad Z = \sum_i \sum_k c_{ik} + \alpha$$

$$\text{st} \quad \alpha \ge f(x^l) + \nabla f(x^l)^T (x-x^l) \; l=1, \ldots, K$$

$$h(x^k) + \nabla h(x^l)^T (x-x^l) \le 0 \qquad \text{(MDP2)}$$

$$\bigvee_{i \in D_k} \left[\begin{array}{c} Y_{ik} \\ g_{ik}(x^l) + \nabla g_{ik}(x^l)^T (x \pm x^l) \le 0 \;\; l \in L(ik) \\ c_{ik} = \gamma_{ik} \end{array}\right] \quad k \in SD$$

$$\Omega(Y) = True$$

$$\alpha \in R, \; x \in R^{n}, \; c \in R^{m}, \; Y \in \{true, false\}^{m}$$

It should be noted that before applying the above master problem it is necessary to solve various subproblems so as to produce at least one linear approximation of each of the terms in the disjunctions. As shown by Turkay and Grossmann (1994) selecting the smallest number of subproblems amounts to the solution of a set covering problem. The above problem (MDP2) can be solved by any of the methods described for the linear case. It is also interesting to note that for the case of flowsheet synthesis problems Turkay and Grossmann (1994) have shown that the above solution method becomes equivalent to the modelling/decomposition strategy by Kocis and Grossmann (1988) if the master problem (MDP2) is converted into MILP form using a convex hull representation. Also, these authors have shown that while a logic-based GBD method cannot be derived as in the case of the OA algorithm, one can nevertheless determine for the MILP version of the master problem (MDP2) one Benders iteration which then yields a sequence similar to the GBD method for the algebraic case.

Computational Experience

From the methods described in the previous section the symbolic integration of logic both in DNF and CNF form have been automated in a special version of OSL, the MILP solver from IBM (Raman and Grossmann, 1993a). Also systematic methods have been developed to automate the generation of logic cuts in process networks (Raman and Grossmann, 1993a; Hooker et al., 1994). Work is also currently under way to automate the logic version of the OA and GBD algorithms.

In order to appreciate the potential impact of integrating logic in discrete optimization problems numerical results on selected examples are given in Table 3. Example (a) deals with an MILP for the synthesis of separation sequences involving 6 components (see Raman and Grossmann, 1992). Applying the standard version of Benders decomposition convergence is not achieved after several hours and more than one hundred iterations on an older Vax computer. In contrast, adding inequalities for the logic cuts in (DP4) convergence is achieved in only 13 iterations, and this despite the fact that the number of constraints is doubled. Note that the integrated master with heuristics is not as effective in this case. Example (b) deals with a small MINLP planning problem in which similar trends are observed when adding the logic cuts. The examples in (c) deal with the symbolic and hybrid integration of logic using branch and bound (see Raman and Grossmann, 1993). Note that for the MILP for the separation of 6 components the reduction in number of nodes enumerated is significant. The more impressive results, however, are with the heat integrated model which corresponds to the motivating example. Adding the inequalities for the logic cuts the problem is solved to optimality in only 8 seconds! And this is accomplished by almost doubling the number of constraints. With the symbolic integration of logic with DNF the time is even further reduced to less than 3 seconds! The reason for the reduction is that in the symbolic integration there is no need to handle the inequalities for the logic cuts. It should be noted that the DNF logic involved 194 disjunctive terms. Therefore, theoretically it is possible to guarantee that the number of nodes in this type of enumeration will not exceed 387 nodes. In actual fact only 20 were needed. Finally, the example in (d) illustrates a problem in which a process network was initially formulated as the generalized disjunctive program (DP2) (see Raman and Grossmann, 1994). Converting it all into MILP form requires more than 1 hour of solution time with OSL. If instead the problem is formulated as in (DP3) in which disjunctions are identified that are not w-MIP representable the modified branch and bound method requires less than 10 minutes of CPU time. Fig. 6 presents the tree searches for a very small version of this problem. Note that even in this case the logic-based branch and bound for the disjunctive model (DP3) requires only 4 nodes as opposed to the 16 that are needed when the model is posed entirely as an MILP and solved with standard branch and bound methods.

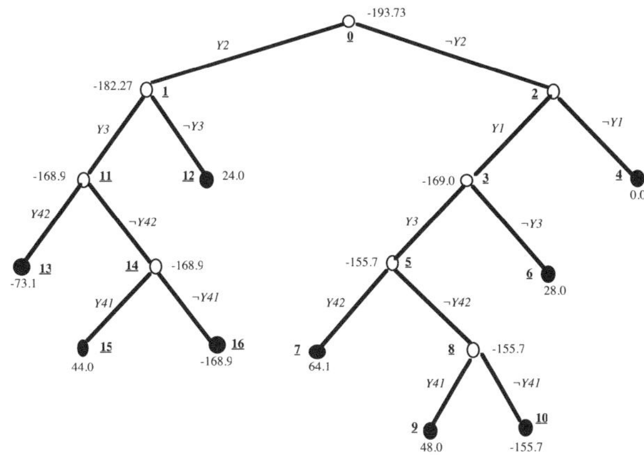

a. Branch and bound standard MILP model.

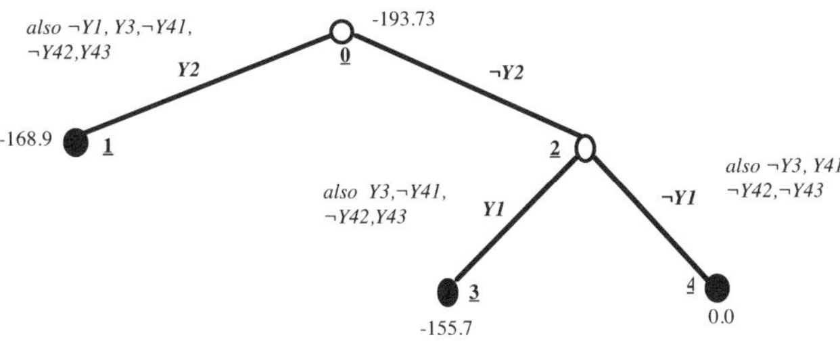

b. Logic based branch and bound for disjunctive model (DP3).

Figure 6. Comparison of tree searches with standard and logic based branch and bound.

Table 3. Computational Results on Selected Example Problems

a. MILP model 6 component separation. *Benders decomposition.*

	Original Model (DP1)	Model with Logic (DP4)	Integrated Master (M2)
Heuristic Constraints			187
Logic constraints		70	70
Other constraints	86	86	86
Iterations	> 100	13	43
CPU time*	> 1000	11.99	338.7

*minutes Micro-Vax II (SCICONIC)

b. MINLP model planning problem. *Generalized Benders Decomposition.*

	Original Model (DP1)	Model with logic (DP4)	Integrated Master (M2)
Heuristic constraints			5
Logic constraints	1	8	8
Other constraints	9	9	9
Iterations	7	3	4
CPU time*	28.20	11.7	18.8

*seconds Micro-Vax II (SCICONIC/MINOS)

c. MILP models. *Branch and bound.*

	Original Model (DP1)	Model with logic (DP4)	DNF based approach	Hybrid DNF approach
Six components				
Logic constraints	0	70	0	11
No. nodes	141	8	18	5
CPU time*	3.46	1.18	1.06	0.7
Heat Integrated Distillation				
Logic constraints	0	215	0	4
No. nodes	>100,000	74	20	17
CPU time*	>5,000	8.37	2.76	2.62

*seconds IBM-RS6000 (OSL)

d. MILP Process Network with semi-continuous demands.

	MILP model (DP1)	Disjunctive Model (DP3)
Constraints	1382	1382
Variables	1326	1326
Binary	73	73
Nodes	16,532	1,771
CPU time*	76.2	8.3

*minutes IBM-RS6000 (OSL)

Global Optimization

Background

A significant effort has been expended in the last five decades toward theoretical and algorithmic studies of **local** optimization algorithms and their computational testing in applications that arise in Process Synthesis, Design and Control. Relative to such an extensive effort that has been devoted to **local nonlinear** optimization approaches, there has been much less work on the theoretical and algorithmic development of **global** optimization methods. In the last decade the area of global optimization has attracted a lot of interest form the Operations Research and Applied Mathematics community, while in the last five years we have experienced a resurgence of interest in Chemical Engineering for new methods of global optimization as well as the application of available global optimization algorithms to important chemical engineering problems. This recent surge of interest is attributed to three main reasons. First, a large number of process synthesis, design and, control problems are indeed global optimization problems. Second, the existing local nonlinear optimization approaches (e.g. generalized reduced gradient and successive quadratic programming methods) may either fail to obtain even a feasible solution or are trapped to a local optimum solution which may differ in value significantly form the global solution. Third, the global optimum solution may have a very different physical interpretation when it is compared to local solutions (e.g. in phase equilibrium a local solution may provide incorrect prediction of types of phases at equilibrium, as well as the components' composition in each phase).

The existing approaches for global optimization are classified as deterministic or probabilistic. The deterministic approaches include: (a) Lipschitzian methods (e.g. Hansen et al. 1992 a, b), (b) Branch and Bound methods (e.g. Al-Khayyal and Falk1983; Horst and Tuy, 1987; Al-Khayyal 1990), (c) Cutting Plane methods (e.g. Tuy et al. 1985), (d) Difference of Convex (D.C.) and Reverse Convex methods (e.g. Tuy 1987 a, b), (e) Outer Approximation methods (e.g. Horst et al. 1992), (f) Primal-Dual methods (e.g. Shor 1990; Floudas and Visweswaran 1990, 1993; Ben-Tal et al. 1994), (g) Reformulation-Linearization methods (e.g. Sherali and Alameddine, 1992; Sherali and Tuncbilek 1992), and (h) Interval methods (e.g. Hansen 1979). The probabilistic methods include (i) random search approaches (e.g. Kirkpatrick et al. 1983), and (ii) clustering methods (e. g. Rinnoy Kan and Timmer 1987). Recent books for global optimization that discuss the above classes are available by Pardalos and Rosen (1987), Torn and Zilinskas (1989), Ratschek and Rokne (1988), Horst and Tuy (1990) and Floudas and Pardalos (1992).

Contributions from the chemical engineering community to the area of global optimization can be traced to the early work of Stephanopoulos and Westerberg (1975), Westerberg and Shah (1978), and Wang and Luus (1978). Renewed interest in seeking global solution was motivated form the work of Floudas et al. (1989). The first exact primal-dual global optimization approach was proposed by Floudas and Visweswaran (1990), (1993) and its features were explored for quadratically constrained and polynomial problems in the work of Visweswaran and Floudas (1992), (1993). At the same time Swaney (1990) proposed a branch and bound global optimization approach and more recently Quesada and Grossmann (1993) combined convex underestimators in a branch and bound framework for fractional programs. Manousiouthakis and Sourlas (1992) proposed a reformulation to a series of reverse convex problems, and Tsirukis and Reklaitis (1993 a, b) proposed a feature extraction algorithm for constrained global optimization. Maranas and Floudas (1992, 1993, 1994 a,b) proposed a novel branch and bound method combined with a difference of convex functions transformation for the global optimization of atomic clusters and molecular conformation problems that arise in computational chemistry. Vaidyanathan and El-Halwagi (1994) proposed an interval analysis based method and Ryoo and Sahinidis (1994) proposed reduction tests for branch and bound based methods.

In this review paper, we will focus, on deterministic global optimization methods since they provide a rigorous framework for exploiting the inherent structure of process synthesis models. In particular, we will discuss decomposition based primal-dual methods and branch and bound with difference of convex functions global optimization approaches developed in the Computer-Aided Systems Laboratory, CASL, of the Department of Chemical Engineering of Princeton University.

Decomposition Methods

Floudas and Visweswaran (1990, 1993) proposed a deterministic primal-relaxed dual global optimization approach, **GOP,** for solving several classes of non-convex optimization problems for their global optimum solutions. These classes are defined as:

Determine a globally ε–optimal solution of the following problem:

$$\begin{aligned} &\min_{x,\,y} f(x, y) \\ \text{st}\quad &g(x, y) \leq 0 \\ &h(x, y) = 0 \\ &x \in X \\ &y \in Y \end{aligned} \tag{P1}$$

where X and Y are non-empty, compact, convex sets, $g(x, y)$ is an m-vector of inequality constraints and $h(x, y)$ is a p-vector of equality constraints. It is assumed that the functions $f(x, y)$, $g(x, y)$ and $h(x, y)$ are continuous, piecewise differentiable and given in analytical form over X x Y. The variables y are defined in such a way that:

a. $f(x, y)$ is convex in x for every fixed y, and convex in y for every fixed x;
b. $g(x, y)$ is convex in x for every fixed y, and convex in y for every fixed x and
c. $h(x, y)$ is affine in x for every fixed y, and affine in y for every fixed x.

Examples of process synthesis problems with this structure are superstructures for separation systems, and heat exchanger networks in which balance equations involve bilinear terms, as well as phase equilibrium problems that can be transformed so as to exhibit the bi-convex characteristics of the above conditions.

Making use of duality theory along with several new theoretical properties, a global optimization algorithm, **GOP,** has been proposed for the solution of the problem through a series of **primal** and **relaxed dual** problems that provide valid upper and lower bounds on the global solution. The **GOP** algorithm decomposes the original problem into **primal** and **relaxed dual** subproblems. The primal problem is solved by projecting on the y variables, and takes the form:

$$
\begin{aligned}
v(y^k) = \min_{x} \; & f(x, y^k) \\
\text{st} \quad & g(x, y^k) \le 0 \\
& h(x, y^k) = 0 \\
& x \in X
\end{aligned}
\qquad \text{(P2)}
$$

A feasible solution x^k of the primal problem (P2) with objective $v(y^k)$ represents an upper bound on the global optimum (i.e. Upper Bound = $v(y^k)$) solution of (P1), and at the same time it provides the Lagrange multipliers λ^k, μ^k for the equality and inequality constraints respectively.

The Lagrange multipliers (λ^k, μ^k) are subsequently used to formulate the Lagrange function $L(x, y, \lambda^k, \mu^k)$ which is used in the dual problem. Invoking the dual of (P1) and making use of several properties of the problem structure, the GOP algorithm solves a relaxation of the dual problem through a series of **relaxed dual** subproblems. The y-space is partitioned into subdomains and each relaxed dual subproblem represents a valid underestimation of (P1) for a particular subdomain. Each relaxed dual is associated with a combination of bounds B_p of the x variables which appear in bilinear x-y products in the Lagrange function, and takes the forms:

$$
\min \quad \mu_B
$$

$$
\text{st}
$$

$$
\left.
\begin{aligned}
&\mu_B \ge L\left(x^{Bj}, y, \lambda^k, \mu^k\right)\Big|_{x^k}^{lin} \\
&\nabla_{x_i} L\left(x, y, \lambda^k, \mu^k\right)\Big|_{x^k} \le 0 \;\; if \;\; x_i^{Bj} = x_i^U \\
&\nabla_{x_i} L\left(x, y, \lambda^k, \mu^k\right)\Big|_{x^k} \ge 0 \;\; if \;\; x_i^{Bj} = x_i^L
\end{aligned}
\right\} k = 1, 2 \ldots, (K-1)
\qquad \text{(P3)}
$$

$$
\left.
\begin{aligned}
&\mu_B \ge L\left(x^{Bp}, y, \lambda^K, \mu^K\right)\Big|_{x^K}^{lin} \\
&\nabla_{x_i} L\left(x, y, \lambda^K, \mu^K\right)\Big|_{x^K} \le 0 \;\; if \;\; x_i^{Bp} = x_i^U \\
&\nabla_{x_i} L\left(x, y, \lambda^K, \mu^K\right)\Big|_{x^K} \ge 0 \;\; if \;\; X_i^{Bp} = x_i^L
\end{aligned}
\right\} \begin{matrix} current \\ iteration \\ K \end{matrix}
$$

The first three sets of constraints of (P3) correspond to the previous (K-1) iterations with the first one denoting the Lagrange underestimating cuts and the second and third defining the partitioning of the y-space. In the current iteration K the bounds B_j of the previous iterations are fixed while the current combinations of bounds B_p need to be considered. The last three sets of constraints, which change as B_p change, are the underestimating cuts for the partitioned subdomain under consideration. Hence, the relaxed dual problems at the current iteration K are equivalent to setting the x-variables to a combination of their bounds B_p, and solving for a corresponding domain of the y-variables. After solving (P3) for all combinations of bounds B_p, we select the minimum of these solutions and the solutions of the previous iterations. This will provide the new y to be considered in the primal problem (P2) and its corresponding solution is guaranteed to be a valid lower bound on (P1). Solving the primal problem (P2) and updating the upper bound as the minimum solution found, a monotonically non-increasing sequence of updated upper bounds is generated. Solving the relaxed dual problems (P3), a monotonically non-decreasing sequence of valid lower bounds is generated due to the accumulation of previous constraints. As a result, the GOP algorithm attains finite convergence to an ε-global solution of (P1) through successive iteration between the primal and relaxed dual problems.

The GOP algorithm along with its primal problem (P2) and its relaxed dual problems (P3) have an interesting geometrical interpretation. Figures 7a, 7b and 7c illustrate graphically the GOP applied to the motivating pooling/blending problem discussed earlier. For a starting point of p = 2, the primal problem corresponds to point A of Figure 7a. Note that for p = 2 the primal problem is a linear programming problem with objective equal to zero. The y-space, which is $1 \le p \le 3$, is partitioned into 2 sub-domains, one for $1 \le p \le 2$ and the other for $2 \le p \le 3$, and one relaxed dual problem is solved for each sub-domain. Figure 7a shows the linear underestimator AB for $1 \le p \le 2$, and the underestimator AC for $2 \le p \le 3$. Note that the underestimators are linear since the relaxed dual problems are linear in p and the points B and C correspond to the solutions of the corresponding relaxed dual problems. Also note that the underestimator AB passes through the global optimum (p = 1.5, -750). At the end of the first iteration we have an upper bound of zero and a lower bound of -1500. Since -1500 < -350, the next point under consideration for p is p = 1. For p = 1 the primal problem has as solution point D with objective value of -700. Since point D is in the boundary of the range of p, there is only one relaxed dual problem and hence one underestimator, shown as DE in Figure 7b, where point E is the solution of the relaxed dual problem.

At the end of the second iteration, we have an upper bound of -700 and a lower bound of -884.61. Since -884.61 < -350, the next p under consideration is p = 1.41. Figure 7c shows the underestimating function after three iterations of the GOP algorithm. Note that we have a piece-wise linear underestimating function. Also note that since the primal problem for p = 1.41 has lower value than -350 we can eliminate the domain $2 \le p \le 3$. The GOP algorithm has quickly identified the region of the global optimum by providing tight upper and lower bounds, and converges to the global solution in 6-7 iterations.

Visweswaran and Floudas (1990) demonstrated that the *Global Optimization Algorithm, GOP,* can address several classes of **non-convex** mathematical problems that include:

i. Bilinear, negative definite and indefinite quadratic programming problems.
ii. Quadratic programming problems with quadratic constraints.
iii. Unconstrained optimization of polynomial functions.
iv. Optimization problems with polynomial constraints.
v. Constrained optimization of ratios of polynomials.

Visweswaran and Floudas (1992) studied the class of polynomial functions of one variable in the objective and constraints of problem (P1) and showed that the primal problem reduces to a single function evaluation while the relaxed dual problem is equivalent to the simultaneous solution of two linear equations in two variables. The resulting global optimization approach was demonstrated to perform favorably compared to other algorithms.

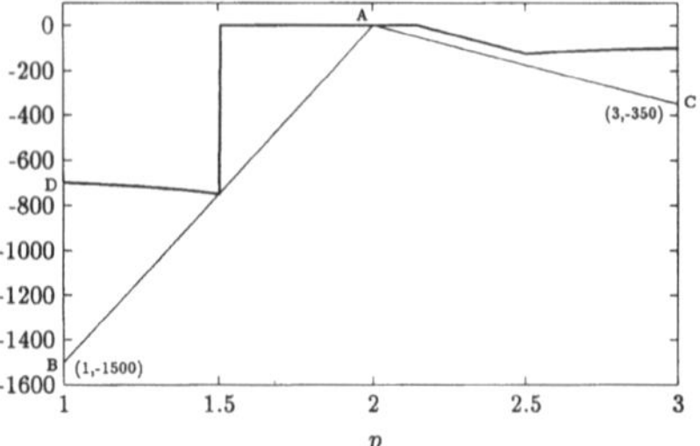

Figure 7a. Iteration I.

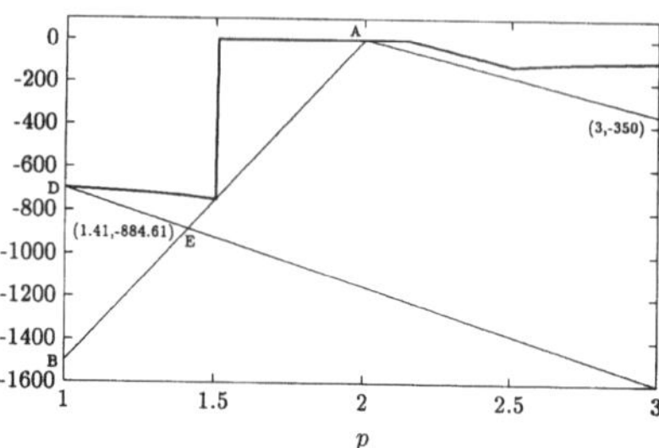

Figure 7b. Iteration II.

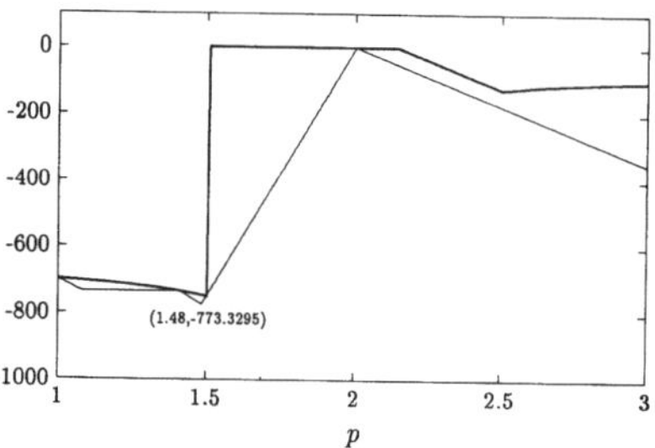

Figure 7c. Underestimator after iteration III.

Visweswaran and Floudas (1993) proposed new theoretical properties that enhance significantly the computational performance of the GOP algorithm. These properties exploit further (i) the structure of the linearized Lagrange function around x^k, which contains bilinear terms in x and y, linear terms in x, and either linear or convex terms in y, and (ii) the gradients of linearized Lagrange function around x^k, which are linear functions of only the y variables. The first property identifies the combinations of bounds that need not be considered if the gradients of the linearized Lagrange function maintain the same sign. The second property shows that if the gradient of the linearized Lagrange function with respect to x_i is zero, then we can set x_i to either its lower or upper bound. The third property allows for updates of the bounds on the x variables at each iteration. Properties 1 and 2 reduce significantly the number of combinations of bounds of the x variables, and hence reduce the number of relaxed dual problems that needed to be solved at each iteration. Property 3 results in tighter underestimators for each of the partitioned subdomains, which in turn results in faster convergence of the upper and lower bounding sequences. The effect of the new properties is illustrated through application of the

GOP algorithm to a difficult indefinite quadratic problem, a multiperiod tankage quality problem that occurs frequently in the modeling of refinery processes, and a set of pooling/blending problems from the literature. In addition, extensive computational experience is reported for randomly generated concave and indefinite quadratic programming problems of different sizes. The results show that the properties help to make the algorithm computationally efficient for fairly large problems. Visweswaran and Floudas (1994) presented a (MILP) formulation for all relaxed duals at each iteration of the GOP algorithm. This is based on a branch and bound framework for the GOP and allows for implicit enumeration of the partitioned subdomains.

A very important advance on the GOP approach has been recently made by Liu and Floudas (1993). It is shown that the GOP approach can be applied to very general classes of nonlinear problems defined as:

$$\min_x \quad F(x)$$
$$G_i(x) \le 0 \qquad i = 1, 2, \ldots, m \qquad \text{(P4)}$$
$$x \in X$$

where X is a non empty, compact, convex set in R^n, and the functions F(x), $G_i(x)$ are C^2 continuous on X. This result, even though it is an existence theorem, is very significant because it extends the classes of mathematical problems that the GOP can be applied to from polynomials or ratios of polynomials to arbitrary nonlinear objective function and constraints that may include exponential terms and trigonometric terms with the only requirement that these functions have continuous first and second order derivatives. Based on this result, it is clear the GOP approach is applicable to very broad mathematical problems.

Branch and Bound Methods with (D.C.) Transformation

A novel branch and bound global optimization approach which combines a special type of difference of convex functions' transformation with lower bounding underestimating functions was recently proposed by Maranas and Floudas (1994 a,b). This approach is applicable to the broad class of optimization problems stated in (P4), and this special type of (D.C.) transformation is the basis of the result reported in Liu and Floudas (1993). In the sequel, we will discuss the essential elements of this approach by considering the problem of:

$$\min_x \quad F(x) \qquad \text{(P5)}$$

$$\text{st} \quad x \in X \equiv \left\{ x_i \,\middle|\, x_i^\ell \le x_i \le x_i \le x_i^u, i = 1,2,\ldots,n \right\}$$

where X is a nonempty, compact, convex set in R^n, and the objective function F(x) is C^2 continuous on X.

Adding a separable quadratic term to F(x), introducing new variables $x_i' = x_i$, and subtracting the same term from F(x) we have:

$$\min_{\substack{x_i^\ell \le x_i \le x_i^u \\ (x_i')^\ell \le x_i' \le (x_i')^u}} F(x) + \alpha \sum_{i=1}^{n} \left[x_i^2 - x_i \cdot x_i' \right] \qquad \text{(P6)}$$

$$\text{st} \qquad x_i - x_i' = 0 \qquad i = 1,2,\ldots,n$$

The key idea is to employ eigenvalue analysis and define the nonnegative parameter α in such a way that the following term:

$$\phi(x) \equiv F(x) + \alpha \sum_{i=1}^{n} x_i^2$$

becomes convex. Then, (P6) takes the form

$$\min_{\substack{x_i^\ell \le x_i \le x_i^u \\ (x_i')^\ell \le x_i' \le (x_i')^u}} \phi(x) - \alpha \sum_{i=1}^{n} x_i x_i' \qquad \text{(P7)}$$

$$\text{st} \qquad x_i - x_i' = 0$$

which has as objective a difference of two convex functions out of which the one that is subtracted is separable quadratic. Formulating the dual of (P7) and applying the KKT conditions, Maranas and Floudas (1994.a) showed that the dual of (P7) is equivalent to (P8) (see Appendix A.3 of that paper)

$$\min_{x_i^\ell \le x_i \le x_i^u} L(x) = \left[F(x) + \alpha \sum_{i=1}^{n} \left(x_i^\ell - x_i\right)\left(x_i^u - x_i\right) \right] \qquad \text{(P8)}$$

where α is a nonnegative parameter which is greater or equal to the negative one half of the minimum eigenvalue of the Hessian of F(x) over the box $x_i^l \le x_i \le x_i^u$, $i = 1, 2, \ldots, n$

$$\left(\text{i.e.,} \quad \alpha \ge \max \left\{ 0, -\frac{1}{2}\lambda_{min} \right\} \right).$$

Note that the term added to F(x) has the effect of overpowering the nonconvexity characteristics of F(x) with the addition of the term (2 α) to all of the eigenvalues of its Hessian. The smaller the value of α, the tighter the underestimator L(x) is for F(x) which may imply less total number of iterations for convergence. Hence, one would ideally desire the non negative parameter α to be exactly equal to $-1/2\ \lambda_{min}$.

Note however that in the branch and bound with difference of convex function transformation it suffices to find an upper bound on $-1/2\ \lambda_{min}$ and select α as equal to this upper bound. In this case we add more convex terms than needed and do not produce the tightest underestimator, but we satisfy the required conditions for convergence.

Given F(x), the selection of the nonnegative parameter α may involve (i) the derivation of analytical expressions for the eigenvalues of its Hessian, or (ii) the development of bounds on the eigenvalues of the Hessian of F(x). Maranas and Floudas (1992), (1993) studied alternative (i) for a variety of atomic/molecular clusters. They derived analytical expressions for the eigenvalues for any potential function which is a function of only the distance between atoms (e. g. Lennard-Jones, Coulomb, Mie, Morse, Gaussian, Born-Mayer, Buckingham). Maranas and Floudas (1994. a,b) proposed a number of ways of obtaining bounds on the eigenvalues of the Hessian of F(x). One general way is via the use of the measure of a matrix, a concept recently applied to the stability of reactor networks at the process synthesis level (see Kokossis and Floudas, 1994). If a denotes the Hessian of F(x), then the measure of the matrix (-A), denoted as μ (-A), provides an upper bound on $(-\lambda_{min})$. Its formulation is a convex problem, and we can use either the 1 or ∞ norm. Appendix A.2 of Maranas and Floudas (1994.a), describes such a formulation.

It should be pointed out however that if λ_{min} goes to $(-\infty)$, then this represents a case in which we cannot create $\phi(x)$ convex. A sufficient condition which excludes such a possibility is when the elements of the Hessian matrix have finite values. This can be seen easily using the measure of a matrix concept. One instance of λ_{min} tending to $(-\infty)$ is reported in the Weber's facility location problem (see Maranas and Flouds, 1994c)

The function L(x) is a lower bounding function of F(x), and exhibits the following important properties:

Property 1: L(x) is always a valid underestimator of F(x) inside the box $[x_i^l, x_i^u]$, that is $L(x) \le F(x)$.

Property 2: L(x) matches F(x) at all corner points of the box.

Property 3: L(x) is convex in the box $[x_i^l, x_i^u]$.

Property 4: The maximum separation between L(x) and F(x) is bounded and is proportional to α and to the square of the diagonal of the box $[x_i^l, x_i^u]$, that is

$$\max_{x_i^\ell \le x_i \le x_i^u} \big(F(x) - L(x)\big) = \frac{1}{4}\alpha \sum_{i=1}^{n} \left(x_i^u - x_i^\ell\right)^2$$

Property 5: The underestimator L(x) constructed over a sub-box of the current box is always tighter than the underestimator of the current box.

In summary, the properties show that L(x) is a convex, lower bounding function of F(x), L(x) matches F(x) at all corner points of the box constraints inside which it has been defined. The values of L(x) at any point, if L(x) is constructed over a tighter box of constraints each time, define a nondecreasing sequence. Also note that Property 4 answers the question of how small the sub-boxes must become before the upper and lower bounds of F(x) are within ε. If δ is the diagonal of the sub-box, and ε is the convergence tolerance, we have:

$$\delta < \sqrt{\frac{4\varepsilon}{\alpha}}$$

Note that δ is proportional to the square root of ε and inversely proportional to the square root of α. As a result, the smaller value of α the faster the convergence rate becomes.

These properties of the lower bounding function, L(x), coupled with an efficient partitioning scheme resulted in a branch and bound global optimization approach that is guaranteed to converge to an ε-global solution in a finite number of iterations. Maranas and Floudas (1994a) analyzed the structure of the branch and bound tree resulting from the subdivision process and developed formulas for finite upper and lower bounds on the total number of iterations required for ε convergence. The maximum number of iterations is exponential in the total number of variables while the minimum number of iterations depends linearly on the total number of variables. Computational experience with molecular conformation problems indicated that the total number of iterations is much close to the minimum one.

Figure 8 provides the geometrical interpretation of the lower bounding scheme for a function F(x) of one variable x in a box $[x^L, x^U]$. Starting at a point x^o we partition the original box into two intervals $[x^L, x^o]$ and $[x^o, x^U]$, while $F(x^o)$ is the current upper bound. For each interval we solve the corresponding convex lower bounding problem and obtain their respective minima at x^1, $L(x^1)$ and x^2, $L(x^2)$ respectively. Note at this point the underestimating functions shown with non-solid lines.

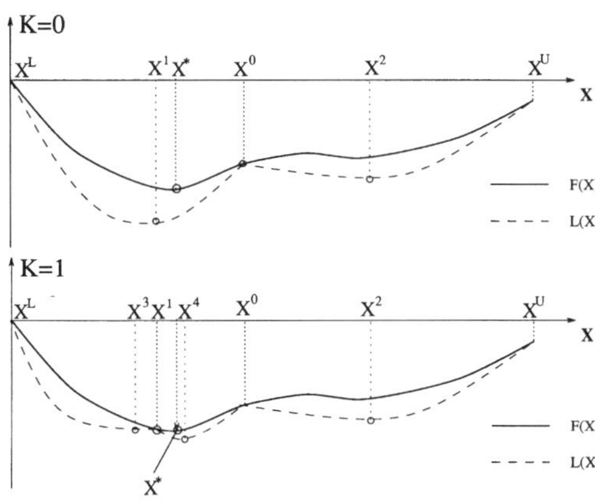

min $F(X)$, single variable problem in X

$$L(X) = F(X) + \alpha(X^{LBD} - X)(X^{UBD} - X)$$

Figure 8. Geometric interpretation of branch and bound with (D.C.).

Global Optimization Tools and Computational Experience

Global optimization tools have been recently developed in the Computer Aided Systems Laboratory, CASL, of the Department of Chemical Engineering at Princeton University for the primal-relaxed dual algorithm, GOP, and the branch and bound approach that combines (D.C.) transformation and a special type of lower bounding function. These tools are denoted as cGOP and αBB for the decomposition and branch and bound global optimization algorithms respectively. Both cGOP and αBB are written entirely in C and make use of MINOS, NPSOL, CPLEX for linear subproblems; MINOS, NPSOL for nonlinear programming subproblems. They have been implemented as a library of subroutines with emphasis on modularity and expandability, the subroutines for the same task have the same interfaces, and modifications in the problem data are allowed at any stage. Both cGOP and αBB have a user specified function capability which allows for connection to any external subroutine that can be treated as a black box. The current versions of cGOP and αBB can be either standalone or can be called as subroutines.Computational experience with cGOP and αBB is shown in Table 4 and Table 5 for a wide variety of applications, that include: pooling/blending problems, heat exchanger network synthesis problems, nonsharp separation synthesis, problems with quadratic objective and box constraints, concave programming problems, bilevel linear optimization problems, minimization of the Gibbs free energy with NRTL and UNIQUAC in phase and chemical reaction equilibrium, tangent plane stability criterion in phase equilibrium, clusters of atoms and molecules, molecular structure determination problems, and financial planning problems. The first three and the last pooling problems correspond to the Haverly problem and the multiperiod tankage problem and are described in Floudas and Visweswaran (1990) and Visweswaran and

Floudas (1993). The fourth and fifth pooling problems are described in Ben-Tal et al. (1994). The first two heat exchanger problems are taken from Floudas and Ciric (1989) while the last three are described in Ben-Tal et al. (1994). The first two heat exchanger problems are taken from Floudas and Ciric (1989) while the last three are described in Quesada and Grossmann (1993). The separations problem is described in Aggarwal and Floudas (1990). The minimization of Gibbs free energy problems are discussed in McDonald and Floudas (1994a). The tangent plane stability criterion problems are presented in McDonald and Floudas (1994b). The quadratic objective with box constraints, concave objective with linear constraints, and indefinite quadratic problems are discussed in Visweswaran and Floudas (1993). The Lennard Jones clusters of atoms problems are discussed in Maranas and Floudas (1993). The molecular structure determination problems are presented in Maranas and Floudas (1994a,b). The molecular structure determination problems are presented in Maranas and Floudas (1994a,b). The financial planning problems are described in Maranas et al. (1994). As Tables 4, 5 illustrate, small medium, and in certain cases large global optimization problems can be solved within a modest computational effort.

Table 4. Computational Results with cGOP.

Problem Type	Problem Name	N_x	N_y	N_c	N_I	CPU (sec)
	HAV1	8	1	6	13	0.22
	HAV2	8	1	6	14	0.21
Pooling	HAV3	8	1	6	7	0.10
	BTP1	8	2	7	7	0.15
	BTP2	15	4	19	41	5.80
	MTP1	33	6	22	11	0.50
	HEX1	8	8	13	11	0.47
	HEX2	12	15	19	39	4.35
Heat	II1	4	8	13	4	0.08
Exchanger	II2	5	6	9	3	0.03
Networks	II3	8	18	30	7	0.14
Separations	SEP1	28	18	32	17	3.84
Minimization	BAW2	8	4	2	107	1.23
of Gibbs	TW2	8	4	2	7	0.07
free energy	TWA3	18	6	3	74	2.04
	PBW3	18	6	3	313	6.27
	PBW3P	©18	6	3	1490	29.95
	BAW2L	2	2	1	32	0.15
	BAW2G	2	2	1	36	0.16
Tangent	TWA3T	6	3	2	16	0.18
Plane	TWA3G	6	3	2	85	0.94
Stability	PBW3T1	6	3	2	53	0.62
Criterion**	PBW3G1	6	3	2	213	2.37
	PBW3T6	6	3	2	549	4.98
	PBW3G6	6	3	2	757	7.09
Quadratic	QBR1	10	300	—	2	6.45
Objective	QBR2	20	300	—	2	46.01
Box	QBR3	30	160	—	2	345.83
Constraints	QBR4	30	300	—	2	411.016
	CLR1	50	50	50	2	1.62
Concave	CLR2	100	100	100	2	22.95
Objective	CLR3	50	150	100	2	0.65
Linear	CLR4	50	200	100	2	2.73
Constraints	CLR5	50	250	100	2	10.47
	CLR6	100	250	100	2	47.5
Indefinite	IND1	100	100	100	2	11.53
Objective	IND2	50	50	50	2	0.71
Linear	IND3	100	50	50	2	4.35
Constraints	IND4	50	100	50	2	1.28
	IND5	50	200	50	3	15.17

N_x = number of x-variables
N_y = number of y-variables
N_c = number of constraints
N_I = number of iterations
CPU = seconds in HP-730
** = using GLOPEQ (McDonald and Floudas, 1994)

Table 5. Computational Results with αBB.

I. *Clusters of Atoms/Molecules*

Problem Name	TV	NCV	RT	N_I
LJ8	18	3	1%	12
LJ13	33	3	1.5%	15
LJ18	48	3	1.5%	20
LJ22	60	3	1.5%	16
LJ24	66	3	1.5%	19

II. *Molecular Structure Determination*

Problem Name	TV	NCV	RT	N_I
PRO	21	2	0.01%	400
APRO	27	2	0.01%	200
ABUT	51	3	0.01%	1000
IBUT	54	3	0.01%	100
NPEN	90	4	0.01%	1000

III. *Financial Planning*

Problem Name	TV	NCV	RT	N_I
FM100	8	8	1	2
FM300	8	8	1	2
FM500	8	8	1	3
FM1000	8	8	1	6
FM10000	8	8	1	6
FMC100	8	8	1	2
FMCTX100	8	8	1	7

TV = total number of variables
NCV = nonconvex variables
RT = relative tolerance

Conclusions

This paper has attempted to present an overview of two major emerging areas in algorithmic synthesis: logic and global optimization. As indicated at the beginning of the paper these areas have been motivated by the need to improve the modelling in discrete optimization techniques, reduce the combinatorial search and avoid getting trapped into poor suboptimal solutions. In the next two subsections we briefly discuss some future directions for research.

Current and Future Directions for Logic Based Optimization

Comparing the review on MINLP given by Grossmann (1990a) at the previous Snowmass meeting, it is apparent that the work on logic based optimization has provided a new direction to address the need of integrating qualitative knowledge into mixed-integer optimization models for synthesis (see also Rippin, 1989). As has been shown by developing new models and branch and bound methods that effectively incorporate logic, order of magnitude reductions can be achieved in the combinatorial search involved in these problems. Furthermore, another very important aspect has been to achieve a better understanding of some fundamental issues related to the modelling of discrete optimization problems. In particular, the concept of w-MIP representability has proved to be a useful theoretical concept for characterizing the nature of discrete constraints. While significant progress has been made, it is clear that a number of major issues and challenges still remain for future research. These include the following:

1. The handling of temporal and modal logic is challenging and should prove to be very useful for a wide range of problems in process scheduling.
2. Other kinds of logic cuts should be investigated apart from the logic relating units in a superstructure. The cuts affect the solution efficiency considerably and also allow one to better understand the modelling of discrete programming problems. One possibility for logic cuts are constraints that prevent multiple mathematical representations for the same design configuration within a superstructure.
3. Most of the work on integration of logic has been directed to discrete linear problems. Much work remains to be done in the integration of logic for nonlinear problems. In addition, there is the issue of integration with new cutting plane methods such as the one by Balas et al. (1993).
4. The problem of developing techniques to efficiently model and solve superstructures of large scale process flowsheet problems is another major issue. The use of disjunctions should serve to reduce the level of nonlinearity present in a mixed-integer representation, and allow for a systematic scheme for generating efficient models for these problems.
5. Further study is needed on the representability of disjunctive constraints as mixed-integer constraints. Our work on w-MIP representability can only be regarded as preliminary work in the area and has just demonstrated the potential for research in this problem. A better understanding of representability issues could

lead to the development of modelling languages for generating efficient discrete optimization models.

6. The development of computer software that efficiently automates the various approaches based on logic and their more extensive testing on large scale problems is still required.
7. The integration with other design methodologies should be explored in which logic information can be generated from a preliminary screening. Example of this are the work by Friedler et al. (1991) and the work by Daichendt and Grossmann (1994a,b).
8. The ultimate objective is to provide a solid foundation to new classes of hybrid optimization models which are expressed in terms of equations and logic relations.

Progress and better understanding in the above problem will undoubtedly lead to a new generation of discrete optimization models and solution methods. Furthermore, it is clear that these efforts can complement advances in global optimization.

Current and Future Directions in Global Optimization

In the global optimization section we have attempted to present an overview of global optimization methods which are based on the concepts of decomposition and branch and bound coupled with a (DC) transformation. From this review, it is apparent that we have experienced a significant progress in the area of global optimization and its applications in Chemical Engineering over the last five years. New theoretical results and algorithms have emerged and their application to a number of Process Synthesis, Design, and Control problems has already resulted in encouraging results. At the same time applications in the area of computational chemistry, facility location, and financial planning demonstrate clearly the potential impact of global optimization in the design of new materials and biological systems, the design of process layout, and the design of financial management systems. It is also worth noting that it is the first time that the progress in the area of global optimization is reviewed in a FOCAPD meeting, which is indicative of the recent advances, the potential usefulness, and the growth of this area. Global optimization, as a new area, however has a number of important challenges and several open problems which will be the subject of current and future research work. These challenges include:

1. New global optimization approaches for nonconvex (MINLP) models arising in Process Synthesis.
2. Global optimization methods for generalized geometric programming problems (e.g. signomials) which arise in many design and robust control applications.
3. New global optimization methods for nonconvex models with trigonometric and exponential functions that arise in Computational Chemistry, Biology and classical reaction engineering applications.
4. Global optimization methods which can determine all solutions of nonlinear systems of equations that arise in phase equilibrium, azeotropic distillation, and reaction engineering.
5. Global optimization methods for bilevel and multilevel linear and nonlinear models that appear in planning problems, flexibility analysis, and optimal control approaches in batch distillation.
6. New global optimization approaches which can address implicitly defined functions; and
7. Distributed computing methods for global optimization with the aim at addressing medium to large scale optimization problems.

Even though the above challenges represent undoubtedly formidable tasks, we should see exciting developments over the next decade.

Acknowledgments

Ignacio Grossmann is grateful to the Engineering Design Research Center at Carnegie Mellon and to Eastman Chemicals Co. for supporting the work on logic optimization. Chris Floudas is grateful to the Air Force and to the National Science Foundation for supporting the work on global optimization.

References

Aggarwal A. and C.A. Floudas (1990). Synthesis of General Distillation Sequences-Nonsharp Separations. Computers and Chemical Engineering, **14**, 6, 631-653.

Al-Khayyal F.A. and J.E. Falk (1983). Jointly Constrained Biconvex Programming. *Math Opers. Res.*, **8**.

Al-Khayyal, F.A. (1990). Jointly Constrained Bilinear Programs and Related Problems: An Overview. *Computers in Mathematical Applications*, **19**, 53-62.

Andrecovich, M.J. and A.W. Westerberg (1985b). An MILP Formulation for Heat-Integrated Distillation Sequence Synthesis. *AIChE J.*, **31**, 1461.

Bagajewicz, M. and V. Manousiouthakis (1991). On the Generalized Benders Decomposition. *Comput. Chem. Engng.*, **15**, 691.

Balas, E. (1974). Disjunctive Programming: Properties of the convexhull of feasible points. MSRR #348, Carnegie Mellon University, Pittsburgh, PA.

Balas, E. (1985). Disjunctive Programming and a hierarchy of relaxations for discreteoptimization problems, *SIAM J. Alg. Disc. Meth.*, **6**, 466-486.

Balas, E., Ceria, S. And Cornuejols, G. (1993). A Lift-and-Project Cutting Plane Algorithm for Mixed 0-1 Programs, *Mathematical Programming*, **58**, 295-324.

Beaumont, N. (1991). An Algorithm for Disjunctive Programs. *European Journal of Operations Research*, **48**, 362-371.

Benders, J.F. (1962). Partitioning Procedures for Solving Mixed-variables Programming Problems. *Numerische Mathematik* , **4**, 238-252.

Ben-Tal A., G. Eiger, and V. Gershovitz (1994). Global Minimization by Reducing the Duality Gap. *Mathematical Programming*, in press.

Cavalier, T.M. And Soyster, A.L. (1987). Logical Deduction via Linear Programming. IMSE Working Paper 87-147, Department of Industrial and Management Systems Engineering, Pennsylvania State University.

Cavalier, T.M., Pardalos, P.M. And Soyster, A. L. (1990). Modeling and Integer Programming techniques applied to Propositional Calculus. *Computers and Operations Research*, **17**(6), 561-570.

Chandru, V. And Hooker, J.N. (1989). Extended Horn Sets in Propositional Logic. Working Paper 88-89-39, Graduate School of Industrial Administration, Carnegie Mellon University, Pittsburgh.

Clocksin, W.F. And Mellish, C.S. (1981). *Programming in Prolog*. Springer-Verlag, New York, NY.

Cook, S.A. (1971). The complexity of theorem proving procedures. Proceedings of the 3rd ACM Symp. on the Theory of Computing, 151-158.

Daichendt, M.M. and I.E. Grossmann (1994a). Preliminary Screening for the MINLP Synthesis of Process Systems I: Aggregation and Decomposition Techniques. *Comput. Chem. Engng.*, **18**, 663.

Daichendt, M.M. and I.E. Grossmann (1994b). Preliminary Screening for the MINLP Synthesis of Process Systems II: Heat Exchanger Networks. *Comput. Chem. Engng.*, **18**, 679.

Davis, M. And Putnam, H. (1960). A computing procedure for quantification theory. *J. ACM*, **8**, 201-215.

Denenberg, L. And Lewis, H.R. (1983). Logical Syntax and Computational Complexity. *Proceedings of the Logic Colloquium* at Aachen, Springer Lecture Notes in Mathematics, 1104, 109-115.

Douglas, J.M. (1985). A Hierarchical Decision Procedure for Process Synthesis. *AIChE J.,* **31**, 353.

Dowling, W.F. And Gallier, J.H. (1984). Linear-time algorithms for testing the satisfiability of propositional Horn formulae. *Logic Programming*, **3**, 267-284.

Duran, M.A. and I.E. Grossmann (1986). An Outer-Approximation Algorithm for a Class of Mixed-integer Nonlinear Programs. *Math Programming*, **36**, 307.

Floudas, C.A. and A. Aggarwal (1990). A decomposition strategy for global optimum search in the pooling problem. *Opers Res. J. Comput.*, **2**(3).

Floudas, C.A., A. Aggarwal and A.R. Ciric (1989). Global optimum search for nonconvex NLP and MINLP problems. *Comput. Chem. Engng.*, **13**, 1117.

Floudas, C.A. and A.R. Ciric (1989). Strategies for Overcoming Uncertainties In Heat Exchanger Network Synthesis. *Computers and Chemical Engineering*, **13**, 10, 1133-1152.

Floudas, C.A. and G.E. Paules IV (1988). A Mixed-Integer Nonlinear Programming Formulation for the Synthesis of Heat-Integrated Distillation Sequences. *Comput. Chem. Engng.*, **12**, 531.

Floudas, C.A., and Pardalos, P. M. (1990). A Collection of Test Problems for Constrained Global Optimization Algorithms. *Lecture Notes in Computer Science*, Springer-Verlag, Berlin, Germany, V. 455.

Floudas, C.A., and Pardalos, P.M. (1992). Recent Advances in Global Optimization. Princeton University Press, Princeton, New Jersey.

Floudas, C.A. and V. Visweswaran (1993). A Primal-Relaxed Dual Global Optimization Approach. *Journal of Optimization, Theory, and its Applications*, **78**, 2, 87-225.

Floudas, C.A. and V. Visweswaran (1990). A Global Optimization Algorithm (GOP) for Certain Classes of Nonconvex NLPs — I. Theory. *Computers and Chemical Engineering*, **14**, 12, 1397-1417.

Friedler, F., K. Tarjan, Y.W. Huang and L.T. Fan (1991). An Accelerated Branch and Bound Method for Process Synthesis. Presented at the 4th World Congress of Chemical Engineering, Karlsruhe.

Friedler, F., K. Tarjan, Y.W. Huang and L.T. Fan (1993). Graph-Theoretic Approach to Process Synthesis: Polynomial Algorithm for Maximal Structure Generation. *Comput. Chem. Engng.*, **17**, 929-942.

Grossmann, I.E. (1985). Mixed-Integer Programming Approach for the Synthesis of Integrated Process Flowsheets. *Comput. Chem. Engng.*, **9**, 463.

Grossmann, I.E. (1990a). MINLP Optimization Strategies and Algorithms for Process Synthesis. in *Foundations of Computer-Aided Design*, J.J. Siirola,I.E. Grossmann and G. Stephanopoulos (Eds.), Cache-Elsevier, Amsterdam.

Grossmann, I.E. (1990b). Mixed-Integer Nonlinear Programming Techniques for the Synthesis of Engineering Systems. *Res. Eng. Des.*, **1**, 205.

Grossmann, I.E. and M.M. Daichendt (1994). New Trends in Optimization-based Approaches for Process Synthesis. To appear in *Proceedings of Process Systems Engineering*, Korea.

Geoffrion, A.M. (1972). Generalized Benders Decomposition. *Journal of Optimization Theory and Applications*, **10**(4), 237-260.

Gundersen, T. and L. Naess (1988). The Synthesis of Cost Optimal Heat Exchanger Networks. An Industrial Review of the State of the Art. *Comput. Chem. Engng.*, **12**, 503.

Hendry, J.E., D.F. Rudd and J.D. Seader (1973). Synthesis in the Design of Chemical Processes. *AIChE J.,* **19**, 1.

Hansen, P., B. Jaumard and S. Lu (1992a). Global Optimization of Univariate Lipschitz Functions: I. Surrey and Properties. *Mathematical Programming*, **55**, 251-272.

Hansen, P., B. Jaumard and S. Lu (1992b). Global Optimization of Univariate Lipschitz Functions: New Algorithms and Computational Comparison. *Mathematical Programming*, **55**, 273-292.

Hansen, E.R. (1980). Global Optimization Using Interval Analysis: TheMulti-dimensional Case. *Numerische Mathematik*, **34**, 247-270.

Haverly, C.A. (1978). Studies of the behaviour of recursion for the pooling problem. *SIGMAP Bull*, **25**, 19.

Hlavacek, V. (1978). Synthesis in the Design of Chemical Processes,. *Computera and Chemical Engineering*, **2**, 67-75.

Hooker, J.N. (1988). Resolution vs. Cutting plane solution of inference problems: some computational experience. *Optns. Research Letters*, **7**(1), 1.

Hooker, J.N., H. Yan, I.E. Grossmann, and R. Raman (1994). Logic Cuts for Processing Networks with Fixed Charges. *Computers and Operations Research*, **21**, 265-279.

Horst, R. and H. Tuy (1987). On the convergence of global methods in multiextremal optimization. *J. Optimization Theory Applic.*, **54**, 253.

Horst, R., Thoai, N.V., and De Vries, J. (1992). A New Simplicial Cover Technique in Constrained Global Optimization. *Journal of Global Optimization*, **2**, 1-19.

IBM (1991). *OSL User Reference Manual*. IBM Corp, Kingston, New York.

Jeroslow, R.G. And Lowe, J.K. (1984). Modelling with Integer Variables. *Mathematical Programming Study*, **22**, 167-184.

Jeroslow, R.E. and Wang, J. (1990). Solving propositional satisfiability problems. *Annals of Mathematics and AI*, **1**, 167-187.

Karp, R.M. (1972). Reducibility among combinatorial problems. *Complexity of Computer Calculations* (Miller, R. E. and Thatcher, J. W., eds) Plenum, New York, 85-104.

Kocis, G.R. and I.E. Grossmann (1987). Relaxation Strategy for the Structural Optimization of Process Flow Sheets. *Ind. Eng. Chem. Res.*, **26**, 1869.

Kocis, G.R. and I.E. Grossmann (1989b). A Modeling and Decomposition Strategy for the MINLP Optimization of Process Flowsheets. *Comput. Chem. Engng.* **13**, 797.

Kokossis, A.C. and C.A. Floudas (1994). Stability in Optimal Design: Synthesis of Complex Reactor Networks. *AIChE J.*, **40**(5), 849-861.

Lien, K. M. And Wahl, P. E. (1991). If you can't beat them, join them. Combine Artificial Intelligence and Operations Research Techniques in Chemical Process Systems Design. *Proceedings of PSE'91*, **4**, 11-15, Montebello, Canada.

Liu, W.B. and C.A. Floudas (1993). A Remark on the GOP Algorithm for Global Optimization. *Journal of Global Optimization*, **3**, 4, 519-522.

Loveland, D.W. (1978). *Automated Theorem Proving: A Logical Basis*. North Holland, Amsterdam.

Mahalec, V. and Motard, R.L. (1977). Procedures for the initial design of chemical processing systems. *Computers and Chemical Engineering*, **1**, 57-68.

Manousiouthakis, M. and D. Sourlas (1992). A Global Optimization Approach to Rationally Constrained Rational Programming. *Chem. Eng. Comm.*, **115**, 127-147.

Maranas, C.D. and C.A. Floudas (1992). A Global Optimization Approach for Lennard-Jones Microclusters. *J. Chem. Phys.*, **97**(10), 7667-7678.

Maranas, C.D. and C.A. Floudas (1992). Global Optimization for Molecular Conformation Problems. *Annals of Operations Research*, **42**, 85-117

Maranas, C.D. and C.A. Floudas (1994a). Global Minimum Potential Energy Conformations of Small Molecules. *Journal of Global Optimization*, **4**, 2, 135-170.

Maranas, C.D. and C.A. Floudas (1994b). A Deterministic Global Optimization Approach for Molecular Structure Determination. *Journal of Chemical Physics,* **100**, 2, January 15, 1247-1261.

Maranas, C.D. and C.A. Floudas (1994c). A Global Optimization Method for Weber's Problem with Attraction and Repulsion. *Large Scale Optimization: State of the Art* . W. W. Hager, D. W. Hearn, and P. M. Pardalos, Eds., Kluwer Academic Publishers, 265-300.

Maranas, C.D., I.P. Androulakis, C.A. Floudas, A.J. Berger, and J.M. Mulvery (1994). Solving Stochastic Control Problems in Finance via Global Optimization. Submitted for publication.

McDonald, C.M. and C.A. Floudas (1994a). Global Optimization for the Phase and Chemical Equilibrium Problem: Application to the NRTL Equation. *Computers and Chemical Engineering*, accepted for publication.

McDonald, C.M. and C.A. Floudas (1994b). Global Optimization for the Phase Stability Problem. *AIChE J.*, accepted for publication.

Mendelson, E. (1987). *Introduction to Mathematical Logic*. Van Nestrand, New York.

Nemhauser, G.L. And Wolsey, L.A. (1988). *Integer and Combinatorial Optimization*. Wiley-Interscience, New York.

Nishida, N., G. Stephanopoulos and A.W. Westerberg (1981). A Review of Process Synthesis. *AIChE J.*, **27**, 321.

Papoulias, S.A. and I.E. Grossmann (1983a). A Structural Optimization Approach in Process Synthesis. Part I: Utility Systems. *Comput. Chem. Engng.*, **7**, 695.

Papoulias, S.A. and I.E. Grossmann (1983b). A Structural Optimization Approach in Process Synthesis. Part II: Heat Recovery Networks. *Comput. Chem. Engng.*, **7**, 707.

Papoulias, S.A. and I.E. Grossmann (1983c). A Structural Optimization Approach in Process Synthesis. Part III: Total Processing Systems. *Comput. Chem. Engng.*, **7**, 723.

Pardalos, P.M. and Rosen, J.B. (1987). Constrained Global Optimization: Algorithms and Applications. *Lecture Notes in Computer Science*, Springer-Verlag, Berlin, Germany, **268**.

Post, S. (1987). Reasoning with Incomplete and Uncertain Knowledge as an Integer Linear Program. *Proceedings Of Avignon 87: Expert Systems and their Applications*. Avignon, France.

Quesada, I., and I.E Grossman (1993). Global Optimization Algorithm forHeat Exchanger Networks. *Ind. Eng. Chem. Res.*, **32**, 487-499.

Quesada, I. and I.E. Grossmann (1993a). A Global Optimization-Algorithm for Linear Fractional and Bilinear Programs. Presented at TIMS/ORSA Meeting, Chicago.

Quesada, I. and I.E. Grossmann (1993b). Global Optimization Algorithm for Heat Exchanger Networks. *Ind. Eng. Chem. Research*, **32**, 487.

Quesada, I. and I.E. Grossmann (1993c). Global Optimization Algorithm of Process Networks with Multicomponent Flows. presented at AIChE Meeting, St. Louis.

Raman, R. and I.E. Grossmann (1991). Relation Between MILP Modelling and Logical Inference for Chemical Process Synthesis. *Computers and Chemical Engineering*, **15**, 73.

Raman, R. and I.E. Grossmann (1992). Integration of Logic and Heuristic Knowledge in the MINLP Optimization for Process Synthesis. *Computers and Chemical Engineering,* **16**, 155-171.

Raman, R. and I.E. Grossmann (1993a). Symbolic Integration of Logic in Mixed Integer Linear Programming Techniques for Process Synthesis. *Computers and Chemical Engineering*, **17**, 909.

Raman, R. and I.E. Grossmann (1993b). Symbolic Integration of Logic in MILP Branch and Bound Methods for the Synthesis of Process Networks. *Annals of Operations Research,* **42**, 169-191.

Raman, R. and I.E. Grossmann (1994). Modeling and Computational Techniques for Logic Based Integer Programming. *Computers and Chemical Engineering*, **18**, 563.

Rardin, R. L. And Choe, U. (1979). *Tighter Relaxations of Fixed Charge Network Flow Problems*. Georgia Institute of Technology, Industrial and Systems Engineering Report Series, #J-79-18, Atlanta.

Ratschek, H., and Rokne, J. (1988). *New Computer Methods for Global Optimization*. Halsted Press, Chichester, Great Britain.

Rinnoy Kan A.H.G. and G.T. Timmer (1987). Stochastic global optimization methods. Part I: clustering methods. *Math. Program*, **39**, 27.

Rippin, D.W.T. (1990). Introduction: Approaches to Chemical Process Synthesis. in *Foundations of Computer-Aided Design*, J.J. Siirola, I.E. Grossmann and G. Stephanopoulos (Eds.), Cache-Elsevier, Amsterdam.

Ryoo H.S. and N.V. Sahinidis (1994). Global Optimization of Nonconvex NLPs and MINLPs with Applications in Process Design. *Computers and Chemical Engineering*, submitted for publication.

Sahinidis, N.V. and I.E. Grossmann (1991). Convergence Properties of Generalized Benders Decomposition. *Computers and Chemical Engineering*, **15**, 481.

SCICON (1986). *SCICONIC / VM User Guide*. SCICON Ltd., London.

Sherali, H., and Tuncbilek, C. H. (1992). A Global Optimization Algorithm for Polynomial Programming Problems Using a Reformulation-Linearization Technique. *Journal of Global Optimization*, **2**, 101-112.

Sherali, H. D. and A. Alameddine (1992). A New Reformulation Linearization Technique for Bilinear Programming Problems. *J. Of Global Optimization*, **2**, 4, 379-410.

Shor, N.Z. (1990). Dual Quadratic Estimates in Polynomial and Boolean Programming. *Annals of Operations Research*, **25**, 163-168.

Stephanopoulos, G. and A.W. Westerberg (1975). The Use of Hestenes' Method of Multipliers to Resolve Dual Gaps in Engineering System Optimization. *Journal of Optimization Theory and Applications*, **15**(3), 285-309.

Swaney, R.E. (1990). Global Solution of Algebraic Nonlinear Programs. AIChE Annual Mtg. Chicago, Il.

Torn, A., and A. Zilinskas (1989). Global Optimization. *Lecture Notes in Computer Science*, **350**, Springer-Verlag, Berlin.

Tsirukis, A.G. and G.V. Reklaitis (1993). Feature Extraction Algorithms for Constrained Global Optimization I. Mathematical Foundation. *Annals of Operations Research*, **42**, 229.

Tsirukis, A.G. and G.V. Reklaitis (1993). Feature Extraction Algorithms for Constrained Optimization II. Batch Process Scheduling Application. *Annals of Operations Research*, **42**, 275.

Turkay, M. and I.E. Grossmann (1994). A Logic Based Outer-Approximation Algorithm for MINLP Optimization of Process Flowsheets. Presented at AIChE Annual Meeting, San Francisco.

Tuy, H., Thieu, T.V., and Thai, N.Q. (1985). A Conical Algorithm for Globally Minimizing a Concave Function over a Closed Convex Set. *Mathematics of Operations Research*, **10**, 498-514.

Tuy, H. (1987). Global Minimum of a Difference of Two Convex Functions. *Mathematical Programming Study*, **30**, 150-182.

Vaidyanathan R. and M. El-Halwagi (1994). Global Optimization of Nonconvex Nonlinear Programs via Interval Analysis. *Computers and Chemical Engineering*, in press.

Viswanathan, J. and I.E. Grossmann (1990). A Combined Penalty Function and Outer-Approximation Method for MINLP Optimization. *Comput. Chem. Engng.*, **14**, 769.

Visweswaran, C., and C.A. Floudas (1990). A Global Optimization Algorithm (GOP) for Certain Classes of Nonconvex NLPs — II. Application of Theory and Test Problems. *Computers and Chemical Engineering*, **14**(12), 1419-1434.

Visweswaran, V., and Floudas, C.A. (1992). Unconstrained and Constrained Global Optimization of Polynomial Functions in One Variable. *Journal of Global Optimization*, **2**, 73-100.

Visweswaran, V. and C.A. Floudas (1993). New Properties and Computational Improvement of the GOP Algorithm for Problems with Quadratic Objective Function and Constraints. *Journal of Global Optimization*, **3**, 4, 439-462.

Visweswaran V. and C.A. Floudas (1994). An MILP Reformulation of the Relaxed Duals in the GOP Algorithm. manuscriopt in preparation.

Westerberg A.W., and J.V. Shah (1978). Assuring a Global Minimum by the Use of an Upper Bound on the Lower (Dual) Bound. *Comput. Chem. Engng.*, **2**, 83.

Williams, H.P. (1988). *Model Building in Mathematical Programming*. John Wiley, Chichester.

AN INDUSTRIAL PERSPECTIVE ON PROCESS SYNTHESIS

Jeffrey J. Siirola
Eastman Chemical Company
Kingsport, TN 37662

Abstract

Systematic approaches for the invention of conceptual chemical process designs have been proposed and discussed for more than twenty-five years. A number of process synthesis approaches, methods, and tools have now been developed to the point of industrial application. Described here are two case studies of the application of such systematic process synthesis techniques to the conceptual design of total process flowsheets and heat-integrated separation schemes for nonideal systems. Compared to conventional design practice, energy savings of 50% and net present cost reductions of 35% have been typically achieved.

Keywords

Conceptual design, Systematic generation, Means-ends analysis, Property hierarchy, Task identification, Task integration, Hierarchical synthesis, Opportunistic synthesis, Total flowsheet synthesis, Separation scheme synthesis.

The Innovation Process and Conceptual Design Synthesis

The industrial chemical *Innovation Process* is an organized multistage procedure which leads from the identification of a customer's need to an operating facility producing a substance believed to address that need. Along the way, a chemical route must be found from available raw materials that produces the desired product, and a facility that implements this chemistry must be conceived, designed, constructed, started, operated and maintained (Fig. 1).

Each stage within the innovation process might be approached by a generalized four-step design paradigm (Fig. 2). In the first *Formulation* step, the goals and specifications for the particular stage of the innovation process are specified. This is then followed by an iteration of three steps consisting of *Synthesis* (generation of an alternative), *Analysis* (determination of the behavior of the alternative generated), and *Evaluation* (comparison of the performance of the proposed alternative against the goals and specifications from the formulation step). If the performance of the proposed solution is judged to be satisfactory, the stage is completed; otherwise a new alternative must be generated, analyzed and evaluated. In general, alternatives are not generated blindly, but are guided by some analysis internal to the synthesis step (sometimes encoded in the problem representation) so that alternatives are generated with anticipated behavior close to the desired behavior. This general design strategy requires that alternative generation methods, analysis methods, and evaluation methods be available for each stage of the process.

The stages within the innovation process interact. Alternatives available at any stage are constrained by decisions made in previous stages. However, optimal selection among alternatives depends in part on how each alternative would be implemented by all of the stages which follow. Because of this interacting nature, the innovation process stages are often executed iteratively with stages revisited a number of times, often at increasing levels of detail. The greatest costs are actually incurred in the equipment procurement, plant construction, and operation stages of the innovation process, but these costs are largely determined by decisions made much earlier in the chemistry and engineering design stages.

- **Need Identification**
- **Manufacturing Decision**
- **Basic Chemistry**
- **Detailed Chemistry**
- **Task Identification**
- **Unit Operations**
- **Basic Plant Engineering**
- **Detailed Engineering**
- **Vendor Specifications**
- **Component Acquisition**
- **Construction Plan and Schedule**
- **Plant Construction**
- **Operating Procedures**
- **Commissioning and Start-up**
- **Production Plan and Schedule**
- **Plant Operation and Maintenance**

Figure 1. Innovation process sequence.

Process Synthesis is the invention of flowsheet alternatives at conceptual design stage of the innovation process. The goal of process synthesis is to develop one or more schemes to implement the conversion of available raw materials into desired product materials, fit for use, in the desired amount and on the schedule required, safely, responsibly and economically. This flowsheet will be refined, optimized and implemented in subsequent basic and detailed engineering and construction stages of the innovation process. It is desirable that this conceptual process design prove to be a better starting point for the remainder of the innovation process than other possible flowsheets.

Three fundamental approaches for the synthesis of chemical process flowsheets have been developed. The first, *Systematic Generation*, builds the flowsheet from smaller, more basic components strung together in such a way that raw materials become eventually transformed into the desired products. The second, *Evolutionary Modification*, starts with a existing flowsheet for the same or a similar product and then make structural or operational changes as necessary to adopt the design to meet the objectives of the specific case at hand. The third approach, *Superstructure Optimization*, views synthesis as an optimization over structure, and starts with a larger superflowsheet which contains embedded within it many alternatives and redundant interconnections and then strips parts of the superstructure away while simultaneously optimizing other design parameters.

In the absence of the exhaustive generation of all alternatives, systematic generation can not guarantee structural optimality. Likewise, the quality of solutions generated by evolutionary modification depends critically on the starting flowsheet as well as on the methods used to modify it. The superstructure approach offers the promise of simultaneous optimization of structural as well as other design parameters. However, it requires the starting superstructure from somewhere (which for some problems may be implicit in the formulation), as well as very extensive computational capability.

Both heuristic and algorithmic methods have been proposed for systematic generation and evolutionary modification approaches, whereas mathematical programming and probabilistic optimization methods are obvious tools for the superstructure optimization approach. Evolutionary modification seems to be the approach most traditionally used by conceptual design engineers, particularly those with a large repertoire of known good existing flowsheets. Systematic generation, on the other hand, has been the basis for many of the computer-aided and other systematic process synthesis approaches which have found industrial application. Superstructure optimization, while limited at the present time, offers tremendous potential for the future.

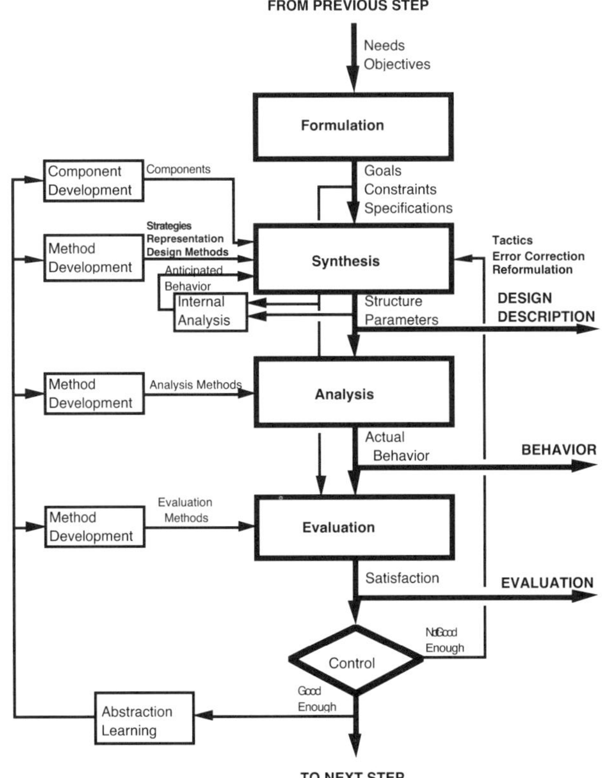

Figure 2. General design paradigm.

Systematic Generation: Means-ends Analysis and the Property Hierarchy

In dealing with the manufacture of substances, there seems to be a number of physical and chemical properties of materials and streams of materials of continuing interest. These properties include the molecular identity of the material, the amount involved, the composition or purity of the material, its thermodynamic phase, temperature, pressure, and possibly size, shape and other physical characteristics. Available raw materials, intermediates and the desired product materials are all characterized by these parameters of identity, amount, concentration, phase, temperature, pressure, form and so on.

In one formulation of the process synthesis problem, the raw materials are considered as an initial state and the desired product considered a goal state. If the value of a particular property of a raw material is different from the desired value of the corresponding property in the desired product, a property difference is detected. The purpose of an industrial chemical process is to apply technologies in sequence such that these property differences are systematically eliminated, and the raw materials become thereby transformed into the desired product. The systematic detection of state differences, the proposed elimination of such differences through the specification of appropriate corrective tasks, and the identification of methods to accomplish these tasks are the essence of a general problem-solving paradigm known as *Means-Ends Analysis* (Simon, 1969).

In many cases there are fairly obvious technologies for reducing or eliminating property differences. Familiar examples include chemical reaction to change molecular identity, mixing and splitting (and purchase) to change amount, separation to change concentration and purity, enthalpy modification to change phase, temperature, pressure, etc. Sometimes there is such a close relationship between a property changing task and an obvious unit operation to accomplish the task that many designers think directly in terms of equipment when developing conceptual process flowsheets.

Generally, property-changing methods are applied to existing streams to produce new streams closer in properties to the desired products. Usually these methods are applied and the conceptual design is developed in a forward or *opportunistic* direction, that is in the same direction as material flow, from the source of raw material through any reactions and toward final products, with recycles considered as an afterthought. This is in contrast to the generally *retrosynthetic* (or backward, from the desired product) approach taken by most chemists in the development of organic reaction sequences. In conceptual design, both the starting materials and the desired products are known, while in reaction sequence development oftentimes only the desired product is known. Opportunistic approaches may be more comfortable since any point in a partial design is a feasible consequence of the raw materials and the property-changing methods so far chosen. Other approaches including hierarchical approaches which are neither opportunistic nor retrosynthetic are also possible.

One complication is that many property-changing methods can only be applied to a stream when certain other properties of the stream are within specified values, which may not be true at the time. For example, a method expecting to exploit relative volatility differences to effect a separation can only be applied if enthalpy conditions are such that simultaneous liquid and vapor phases exist. If conditions for the immediate application of an method believed to be useful do not exist, a new design subproblem may be formulated whose objective is to reduce property differences between the initial stream and the conditions necessary for the application of the method. This recursive strategy is a common feature of the means-ends analysis paradigm.

Alternative solutions are generated when more than one method, for example exploiting different phenomena, are identified that can be applied to accomplish a property difference elimination task. The selection of which method to choose might be made on the basis of some evaluation at the time the methods are being examined. Alternatively, each may be chosen separately and the consequences of doing so followed separately (leading to alternative design solutions) and then each final solution may be evaluated and compared. One additional possibility is that a number of feasible alternative methods are selected and applied in parallel leading to a redundant design or superstructure. At the end of the design process, the superstructure is reevaluated in its entirety, and the less economical redundant portions eliminated.

Another problem is that a prospective method may not completely eliminate a property difference. In such a case, another follow-up method for the same property difference may need to be specified. Filtration followed by drying is such an example. Furthermore, the application of a method may not exclusively change only a single property. The side effects of a difference reduction method may create, increase, or decrease other property differences as well. Other problems include the possibility of not finding a method in the repertoire that can eliminate a property difference. If this occurs in a recursive subproblem (that is, while attempting to meet the preconditions for the application of another method), it may not be fatal if an alternative method at the higher level is available. The paradigm would fail, however, if no method can be found for a property correction task at the topmost level.

Alternative designs may also result when more than one property difference exists between a stream and its goal(s) and these differences are eliminated in different orders. There does, however, seem to be a natural hierarchy among property differences in the chemical process domain, and a natural order to selecting property differences for resolution. The hierarchy is the same as the order in which the differences were previously mentioned: identity first, then amount, then concentration, then phase, then temperature and pressure, then form. The hierarchy arises because properties lower in the hierarchy are often more readily manipulated in order to satisfy the preconditions for the application of difference elimination methods for property changing tasks higher in the hierarchy. Alternative solutions are still possible when multiple differences at the same level in the hierarchy exist, for example when both temperature and pressure need to be changed, or when a mixture of more than two components is to be separated and sent to different destinations.

Systematic Flowsheet Synthesis Algorithm

One approach to the systematic generation of a conceptual designs given available raw materials and the desired products using the means-ends analysis paradigm would be to attack in a hierarchical manner identity differences first (resolved with reaction methods), then amount (resolved with mixing, splitting or purchase methods), then composition (resolved with mixing or separation methods), then phase, temperature and pressure (resolved with enthalpy-changing methods), then form properties (size, shape, etc.). Preconditions are met by adjusting properties lower in the property hierarchy as necessary. The conceptual designer thinking directly in terms of equipment would consider reactors first, then mixers, then columns, decanters, filters, dryers and the like, then heat exchangers, pumps, compressors, valves, etc. to satisfy goal properties for the desired product, and subgoals which arise as preconditions for the application of unit operations higher in the equipment hierarchy. Alternative designs result when alternative unit operations can be used to resolve the same a property difference, or when multiple property differences at the same level in the property hierarchy are resolved in different orders.

The means-ends analysis paradigm, although recursive in nature, is not iterative. However, for efficiency, many chemical processes exploit recycle of mass and energy. Recycle introduces a very serious complication since even the existence, much less the properties, of the mass or energy to be recycled may not be known at the time property differences at the point of recycle were resolved. Sometimes the interactions are minimized by the indirect way in which recycle tasks are implemented in equipment, for example energy recovery with a heat exchanger. Other times the interactions are more severe as in mass recycle by mixing. Recycles are generally considered at the end of one pass through the conceptual design. If the effects of recycle are severe, particularly with respect to structural validity, the design may need to be repeated from the beginning but this time assuming the existence of the potential recycle, until the design process converges.

One particular implementation of this hierarchical flowsheet synthesis procedure (Siirola and Rudd, 1971) consists of *Reaction Path* (chemistry identification), *Species Allocation* (target mass flows among raw materials, reactions, products and wastes including possible recycle of incompletely converted reactants), *Task Identification* (property difference detection and elimination method identification), *Task Integration* (association of methods with actual unit operations, also including combining adjacent tasks and combining complementary tasks), *Utilities System* (basic infrastructure design and preliminary operating cost estimation), and *Equipment Design* (target level equipment design and preliminary capital cost estimation). With characteristics similar to the overall innovation process of which conceptual design is a part, the initial stages of the flowsheet synthesis hierarchy are less well defined, but have a greater impact on the overall outcome of the design. Later stages are more constrained and better defined, but have comparably less potential for economic impact. The stages clearly interact with each other. Alternatives available at any stage are constrained by choices made at earlier stages, but optimal choices are influenced by how all the subsequent stages prove to be executed. Similar to the innovation process, better solution schemes iterate among the stages at different levels of detail. Better yet are schemes which solve several stages simultaneously. A key feature of the procedure is that *task identification*, *task integration*, and *equipment design* are discrete and separate activities. Specifically, unlike the common conceptual design approach, property differences are not resolved directly in terms of equipment.

In the original automated implementation of this synthesis procedure (Siirola, 1970), preconditions for property changing methods were met, if necessary, by adjusting properties only lower in the property hierarchy. For example, separations might be identified in order to get appropriately pure feed conditions for reaction tasks, and temperatures may be adjusted in order to get better conditions for executing separation tasks. However, the procedure would never consider changing the identity of a species to get better conditions for a separation, as that would have violated the property hierarchy. Multiple design alternatives were generated by applying different methods for task resolution (for example by exploiting different physical phenomena to effect a separation), and by attacking multiple differences at the same level of the property hierarchy in different orders.

A Total Flowsheet Example

Consider the following industrial example of a process for the production of methyl acetate. A process will first be synthesized in the traditional conceptual process design manner, then modified using evolutionary approaches, and finally resynthesized using the systematic means-ends analysis procedure both with and without adherence to the property hierarchy. For simplicity here, only identity, amount, and composition differences will be resolved (that is, temperature and pressure change will be ignored).

Conventional Approach

Methyl acetate may be produced by the elementary equilibrium-limited acid-catalyzed esterification of methyl alcohol and acetic acid which also byproduces water. The reaction is nearly athermic, with an equilibrium constant on the order of unity.

Although acetates generally form azeotropes with both their corresponding alcohols and water, the azeotrope with water is generally heterogeneous and lower-boiling. Therefore, acetates are typically produced from acetic acid and an alcohol with a textbook flowsheet involving a reactor fitted with a fractionating column which removes the acetate-water heterogeneous azeotrope overhead while forcing unreacted acetic acid and alcohol back to the reactor. The product acetate and water are separated in a decanter and purified additionally as necessary (Witzeman and Agreda, 1993). However, this textbook flowsheet is not applicable to the production of methyl acetate because the methyl acetate-methanol azeotrope boils lower than the methyl acetate-water azeotrope and further, the methyl acetate-water azeotrope is unique among acetates in that its composition is outside the immiscible region. Therefore an alternative conceptual process design must be generated.

If methanol and acetic acid are available as raw materials, and methyl acetate is the desired product, according to the property difference hierarchy, an identity difference is first detected between the desired product and either of the raw materials. A known chemical reaction method, namely esterification, can be applied to a mixture of the raw materials brought to the proper conditions to produce methyl acetate and eliminate the identity difference between the reaction effluent and the desired product. Thinking directly in terms of equipment, this task might be immediately implemented as a stirred tank reactor.

The effluent of the reactor contains product methyl acetate, byproduct water and significant unreacted methanol and acetic acid. The species allocation is trivial: methyl acetate is directed toward the product, water to wastewater treatment, and both methanol and acetic acid recycled back to the reactor in order to improve feedstock usage efficiency. Since the reactor effluent does not meet the purity specifications of any desired destination, separation tasks need to be specified to eliminate composition differences between the reactor effluent and the various destinations of its various components.

The species are all liquids at ambient conditions with normal boiling points between about 50C and 120C. It may be possible to exploit volatility difference to effect the required separations. Water and methyl acetate form a low-boiling azeotrope, and also have a limited immiscible region but which is unusual in that it does not include the azeotrope. Methanol and methyl acetate form an azeotrope boiling even lower. There are no other binary or ternary azeotropes in the system, although water and acetic acid almost form an azeotrope, exhibiting a tangent pinch on the water end (and acetic acid dimerizes in the vapor phase).

One common separation sequencing heuristic is to perform the easiest split first. Considering the boiling point distribution of the components and azeotropes, and thinking in terms of equipment, an opportunistic first separation is a distillation column in which all of the methyl acetate, all of the methanol, and as much water as azeotropes with the methyl acetate is taken overhead, and all of the acetic acid and the remaining water is taken as underflow. At first this may not seem like much progress since neither column effluent stream is directly suitable for any destination. However, the bottom stream is essentially a binary mixture of acetic acid and water, and the separation of that mixture is a problem which has been solved before!

The separation of acetic acid-water mixtures is a feature of a number of large-scale industrial processes including manufacture of cellulose acetate, terephthalic acid and even acetic acid itself. Simple binary distillation is feasible but very expensive because of the tangent pinch. The literature contains a number of flowsheets for this separation. Most involve solvent extraction and/or azeotropic distillation to circumvent the tangent pinch, and differ in the acetic acid-water composition addressed and the nature of the extraction solvent or azeotropic entrainer. One particular textbook flowsheet (Lodal, 1993) can be immediately applied to the acetic acid-water composition from the underflow of the splitting column in this example. The design consists of a solvent extractor, an azeo column and decanter, flash columns for solvent recovery from the extractor raffinate and the decanter aqueous layer, and a color column for purification of the azeo column underflow. All of the acetic acid appears in the color column overhead suitable for recycle back to the reactor. Water exits from the two flash column underflows suitable for discharge to the wastewater treatment system. Ethyl acetate in this case is used for both the extraction solvent and the azeo column entrainer, and is totally recycled within the flowsheet.

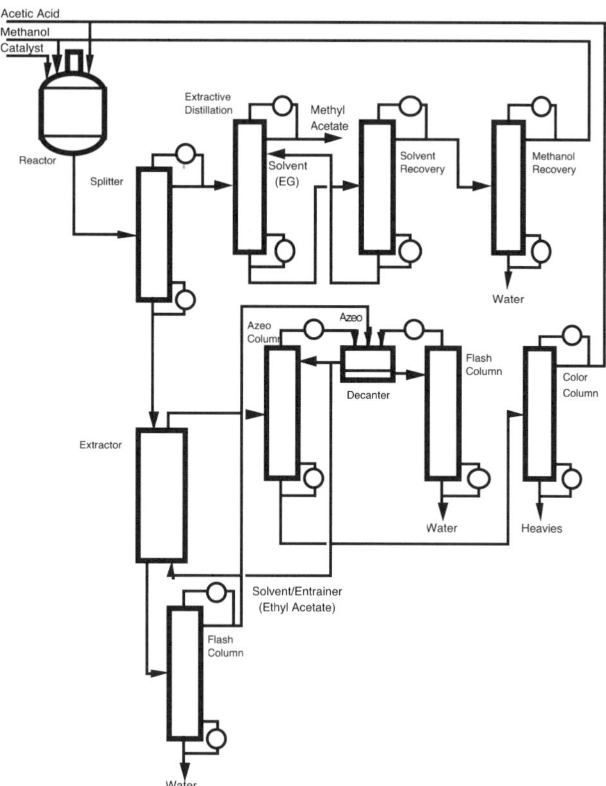

Figure 3. Methyl acetate: Conventional approach.

This now leaves the overhead from the first splitting column. The homogeneous azeotropes in this mixture are a significant problem. However, a well-known pattern for breaking homogeneous azeotropes involves extractive distillation with a higher-boiling solvent such as for this case, ethylene glycol. A standard textbook configuration consisting of an extractive distillation column followed immediately by a solvent recovery column produces a pure methyl acetate product as an overhead from a first column and the methanol-water mixture as an overhead from the second. The methanol-water distillate may be separated in one final binary distillation column producing methanol suitable for recycle back to the reactor and water suitable for discharge to the wastewater treatment system. Composition specifications of the product methyl acetate and the byproduct water are met, and all unreacted methanol and acetic acid have been recycled to the reactor. No additional identity, amount, or concentration differences exist and the conceptual design is complete. The flowsheet includes one reactor, one extractor, one decanter, eight distillation columns, and employs two separate mass separation agents (Fig. 3).

Evolutionary Modification

The flowsheet generated by the conventional conceptual process design approach using systematic generation and parts of textbook flowsheets and standard patterns wherever possible seems especially complicated for such simple chemistry involving so few components. This first flowsheet might be a candidate for evolutionary modification.

Some evolutionary strategies search for thermodynamic inefficiencies, for example through exergy analyses. Another strategy is to search for structural inefficiencies like design redundancies. In our first flowsheet, both flash columns have similar feeds, both overheads go to the same decanter and both underflows go to wastewater treatment. Therefore, these two columns can probably be combined. Secondly, the acetic acid color column, which was a feature of the textbook flowsheet taken for this part of the design may not be required. Upon investigation, it is found that the original flowsheet was designed to handle dirty acetic acid-water mixtures containing minor unspecified high-boiling contaminants. In the present case, it is determined that the acetic acid composition from the underflow of the azeo column is sufficiently pure to be recycled to the reactor directly, and thus another column is eliminated. Enthalpy-changing tasks were not considered in this example, but if they were, a number of energy recovery opportunities might also be evaluated using heat integration techniques.

Another point of concern is the two external mass separating agents. Both were chosen from textbook patterns and flowsheets. Although both are readily recovered and recycled within their respective sections of the process, a common simplifying heuristic in conceptual design suggests that if possible, mass separation agents should be chosen from among components already within the system. Considering first the agent that must serve the dual role of extraction solvent as well as azeotropic entrainer, the only possible component in the original system that might work is methyl acetate. However, as previously noted, the region of immiscibility is not large, and depending on the feed composition, it might not be a feasible extraction solvent at all. Moreover, the azeotrope with water is not homogeneous, so the azeotropic distillate would not decant. Methyl acetate is totally unsuitable as a mass separating agent in the present design.

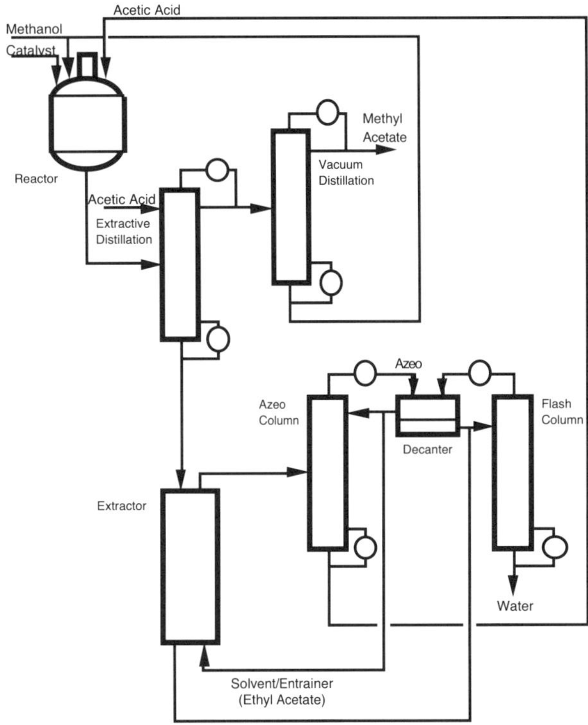

Figure 4. Evolutionary modification.

On the other hand, acetic acid is a candidate for the extractive distillation solvent as it can break the methyl acetate-water azeotrope. If it were directly substituted for the ethylene glycol, methyl acetate and methanol would be produced overhead, and an acetic acid-water mixture would be produced in the underflow. The acetic acid-water mixture is sent directly to the lower part of the flowsheet with the underflow of the first splitting column, thus saving one additional column. Furthermore, the splitting column and the extractive distillation column might also be combined since one is fed from the overhead of the other and both underflows go to the same place. The methyl acetate-methanol overhead (which also forms a homogeneous azeotrope) may be separated into methanol and sufficiently pure methyl acetate to meet the desired product specification either by binary distillation at reduced pressure (which shifts the azeotropic composition), or by pervaporation, or again by extractive distillation with acetic acid. The final evolved design contains four fewer columns than the original design (Fig. 4).

The relative performances of the original and evolved designs can only be determined by analysis. The evolved design has fewer columns, but has the added complication of either refrigeration or membranes, and the flow of acetic acid which must be separated from water is much increased. The results of detailed simulations indicate in fact that combination of the flash columns and elimination of the color column is advantageous, but that using acetic acid for the extractive mass separation agent unfortunately is not.

Means-ends Analysis Adhering to the Property Hierarchy

If the hierarchical means-ends analysis synthesis procedure is applied to the methyl acetate problem, the task identification, task integration, and equipment design stages are kept completely separate. Employing the property difference hierarchy, a reaction task (Task A) is identified first as before. When examining the differences between the reaction task effluent and the product methyl acetate and byproduct water destinations, two sets of differences are detected. For the methyl acetate destination, acetic acid, water, and methanol must all be essentially completely removed. For the water destination, methyl acetate, methanol, and acetic acid must all be essentially completely removed. The obvious method to resolve all of these concentration differences is to exploit relative volatility. Considering the methyl acetate destination first, the separation of acetic acid should be relatively easy because of the large boiling point difference. However, removal of the water and methanol may be difficult because of homogeneous azeotropes. Turning to the water destination, removal of both methyl acetate and methanol should be easy because of the large boiling point differences, but removal of the acetic acid may be difficult because of the tangent pinch.

Following the outline from the evolutionary discussion above, the acetic acid and most of the water may be removed from the methyl acetate destination by a distillative separation (Task B1). The remaining methanol and the water can also be removed from the methyl acetate (despite the azeotropes) by solvent enhanced distillative separation using acetic acid (Task C). For the water destination, the methyl acetate and methanol may be removed by simple distillative separation (Task B2, perhaps the same task that separated the acetic acid from the methyl acetate destination), while in principal acetic acid can also be completely separated from water also by simple distillative separation (Task D) (Fig. 5).

The tasks thus identified may be integrated into actual equipment in a number of ways. One possibility is to recognize that Tasks B1 and B2 are the same task, Task B. Further, the reaction Task A, the first distillative separation Task B, and the solvent-enhanced distillative separation Task C may all be integrated into a single shell. The remaining acetic acid-water separation Task D may then be implemented as a separate column (Fig. 6). An analysis of this scheme indicates that while feasible, the single-column implementation of the acetic acid-water separation task is quite expensive because of the tangent pinch.

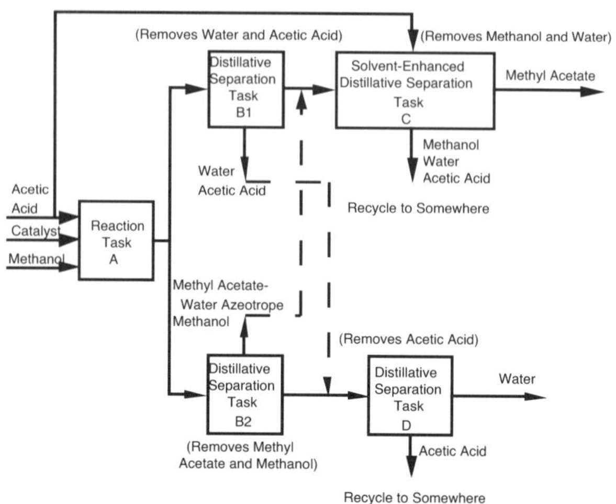

Figure 5. Task identification with property hierarchy.

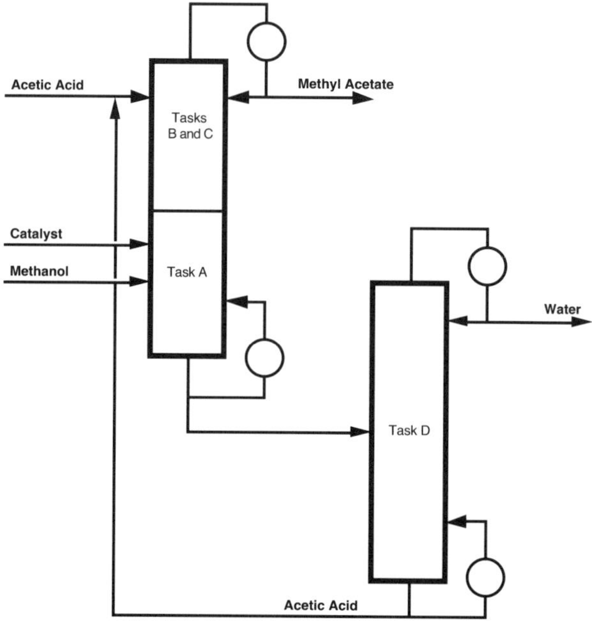

Figure 6. Task integration into equipment.

Ignoring Strict Adherence to the Property Hierarchy

Relative volatility is not the only property that might be exploited to address composition differences. For example, if the property hierarchy is ignored, then an identity-changing esterification reaction task with acetic acid might be chosen to eliminate methanol from the methyl acetate stream (Task E). Similarly, an esterification reaction task with methanol might be chosen to eliminate acetic acid from the water stream (Task F). Furthermore, similar to the previous example, solvent-enhanced distillative separation using acetic acid might be used to break the methyl acetate-water azeotrope (Task G), and conventional distillative separation might be used to separate both methyl acetate (Task H) and methanol (Task I) from water, and acetic acid from methyl acetate (Task J). What results then are three reaction tasks (one to produce the desired product and two to remove offending species from destination streams), one solvent-enhanced distillative separation task (using acetic acid), and three conventional distillative separation tasks (Fig. 7).

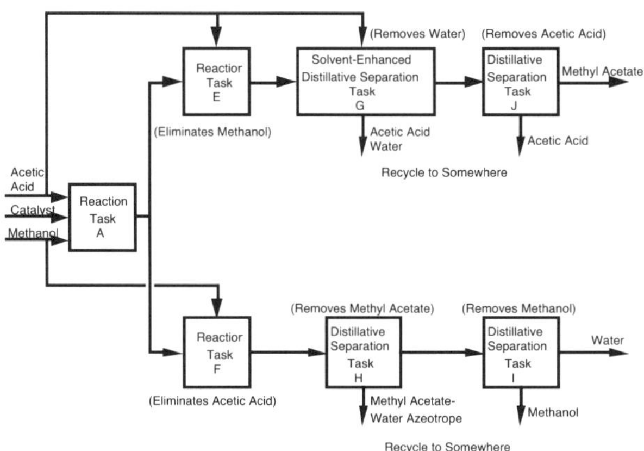

Figure 7. Task identification ignoring hierarchy.

There are a number of ways in which these tasks may be integrated into actual processing equipment. It turns out in this case that considering hydraulics, material flows, and energy requirements, all seven tasks may be integrated into a single shell (Fig. 8). All three reactive tasks are integrated into the center of the column with Task A in the middle and Task E above and Task F below. Extractive distillation Task G sits above that with solvent removal Task J above that. Low boiler separation Tasks H and I are combined and sit below the reaction zones. Proper operation requires that an extractive zone must be above the reactive zone (otherwise the azeotrope of water and methyl acetate formed at the top of the reaction zone can not be broken), which requires that either the catalyst (if homogeneous) be introduced between zones (G) and (E), or else (if heterogeneous) be only installed below zone (G). The resulting capital cost and operating cost of the single-column reactive-extractive distillation design for methyl acetate production are both just one-fifth of that for the optimized (color column and extra flash column eliminated) alternative generated by the conventional design and evolutionary modification approach (Agreda and Partin, 1984).

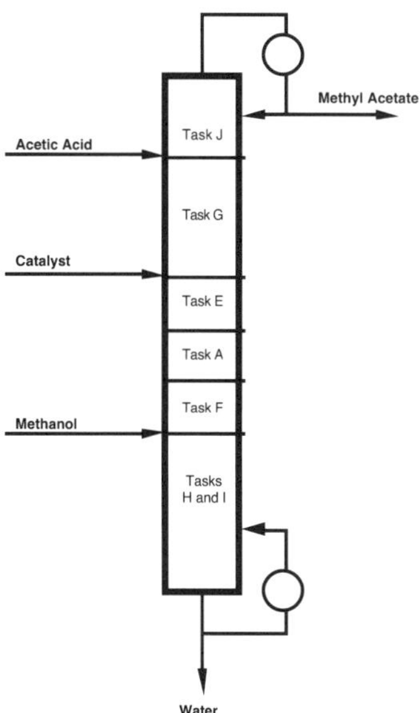

Figure 8. Reactive/extractive distillation process.

Another Look at Acetic Acid Dehydration

In the last example, consideration of task identification separately from task integration into equipment lead to an elegant single-column reactive extractive distillation process for methyl acetate. Process simplification appears to have advantages.

Now, let us return to the simple binary separation of acetic acid-water as might be encountered in the manufacture of cellulose acetate, which as mentioned previously can be feasibly separated in one single column, but generally is not because of the tangent pinch on the water end. Where did the textbook separation scheme for acetic acid-water previously mentioned come from, and how might such alternatives to a simple distillative separation be systematically generated?

The synthesis tools required to propose property difference resolution methods depend in part on the nature of the system. In the case of distillative separations for resolving compositional differences, if the system is relatively ideal, then a simple volatility ordering may be sufficient to represent the solution thermodynamics. Alternative separation trains can be synthesized at the targeting level of detail using list-processing techniques. However, if the system exhibits more nonideal thermodynamics involving azeotropes and regions of immiscibility, a more complex representation and synthesis procedure may be necessary.

- **Define Goal Specifications**
- **Construct RCM Representation**
- **Identify Critical Features**
- **Identify Interesting Compositions**
- **Address Critical Features First**
- **Resolve Remaining Differences Opportunistically**
- **Pursue All Resolution Alternatives**
- **Pursue Both Distillative Separation Alternatives**
- **Avoid Separating Ternary Mixtures Simultaneously into Lightest Overhead and Heaviest Underflow**
- **Beware of Reflux Composition Different from Overhead**
- **Recycle Opportunistically**
- **Check Recycle Mass Balance**
- **Special Consideration for Pressure-Shifting Boundaries, Exploiting Boundary Curvature, Reaching Saddle Compositions, and Solvent-Enhanced Separations**
- **Eliminate Redundant Tasks**
- **Integrate Tasks into Equipment**
- **Evaluate Each Design Alternative**

Figure 9. Separation scheme synthesis method.

An industrial method has been developed for separation scheme synthesis exploiting vapor-liquid and liquid-liquid phase equilibria (Fig. 9). Key features of this method include the *Residue Curve Map* representation of solution thermodynamics overlaid with pinched regions and liquid-liquid equilibria data (Foucher, Doherty, and Malone, 1991); identification of thermodynamic *Critical Features* to be avoided (pinched regions), overcome (distillation boundaries), or exploited (liquid-liquid tie lines); a *Strategic Hierarchy* to address issues raised by the critical features first; the identification of *Interesting Compositions* (useful as mass separating agents but which must be regenerated and recycled within the process); *Opportunistic Resolution* of remaining concentration property differences by mixing, decant, or distillative separation as permitted by distillation region constraints; and pursuit of *Multiple Alternatives* (heaviest underflow as well as lightest overhead) for every multicomponent distillative separation.

The tangent pinch is a critical feature of the binary acetic acid-water system and simple distillation through the pinched region is to be avoided. In this example, this is accomplished by exploitation of a third component as a mass separation agent. One such class of such agents is partially immiscible with water with which it also forms a minimum-boiling azeotrope. Such an agent is ethyl acetate which boils lower than water. The resulting system contains a Type I immiscibility region critical feature but no distillation boundaries. Interesting compositions for potential recycle include the pure components, the water-ethyl acetate azeotrope, and the compositions of the two liquid phases into which the heterogeneous azeotrope separates.

Following the hierarchy to avoid the pinch critical feature and to exploit the immiscibility region critical feature, it is seen that an extractive separation method may get around the relative volatility pinch and produce a raffinate arbitrarily free of acetic acid. Such a method may be applied directly to the feed composition in the present example using either pure ethyl acetate or the organic layer from the azeotrope as the mass separation agent (solvent), and each choice leads to a different design. Application of an extractive method using water saturated ethyl acetate produces an ternary extract stream and an binary raffinate stream as in Fig. 10.

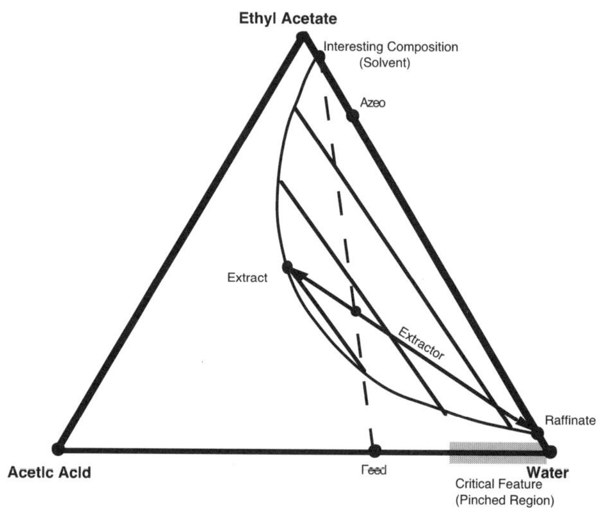

Figure 10. Acetic acid dehydration: Extractor.

For the extract stream, since there are no distillation boundaries, the distillation region is the entire composition space. Of all the distillative separation tasks that could be applied opportunistically to this stream, two especially noteworthy include one which produces the lowest-boiling composition (the ethyl acetate-water azeotrope) as overhead and an underflow water acetic acid mixture determined by mass balance, and a second which produces the highest-boiling composition (acetic acid) as underflow and an overhead water ethyl acetate mixture determined by mass balance. Again, each choice leads to a different design.

Since the water end of this separation is still somewhat pinched, it is desirable to make this distillative separation as easy as possible. The separation might be enhanced by adjusting the feed to be exactly collinear with acetic acid and the azeotrope so that it would be possible to simultaneously take the lowest-boiling (azeotropic) composition overhead and the highest-boiling (acetic acid) composition under (thus maximizing the temperature difference across the column and presumably minimizing the acetic acid-water separation difficulty). The feed could be so adjusted by adding either of the mass separation agent compositions previously identified for the extractive separation task.

In the integration of this separation task into actual equipment, the added composition (in this case called the azeotropic entrainer) may be directly mixed with the extract composition feed and introduced together into the middle of the column with reflux provided by condensed overhead in the conventional manner. Alternatively, since the azeotropic entrainer has a lower boiling point than the extract, it might better be introduced separately in the upper part of the column. Alternatively again, depending upon the staging and reflux requirements, the amount of azeotropic entrainer may enough to provide most or even all of the required column reflux if introduced at the very top of the column. In this last case, the staging may be adjusted so that the amount of azeotropic entrainer required by mass balance considerations (so that the net feed is exactly collinear with acetic acid and the azeotrope) is also the amount of reflux appropriate for the number of stages to produce the desired purities (Fig. 11).

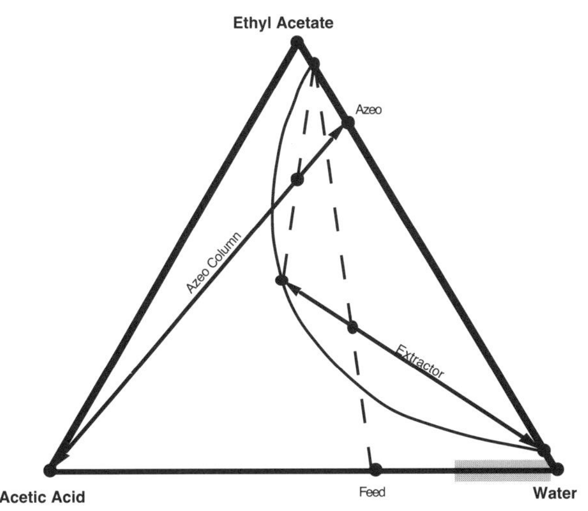

Figure 11. Simultaneous azeo and acetic acid separation.

Since the overhead binary azeotrope is heterogeneous, it may be opportunistically decanted producing the exact composition in the organic layer previously assumed for the azeotropic entrainer and the extraction solvent, and also an aqueous layer as determined by the liquid-liquid tie line through the azeotrope. By mass balance it can be shown that the amount of mass separation agent assumed to be available for both the extraction solvent and the azeotropic entrainer has not quite yet been generated by the decant, so the recycle problem is not yet completely solved.

The aqueous layer of the decant task and the raffinate from the extraction task are both saturated with ethyl acetate which may opportunistically separated by a distillative method to again produce the azeotrope overhead and pure water underflow. With the overhead recycled to the decant task, mass balance now indicates that the full amount of mass separating agent assumed to be available has now been regenerated. The tasks specified accomplish the composition goals producing pure water and pure acetic acid, while avoiding the tangent pinch. The mass separating agent remains totally within the system. The subsequent integration into equipment involves an extractor, two columns, and one decanter (Fig. 12) and (minus an additional distillative separation to separate acetic acid from unspecified high boilers) is the textbook flowsheet noted previously.

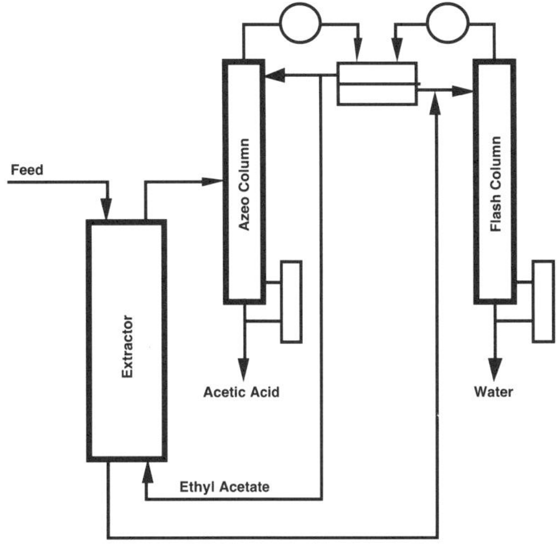

Figure 12. Ethyl acetate flowsheet.

In the above example, the ternary feed composition in the azeo column was adjusted such that simultaneously the most volatile composition in the region was overhead while the least volatile composition was underflow, thereby reducing the distillative separation task to a single alternative (with apparently maximum separation potential). This is not necessarily as clever as it may appear. Since in distillation the overhead and bottom composition must nearly lie on the same residue curve, and all residue curves for this system start at the azeotrope and end at acetic acid, there are many designs which will be feasible. This can be a particular problem in designs which reflux compositions different from the overhead (as in this case). Slight changes in reflux lead to overall feed composition shifts in a direction orthogonal to the column mass balance line (which does not happen in normal columns which reboil the underflow and reflux the overhead). In this case, slight shifts in the mass balance line can result in major shifts in the actual column composition profile, so much so that the major contaminant in the acetic acid product may change from water to ethyl acetate or vice versa. If there are excessive stages in the top of the column, the overhead composition will be pegged to the azeotrope and in response to reflux perturbations, the bottom composition may swing from the acetic acid-water face of the composition triangle to the acetic acid-ethyl acetate face, so that the acetic acid purity specification may not be met. It may appear that the problem may be resolved by relatively increasing the staging in the bottom of the column, pegging that composition to pure acetic acid, and letting the top binary composition float on either side of the azeotrope. As long as the top composition remains in the heterogeneous region, the flowsheet pretty much works as designed. However, even in such a case, if the mass balance line moves from one side of the azeotrope to another, the column composition profile will also shift dramatically and in the bottom of the column will approach the acetic acid vertex from one side of the composition triangle or the other. Although the acetic acid purity specification may still be met, the identity of the next plentiful contaminant will change from water to entrainer or vice versa, which may be entirely unacceptable. Special arrangements, for example the addition of additional minor feeds may be required to reliably start-up and operate such a column on the desired composition profile.

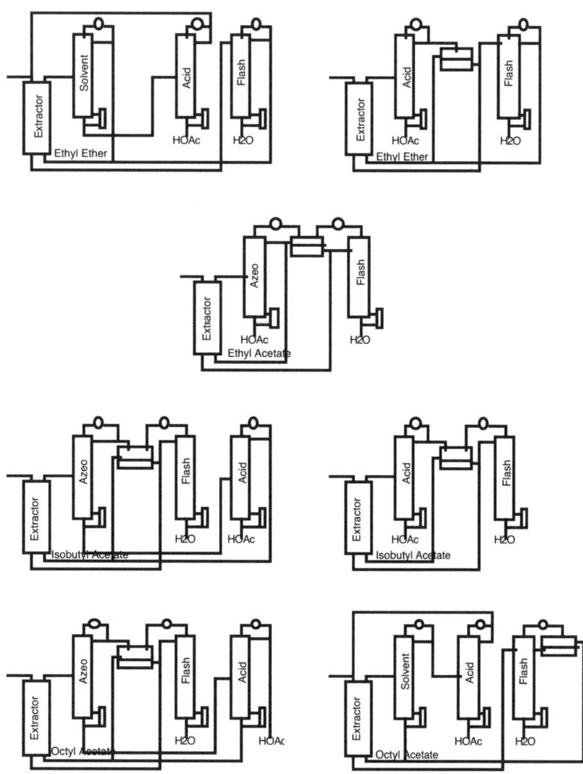

Figure 13. Alternative flowsheets.

Many mass separation agents besides ethyl acetate may be used to facilitate the separation of acetic acid and water. In each case, the first separation method is an extraction task to remove as much water as possible without exploiting relative volatility. This is followed by opportunistic distillative separations designed to avoid the pinched region and recover the mass separating agent. Figure 13 shows additional flowsheets which result from this systematic separation synthesis method depending on the boiling point of the mass separation agent and whether the column treating the extract removes the lowest-boiling composition overhead, or the highest-boiling species under. Which flowsheet is best can only be determined by simulation and evaluation of the each flowsheet, which is left as an exercise for the reader. Several have been used industrially.

Previous research has indicated that there can be a significant benefit to tightly coupling task specification and task integration into equipment. One example is heat-integrated distillation sequences with energy recovery either within the sequence or by the utility system, especially when column pressures are adjusted to allow recovery of latent heat (Siirola, 1981). For processes that contain a single or a particularly dominant distillation such as occurs in the present example, significant energy recovery by such an approach may not possible. However, the implementation of a single distillative separation may be split into two columns in such a way that energy rejected from one might be recycled to the other, approximately halving the net utility requirement. This may be done by operating the columns at different pressures, possibly assisted by the degree of separation designed for each column.

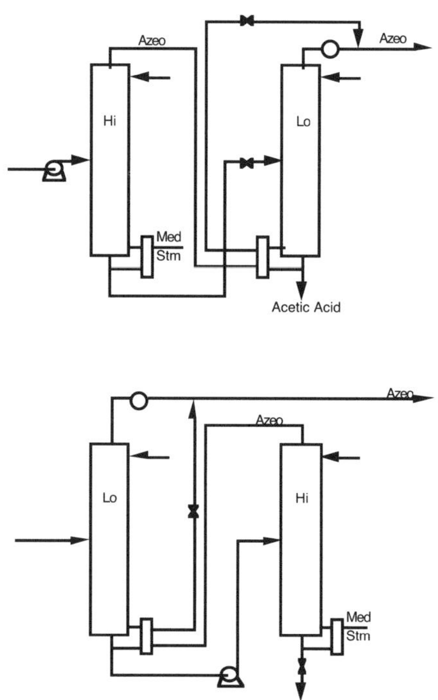

Figure 14. Multiple effect azeo distillation.

In one parametric study (Blakely, 1984), five different two-column multiple effect distillation designs were compared against a single column as a function of feed flowrate, feed composition, and relative volatility. The results of that study predict that for the case of ethyl acetate as the mass separating agent, either of the multiple effect designs for the azeo column in Fig. 14 (which differ in which column is operated at higher pressure) is expected to be significantly superior to a single column implementation. In addition, a number of conventional heat integration techniques may be applied to either design. The resulting final two designs (Fig. 15), which interestingly implement the color column function differently because of the location of composite curve pinches, both accomplish exactly the separation which could have been achieved in one simple column, but do so with 50% less energy and at 30% lower net present cost than even the textbook acetic acid dehydration flowsheet. In this case, simplicity isn't necessarily best.

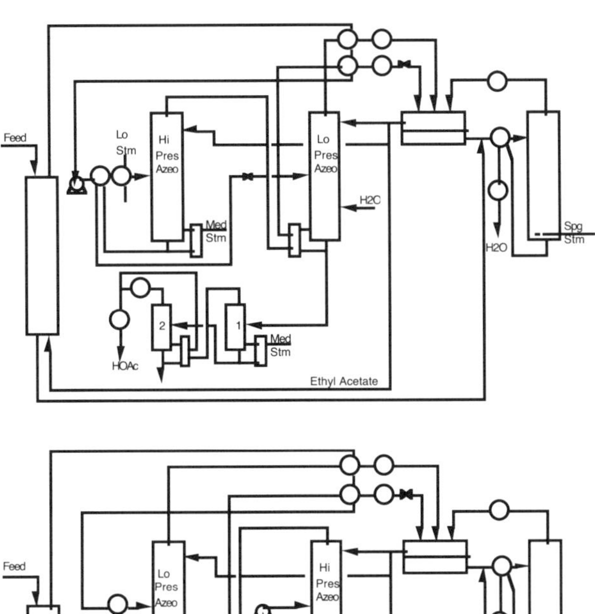

Figure 15. Heat-integrated ethyl acetate flowsheets.

Conclusions

Systematic approaches to process synthesis have begun to have industrial impact. Features of these systematic approaches include architectures which are hierarchical and iterative in both scope and level of detail, means-ends analysis as a recursive design paradigm, hierarchical as well as opportunistic goals attacked by an iterative formulation-synthesis-analysis-evaluation design strategy, consideration of tasks to be accomplished separately from how these tasks may be implemented in actual equipment, exploiting properties outside of the conventional hierarchy to accomplish identified tasks, the importance of representations such as residue curve maps that encapsulate analysis, and the advantages of solving synthesis problems simultaneously. With these techniques — some automated and some not — an increased number of higher value, lower energy, and even novel designs have been generated and actually implemented.

References

Agreda, V.H. and L.R. Partin (1984). U.S. Patent 4,435,595 (to Eastman Kodak Company).

Blakely, D.M. (1984). *Cost Savings in Binary Distillation through Two-Column Designs*. M.S. dissertation, Clemson University, Clemson.

Foucher, E.R., M.F. Doherty, and M.F. Malone (1991). Automatic Screening of Entrainers for Homogeneous Azeotropic Distillation. *Ind. Eng. Chem. Res.*, **30**, 760-772.

Lodal, P.N. (1993). Production Economics. In V.H. Agreda and J.R. Zoeller (Eds.), *Acetic Acid and its Derivatives*, Marcel Dekker, New York. 61-69.

Siirola, J.J. (1970). *The Computer-Aided Synthesis of Chemical Process Designs*. Ph.D. dissertation, University of Wisconsin, Madison.

Siirola, J.J. and D.F. Rudd (1971). Computer-Aided Synthesis of Chemical Process Designs. *Ind. Eng. Chem. Fundam.*, **10**, 353-362.

Siirola, J.J. (1981). Energy Integration in Separation Processes. In R.S.H. Mah and W.D. Seider (Eds.), *Foundations of Computer-Aided Process Design*, Vol. II. Engineering Foundation, New York. 573-594.

Simon, H.A. (1969). *Science of the Artificial*. MIT Press, Cambridge.

Witzeman, J.S. and V.H. Agreda (1993). Alcohol Acetates. In V.H. Agreda and J.R. Zoeller (Eds.), *Acetic Acid and its Derivatives*, Marcel Dekker, New York. 257-284.

CONTRIBUTED PAPERS

G. V. Reklaitis
Purdue University
West Lafayette, IN

Commentators:
Truls Gundersen[1], Malcolm Preston[2], W. David Smith[3]
[1]Telemark Institute of Technology, Porsgrunn, Norway
[2]ICI Engineering, Runcorn, England
[3]DuPont Central Research, Wilmington, DE

Introduction

In this session, thirty one contributed papers were presented in a poster format, with overviews and assessments provided by a team of three commentators. The contributions were divided into three topical clusters: Design and Synthesis, Modeling and Analysis, and Computing Methods and Tools.

Design and Synthesis

Commentary on the ten papers in this cluster (the first ten in the order listed in this volume) was provided by Truls Gundersen. He first of all noted that these papers all constituted work by established researchers in the field and were generally of very good quality. He pointed out that two papers in the Modeling and Analysis grouping, those by Krajnc and Glavic and by Omtveit and Lien might also have been included in this grouping because of their significant design orientation. Gundersen observed that eight of the papers employed optimization technology to solve design related problems, reflecting both the theme of the conference (computer-aided) and the current trend in academic process design research. Two of the ten papers employed graphical construction techniques to gain physical insights, along the lines of pinch analysis for process integration. He suggested that it is now about time for industry to start exploring some of these new optimization based approaches for the design and operation of their plants. Another trend evident from the papers is the change in focus away from continuous bulk production to batch-wise production of specialty chemicals. Indeed four of the papers in the session addressed batch design issues. He suggested that perhaps chemical engineers may further have to shift their emphasis from process engineering to product engineering, as proposed in the papers by Sunol and by Venkatasubramanian et al.

Of the eight optimization based papers, one employs stochastic search techniques: the paper by Venkatasubramanian et al. advances the use of genetic algorithms to conduct an evolutionary search for desirable molecular structures that can meet product performance requirements. The paper uses design examples from the polymers and refrigerant domains to demonstrate some success in overcoming the combinatorial explosion of structural choices. The seven other papers in this class employ mathematical programming constructions. The Sunol paper uses a hierarchical and hybrid approach to examine life cycle design for a composite material manufactured using supercritical fluid processing. The papers by Kokossis and Floudas and by Friedler et al. direct their attention to reactor network synthesis. The former proposes a mixer-settler extractor based superstructure for a family of two phase complex reactions, while the latter approaches the reactor network problem using a decision-mapping procedure to handle the combinatorial complexity associated with the underlying MINLP's. While interesting, the latter paper does not provide enough details to allow full appreciation of the power of the approach.

The remaining four mathematical programming based contributions are directed at batch processing applications. Barbosa-Povoa and Macchietto discuss the detailed design of multipurpose batch plants, including consideration of aspects of plant lay-out, intermediate storage, and other operational constraints. While a nice step forward, the approach remains subject to limitations and simplifications such as time discretization, linear cost relations, simplified process models, and single campaigns with fixed product slates. The paper by Bassett et al. examines computational strategies for solving the very large scale MILP models associated with integrated batch scheduling, planning, and design applications under market uncertainties. A hierarchical model described in earlier work is extended to handle the multiple time scales inherent to the scheduling-planning-design levels. Uncertainty is handled via the concept of scenario. Computing strategies are discussed for exploiting the parallelism inherent in the decomposition and illustrated with case studies involving several thousand integer variables. The Rotstein paper discusses improved estimation of the overall reliability of multipurpose batch plants subject to equipment failures. The proposed approach decomposes networks into "minimal cuts" and disaggregates the cuts into relevant units. Three MILP models are presented that gradually improve reliability estimates, at the expense of increasing computing time. Finally, the paper by Thomaidis and Pistikopoulos presents a stochastic MINLP model for the incorporation of flexibility, reliability, availability, and maintenance in process design. The key features of the proposed model include a maintenance superstructure, explicit consideration of uncertain continuous parameters, and the determination of the duration and expected revenue of operable states. Application is confined to a simple multiproduct plant example.

The remaining two papers, Dye et al. and Richburg and El-Halwagi, propose graphical construction based design methods. The Dye paper treats the synthesis of crystallization based separation schemes, a separation technology that plays an increasingly important role in specialty chemicals production but has received little systematic process design attention. An approach in the spirit of process synthesis is proposed using phase diagrams and related projections to draw conclusions about the process structure. The Richburg paper discusses the application of graphical, pinch analysis like, constructions as short cut methods for designing refrigeration driven condensation systems for recovering volatile organic compounds (VOC's). Condensation networks provide an interesting contrast to mass exchange and membrane systems which have received the largest attention in VOC recovery. An application is discussed involving recovery of vinyl chloride from PVC plants.

Modeling and Analysis

The ten contributions in this cluster, which were highlighted by David Smith, can be grouped into three categories: process modeling and analysis (six papers), process dynamics (2 papers), and analysis techniques (2 papers). The first and largest group of papers includes four which address issues in the modeling of separation systems. The paper by Hytoft et al. presents the components of a computer package for the simulation, design, and analysis of supercritical extraction processes. Key computational issues associated with the steady state and dynamic modeling of fluid-liquid extraction at elevated pressure and of solvent recovery at low pressure as well as with the determination of phase transitions are discussed. Examples are presented involving chemical, biochemical, and food processing.

The papers by Zhu and Shen and Pekkanen address issues in reactive distillation. The former paper describes a steady-state nonequilibrium stage column model, assuming no reactions in the vapor phase. Solution is proposed via a homotopy continuation method. Some comparisons between calculations and experimental data are reported for an ethanol-acetic acid esterification but no comparisons are offered to work in the literature, such as that of Taylor and co-workers. Pekkanen reports theoretical conditions for the possible existence of a reactive pinch in reactive distillation but offers no physical examples in which these conditions are realized. Zitney and coworkers discuss the results of exploiting sparse matrix technology, vectorized residual calculations and parallel asynchronous data transfer to attain computational speed-ups using a supercomputer in large scale distillation applications. Interesting results are reported for a large scale industrial dynamic simulation study involving multiple coupled columns and practical implications are discussed for process operations.

Shyamkumar et al. describe a flowsheeting case study directed at exploring waste reduction alternatives in the sulfolane process. While no conclusions could be drawn concerning a general design approach to waste reduction, the study reaffirms the power of process modeling and flowsheet optimization when combined with good engineering judgment. Finally, Omtveit and Lien report a computational strategy for optimizing the temperature profile for a class of reaction problems with fixed inlet compositions and profiles defined by two composition degrees of freedom. The approach has design implications since it points to the establishment of performance targets for reactor systems for this limited class of problems.

Two of the papers, those by Braatz et al. and Ciric, address issues in process dynamics. The former discusses at the conceptual level the advantages of introducing autoregulatory feedback elements into process designs so as to both provide stability and improve open loop performance. Two simple examples: a flow divider valve and a compressor system are described and some simulated performance results are reported for the latter case. The Ciric paper presents a computationally attractive approach to determining the dynamic stability of a process. The test requires the solution of a constrained optimization problem derived from a Lyapunov function criterion. The potential of this interesting approach is demonstrated using the fold bifurcation and a binary distillation problem.

Finally, Krajnc and Glavic report on the impact of parameter variations on the comparative effectiveness of pinch technology and optimization based methods for heat integration. Shacham et al. reemphasize the interactive use of data visualization and linear and nonlinear data regression executed via contemporary numerical/statistical packages to evaluate process data, discriminate between alternative models, and determine model parameters.

Computing Methods and Tools

As noted by Malcom Preston, the eleven papers in this cluster (the last eleven of the group of thirty-one in this section of the proceedings volume) can be classified into four gradations: people/tools (two contributions), tools/techniques (three contributions), techniques/methodology (three contributions), and methodology/philosophy (three papers). He noted that to the practicing engineer tools were the most useful things if correctly applied to the right problem. He also observed that the longer the word describing the classification, the more remote from the real world it appeared to become and the more the engineer needed to beware.

In the people/tools class, the paper by van Klavern et al. demonstrated that computer simulations are emerging as bona fide educational tools. Sim Refinery, based on the popular computer game, Sim City, shows that disasters can be fun and that perhaps expensive high fidelity training simulators are not necessary to transfer important lessons. A second paper in this class, that by Partin, shows how benefits can be gained from the recent advances in spreadsheet capabilities. However, in extolling the advantages of spreadsheets, the paper does not discuss the important issues of validation and verification of the resulting computational procedures.

The three tools/techniques oriented papers, those by Stadtherr and Schnepper, Book and Harney, and Seader and Balaji, all embraced numerical methods for solving the sets of nonlinear equations found in many chemical engineering problems, particularly those arising in physical properties evaluations. These techniques and details are normally hidden from the engineering end-user and thus it is important that robust and reliable methods be implemented which can find all the correct solutions with integrity.

Of the techniques/methodology class of papers, the one by Cassata et al. is perhaps the best in terms of relating to industrial applicability. In using an object oriented approach to provide an environment for several activities relating to FCC, it offers a good demonstration of how much better integrated design is likely to become in the future. The paper by Preisig is also an exemplar of the object-oriented paradigm, with more of a focus on representational issues than immediate use in a design environment. Finally, the Schembecker et al. paper describes a useful contribution to the important area of chemical reactor synthesis in early stage design. The software package combines a "classical" rule-based expert system with algorithmic calculations into a hybrid consulting system.

The first of the three papers in the methodology/philosophy class, Fielding et al., prompted the observation that the area of data modeling and data exchange was in danger of drifting from methodology to philosophy rather than towards the direction of more useful techniques and tools. The paper appears to do little to reverse this view, although the domain of activity modeling is an exciting and important one. Although the work on activity modeling described by Debelak and De Carin is perhaps a rearticulation of several systems engineering truths revealed by Rippin and co-workers in papers in the late 60's and early 70's, it is as previously stated a very exciting area. Much fruitful work will be borne of this burgeoning area in looking at total activities and their subsystems in the context of problem solution rather than methodology obsession. Finally, the paper by Banares-Alcantara et al. is a thought provoking preview of how process design environments might be 10+ years from now. In researching these complex issues, the authors admit that much is not currently implementable. Nevertheless, they offer a glimpse of a pinnacle at which aim should be taken.

Collectively, the papers in this part of the FOCAPD conference offered a stimulating snap-shot of the range of research topics currently under investigation in the vigorous computer aided design field. The authors and commentators are to be commended for their willingness to share their work and insights.

SYNTHESIS OF CRYSTALLIZATION-BASED SEPARATION SCHEMES

Susan R. Dye, David A. Berry and Ka M. Ng
Department of Chemical Engineering
University of Massachusetts
Amherst, MA 01003

Abstract

We present in this paper systematic methods for overcoming the eutectic limitations in crystallization, thus leading to processes for the complete separation of a multicomponent mixture. The solid-liquid phase behavior is represented by various phase diagrams and the related projections. We show that only a limited number of flowsheet structures are needed to handle a wide variety of mixtures.

Keywords

Process synthesis, Extractive crystallization, Selective crystallization, Selective dissolution, Crystallization of reciprocal tonic systems.

Introduction

Crystallization is widely used for the separation of organic and inorganic high boilers, and heat sensitive biochemicals. It is also the method of choice for separating close-boiling isomers. With the gradual shift of emphasis from commodity to specialty chemicals in the chemical industry of developed countries, crystallization is expected to play an increasingly significant role in separation processes.

Relatively little has been done to integrate crystallization into process synthesis. In particular, there is need for systematic design methods for completely separating multicomponent mixtures. For example, consider a plant for the manufacture of adipic acid. A significant fraction of adipic acid is lost during separation from the byproducts, glutaric and succinic acids. Another example is the crystallization of para xylene from a mixture of para and meta-xylenes formed by catalytic reforming of naphtha. Current processes do not allow recovery of meta-xylene even though it is a valuable precursor for isopolyester resins.

The problem that must be overcome is how to handle multiple-saturation troughs and surfaces which cause cocrystallization. The current practice in chemical processing avoids direct confrontation of this problem; thus some product is always lost in the effluent mother liquor (Barnicki and Fair, 1990). This paper introduces a number of schemes which completely separate multicomponent mixtures by extractive crystallization, selective crystallization and dissolution, and crystallization of reciprocal ionic systems. Purification of solid solutions is not considered.

Systematic Separation Schemes

Extractive Crystallization

In extractive crystallization, an additional component (or components) is added to a multi-solute system to avoid cocrystallization at eutectics, thus achieving complete separation (Rajagopal et al., 1991). Consider the phase behavior of a four-component system with 3 solutes (A, B, C) and a solvent (S). Solid-liquid equilibria can be depicted with a tetrahedral phase diagram (Figure 1). There are 6 binary eutectics, 4 ternary eutectics, 1 quaternary eutectic, 12 double-saturation troughs, 4 triple-saturation troughs and 6 double-saturation surfaces in a simple eutectic system. A two-dimensional representation called a Jänecke projection can be obtained by compressing the tetrahedral diagram onto the base as viewed from the apex. There is a saturation compartment for each component. Cooling a liquid stream with a composition located in any of the 4 compartments results in the crystallizing of a pure product.

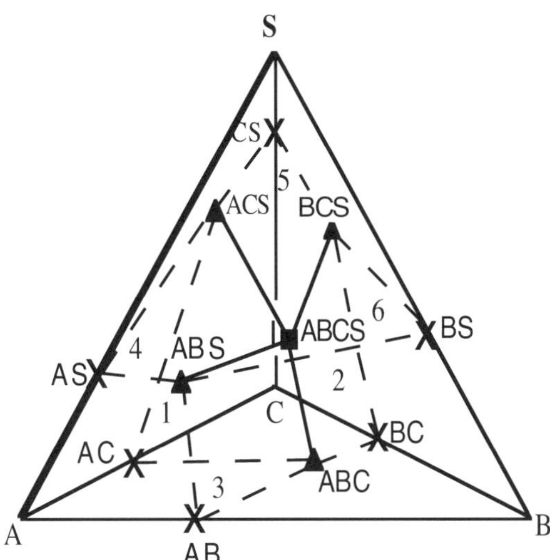

Figure 1. Tetrahedral phase diagram with simple eutectics.

Figure 2.a shows the equipment configuration for the separation of three solutes, A, B and C, which exhibit the phase behavior depicted in Figure 2.b. The stream numbers on the flowsheet correspond to the composition numbers on the phase diagram. There are 3 double-saturation surfaces in Figure 2.b: (1) AC/ACS/ABCS/ABC, (2) BC/BCS/ABCS/ABC, and (3) AB/ABS/ABCS/ABC. Part of surface 1 is on top of part of surface 2.

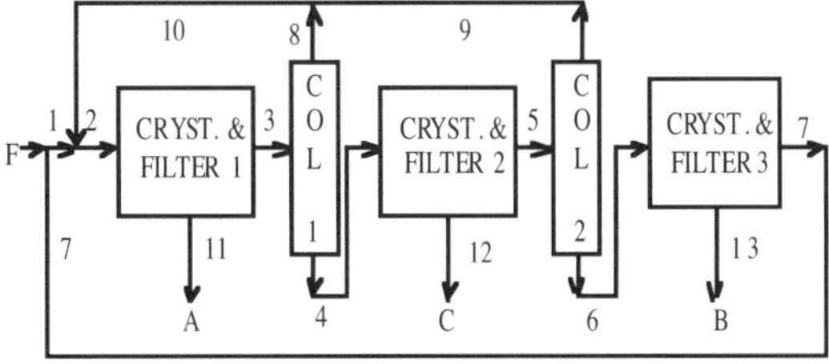

Figures 2.a. Extractive crystallization of three solutes.

The process works as follows. The feed, F, is mixed with recycle stream 7 to form stream 1. Recycled solvent is added to stream 1 to form stream 2. The composition of stream 2 is in compartment A. The liquid mixture is cooled to produce crystals of pure A. The mother liquor, or stream 3, is sent to flash unit 1 where a fraction of the solvent is removed. As a result, point 4 is below surface 1 and is in compartment C. Cooling stream 4 leads to the recovery of component C. All the solvent in the effluent mother liquor from crystallizer 2 is removed in flash unit 2. Depleted of solvent, the composition of stream 6 is on the base of the tetrahedron and is in compartment B. Pure B is recovered by cooling stream 6. The remaining mother liquor, stream 7, consisting of components A, B and C, is recycled.

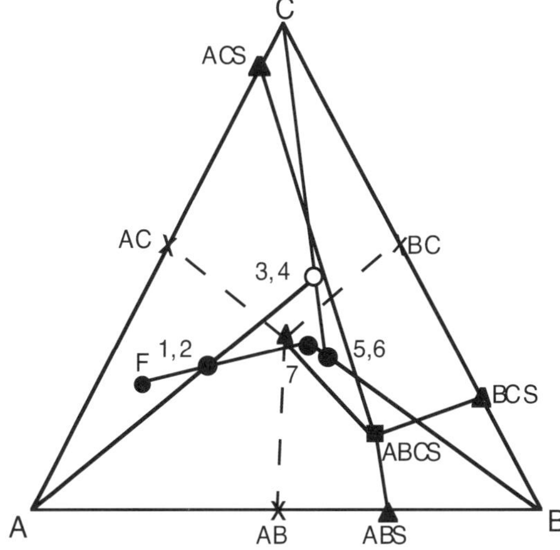

Figures 2.b. Phase diagram and process stream compositions for extractive crystallization.

Selective Crystallization and Dissolution

Mixtures of solids may be separated by use of selective crystallization and dissolution (Ng, 1991). There are two basic schemes: the conventional and the alternate processes. Solids with reverse joint solubilities (Figure 3a) can be separated using the conventional process, which consists of a dissolver-filter-crystallizer-filter train (Figure 3.b).

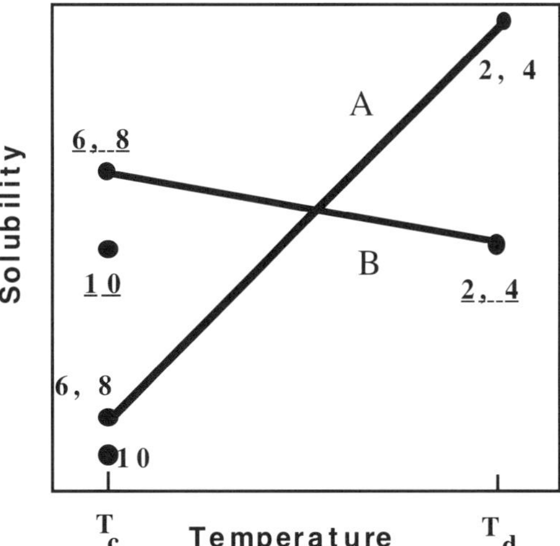

Figure 3.a. Joint solubilities as a function of temperature.

This process is governed by the following equations:

$$(W_r+W_m)S_A(T_d) = W_rS_A(T_c)+M_A \tag{1}$$

$$(W_r+W_m)S_B(T_d) = W_rS_B(T_c) \tag{2}$$

Equation (1) indicates that all the A in the feed is dissolved in the dissolver. Equation (2) implies that none of the B in the feed dissolves in the dissolver.

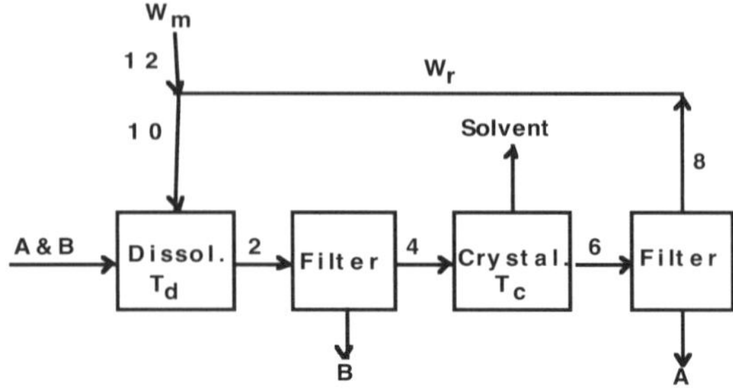

Figure 3.b. Process flowsheet for the separation of components with reverse joint solubilities.

Figure 4 shows the alternate flowsheet, required when the joint solubilities increase with temperature.

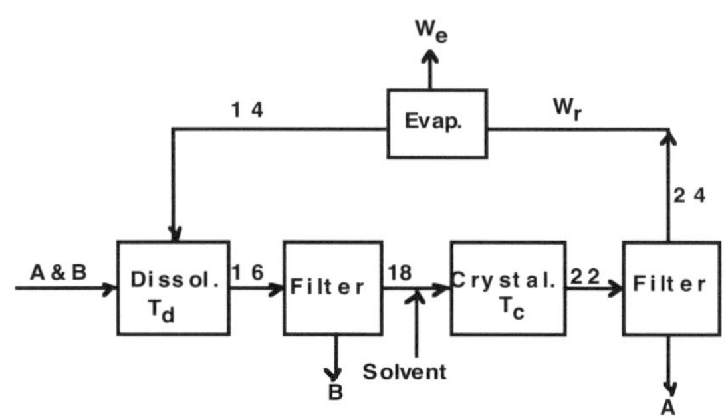

Figure 4. Process flowsheet for the separation of components whose joint solubilities increase with increasing temperature.

A sequence of (n-l) loops can be used to separate an n-component mixture. For example, Figure 5 shows the cascade required to separate A, B and C, where $S_A(T_d)/S_A(T_c) > S_B(T_d)/S_B(T_c) > S_C(T_d)/S_C(T_c)$. The alternate process is used for the first loop to recover A and the conventional process is used to separate B and C in the second loop.

Crystallization Involving Reciprocal Ionic Systems

In a hypothetical system composed of ions A+, B+, X^- and Y^- and solvent, the following reaction occurs:

$$AX + BY \leftrightarrow AY + BX \tag{3}$$

Figure 6.a shows a Jänecke projection of this system's phase diagram at three temperatures: Tc, Ti, and Th. The surface formed by the Tc isotherm contains more solvent than all points forming the Th isothermal surface. The Ti isotherm contains an intermediate amount of solvent relative to the T_c and T_h isotherms. Each of the 4 saturation surfaces defined by the projection is in actuality the top surface of a saturation compartment in a tetrahedral phase diagram. A pure salt may be crystallized by removing solvent so long as the solution remains completely within one compartment. The projection also shows the stream compositions, the feed composition, and the process path for separating pure salts BX and AY from an equimolar feed of BY and AX.

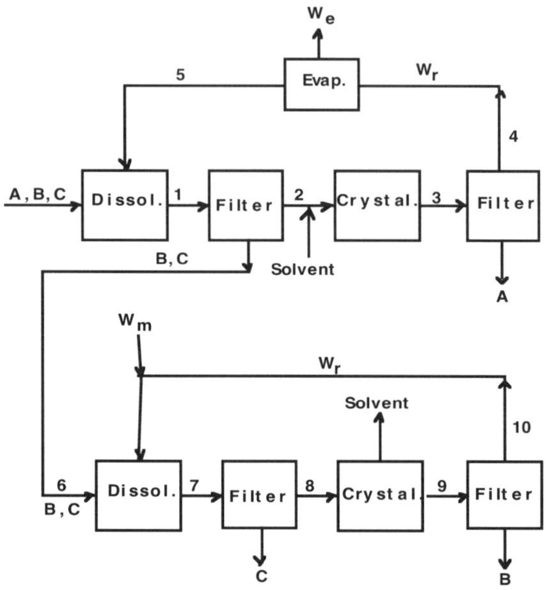

Figure 5. The use of two loops to separate a three-component mixture.

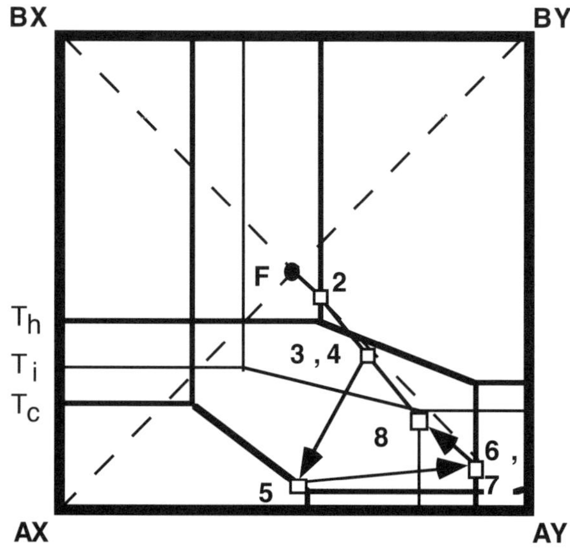

Figure 6.a. Projection process stream compositions for the recovery of AY and BX.

Figure 6.b shows the accompanying equipment configuration for the path drawn on the projection. Stream 1 is formed by combining a liquid feed, F and recycle stream 13 in a dissolution tank D_s1. The composition of stream 1 lies above the BX saturation surface at T_h. By removing solvent at the crystallizer C1, pure BX crystallizes. Stream 3 is formed by combining stream 2 and stream 8. Cocrystallization in the next crystallizer, C2, is avoided by diluting stream 3 at D_L2 which is held at some temperature T_{d2} such that stream 4 remains unsaturated. When the temperature is changed to T_c at C2, the composition of stream 4 lies in the BY saturation compartment and pure BY crystallizes. When the temperature is changed to T_h, stream 5 lies within the AX compartment. Thus, AX can be recovered by solvent removal. To avoid cocrystallization in C4, stream 6 is diluted at D_L3 so that stream 7 is unsaturated. Cooling stream 7 crystallizes pure AY. The effluent is recycled to stream 2.

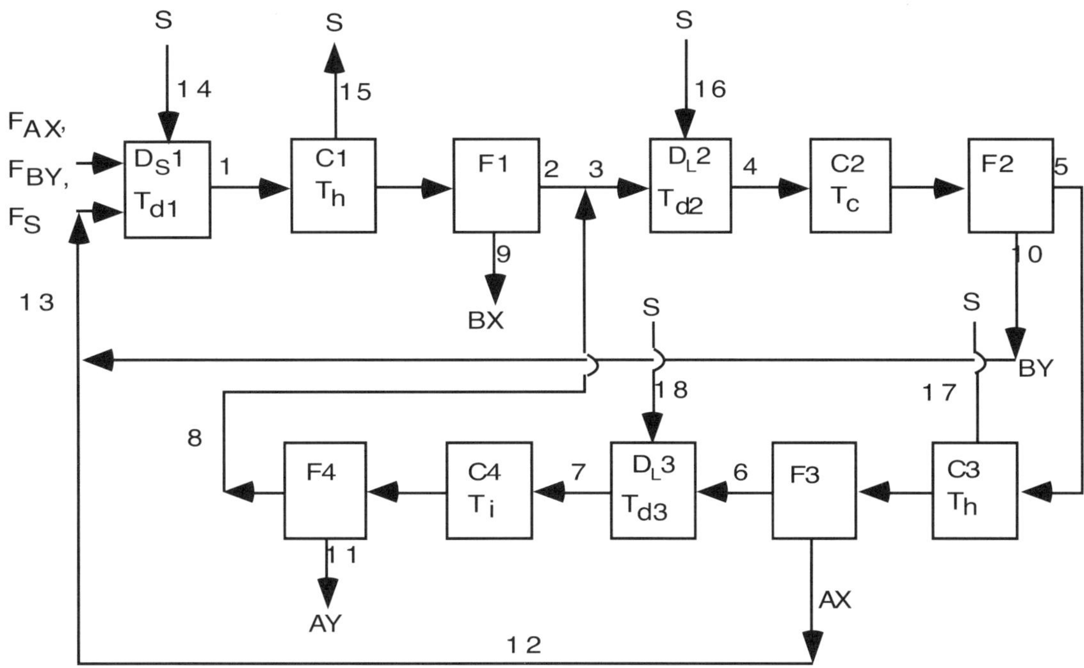

Figure 6.b. Process flowsheet for the recovery of AY and BX from an equimolar feed of AX and BY.

Conclusions

We have presented a brief description of three crystallization-based separation techniques. Given the phase diagram, it is possible to develop the appropriate flowsheets using systematic design methods. Process synthesis issues such as solvent selection, choice of operating temperatures, sensitivity of the process to design variables and process operability have been considered elsewhere.

Acknowledgment

We express our appreciation to the National Science Foundation, Grant No. CTS-9211673, for support of this research.

Nomenclature

M = mass flow rate, kg/s

S = solubility, kg solute/kg solvent

T_c, T_d = crystallizer and dissolver temperature, espectively, °C

W_m, W_r = makeup and recycled solvent flow rates, respectively, kg/s

References

Barnicki, S.D., and J.R. Fair (1990). Separation System Synthesis: A Knowledge-Based Approach. 1. Liquid Mixture Separations. *Ind. Eng. Chem. Res.*, **29**, 421-432.

Ng., K.M. (1991). Systematic separation of a multicomponent mixture of solids based on selective crystallization and dissolution. *Separations Tech.*, **1**, 108-120.

Rajagopal, S., K.M. Ng, and J.M. Douglas (1991). Design and Economic Trade-Offs of Extractive Crystallization Processes. *AIChE J.*, **37**, 437-447.

SYNTHESIS AND OPTIMIZATION OF TWO-PHASE REACTION PROCESSES

A. C. Kokossis
Department of Process Integration, UMIST
Manchester M60 1QD, UK

C. A. Floudas
Department of Chemical Engineering, Princeton University
Princeton, NJ 08544-5263

Abstract

This paper studies the synthesis problem of a chemical process that involves a complex reaction mechanism and two liquid phases. First the reaction mechanism, the process and the assumptions made are discussed. Based upon the specifics of the problem, a representation is generated which can accommodate discontinuous aspects related to the operational and/or the structural alternatives of the chemical process. The discontinuous aspects are described by discrete variables and the synthesis problem is formulated as a Mixed-Integer Nonlinear Programming (MINLP) problem. The potential of the approach is illustrated with an example problem.

Keywords

Process synthesis, Two-phase reaction processes, Discrete-continuous optimisation.

Introduction

The course of a chemical reaction in a multi-phase system is generally described by a combination of chemical kinetics and mass phenomena. In terms of the solid-fluid, gas-liquid or liquid-liquid reactors involved, one must in principle rely on pilot plant and plant experience with various reactor arrangements. Generalisations are not as reliable as in single-phase systems and specific reactor styles must be considered separately, rather than as members of a larger class. Furthermore, since the design parameters of a multiphase chemical reaction vary to a significant degree, the proposition of superstructure schemes, like the ones presented by Kokossis and Floudas (1990a, 1990b), should concentrate on the unique elements of each particular process. Hence, the proposition of the synthesis structure should be based upon the alternatives defined by the particular process, reflect the peculiarities of the chemical system and accommodate for the specific objectives of the problem. The appropriate methodology suggests a synthesis approach based upon a case dependant rather than a general purpose structure. The application of such an approach is further discussed for a two-phase liquid-liquid chemical process.

In a number of liquid-liquid processes, such as nitration or sulfonation, it is advantageous to distribute the reactants between the two phases. In others, the liquid reactants are only partially miscible in one another so that the reaction process must, of necessity, be a two phase reaction. The reaction may occur in one phase or in both phases simultaneously, but generally the solubility relationships are such that the extent of chemical reaction in one of the phases is so small that it can be neglected in the analysis of this kind of reactor. Two phase batch reactions are generally carried on in reaction kettles or autoclaves while two phase continuous reactions are carried out in mixer-settler extractors. In this work, a two-phase complex reaction process is studied. The reaction mechanism is described first. It involves six reaction species and five reversible and irreversible reactions. The synthesis objectives along with the design options determine the framework on which the synthesis structure is generated.

The mathematical formulation of this structure gives rise to a mixed integer nonlinear programming (MINLP) problem whose solution reveals the optimal structure and operation of the process.

The Chemical Process

The reaction species are processed in a two liquid phase system consisted of an organic and an aqueous phase. The reactions occur in the organic phase and are catalysed by completely dissociated species. The organic phase consists of inert propanol or isopropanol and an aqueous phase in enticed by the successive accumulation of one of the components. No reactions take place in this second phase which affects the process only by selectively extracting products and reactants from the organic phase.

In its original from the reaction mechanism involves ten reacting species denoted by HA, HF, HG, A^-, B, C^-, D, E^-, F^- and G^-. Two feed streams are furnished to the system: one is composed by inert solvent and components HA and B; the second is made up entirely by E^-. In terms of the developing reactions, HA is in rapid equilibrium with A^-: $F^- + HA \rightarrow HF + A^-$, while A^- reacts irreversibly with B to form C^-: $A^- + B \rightarrow C^-$. The produced C^- is reversibly decomposed to E^- and D (equilibrium to the right): $C^- \leftrightarrow D + E^-$. Component D represents the desire product and is further consumed by an irreversible and very slow reaction with A^-: $D + A^- \rightarrow G^-$.

Feed components B and E^- reversibly react to form F- (rates comparable but slow): B + E- $\leftrightarrow$ F-, while G^- is in rapid equilibrium with A^-: $G^- + HA \leftrightarrow HG + A^-$. As the reaction proceeds, component E- is produced and accumulated in the organic phase. At a critical concentration, E^- entices the generation of a second phase in which E- has a strong preference in contrast with B which retains a preference in the organic phase. Their phase partitioning coefficients are independent of the concentration and they are only functions of the temperature.

The rest of the reactants have a strong preference in the organic phase and their presence in the extractive phase can be neglected. The process is virtually isothermal and in commercial scale it becomes adiabatic. The rapid equilibrium reactions are associated with constants in the order of 10 with the forward reactions being 10-100 times faster than the backward ones. By lumping components HF and F- in the velocity constants and assuming that species HA and HG are 100% dissociated into A- and G- respectively, the reaction mechanism can be simplified as follows:

$$A^- + B \rightarrow C^- \tag{1}$$

$$C^- \leftrightarrow D + E^- \tag{2}$$

$$D + A^- \rightarrow G\text{-} \tag{3}$$

$$D + A^- \rightarrow G\text{-} \tag{4}$$

$$B + E^- \leftrightarrow A\text{-} \tag{5}$$

The above system consists of five reactions and six species. All reactions are of first order and, based upon the qualitative arguments made for the original mechanism, the kinetic constants of the assumed reaction system are given by Table 1.

Table 1. Kinetic Constants for the Reaction Process.

Reaction	Forward reaction (gmol/min)	Backward reaction (gmol/min)
1	0.1	—
2	1.0	0.5
3	0.2	—
4	0.2	0.05
5	2.0	2.0

Problem Description

A number of different objectives is associated with the system described in the above section. The synthesis problem is focused on developing the appropriate structure and operation that processes the reaction mechanism in question and accounts for the economically attractive alternatives. Given is a cascade of CSTR's, a number of potential settlers and the specifications of the feed streams. Also given are economic data pertaining to the process units, the raw materials and products, as well as the design parameters of the overall venture. The objectives of the problem consist in three primal synthesis targets which are described by:

a. the maximisation in the production of the desired component D;
b. the minimisation of the total amount of B consumed in the reaction process and/or absorbed by the extractive phase; and
c. the minimisation of the annual venture cost of the plant.

The need for an appropriate weighting of these targets suggests the venture profit as the objective function of the problem. At the synthesis level, the existing alternatives are described by: (i) the distribution of the feed along the cascade; (ii) the sizes of the reactors; and (iii) the allocation of the settlers in the cascade. The deliberate exploitation of the above choices implies a complete and exhaustive search among the different synthesis scenarios and assumes the proposition of a synthesis structure which should not only embed the existing alternatives but also accommodate an effective formulation of the objectives.

The main assumptions made in this work are as follows: (a) the process is isothermal and no reaction takes place in the extractive phase; (b) in both the reactive and the extractive phase the mixing is complete and the concentration of the reactants are uniform; (c) between the two phases, a linear equilibrium always holds and the partition equilibrium is established; and (d) the operating cost of the reactors and settlers is negligible while the capital cost of each settler consists of a fixed charge cost.

The Synthesis Structure

The chemical system consists of a cascade of continuous stirred tank reactors and a number of settlers. The reactors process single or two-phase streams while the settlers separate two phase streams into the extractive phase, which is discarded from the process, and the organic phase which is forwarded to a subsequent stage. Since the settlers may or may not appear in the optimal configuration, they define a set of structural parameters for the problem. Furthermore, once their distribution along the cascade is considered among the synthesis objectives, potential positions of settlers are to be assumed at all reactor outlets. Associated with the optimal distribution of the feed, a set of feed stream splitters are considered at each reaction stage.

Since, at the synthesis level, it is not known whether a particular CSTR in the cascade processes a single or a two-phase stream, a synthesis structure based upon the above set of structural parameters is incomplete and embeds a number of unrealistic scenarios (i.e. settlers processing single phase streams). Should the optimal allocation of settlers be considered separately from the other objectives of the problem, the alternatives discussed in the previous section will not be properly explored. In order to handle this difficulty, one has to differentiate between single and two phase CSTR's as two different types of reactors. Only one of these reactors is to appear at each stage and the structural units of the synthesis problem are, therefore, represented by the: (i) fresh feed splitters FS_k; (ii) single phase CSTR's (SPC's); (iii) splitters SP_k which process single phase streams; (iv) two-phase CSTR's (TPC's); (v) splitters TP_k which process two-phase streams; and (vi) settlers (STL's) following each two-phase CSTR.

At each stage, a set of interconnections among the structural units is established by applying the following rules:

1. The inlet streams of each two-phase reactor consists of substreams originating from all three splitters FS_k, SP_k and TP_k. The outlet from the reactor is feeding the splitters TP_{k+1} and the settler.
2. The inlet streams of each single-phase consists of substreams originating from splitters SP_k and FS_k. The entire reactor outlet is directed to the splitters SP_{k+1} of the next stage.
3. The two phase inlet stream of a settler is a substream of the two-phase outlet. The organic phase is feeding the SP_{k+1} splitters while the rest is extracted and removed from the process.

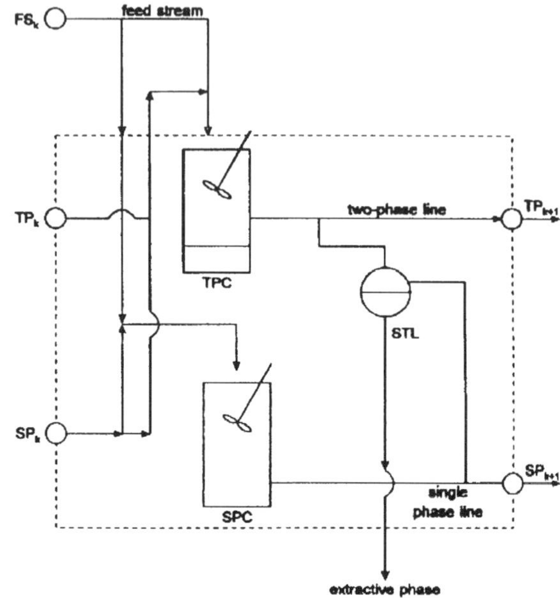

Figure 1. The kth stage of the process.

The resulting stream network is illustrated in Fig. 1. The dotted box in the figure represents a single stage of the overall process. The adopted synthesis structure, which is shown in Fig. 2, is made up of serial combination of these stages and contains all possible scenarios for processing the chemical system in question.

Illustration Example

The synthesis problem described in the previous section has been formulated and solved as a mixed integer non-linear programming (MINLP) problem. The separability and linearity in terms of the integer variables, suggests a projection of the problem upon integer variables and a decomposition of the original formulation into a nonlinear programming problem (NLP) and a mixed integer linear programming problem (MILP). The solution algorithm then consists in the successive solution of these problems and iteration based upon the Generalised Benders Decomposition algorithm. The implementation of the GBD algorithm in the particular problem is proved successful and provided consistent results.

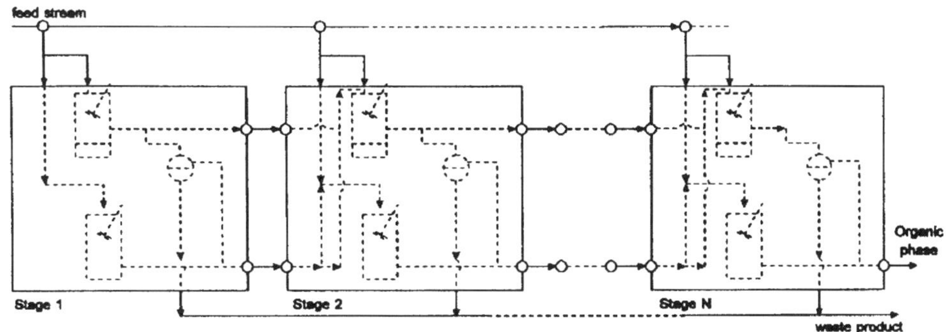

Figure 2. The synthesis structure.

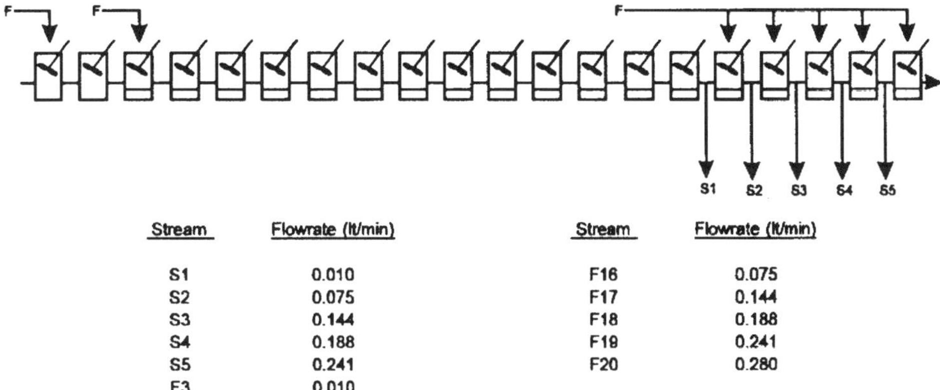

Stream	Flowrate (lt/min)	Stream	Flowrate (lt/min)
S1	0.010	F16	0.075
S2	0.075	F17	0.144
S3	0.144	F18	0.188
S4	0.188	F19	0.241
S5	0.241	F20	0.280
F3	0.010		

Figure 3. Optimal solution for Case II.

The parameters of the optimisation problem include the specifications of the feed streams, the equilibrium constants, and the economic parameters of the process. The undistributed organic feed is furnished at a rate of 100 lt/min and contains 1.0 gmol/lt of A^- and 5.0 gmol/lt of B.

The distributed feed contains 1.0 gmol/lt of E^- and the extractive phase appears beyond a critical concentration of 0.5 gmol/lt of E^-. The phase partioning coefficients are set to 10 and 0.2 for E^- and B respectively and they have been considered negligible for the rest of the components. Product D can be sold at \$2.25/gmol while feed components A^- and B can be purchased at \$0.01 gmol/lt. Reactant E^-, is much cheaper and can be purchased at \$0.01 gmol/lt. The purchase and instalment cost of each settler is \$360 and the operating time of the process is 720 hrs/yr.

The reactor system in this case consists of 20 CSTR's and 5-10 settlers. The larger optimisation problem has 564 equations, 481 continuous variables and 60 integer variables. The optimal configuration for this case is shown in Fig. 3. All but the 1st and 2nd CSTR's are two-phase reactors and the catalyst stream is feeding the 3rd, 16th, 17th, 18th, 19th and 20th CSTR. The five settlers found in the optimal solution are located in front of the last five CSTR's. The solution algorithm consumed 4.07 CPU sec/iteration and the maximum profit is found to be equal to \$982,454/yr.

Conclusions

The synthesis problem of a two phase reaction system has been studied in this work. The presented approach considers two types of design parameters: (i) stochastic parameters which describe the structural alternatives of the process, and (ii) operational parameters which account for the sizes analysis results to the optimal flowsheet of the process of the units and the levels of the various inputs and outputs of the system. As a result, the optimisation problem based upon the proposed synthesis structure involves both integer and continuous variables and gives rise to a Mixed-Integer Nonlinear Programming (MINLP) problem. The various steps and the potential of the approach are illustrated with a complex reaction mechanism for which analysis results to the optimal flowsheet of the process.

References

Kokossis, A.C. and Floudas, C.A. (1990a). Optimization of Complex Reactor Networks — I: Isothermal Operation. *Chem. Engr. Science*, **45**, 595-614.

Kokossis, A.C. and Floudas, C.A. (1990b). Synthesis of Isothermal Reactor Separator Recycle Systems. *Chem. Engr. Science*, **46**, 1361-1383.

DECISION-MAPPING FOR DESIGN AND SYNTHESIS OF CHEMICAL PROCESSES: APPLICATION TO REACTOR-NETWORK SYNTHESIS

F. Friedler[1,2], J. B. Varga[2], and L. T. Fan[1]
[1]Department of Chemical Engineering
Kansas State University
Manhattan, KS 66506

[2]Research Institute of Chemical Engineering, Hungarian Academy of Sciences
Veszprém, Pf. 125, H-8201, Hungary

Abstract

The extreme complexity of decisions involved in process synthesis can be attributed to the fact that such decisions are concerned with specification or identification of highly interconnected systems, i.e., process structures, often containing a multitude of recycling loops. Apparently, a rigorous technique capable of exactly representing and efficiently organizing the system of decisions is hitherto unavailable for process synthesis. In order to fill this void, a new mathematical notion, decision-mapping, has been introduced in the present work. The technique derived from this notion enables us to make consistent and complete decisions in process design and synthesis. The terminologies necessary for decision-mappings have been defined based on rigorous set theoretic formalism, and the important properties of decision-mappings have been described. The efficacy of the technique is demonstrated by synthesizing a reactor-network.

Keywords

Process synthesis, Process structure representation, Feasible process structures, Systems of consistent decisions, Branch and bound.

Introduction

Some major issues remain unresolved for developing rigorous and efficient mathematical programming methods of process synthesis. These include the establishment of the mathematical foundation for validating the algorithms for optimal process synthesis and the reduction in the complexity of these algorithms.

Two approaches, logical formulation (Raman and Grossmann, 1993) and combinatorics, are available for analyzing and discerning the unique combinatorial properties of process structures. The present work is based on combinatorics. The fundamental combinatorial properties of feasible process structures have been expressed as a set of axioms (Friedler et al., 1992). For the MINLP model of process synthesis, these axioms constrain the set of possible values of the integer variables to the set of combinatorially feasible values, thus reducing the size of the search space by many orders of magnitude. These axioms constitute a rigorous foundation for the combinatorial segment of process synthesis; nevertheless, they do not directly lead to computational algorithms. This is attributable to the fact that the axioms express self-evident facts instead of procedures; hence, they are not in procedural form. The present work introduces a new combinatorial technique, i.e., decision-mapping, that is capable of representing process networks or structures of any type, size, and complexity. Decision-mapping has been developed with rigorous mathematical formalism and validated by resorting to the combinatorial axioms of process synthesis. Consequently, the algorithms developed on the basis of the decision-mapping can also be validated rigorously.

The usefulness of decision-mappings is illustrated by solving a reactor-network synthesis problem. Nevertheless, the decision-mapping is equally applicable to other classes of process synthesis.

Structural Representation: P-graph

An unambiguous structural representation is mandatory for formal analysis of process structures in process synthesis. A directed bipartite graph termed process graph, or P-graph in short, has been proposed for this purpose (see Friedler et al., 1992), which is briefly described in the following.

Let M be a given finite nonempty set of objects, usually material species or materials, that can be transformed in the process of interest. Transformation between two subsets of M is effected in an operating unit of the process. This operating unit must be linked to other operating units through the elements of these two subsets of M.

Let O be the set of operating units to be considered in synthesis; then, $O \subseteq \wp(M) \times \wp(M)$ where $O \cap M = \varnothing$. Suppose that (α, β) is an operating unit, i.e., $(\alpha, \beta) \in O$; then, α is the input set, and β is the output set of this operating unit. Pair (M, O) is defined to be a P-graph with the set of vertices $M \cup O$ and the set of arcs $\{(x,y) : y = (\alpha, \beta) \in O \text{ and } x \in \alpha\} \cup \{(y,x) : y = (\alpha, \beta) \in O \text{ and } x \in \beta\}$.

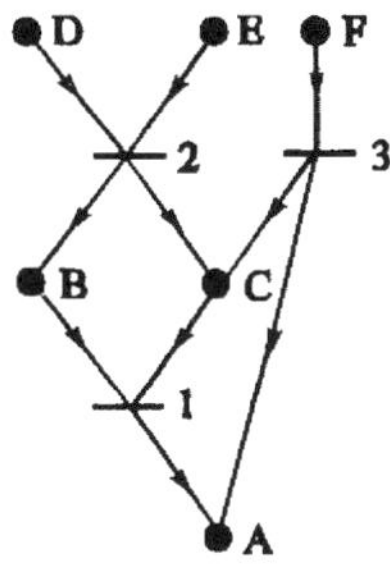

Figure 1. P-graph (M_1,O_1) where A, B, C, D, E, and F are the materials, and 1, 2, and 3 are the operating units.

Example 1. Set M_1 of materials and set O_1 of operating units of P-graph (M_1, O_1), illustrated in Fig. 1, are expressed as $M_1 = \{A, B, C, D, E, F\}$ and $O_1 = \{(\{B, C\}, \{A\}), (\{D, E\}, \{B, C\}), (\{F\}, \{A, C\})\}$. The input and output sets of operating unit 3, i.e., $(\{F\}, \{A, C\})$, are $\{F\}$ and $\{A, C\}$, respectively.

Decision-Mapping: Definitions

Formally, mapping or function f is a subset of the Cartesian product of domain $\mathcal{D}$ and range $\mathcal{R}$. In other words, f is a set of pairs (x, y) where $x \in \mathcal{D}$ and $y = f(x) \in \mathcal{R}$; this set of pairs is denoted by $f[\mathcal{D}]$.

Let the interconnections of the operating units of a synthesis problem be represented by P-graph (M, O). Then, $O \subseteq \wp(M) \times \wp(M)$ holds for the set of materials M and that of operating units O. Let us now induce mapping Δ from M to the set of subsets of O; in other words, $\Delta[M] \subseteq M \times \wp(O)$. This mapping determines the set of operating units producing material X for any $X \in M$; hence, $\Delta(X) = \{(\alpha, \beta) : (\alpha, \beta) \in O \text{ and } X \in \beta\}$.

Definition 1. Let m be a subset of M and X be an element of m, and also let $\delta(X)$ be a subset of $\Delta(X)$ for each $X \in m$. Then, mapping δ from set m to the set of subsets of set O, $\delta[m] = \{(X, \delta(X)) : X \in m\}$, is a *decision-mapping* on m.

Example 2. Let us reconsider Example 1. According to the definition of $\Delta_1(X)$, sets $\Delta_1(A)$ through $\Delta_1(F)$ can be given for P-graph (M_1, O_1) in Fig. 1 as follows: $\Delta_1(A) = \{(\{B, C\}, \{A\}), (\{F\}, \{A, C\})\}$, $\Delta_1(B) = \{(\{D, E\}, \{B, C\})\}$, $\Delta_1(C) = \{(\{D, E\}, \{B, C\}), (\{F\}, \{A, C\})\}$, and $\Delta_1(D) = \Delta_1(E) = \Delta_1(F) = \Delta$. Suppose that $m_1 = \{A, B\}$ is the domain of decision-mapping δ_1; moreover, $\delta_1(A) = \{(\{B, C\}, \{A\})\}$ and $\delta_1(B) = \{(\{D, E\}, \{B, C\})\}$. Hence, $\delta_1[m_1] = \{(A, \delta_1(A)), (B, \delta_1(B))\} = \{(A, \{(\{B, C\}, \{A\})\}), (B, \{(\{D, E\}, \{B, C\})\})\}$ is a decision-mapping of the example.

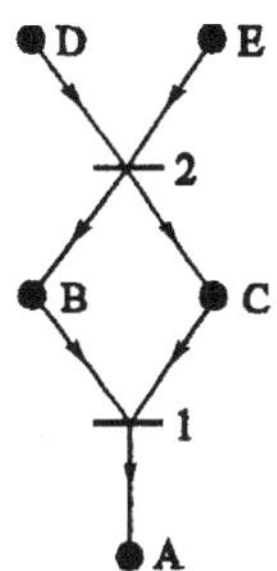

Figure 2. P-graph of decision-mapping d1[m1].

Instruments besides those currently available are needed to precisely define the P-graph of a decision-mapping. Nevertheless, the P-graph of decision-mapping $\delta_1[m_1]$ of Example 2 is given in Fig. 2 for illustration.

A special class of decision-mappings need be introduced for discerning the major properties of process structures.

Definition 2. The *complement* of decision-mapping $\delta[m]$ is defined by $\bar{\delta}[m] = \{(X, Y) : X \in m \text{ and } Y = \Delta(X) \backslash \delta(X)\}$; therefore, $\bar{\delta}(X) = \Delta(X) \backslash \delta(X)$ for $X \in m$. Since $\delta(X)$ is a set of operating units producing material X, $\bar{\delta}(X)$ is the set of operating units producing X, but are excluded from $\delta(X)$.

The consistency of decisions is essential in synthesizing a process. For instance, we may decide independently that an operating unit producing materials X and Y be included in a process for producing material X and be excluded from producing material Y; this leads to a contradiction in the system of decisions. Decision-mappings must be consistent as to circumvent such a contradiction.

Definition 3. Decision-mapping δ[m] for which m≠∅ is consistent if and only if $\delta(X) \cap \bar{\delta}(Y) = \varnothing$ for all X, Y ∈ m.

Example 3. In Example 2, decision-mapping $\delta_1[m_1]$ is consistent; however, decision-mapping $\delta_2[m_2] = \delta_1[m_1] \cup \{(C, \{(\{D, E\}, \{B, C\}), (\{F\}, \{A, C\})\})\}$ is inconsistent since $\delta_2(C) \cap \bar{\delta}_2(A) = \{(\{F\}, \{A, C\})\} \neq \varnothing$. In $\delta_2[m_2]$, operating unit ({F}, {A, C}) is simultaneously included and excluded from consideration, thus resulting in a contradiction in the system of decisions.

To determine which decision-mappings be regarded equivalent is highly important. For example, decision-mapping $\delta_3[m_1 \cup \{C\}] = \delta_1[m_1] \cup \{(C, \{(\{D, E\}, \{B, C\})\})\}$ is different from $\delta_1[m_1]$; however, both yield the same structure. This implies that these two decision-mappings have some equivalence as can be observed in Fig. 2. Such equivalence will be established on the closure of the decision-mapping. For this purpose, let the set of operating units of decision-mapping δ[m] be denoted by op(δ[m]). Moreover, let the set of input and output materials of set o of operating units be denoted by $mat^{in}(o)$ and $mat^{out}(o)$, respectively; thus,

$$op(\delta[m]) = \bigcup_{X \in m} \delta(X), \quad mat^{in}(o) = \bigcup_{(\alpha \cup \beta) \in o} \alpha,$$

$$\text{and } mat^{out}(o) = \bigcup_{(\alpha \cup \beta) \in o} \beta.$$

The union of $mat^{in}(o)$ and $mat^{out}(o)$ will be denoted by mat(o).

Definition 4. For consistent decision-mapping δ[m], let **o** = op(δ[m]), **m** = mat(**o**)∪m, and δ'[**m**] = {(X, Y) : X ∈ **m** and Y = {(α, β) : (α, β) ∈ **o** and X ∈ β}}. Then, δ'[**m**] is defined to be the *closure* of δ[m]. If δ[m] = δ'[**m**], then δ[m] is said to be *closed*. Obviously, the closure of a consistent decision-mapping is closed.

It is necessary to examine if the closure preserves the consistency.

Theorem 1. The closure of a consistent decision-mapping is consistent.

The equivalence of decision-mappings can be established on their closure as stated in the definition given below.

Definition 5. Two consistent decision-mappings are equivalent if their closure is common.

Obviously, a consistent decision-mapping is equivalent to its closure. Moreover, the relation "equivalent" has the reflexive, symmetric, and transitive properties of a mathematically defined equivalence relation.

Example 4. Let us revisit Example 2; the closure of decision-mapping $\delta_1[\mathbf{m}_1]$, $\delta_1'[\mathbf{m}_1]$, follows Definition 4. Domain $\mathbf{m}_1$ is the union of set m_1 and the set of materials involved in the operating units, i.e., $\mathbf{m}_1$ = {A, B, C, D, E}. Thus, $\delta_1'[\mathbf{m}_1]$ = {(A, {({B, C}, {A})}), (B, {({D, E}, {B, C})}), (C, {({D, E}, {B, C})}), (D, ∅), (E, ∅)}. Decision-mapping δ_4[{A, C}]={(A, {({B, C}, {A})}), (C, {({D, E}, {B, C})})} has the same closure as $\delta_1[m_1]$ in Example 2; therefore, they are equivalent; see Fig. 1.

Representation of a P-Graph by a Decision-Mapping

The relationship between the P-graphs and decision-mappings can now be examined. Let P-graph (m, o) be a subgraph of P-graph (M, O); then, m ⊆ M, o ⊆ O, O ⊆ ℘(M) × ℘(M), and o ⊆ ℘(m) × ℘(m).

Definition 6. Suppose that m' ⊆ m and β ∩ m' ≠ ∅ for any (α, β) ∈ o; then, m' is an *active set* of P-graph (m, o).

Definition 7. Let m' be an active set of P-graph (m, o); then, δ[m'] is a *decision-mapping* of *P-graph* (m,o), if δ[m'] = {(X, Y) : X ∈ m' and Y = {(α, β) : (α, β) ∈ o and X ∈ β}}. Obviously, o = op(δ[m]).

Example 5. In Example 2, both {A, B, D} and {A, C} are active sets of the P-graph given in Fig. 2. For these active sets, δ'[{A, B, D}] = {(A, {({B, C}, {A})}), (B, {({D, E}, {B, C})}), (D, ∅)} and δ'[{A, C}] = {(A, {({B, C}, {A})}), (C, {({D, E}, {B, C})})} are two decision-mappings of this P-graph. They have an identical closure; thus, these decision-mappings are equivalent. Nevertheless, it is the case in general, i.e., the different decision-mappings of the same P-graph are always equivalent.

Theorem 2. Let δ[m'] be a consistent decision-mapping; o = op(δ[m']); and m = mat(o). Then, (i) (m, o) is a P-graph; (ii) m' is an active set of P-graph (m, o); and (iii) δ[m'] is a decision-mapping of P-graph (m, o).

The above theorem leads to the following definition:

Definition 8. The P-graph of consistent decision-mapping δ[m'] is (m, o) where o = op(δ[m']) and m = mat(o).

Application of Decision-Mappings for the Accelerated Branch and Bound Algorithm of Process Synthesis

One of the most important applications of the decision-mapping is the acceleration of the branch and bound algorithm in solving the MILP or MINLP model of process synthesis. Suppose that set P of the products, set R of the raw materials, and set O of the operating units are known for set M of the materials given. A synthesis problem is defined by triplet (P, R, O), if none of sets P, R, and O is empty; moreover, P ⊆ M, R ⊆ M, P ∩ R = ∅, M ∩ O = ∅, and O ⊆ ℘(M) × ℘(M). The process structures for synthesis problem (P, R, O) are the subgraphs of P-graph (M, O); nevertheless, the P-graph of a feasible process always needs to be in conformity with certain combinatorial properties expressed as a set of axioms (Friedler et al., 1992). P-graph (m, o) is a *combinatorially feasible-structure* of synthesis problem (P, R, O) if it satisfies the following axioms: (S1) P ⊆ m; (S2) ∀X ∈ m, $op^{in}(\{X\}) = \varnothing$ iff X ∈ R; (S3) o ⊆ O; (S4) $\forall y_0 \in o$, ∃path $[y_0, y_n]$, where $y_n \in P$; and (S5) $\forall X \in m, \exists(\alpha, \beta) \in o$ such that $X \in (\alpha \cup \beta)$.

Global variables: R, Δ(x)(x∈M), U, *best*;
Comment: procedure BOUND determines a lower bound for a subproblem;

```
begin
U:=∞; best:= anything; if P=∅ then stop;
m:=∅; p:=P; ABBS(p, ∅,∅);
if U<∞  then print best
        else print'there is no solution'
end;

procedure ABBS(p, m, δ[m]):
begin
x∈p; C:=℘(Δ(x))\{∅};
for all c∈C do if ∀y∈m,c∩d(y)=∅
                    & (Δ(x)\c)∩δ(y)=∅
        then begin m':=m∪{x};
                δ[m']:=δ[m]∪{(x,c)};
                p':=p∪(mat^in(c))\(R∪m');
                if p'=∅ then begin
                        U:=min(U, BOUND(δ[m']));
                        update best; end
                else if U≥BOUND(δ[m'])
                        then ABBS(p', m', δ[m']);
        end
return
end
```

Figure 3. *Algorithm ABBS*.

The accelerated branch and bound algorithm, algorithm ABB, reduces both the number and size of subproblems to be solved by exploiting the combinatorial properties of feasible process structures represented by the decision-mappings in algorithm ABB. This representation ensures the consistency and completeness of the systems of decisions. Algorithm ABB guarantees the optimal solution of the MINLP or MILP model of a process network synthesis. For illustration, the simplified version of algorithm ABB, i.e., algorithm ABBS, is given in Fig. 3. Additional acceleration can be accomplished by applying rule α(n), stating that the number of operating units producing a material in parallel within a process is limited to n where n is a positive integer.

Example 6. Let us consider operating units given in Example 1 for producing material A from raw materials D, E, and F. The recursive steps of algorithm ABBS are given in Table 1 in the worst case.

Table 1. *Recursive Steps of Algorithm ABBS for Example 6*.

#	p	m	δ[m]
1	{A}	∅	∅
2	{B,C}	{A}	{(A,{1})}
3	{C}	{A,B}	{(A,{1}),(B,{2})}
4	∅	{A,B,C}	{(A,{1}),(B,{2}),(C,{2})}
5	∅	{A}	{(A,{3})}
6	{B,C}	{A}	{(A,{1,3})}
7	{C}	{A,B}	{(A,{1,3}),(B,{2})}
8	∅	{A, B, C}	{(A,{1,3}),(B,{2}),(C,{2,3})}

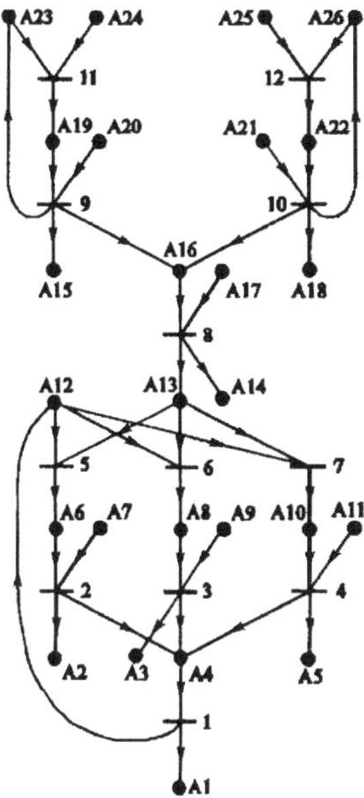

Figure 4. *P-graph of a reactor-network*.

Synthesis of a Reactor-Network

Figure 4 illustrates the mathematically validated super-structure of a commercial reactor-network synthesis problem for producing material A1 from available raw materials through several plausible reaction paths or reactions. This super-structure has been generated by algorithm MSG (Friedler et al., 1993); it contains seven reactors (#5, 6, 7, 9, 10, 11, and 12) and five separators (#1, 2, 3, 4, and 8). Suppose that the mathematical model for each operating unit is available to formulate a MINLP problem for the reactor-network synthesis. This MINLP problem of 12 binary variables can be solved by a basic branch and bound algorithm by generating 991 NLP subproblems in the worst case to obtain the optimal solution; the enumeration tree is given in Fig. 5. On the other hand, algorithm ABB solves only 29 NLP subproblems in the worst case to yield the optimal solution. Each subproblem corresponds to one substructure of the maximal structure given in the form of a decision-mapping, one of which represents the optimal structure; the enumeration tree is given in Fig. 6. Moreover, by applying rule $\alpha(1)$ in solving the problem, the number of subproblems can be reduced to as few as 10.

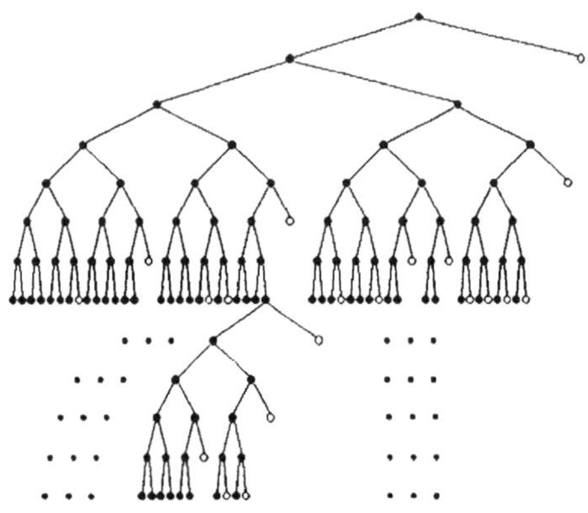

Figure 5. Enumeration tree for the basic branch and bound algorithm which generates 991 subproblems in the worst case.

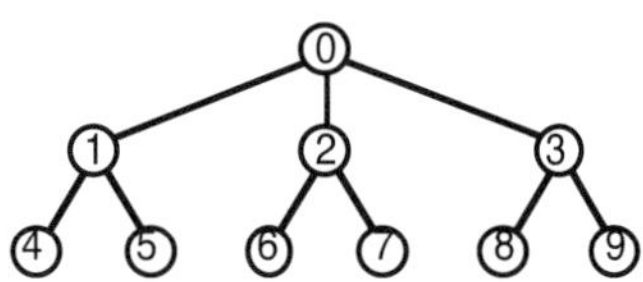

Figure 6. Enumeration tree for the accelerated branch and bound algorithm with rule a(1) which generates 10 subproblems in the worst case.

Conclusions

The complex decision systems involved in process synthesis has been rendered consistent and complete by a novel mathematical notion, decision-mapping. The major properties of the decision-mapping have been identified; moreover, the relationship between the decision-mapping and P-graph has been established. The decision-mapping has been demonstrated through application to be exacting but yet efficient.

References

Friedler, F., K. Tarjan, Y.W. Huang, and L.T. Fan (1992). Graph-theoretic approach to process synthesis: axioms and theorems. *Chem. Engng Sci.*, **47**, 1973-1988.

Friedler, F., K. Tarjan, Y.W. Huang, and L.T. Fan (1993). Graph-theoretic approach to process synthesis: polynomial algorithm for maximal structure generation. *Computers Chem. Engng*, **17**, 929-942.

Raman, R. and I.E. Grossmann (1993). Symbolic integration of logic in mixed-integer linear programming techniques for process synthesis. *Computers Chem. Engng*, **13**, 909-927.

DETAILED DESIGN AND RETROFIT OF MULTIPURPOSE BATCH PLANTS

A. P. F. D. Barbosa-Póvoa and S. Macchietto
Centre for Process Systems Engineering
Imperial College
London SW7 2BY

Abstract

Models are discussed for the optimal design and retrofit of multipurpose batch plants which take into account plant topology (connectivity, layout and service circuits) and generalise the treatment of intermediate storage and operating constraints. The simultaneous economic optimisation of design and operation choices is achieved by solving a MILP problem. An example of retrofit design for capacity and product range expansion demonstrates the realism and detail of the representations used.

Keywords

Batch processing, Plant design, Retrofit, Scheduling, Cleaning, Optimisation, MILP.

Introduction

The design and retrofit of multipurpose batch plants has received some attention lately (Papageorgakis and Reklaitis, 1990, Shah, 1992, Vodouris and Grossmann, 1993, Henning et al., 1994, Papageorgakis and Reklaitis, 1993). The formulations presented address mainly the selection of number, type and size of major processing vessels (*units*) and their operation (campaigns/schedules), but still involve significant simplifications with respect to plant topology (connectivity between units, their layout, design of utility, cleaning-in-place and other service circuits), intermediate storage, important operating constraints (e.g. task precendence, cleaning, etc.) and reuse of equipment in retrofits. All these aspects are often crucial in industrial applications and are addressed in this paper. Only a summary of the ideas and an illustrative example are given here. A full description is given in Barbosa-Póvoa (1994).

Batch Processing Models

Distinct and complementary aspects are modelled, amongst which: i) the transformation processes (product recipes and service operations), ii) the plant (equipment and structure) and iii) any operating constraints.

As in Kondili et al. (1988), transformation processes are described in terms of *State nodes* (representing *materials*) and *Task nodes* (operations transforming one or more input material states into one or more output material states). The directed graph(s) linking input, intermediate and end materials with the transformation tasks (*State-Task-Network(s)* – STNs) define the precedence structure of the *product networks*. Together with material proportions, processing times, utility requirements for each tasks, etc. the STNs can describe product recipes and service operations in an equipment-independent way.

A plant is modelled in the usual manner as a flowsheet, that is an *equipment network* of vessels, processing units, utility sources and sinks, etc. (nodes, characterised by a size) connected by directed arcs (also characterised by a size) representing pipes, air lifts, conveyors, manual or mechanical transportation, etc.

A mapping between the plant and process networks is defined by i) the suitability of each piece of equipment in the equipment network to carry out processing tasks, to store material states and (for connections) to transport material states (with associated maximum and minimum capacity for each state, task, etc.) and ii) the resources (equipment, utilities, operators, etc.) required by each task.

Operations and equipment network are automatically combined into a *maximal State-Task Network* (mSTN) characterised by four types of nodes. This representation was initially developed by Crooks (1992) for multipurpose batch plant scheduling and extended by Barbosa-Povoa (1994) to their design and retrofit. The main advantage of the mSTN is that it unambiguously and explicitly represents all and only the location of material states and the allocations of processing, storage and transfer tasks which are potentially necessary and structurally feasible.

Important constraints arise due to the multipurpose nature of the equipment (e.g. sequence-, time- and frequency-dependent tasks, cleaning requirements, etc.) and to general task preconditions (e.g. feed A before B). Extending the ideas of Crooks (1992), allowable *equipment states* (*eStates*) are explicitly defined for units and connections to represent their condition, which may be modified when an allowable processing, storage or transfer task is executed in it. By appropriate definition of the equipment states and allowable transitions, very complex sequences of operations (finite state machines named *equipment-state Task Networks*, eSTNs.) may be defined and forced on any equipment item. Initial and desired equipment states at specified or final times may also be specified (e.g. start with equipment item j dirty with polymer A, leave it empty and clean). The main advantage of the eSTN is that it unambiguously and explicitly represents all precedence constraints relating equipment states and any processing, storage and transfer tasks which may be assigned to an equipment item, independently from a specific assignment method.

A schematic representation of these modelling tools is given in Figure 1. They are used for generating an efficient MILP mathematical formulation. Additional models (not detailed here) define production requirements, operation mode (e.g. periodic), resource availability profiles, plant/process objectives and all capital and operating costs.

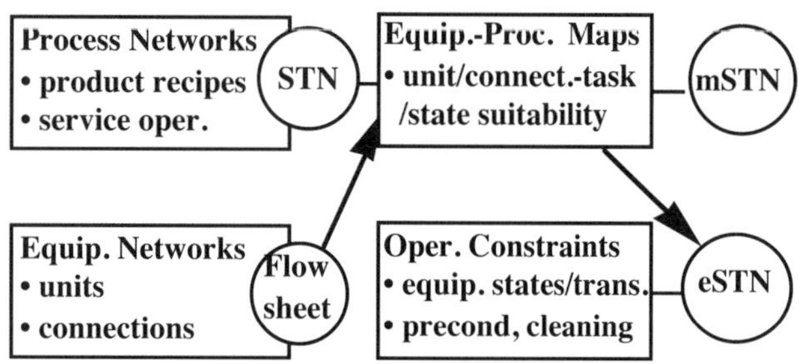

Figure 1. Batch Models structure.

Grassroots Design

The problem is posed as follows.

Given:

- STN descriptions of product and service processes, with associated recipe parameters and resource requirements (equipment, utilities, etc.).
- A plant superstructure (network of units and connections) with associated capacities and suitabilities.
- eSTN descriptions, identifying operation constraints (cleaning, equipment preconditions, etc.).
- Production requirements, a time horizon, operation mode and availability profiles of all resources.
- Capital and operating costs for equipment, materials and utilities.

Determine:

- A plant configuration (equipment network and sizes).
- An operations schedule (sizes, allocation and timing of all batches, storage and transfers)

So as to Optimise an Economic Performance Function.

Minimum capital cost and maximum annualised plant profit (accounting for operation as well as capital performance) have been used in this work.

The mathematical formulation developed (Barbosa and Macchietto, 1994) relies on a discretization of time into a number of intervals of equal duration, fixed a priori, with process events occurring at interval boundaries. Task processing times, operations horizon, etc. are integer multiples of the selected interval. Design and operation decisions are represented by continuous variables (batch sizes, equipment capacities, amounts of materials, etc.) and discrete choices by binary variables (equipment existence, task allocations to equipment and time, etc.). Equipment items (units, dedicated storage and connections) are selected optimally from the defined plant superstructure while operation is optimised so as to satisfy all constraints.

With regards to the operation mode, a single campaign structure with fixed product slate is assumed here, with choice of periodic or non periodic operation. Product requirements are defined for each product as fixed or variable within ranges, in distributed (with arbitrary time profiles) or aggregate (over a cycle, horizon, etc.) form. Demands (and supplies) may be associated to specific equipment. The formulation allows equipment in discrete or continuous size range(s), mixed storage policies, shared intermediate materials, material merging, splitting and recycling, in-phase and out-of-phase operation, etc. in any combination. Equipment costs are approximated by linear functions of equipment size and operating costs as linear functions of batch size. Nonlinear functions can be well approximated by linear ones by utilising a small number of size intervals.

All equations are expressed in linear form, resulting in a MILP problem which is solved by a branch and bound method using a standard package (SCICONIC).

Retrofit Design

The retrofit formulation is obtained as a direct extension of that for grassroots design and thus includes all its features. An existing plant network is modelled as above, with additional data for new equipment (units and connections) eligible for addition (with associated cost parameters, suitability to tasks, etc.) and existing equipment eligible for removal (with a salvage value) or reuse after adaptation (at a retrofit cost). Similarly, changes can be modelled reflecting the revision of utility supply capacities, product slate or recipes, production demands and cost factors. A retrofit mSTN is obtained as for the grassroots design formulation. This superstructure embeds in a non redundant way all feasible retrofit designs and task/equipment allocations resulting from the introduction-deletion-reallocation of equipment and/or tasks and explicitly accounts for all limitations imposed by the structure of the equipment and operations networks.

Layout considerations are introduced by defining plant *zones* of given areas and by associating a footprint attribute to equipment items (area plus associated services, work space, etc.). New binary allocation variables reflect the allowable assignments of equipment to zones and new constraints are formulated to express the unique allocation of any equipment actually installed and the limited area availability in each zone. The space occupancy of new equipment and space recovery by reallocation-removal of old one are taken into account. The cost of connections between two units may depend on their zone allocation. A MILP problem is formulated and solved as before.

Retrofit Example

An existing plant (Table 1 and thin lines in Fig. 3) is to be modified to accommodate new demands for current products S5 and S6 (minimum production rates to be increased from 10 to 13.3 t/h, with a maximum of 15 t/h) and two potential new products (S9 and S10) from a current intermediate (S4) for which demand rates are 0 to 4.17 t/h (S9) and 0 to 20.83 t/h (S10). S9 and S10 will not be manufactured unless it is profitable. Product values and raw materials costs are 0.05 and 0.015 $/kg for S1 and S2, 0.035, 0.045, 0.1 and 0.125 for S5, S6, S9 and S10. The product recipes are shown in Fig. 2 (proportions by mass). Intermediate materials S8 and S3 are highly unstable and must be consumed immediately after being produced, while S4 is stable and may be stored indefinitely.

The existing units are given in Table 1, together with the possible new units: reactors R2, R4, R5 and R6, distillation column D1 and storage vessels V7, V4a, V5a, V6a, V9 and V10. Similar tables (not shown) define existing and potential new connections of interest and equipment eligible for removal or reuse after retrofit. The plant retrofit superstructure (Fig. 3) is characterised by limited connectivity (e.g. no connection between vessel V4a and reactor R3). The *mSTN* for this problem is given in Barbosa (1994). Steam is used in task T4 at a rate proportional to the batch size processed. Its existing supply is limited to 20 t/h and expansion of up to an extra 80 t/h is considered with an associated cost of 0.01 $/te. The problem is to maximise the new plant annualised profit, assuming a total number of 6600 working hours per year, periodic operation with a fixed cycle time of 12 hours and a capital charge factor (Douglas, 1988) of 1/3. The optimal sizes for two cases are presented in Table 2.

Table 1. Unit Characteristics.

Units	Suitability	Capacity [te]		Footprint[a]	Costs[a]
		Min.	Max.	[m^3]-[$/$m^3$]	[$]-[$/Kg]
Existing Units					
R1	T1-T2	-	67.2	-	-
R3	T3-T4	-	112	-	-
V1	S1	-	Unl.	-	-
V2	S2	-	Unl.	-	-
V4	S4	-	45	-	-
V5	S5	-	112	-	-
V6	S6	-	80	-	-
Potential New Units					
R2	T1-T2	20	35	15-0.05	20-1
R4	T3-T4	20	120	10-0.01	35-0.1
R5	T5	40	40	25-0	55-0
R6	T2-T5; S4	40	80	15-0.05	-
D1	T6	30	100	30-0.15	80-1
V4a	S4	20	50	-	100-0.05
V5a	S5	100	100	-	-
V6a	S6	100	100	-	-
V7	S7	-	Unl.	-	-
V9	S9	-	Unl.	-	-
V10	S10	-	Unl.	-	-

a = fixed and unit size (linearly) dependent coefficients

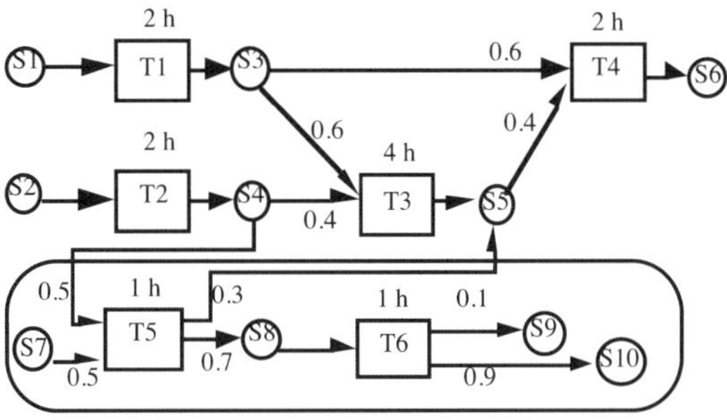

Figure 2. Product recipes (new ones in box).

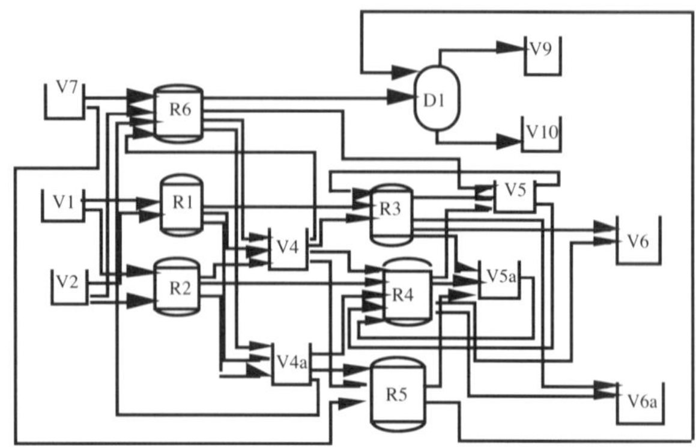

Figure 3. Retrofit plant superstructure.

Case 1

No space constraints are imposed. The new optimal plant is given in Figure 4. No existing units or connections were eliminated. In the optimal schedule (not shown) reactors R1, R2 and R3 perform more than one task (respectively T1/T2, T1/T2 and T3/T4). The remaining units are dedicated to a single task (R4-T4, R5-T5 and D1-T6). Material S4 is stored in V4a, V4 and material S5 in V5, V5a. No expansion of steam capacity is required. The optimal new production capacity is 15 t/h for products S5 and S6, 2.32 t/h for S9 and 20.83 t/h for S10, giving an annualised profit of $19,259,000.

Case 2

Space availability is limited in two main areas: R2 and R4 may be located in area A of 75m^3 and R5, R6 and D1 could be located in area B of 54.5m^3. With size dependent footprints of the units as in Table 1 the design of case 1 cannot be accommodated. A more expensive and larger reactor (R6 instead of R5) and a different connectivity structure are optimal (Fig. 5), leading to only a slightly lower annualised profit of $19,256,000.

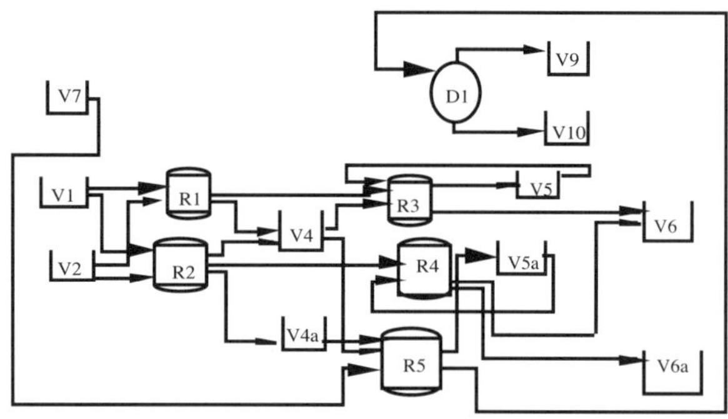

Figure 4. Optimal plant structure — Case 1.

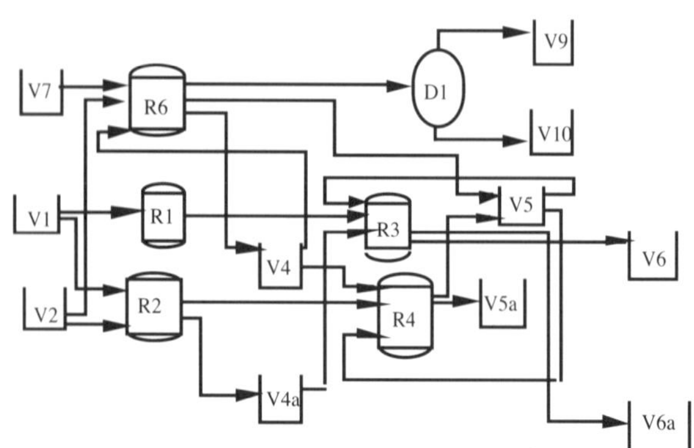

Figure 5. Optimal plant structure — Case 2.

Table 2. Optimal New Units Capacities.

Case	New Units Capacity [t]						
	R2	R4	R5	R6	V4a	V5a	D1
1	35	33.5	40	-	20	100	30
2	35	56.3	-	58.6	50	100	41

No existing units were eliminated but some of the connections were, in particular between R1 and V4, R3 and V5, and V4 and R3. This did not affect the capital cost since existing equipment had zero salvage value. Reactors R2 and R6 are operated in a multipurpose way (R2-T1/T2, R6-T2/T5) whilst R1, R3, R4 and column D1 are dedicated to a single task (R1-T1, R3-T4, R4-T3 and D1-T6). Storage policies, utility capacity and optimal product production rates are as in Case 1.

Case 1 was solved in 273.8 s CPU in a SUN SPARCstation 10 and Case 2 in 679.6 s.

Conclusions

Grassroots and retrofit design can be handled in a common modelling and solution approach. Connectivity and layout aspects can be considered simultaneously with the design of the main processing and storage equipment and with operations scheduling. Trade-offs between design and operation choices can then be made based on an economic objective function. The design/retrofit of utility networks, including Cleaning-In-Place (CIP) circuits is included in the formulation but will be detailed elsewhere due to space reasons. All problems are formulated as MILPs and solved using standard techniques. Some limitations remain with respect to the deterministic nature of all models, discrete time representation and solution efficiency with large problems. However, the method permits the solution of problems of much greater realism than possible so far.

Acknowledgments

This work was supported by an SERC/AFRC grant.

References

Barbosa-Póvoa, A.P.F.D. (1994). *Detailed Design and Retrofit of Multipurpose Batch Plants*. Ph.D. Thesis, University of London, UK.

Barbosa-Póvoa, A.P.F.D. and S. Macchietto (1994). Detailed Design of Multipurpose Batch Plants, *Comput. Chem. Engng.*, **18**(11/12), 1013-1042.

Crooks, C. (1992). *Synthesis of Operating Procedures for Chemical Plants*. Ph.D. Thesis, University of London

Douglas, J.M. (1988). *Conceptual Design of Chemical Processes*. McGraw Hill

Henning, G.P., N.B. Camussi, and J. Cerdá (1994). Design and Planning of multipurpose batch plants involving nonlinear processing networks. *Comput. Chem. Engng.*, **18**, 129-152.

Kondili, E., C.C. Pantelides, and R.W.H. Sargent (1988). A general algorithm for scheduling batch operations. In *Intl. Symp. on Process Systems Engineering*, Sydney, Australia, 62-75.

Papageorgaki, S. and G.V. Reklaitis (1990). Optimal design of multipurpose batch plants. *Ind. Eng. Chem. Res.*, **29**, 2054-2062.

Papageorgaki, S. and G.V. Reklaitis (1993). Retrofitting a general multipurpose batch chemical plant. *Ind. Eng. Chem. Res.*, **32**, 345-362.

Shah, N. (1992). *Efficient scheduling, planning and design of multipurpose batch plants*. Ph.D. Thesis, University of London, UK.

Voudouris, T.V., and I.E. Grossmann (1993). MILP model for scheduling and design of a special class of multipurpose batch plants. In AIChE Annual Meeting, St. Louis.

A GRAPHICAL APPROACH TO THE OPTIMAL DESIGN OF HEAT-INDUCED SEPARATION NETWORKS FOR VOC RECOVERY

Andrea Richburg and Mahmoud El-Halwagi
Chemical Engineering Department
Auburn University
Auburn, AL 36949

Abstract

In this work, we present a short-cut procedure for the optimal design of heat-induced separation systems for recovering volatile organic compounds "VOC's" via condensation. Using phase equilibrium data, the VOC-recovery mass-transfer task is converted into a heat-transfer duty. Then, a two-level targeting approach is developed to identify optimal coolants/refrigerants, their flowrates and system configuration. A case study on recovering vinyl chloride from a gaseous waste is addressed.

Keywords

Volatile organic compounds, Waste recovery, Condensation, Separation.

Introduction

Volatile organic compounds "VOC's" are among the most serious atmospheric pollutants. Recently, several regulations have been enacted to improve the air quality by imposing stringent emission standards on the discharge of VOC's to the environment. An efficient approach towards solving the VOC problem is recovery via separation systems. In this context, three categories of separation processes can be employed; mass exchange (e.g. absorption, adsorption), membrane systems (e.g. gas permeation, pervaporation) and heat-induced separation (e.g. condensation). While much attention has been directed towards the first two categories (e.g. El-Halwagi and Manousiouthaks, 1989; Srinivas and El-Halwagi, 1993), little work has been done in the area of systematically synthesizing optimal VOC-condensation systems. The objective of this work is to develop a graphical approach to the optimal design of VOC-condensation systems. The combination of phase equilibrium and enthalpy data renders an appropriate framework that can be used to identify the minimum operating cost of the system. Next, the heat-transfer driving force is optimized so as to minimize the total annualized cost of the network.

Problem Statement

Given a VOC-laden gaseous stream whose flowrate is G, supply composition of the VOC is y^s and supply temperature is T^s, it is desired to design a cost-effective condensation system which can recover a certain a fraction, α, of the amount of the VOC contained in the stream.

Available for service are several refrigerants. The operating temperature for the jth refrigerant, t_j, is given. For convenience in terminology, these refrigerants are arranged in order of decreasing operating temperature, i.e

$$t_1 \geq t_2 \geq \ldots t_j \ldots \geq t_N \qquad (1)$$

The operating cost of the j(th) refrigerant (denoted by C_j, \$/kJ removed) is known. The flowrate of each refrigerant is unknown and is to be determined through optimization.

Defining Target Composition

At first glance, it may appear that the target composition is calculated via

$$y^t = (1 - \alpha) y^s \tag{2}$$

which is not necessarily optimal. Indeed, one may pass only a fraction, β, of the gaseous stream through the condensation system such that

$$\beta G(y^s - y^t) = \alpha G y^s \tag{3.a}$$

i.e.

$$y^t = (1 - \frac{\alpha}{\beta}) y^s \tag{3.b}$$

since condensation must occur ($y^t < y^S$) the following bounds on β should be met:

$$\alpha < \beta \leq 1 \tag{4}$$

The rest of the gaseous stream, (1 - β)G, is bypassed and the net effect is that a fraction α of the VOC contained in the whole gaseous waste is recovered. Hence, the identification of the optimum value of β is part of the system optimization. It is worth pointing out that additional condensation can be accomplished by mixing the cooled stream with the bypassed portion. This scenario is beyond the scope of this paper but is discussed elsewhere (El-Halwagi et al., 1994).

Conversion of VOC-Recovery Task into a Heat-Transfer Duty

The first step in synthesizing a VOC condensation system is to convert the waste-recovery task from a mass-transfer problem to a heat-transfer duty. This can be accomplished by relating the composition of the VOC to the temperature of the gaseous waste. When a VOC-laden gas is cooled, the composition of the VOC remains constant until condensation starts at a temperature Tc, defined by

$$p^S = p^o(T^c) \tag{5}$$

where ps is the supply (inlet) partial pressure of the VOC, po(T) is the vapor pressure of the VOC expressed as a function of the gaseous-waste temperature, T. Hence, for a dilute system, the molar composition of the VOC in the gaseous emission, y, can be described as

$$y(T) = y^S \qquad if\ T > T^c \tag{6.a}$$

$$y(T) = p^o(T)/[P^{total} - p^o(T)] \qquad if\ T \leq T^c \tag{6.b}$$

where y^S is the supply (inlet) mole fraction of the VOC in the waste and P^{total} is the total pressure of the gas. Therefore, for a given target composition of the VOC, y^t, the waste-recovery task is equivalent to cooling the stream to a separation-target temperature, T*, which is calculated via the following equation:

$$y^t = p^o(T^*)/[P^{total} - p^o(T^*)] \tag{7}$$

Having converted the VOC-recovery problem into a heat-transfer task we are now in a position to develop the design procedure.

Design Approach

In the context of designing the VOC-condensation system, it is first necessary to consider the system configuration schematically depicted in Fig. 1. The initial

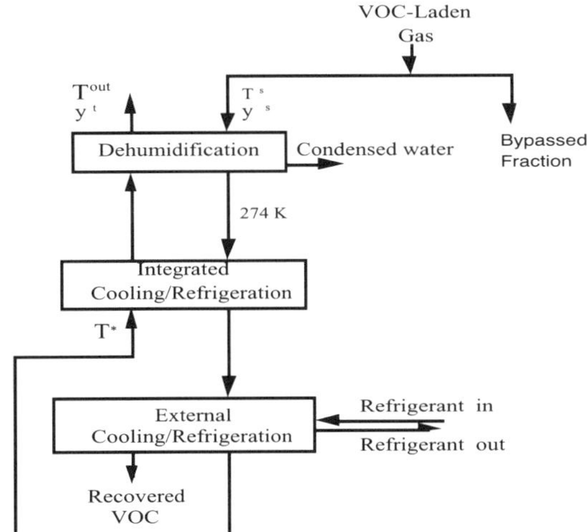

Figure 1. Schematic representation of the VOC-Recovery System

step is to cool the stream to a temperature slightly above that of water freezing so as to dehumidify the gas by condensing the water vapor and prevent the detrimental icing effects in subsequent stages. Next, the stream is cooled to T* to recover the VOC. In order to utilize the cooling capacity of the gaseous stream at T*, it is recycled back to the system for heat integration. The remaining cooling duty is accomplished by a refrigerant.

The design procedure starts by identifying the minimum utility cost for a given heat-transfer driving force. Next, the fixed and operating costs are traded off by iterating over the driving forces until the minimum total annualized cost "TAC" is attained.

Minimum Utility Cost

It is beneficial to develop the enthalpy expressions for the gaseous VOC-laden stream as its temperature is cooled from Ts to some arbitrary temperature, T, which is below Tc. Assuming that the latent heat of the VOC remains constant over the condensation range, the enthalpy change (e.g. kJ/kmole of VOC-free gaseous stream) can be evaluated through

$$h(t) - h(T^S) = \int_{T^S}^{T} C_{p,g}(T)\,dT + \int_{T^S}^{T} y(T)\,C_{p,y}(T)\,dT + [y^S - y(T)]\lambda + \int_{T^C}^{T} [y^S - y(T)]\,C_{P,L}(T)\,dT \quad (8)$$

A similar expression (excluding the latter two terms) can be derived for the enthalpy of the recycled cold gas.

A convenient way of identifying the minimum utility cost of the system is the pinch diagram (Linnhoff and Hindmarsh, 1982). A temperature scale for the gaseous stream to be cooled, T, is related to a temperature scale for the recycled cold gas, t, by using a minimum driving force ($T = t + \Delta T_1^{min}$). Next, Eq. (8) is used to plot the enthalpy of the hot stream (VOC-laden stream) versus T. Similarly, one can plot the enthalpy of the cold gas recycled to the system against t. The cold stream is slid down until it touches the hot stream at the pinch point and, therefore, the minimum cooling requirement, Q^c, can be identified (Fig. 2). For each value of, Eq. (3.b) is used to determine y^t which in turn is employed in Eq. (7) to calculate T^*. As can be seen from Fig. 2, for a given T^* the value of Q^c is determined graphically. In addition, the outlet temperature for the gaseous stream leaving the network, T^{out}, is identified.

The next step is to select the refrigerant that minimizes the utility cost. This step can be accomplished by examining the cost of refrigerants that operate below T^*. Consider two refrigerants u and v whose costs (\$/kJ removed) are C_u and C_v where u > v. It is useful to recall that the refrigerants are arranged in order of decreasing operating temperatures, hence $t_u < t_v$. If $C_u \le C_v$, then refrigerant u is preferred over v. This rationale can be employed to compare all refrigerants below T^* with the result of identifying the one that yields the lowest cooling cost. Finally, tradeoffs between operating and fixed costs must be established. this step is undertaken iteratively. For given values of ΔT_1^{min} and ΔT_2^{min}, the pinch diagram is used to obtain minimum cooling cost and outlet gas temperature. By conducting enthalpy balance around each unit, intermediate temperatures and exchanger sizing can be determined. Hence, one can evaluate the fixed cost of the system. Next, ΔT_1^{min} and ΔT_2^{min} are altered, until the minimum TAC is identified.

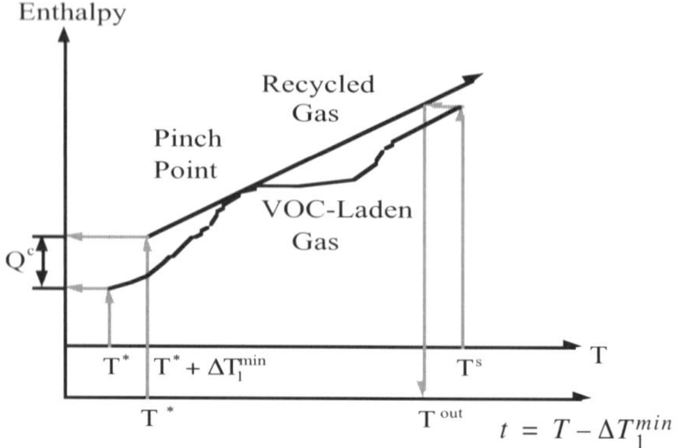

Figure 2. Pinch diagram for the VOC-Condensation System

Special Case: Dilute Waste Streams

A particularly useful special case is the situation of dilute waste streams. This situation typically implies that the latent heat change of the stream is negligible with respect to its sensible heat. In such cases, one can show that the cooling utility is given by

$$Q^c = \beta G \bar{C}_{p,g} \Delta T_1^{min} \quad (9)$$

while the rest of the cooling task is fully integrated with the recycled gaseous stream. By combining Eqs (3.a), (7) and (9), and noting that for dilute streams $p^o(T^*) << P^{total}$, we get

$$Q^c = \alpha G y^s \bar{C}_{p,g} \Delta T_1^{min} / \{ y^s - [p^o(T^*)/P^{total})]\} \quad (10)$$

since $p^o(T^*)$ is a monotonically increasing function of T^*, the lower the value of T^*, the lower the Q^c. For a given refrigerant, j, and a minimum driving force ΔT_2^{min}, the lowest attainable gas temperature is $t_j + \Delta T_2^{min}$. Hence, the optimal value of T^* can be obtained by comparing the costs of refrigerants needed to cool the gas to $t_j + \Delta T_2^{min}$ where j = 1, 2, ..., N. This entails searching over a finite set of at most N temperatures to identify the minimum cooling cost. Once the optimal T^* is determined, Eqs. (3.b) and (7) are used to calculate the optimal β and the optimum network is configured.

Case Study: Recovery of Vinyl Chloride "VC"

In a polyvinyl chloride plant, a continuous air emission leaves a dryer stack. The primary pollutant in this stream is VC. The flowrate of the gas is 0.20 kmole/s, its supply temperature is 338 K and contains 0.5 mol/mol% of

VC. It is desired to recover 90% of the VC in the gaseous waste. Available for service are four refrigerants: SO_2, HFC134, NH_3 and N_2 whose respective operating temperatures are 280, 265, 245 and 140 K and operating costs are 12, 10, 13 and 16 \$/10^6 kJ removed, respectively. The average specific heats of the air, VC vapor and VC liquid are taken as 29, 50 and 85 kJ/(kmole.K), respectively. The latent heat of condensation for VC is assumed to be 24,000 kJ/kmole over the operating range. The vapor pressure of VC is given by (Yaws, 1994):

$$\log p^o(T) = 52.9654 - 2.5016 * 10^3/T - 17.914 \log T + 0.0108 * T - 4.531 * 10^{-14} T^2$$

where $p^o(T)$ is in mmHg and T is in K. The cost of condensers/heat exchangers (\$) is taken as 1,500*(heat transfer area in $m^2)^{0.6}$. The overall heat-transfer coefficients for the dehumidification, the integrated cooling/condensation and external cooling/condensation sections are 0.05, 0.05 and 0.10 kW/(m^2.K), respectively. The fixed cost of refrigeration systems for SO_2, HFC134 and NH_3 is provided elsewhere (US EPA, 1991). For N_2, it is assumed that no refrigeration system is needed; instead liquid nitrogen is purchased and stored in a tank whose cost is \$150,000. For all units, a linear depreciation scheme is used with 10 years of useful life period and negligible salvage value.

By following the proposed procedure, the minimum total annualized cost was found to be \$67,950/year of which \$25,150/year is operating cost and \$42,800/year is annualized fixed cost. The optimal values of ΔT_1^{min} and ΔT_2^{min} were determined to be 5.5 and 11.5 K, respectively. The solution procedure also identified the optimal value of β to be 0.92 which corresponds to a y^t of 0.012 mol/mol%.

Conclusion

A short-cut method has been proposed for the optimal design of VOC condensation systems. The design is based on transforming the mass-recovery task into a heat-transfer duty. A pinch-based graphical approach is used to minimize the operating cost of the system. Subsequently, heat-transfer driving forces are employed to iteratively trade off capital versus operating cost so as to minimize the total annualized cost of the system.

Acknowledgment

The financial support of the National Science Foundation (grant #CTS-9211039) is gratefully acknowledged.

Nomenclature

C_j = cost of the j(th) refrigerant, \$/kJ removed
$C_{p,g}$ = specific heat of the VOC-free gas, kJ/(kmole. K)
$\bar{C}_{p,g}$ = average specific heat of VOC-free gas, kJ/(kmole.K)
$C_{P,L}$ = specific heat of the liquid VOC, kJ/(kmole.K)
$C_{P,V}$ = specific heat of the vapor VOC, kJ/(kmole.K)
G = flowrate of VOC-free gaseous stream, kmole/s
h = specific enthalpy of gaseous stream, kJ/kmole of VOC-free gaseous stream
j = index for refrigerants
N = total number of potential refrigerants
$p^o(T)$ = vapor pressure of the VOC at T, mm Hg
P^{total} = pressure of gaseous stream, mm Hg
Q^c = minimum cooling requirement, kW
t = temperature of cold gas to be heated, K
t_j = operating temperature of the j th refrigerant, K
T = temperature of gaseous stream being cooled, K
T^c = dew temperature for the VOC, K
T^s = supply (inlet) temperature of gaseous stream, K
T^* = separation-target temperature to which the gaseous stream has to be cooled, K
y^s = supply composition of VOC, kmole VOC/kmole VOC-free gaseous stream
y^t = target composition of VOC, kmole VOC/kmole VOC-free gaseous stream

Greek

α recovery fraction of VOC in gaseous stream, kmole VOC recovered/kmole VOC in gaseous waste
β fraction of gaseous stream to be passed through the condensation system
ΔT_1^{min} minimum driving force for the dehumidification and integrated cooling/condensation blocks, K
ΔT_2^{min} minimum driving force for the external cooling/condensation block, K
λ latent heat of condensation for VOC, kJ/kmole

References

El-Halwagi, M.M., B.K. Srinivas and R.F. Dunn (1994). Synthesis of Optimal Heat-Induced Separation Networks. *Chem. Eng. Sci.* (in press).

El-Halwagi, M.M. and V. Manousiouthakis (1989). Synthesis of Mass-Exchange Networks. *AIChE J.*, **35,** 1233-1244.

Linnhoff, B. and E. Hindmarsh (1983). The Pinch Design Method for Heat Exchange Networks, *Chem. Eng. Sci.*, **38**, 745-763.

Srinivas, B.K. and M.M. El-Halwagi (1993). Optimal Design of Pervaporation Systems for Waste Reduction. *Comp. Chem. Eng.*, **17**, 957-970.

U.S. EPA (1991). *Control Technologies for Hazardous Air Pollutants*, EPA/625/6-91/014, 4.55-4.64.

Yaws, C.L. (1994). *Handbook of Vapor Pressure*, Gulf Pub. Co., Houston, **1**, 343.

FLEXIBILITY, RELIABILITY AND MAINTENANCE IN PROCESS DESIGN

Thomas V. Thomaidis and Efstratios N. Pistikopoulos[1]
Department of Chemical Engineering
Imperial College
London SW7 2BY

Abstract

In design and operation of chemical process systems, two categories of uncertainties play an important role, uncertainty with respect to realization of continuous process parameters (product demands, feedstock qualities, reaction kinetics), and uncertainty with respect to discrete states (typically related to equipment availability and maintenance). In this paper, a design and maintenance optimization procedure is proposed for obtaining process systems capable to absorb random process variations and equipment failures/malfunctions at maximum expected profit-while maintenance planning options are systematically considered based on a maintenance superstructure representation. The application of the developed framework is presented via the design of a flexible and reliable multiproduct batch plant example.

Keywords

Process flexibility, Reliability, Maintenance, MI(N)LP, Process design.

Introduction

Reliability, Availability and Maintenance are not traditionally considered in "conceptual process design" practice despite their important role in plant operations and their essential contribution to process economics, safety and effectiveness. Similarly, the existence of continuous process variations (for example, in the feed-stream compositions and flowrates, uncertainty in product demands, etc.) which are very common in plant operations is typically overlooked during the design stage of process systems. Flexibility considerations in the process industry unlike the manufacturing industry have never been perceived as distinct design objectives. However, increasing standards for product quality, safety and the enforcement of strict environmental regulations, have gradually motivated the need for integrating operability in the early design stage, in which the most benefits can be achieved, Thomaidis and Pistikopoulos, 1993; Grievink et al., 1993).

The objective of this paper is to present some recent developments of a methodology for the systematic incorporation of flexibility, reliability, availability and maintenance (FRAM) in process design optimization.

Integration of Design Under Uncertainty and Maintenance Optimization

The problem to be addressed in this section can be defined as follows. Given a process model, statistical data for process variations, reliability and repair rates, cost data, and a set of potential maintenance policies (number of service crews, order of execution of maintenance tasks, calendar-time based preventive maintenance), determine an optimal flexible and reliable process design and an optimal maintenance strategy that *maximize* an *expected profit* accounting for revenues, investment, operating and maintenance costs.

An optimization model has been developed based on the following ideas. First, the operable operating states (due to equipment failures) are identified based on process flowsheet structural information. These states may potentially contribute to the overall expected profit proportional-

[1] Author to whom correspondence should be addressed.

ly to their corresponding state probabilities, P^k. In the presence of continuous uncertainty Θ, the expected revenue R^k of an operable system state k is given by

$$R^k = \int_{\Theta \in FOR^k(d)} r^k(\Theta) * jpdf(\Theta) d\Theta \qquad (1)$$

where r^k is a profit function for state k; **jpdf** denotes the joint probability density function[2] of Θ,***FOR*** is the feasible region of operation of the process design *d, (*see also example problem).

Since maintenance is postulated and state probabilities depend on the (unknown optimal) maintenance policy, a **maintenance superstructure** representation is adopted to embody all possible alternatives. The maintenance superstructure concept can be better illustrated for the batch process shown in Fig. 1,

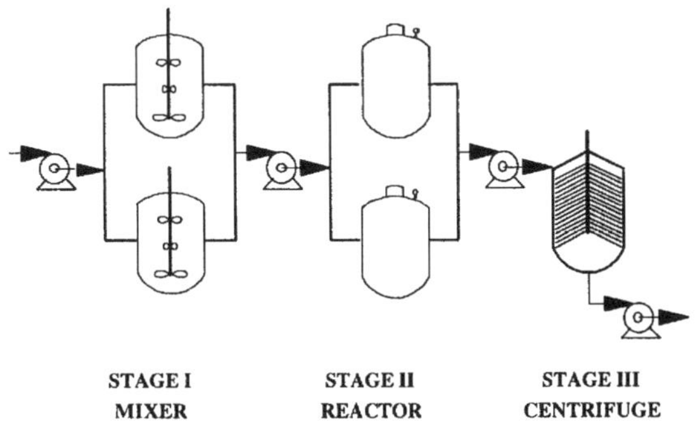

Figure 1. Multiproduct batch plant.

where we further assume a one failure-mode Markovian model (breakdown maintenance), Billinton and Allan (1983). Since the process involves five units the Markovian model comprises thirty-two discrete operating states.However, if identical operating states are properly aggregated (Straub and Grossmann, 1992), then a significant reduction of the state space model can be achieved involving only eighteen unique operating states k ($k \in K$), with an aggregation factor π^k, see also Fig. 2.

Maintenance Superstructure Model

Consider for instance a reference system state k (kŒK) together with its corresponding upgraded and degraded states, M^k and L^k respectively, see Fig. 3.

The transition from an upgraded state m ($m \in M^k$) to a reference state k, due to failure of all or a subset of operable equipment J^m in the upgraded state m (with corresponding failure rates λ_j, $j \in J$), is represented by a fixed arc (straight arrows in Fig. 3). On the other hand, the transition from a reference state k to an upgraded state m due to the repair of the set of units J^k (with repair rates μ_j) depends on the execution (or not) of specified maintenance tasks. A variable arc (depicted by the bold curved arrows in Fig. 3) is thus introduced denoted by a binary (0,1) variable $U^{l,k}$ (i.e. $U^{l,k}=1$ implies that system is upgraded from degraded state l to upgraded reference state k; else 0) associated with a set of maintenance tasks.

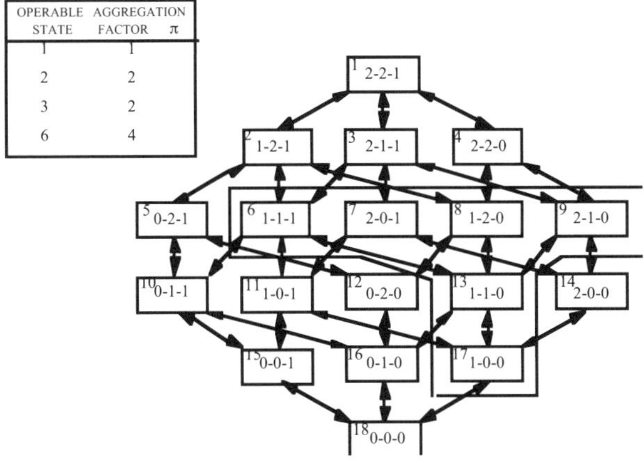

Figure 2. State space diagram.

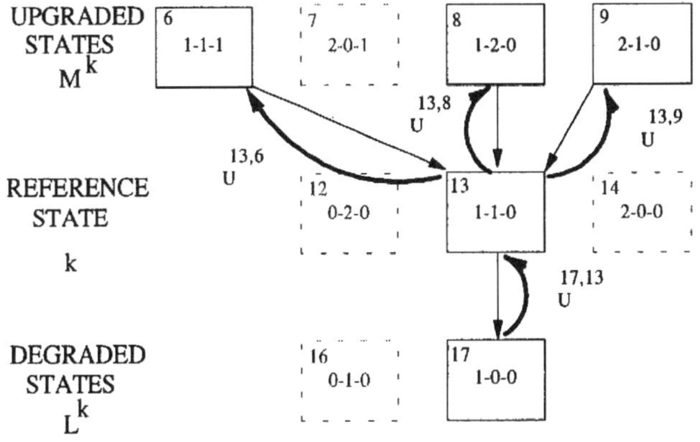

Figure 3. Maintenance superstructure model.

Mathematical Model

By incorporating these aspects into the frequency balance equations for the determination of the state probabilities Pk of the Markovian model, the following conceptual mathematical formulation (P) is proposed for the simultaneous design and maintenance optimization.

max Expected Profit

Expected Profit = *Expected Revenue*
- *Installation Cost*
- *Operating Cost*
- *Maintenance Cost*

s.t. (P)

---

[2] Any probability distributional form can be employed

(A.1) Process Model
Maintenance Superstructure Model
(A.2) Frequency balance equations
(A.3) Normalizing equation
(A.4) Required maintenance resources

The detailed mathematical model is described elsewhere, Thomaidis and Pistikopoulos (1994). Note that the set of constraints (A.1) of problem (**P**) represent the process model where continuous uncertainty is explicitly included (either through a multiperiod multiple-scenario approach or via a stochastic optimization framework); the set of constraints in (A.2), (A.3) and (A.4) correspond to the frequency balance equations of the modified Markovian model based on the maintenance superstructure concept of Fig. 3. Finally, the objective function in (**P**) balances revenues and costs.

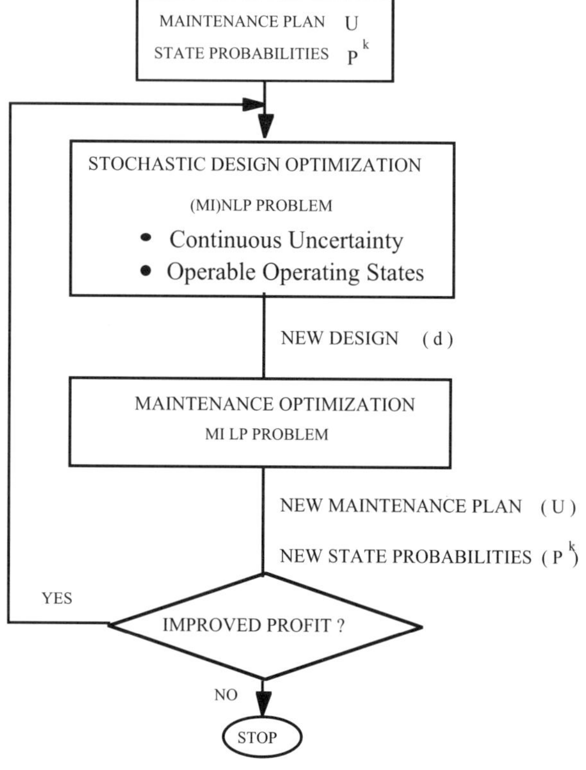

Figure 4. Algorithmic procedure for design-maintenance optimization.

Problem (**P**) corresponds to a Stochastic Mixed Integer Non-Linear Programming (SMINLP) formulation, since it involves (0,1) variables (denoting the discrete maintenance decisions), stochastic parameters (denoting continuous uncertainty) and a nonlinear model.

Algorithmic Procedure

The following decomposition strategy has been developed for the efficient solution of problem (**P**) based on Generalized Benders Decomposition principles. By selecting the (0,1) variables U as complicating variables, the balance equations in (A.2) and (A.3) can be solved yielding the values of the state probabilities P^k; furthermore, the required maintenance resources can also be determined from equation (A.4). The stochastic design optimization (with fixed values of P^k) can then be performed providing the optimal values of the design variables d, the expected revenues R^k for each operable state, and a lower bound on the overall objective function of problem (**P**).

Fixing the design variables d transforms (**P**) to a Mixed Integer Linear Programming (MILP) problem, where the state probabilities P^k and the binary vector U are unknown. Its solution returns a new set of U and P^k variables (and the number of service crews NC), while providing an upper bound on the objective function of (**P**). Successive iterations between the stochastic design subproblem and the maintenance part, as depicted in Fig. 4, result in an optimal flexible and reliable design and an optimal maintenance planning policy that maximize the overall expected profit function of problem (**P**).

A similar procedure can be also developed when calendar-time based preventive maintenance activities are considered (via block and age replacement models), as in Arsenis (1994) and the SPARC decision support tool, Smit et al. (1994), by properly modifying the maintenance optimization subproblem. Aggregation of units in subsystems, truncation of the state space and the evaluation of the expected revenue only for the most significant states (based on a bounding procedure, Straub and Grossmann (1992)), can also be employed to reduce model size and computational complexity. Finally, note that the above optimization procedure can be extended to account for the execution of different maintenance strategies over a time horizon H, by defining a number of time periods. These aspects are currently under investigation.

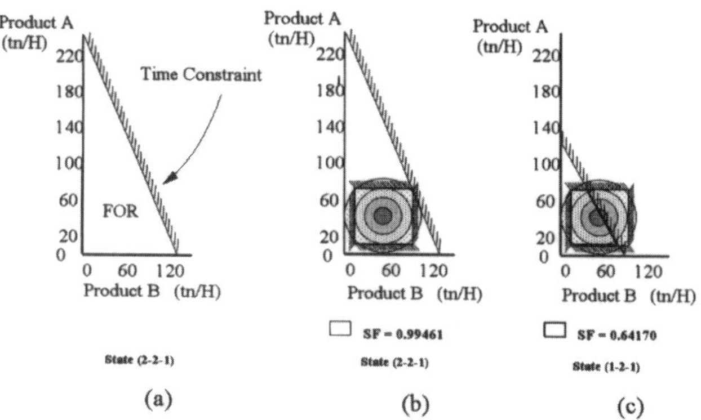

Figure 5. Feasible Operating Region for fully operable and degraded operating states.

Example Problem

Consider the multiproduct batch plant configuration shown in Fig.1 involving three main processing stages (mixing, reaction and purification) for manufacturing two different products (A,B). In the first stage two identical units (M-1, M-2) are used in parallel in order to mix raw materials. The reaction stage involves two identical reactor units (R-1, R-2), for the conversion of raw materials, whereas in the last stage a centrifuge (C-1) is used to purify the intermediate to final products. according to desired specifications. Data for the particular process are given in Table 1. Product demands are considered uncertain following Gaussian probability distribution functions, with means and variances given in Table 2. Repair and failure rates, μ_j, λ_j, together with maintenance cost data, C_{mj} (typically obtained from manufacturers specifications and historical data coupled with parameter-estimation techniques) are also given in Table 2. The expected revenue, R^k, which is a function of the uncertain production rates Q_i and the price of products pi, and the corresponding investment cost function CD, are as follows (with the data of Table 2):

$$R^k = \int_{Q \in FOR^k(V)} p^T Q^* jpdf(Q)\, dQ$$

$$C_D = \sum \alpha_j m_j V_j^{\beta_j}$$

For the batch plant of Fig. 1 and for one failure mode the system resides in four operable aggregated operating states {1,2,3,6} corresponding actually to nine discrete physical operating states, see Fig. 1. Note here that only the operable states are considered in the design optimization problem since only these actually contribute to system profitability. Further reduction techniques for the number of states together with bounding procedures are discussed elsewhere, Thomaidis and Pistikopoulos (1994).

The algorithm converged after four iterations yielding an optimal design with values for the capacities of the five units V1=V2= 500 Lit, V3=V4= 741 Lit and V5= 988 Lit respectively, and the optimal maintenance plan of Fig. 4 involving one service crew and maintenance cost of \$62,041, with a maximum expected profit of \$260,080.

Table 1. Reliability, Maintenance and Cost Data.

	Investment cost factors		Repair rates	Failure rates	Maint. cost	Stage Capacities	
						Min	Max
	$\alpha(Lit^{-1})$	β	$\mu(hr^{-1})$	$\lambda(hr^{-1})$	C_{Mj}	Sizes (Lit)	
1	350	0.6	0.225	0.0070	300	500	2,500
2	350	0.6	0.220	0.0075	250	500	2,500
3	650	0.6	0.090	0.0080	150	500	2,500
4	650	0.6	0.092	0.0082	100	500	2,500
5	450	0.6	0.150	0.0110	450	500	2,500

Time Horizon H = 12,000 hr

Fixed cost per service crew δ = 50,000

Table 2. Demand, Process and Operating Data for the Multiproduct Batch Plant

	Processing times			Size factors L/tn			Demand		Price
stage ⇒	R	M	C	R	M	C	μtn/H	σtn/H	units/tn
A	12	20	4	2,000	3,000	4,000	10	40	460
B	8	12	3	4,000	6,000	3,000	60	15	520

Conclusions

A design optimization procedure has been proposed with which inherently flexible and reliable process systems are obtained at maximum expected profit by identifying the optimal maintenance strategy. The algorithm is based on a maintenance superstructure representation, while continuous uncertain parameters are explicitly considered in process models. The proposed optimization scheme is amenable for considering both corrective and preventive maintenance activities.

A case-study, the design of flexible and reliable multiproduct batch plant, was presented. The methodology presented here is not of course, completed to this end; However, the main objective is to link the proposed technique, which is mainly suitable for process design studies, to existing powerful reliability-maintainability techniques (for example the SPARC tool, Smit et al., (1994)).

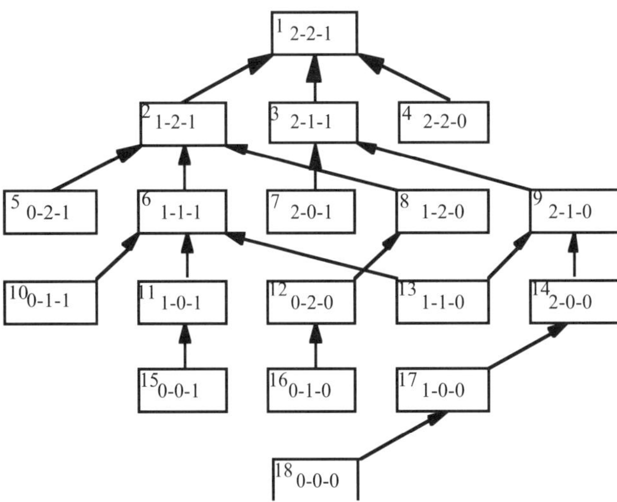

Figure 6. Optimal maintenance plan.

References

Arsenis, S. (1994). *On the reliability and availability of a system of components active, standby and in corrective or preventive maintenance*. ESReDA'94, Chamonix, France.

Billinton, R. and R.N. Allan (1983). *Reliability evaluation of engineering systems. Concepts and techniques*. Plenum Press, N.Y., 1st Edition.

Geoffrion, A.M. (1974). Generalized Benders Decomposition. *Journal of Optimization Theory and Applications*, **10**, 237-260.

Grieving J., R. Decker, K. Smit and C.F. Van Rijn (1993). Managing reliability and maintenance in process industries. Conference on Foundation of Computer Aided Operations, FOCAPO.

Smit, A.C.J.M, C.F.H. Van Rijn and S.G. Vanneste (1994). SPARC: a Comprehensive Reliability Engng. Tool. ESReDA'94, Chamonix, France.

Straub, D.A., and I.E. Grossmann (1992). Evaluation and design optimization of stochastic flexibility in multiproduct batch plants. *Comp. Chem. Engng.*, **16**, 2, 69-87.

Thomaidis, T.V. and E.N. Pistikopoulos (1994). Integration of flexibility, reliability and maintenance in process synthesis and design. ESCAPE-3, *Comp. Chem. Engng.*, **18**, Suppl. S259-S263.

RELIABILITY ESTIMATION AND OPTIMIZATION IN MULTIPURPOSE BATCH PROCESSES

Guillermo E. Rotstein
Centre for Process Systems Engineering
Imperial College
London SW7 2BY

Abstract

Batch scheduling recipes are usually modeled by network representations. The reliability of a schedule can be then related to the reliability of the networks involved in the schedule. This facilitates the estimation of a reliability index within a MILP formulation, in which the activation status of each network is modeled by a single binary variable. However, when the networks share process units, the index values lead to a conservative estimation of reliability. Here, we reformulate the MILP model, decomposing the networks into minimal cuts. The new formulation improves the reliability estimation when the production networks share minimal cuts and when some minimal cut share equipment units.

Keywords:

Batch design, Batch scheduling, MILP modeling, Reliability, Uncertainty.

Introduction

The use of multiperiod MILP models for the scheduling of multipurpose batch plants is well established. The problem of accounting for uncertainties in these models has been addressed only recently (Rotstein et al, 1993). In process design, a combined stochastic flexibility/reliability (*SFR*) index was introduced to quantify the probability of managing the expected uncertainty. It is estimated by summing up the products of the probability of each operating state and its stochastic flexibility index (Straub and Grossmann, 1990 and Pistikopoulos and Thomaidis, 1992). However, in batch scheduling, this approach may lead to a large number of states, since the topology must be considered repeatedly for each time interval. Moreover, in multipurpose plants, several routes for each product may exist. One must then consider the possibility of redirecting intermediates to alternative routes when equipment malfunctions stop normal processing

Recently, (Rotstein et al, 1993) a reliability index (*SR*) for batch schedules was estimated through the evaluation of reliability in a set of associated networks. Even though a lower bound on the index could be evaluated by MILP scheduling techniques, dependence of the networks may lead to conservativeness in the bound. In this paper, reformulations are introduced, decomposing the networks into minimal cuts, and disaggregating the cuts into relevant units. This improves the reliability estimation when the production networks share cuts, and when some cuts share equipment units.

Stochastic Reliability. Mathematical Formulation

A multiperiod MILP scheduling model is written as:

$$\mathbf{A}_1\mathbf{x} + \mathbf{A}_2\,\mathbf{I} \geq \mathbf{b}$$
$$x \in R^n, \mathbf{I} \in B^m \qquad (1)$$

In batch scheduling, the discrete variables **I** typically represent the activation of a particular task instance. Boolean uncertainty is then modeled by associating a reliability, ρ_ι, to each task instance (ρ_ι is the probability of activating a task i, during the considered time horizon, with no failure in any of the required equipment items).

A stochastic reliability index accounts for the probability of equipment failure (availability) and the existence of operating alternatives. For batch schedules, the index can be estimated employing a network representation for batch recipes (Rotstein et. al., 1993). Assuming the availabilities of the equipment are known, independent and constant, the schedule reliability can be estimated by calculating the reliability of the corresponding networks. Define:

Production Route: any path from initial to goal nodes.

Production Policy (Y_i): subnetwork consisting of all production routes shari initial and goal task nodes.

The following methodology is now proposed:

a) Determine all (n_p) production policies.
b) Estimate the reliability of each policy, SR(Y_i).
c) Calculate the schedule reliability, *SR*. This gives the probability of successful schedule completion when rescheduling to the alternative routes is feasible. If gaps in the routes capacities exist or the rescheduling of a task is delayed the completion time may be extended. A lower bound on *SR* is given by the reliability of a system in which all the policies are independent and connected in series. Then:

$$\underline{SR} = \prod_{\substack{i=1 \\ Y_i \text{ is active}}}^{n_p} SR(Y_i) \tag{2}$$

Steps (a) and (b) are performed before the optimization. However, step (c), requires the identification of the policies active in a proposed schedule within the MILP procedure. This is achieved by adding constraints and variables to the original model. These link the activation of tasks to the activation of the corresponding policies. First, a linearized expression for $\underline{\text{SR}}$ is obtained:

$$\log(\underline{SR}) = \sum_{i=1}^{n_p} \mathbf{y}_i \log(SR_i)$$
$$\mathbf{y}_i = \begin{cases} 1 & \text{if } \mathrm{Y_i} \text{ is active} \\ 0 & \text{otherwise} \end{cases} \tag{3}$$

Now, let W_i be the set of binary variables characteristic of policy Y_i (i.e., they represent tasks required only by that policy). The characteristic tasks of a policy can only be active if the corresponding policy is active. This is stated by the following constraints, where N_i is an upper bound on the number of task activations:

$$\sum_{\substack{j,k,t \\ I_{jkt} \in W_i}} I_{jkt} \le N_i y_i \tag{4}$$

Finally, introducing eqs. (3) and (4) in formulation (1)

Model_1

$$\underset{x,I,y}{Max} \log(\underline{SR}) = \underset{x,I,y}{Max} \sum_{k=1}^{n_K} y_i \log(SR_i)$$
$$s.t.$$
$$\sum_{\substack{j,k,t \\ I_{jkt} \in W_i}} I_{jkt} \le N_i y_i$$
$$\mathbf{A_1 x + A_2 I \ge b}$$
$$x \in R^n, \mathbf{I} \in B^m, \mathbf{y} \in B^{n_p}$$

Minimal Cuts Decomposition

A Minimal Cut Set is a minimal set of components whose failure results in the failure of the system. A lower bound on a network reliability can be evaluated by representing it as a serial connection of its minimal cut sets (Gertsbakh, 1989):

$$SR(Y_i) \ge \prod_{j=1}^{l_i} r(C_{ij})$$
$$Y_i = \{C_{i1}, C_{i2}, \ldots, C_{il_i}\} \tag{5}$$

C_{ij} is a cut of the policy Y_i and r(C_{ij}) is its reliability. Combining equations (3) and (5), and linearizing:

$$\log \underline{SR} \ge \sum_{i=1}^{m} \sum_{\substack{j=1 \\ C_{ij} \text{ is active}}}^{l_i} \log\{r(C_{ij})\} \tag{6}$$

When the operating policies share cuts, the lower bound obtained using (6) is conservative. A cut may belong to several policies, but its effect on the reliability should be considered only once. One can improve the reliability estimate by disaggregating the policy variables, into variables representing the activation of distinct cuts:

$$c = \begin{cases} 1 & \text{if cut } \mathbf{K}_k \text{ is active} \\ 0 & \text{otherwise} \end{cases} \tag{7}$$

where $\mathbf{K}_k$ is a minimal cut and n_k is the number of distinct cuts in all the policies. By the definition of a cut, if a policy is active all its cuts must be active:

$$c_k \ge y_i \qquad \forall K_k \in Y_i \tag{8}$$

Thus a tighter lower bound on the reliability of the schedule is obtained by reformulating Model_1 into:

Model_2:

$$\underset{x,I,y,c}{Max}\ \log(\underline{SR}) = \underset{x,I,y,c}{Max} \sum_{k=1}^{n_K} c_k \log\{r(K_k)\}$$

$s.t.$

$$c_k \geq y_i \quad \forall K_k \in Y_i$$

$$\sum_{\substack{j,k,t \\ I_{jkt} \in W_i}} I_{jkt} \leq N_i y_i$$

$$\mathbf{A_1 x + A_2 I \geq b}$$

$$x \in R^n, \mathbf{I} \in B^m, \mathbf{y} \in B^{n_p}, \mathbf{c} \in B^{n_K}$$

Associated Minimal Cuts Decomposition

Cut decompositions into equipment networks can further improve the reliability index estimate. One can:

1) Improve the reliability estimation for each cut, $r(K_k)$. This is done before the optimization.
2) Consider that some cuts are associated (i.e. share equipment units). Now, model reformulation is required. Here notation is simplified by considering that there is only one set of associated minimal cuts. The extension to several sets is straightforward. We decompose SR into:

$$SR(Y_i) = \alpha \prod_{j \in A} r(K_j) \prod_{j \in A} r(k_j) \qquad (9)$$

where A is a set of associated cuts, and α is a correcting factor to account for that association. The value of α depends on the subset of A activated within a schedule. For instance: if $A = (K_2, K_3, K_4)$, $r(K_i) = \rho\rho_i$ (i.e. they share a unit of reliability ρ) and only cuts K_2 and K_4 are active within a schedule then $\alpha = 1/\rho$ (i.e. the reliability of the unit is considered only once). However, if all the cuts are active then $\alpha = 1/\rho^2$. The reformulation of the model is done in two stages:

a) Estimate the potential a values by decomposing the associated set into $A = \bigcup_{i=1}^{n_\alpha} S_i$. Each S_i is a different combination of associated cuts. Then, estimate the value of a for each of the combinations:

$$\alpha_i = \frac{r(S_i)}{\prod_{K_j \in S_i} r(K_j)} \qquad (10)$$

b) Estimate the actual α value as a function of the active cuts. This corresponds, for a given schedule, to the subset S_i for which all the cuts are active ($a_i = 1$) and there is no other S_i for which all the cuts are active and $S_i \subset S_{i'}$:

$$\log \alpha = \sum_{i=1}^{N_\alpha} a_i \log \alpha_i$$

where

$$a_i = \begin{cases} 1 & \text{if } a_{i'} = 0, \forall i' \text{ s.t. } S_i \subset S_{i'}, c_k = 1, \forall k \text{ s.t. } K_k \in S_i \\ 0 & \text{otherwise} \end{cases}$$

The final MILP formulation obtained is:

Model_3:

$$\underset{x,I,y}{Max}\ \log(\underline{SR}) = \underset{x,I,y,c}{Max} \sum_{k=1}^{n_K} c_k \log\{r(K_k)\} + \sum_{i=1}^{N_\alpha} a_i \log\alpha_i$$

$s.t.$

$$c_k \geq y_i \quad \forall K_k \in Y_i$$

$$\sum_{\substack{j,k,t \\ I_{jkt} \in W_i}} I_{jkt} \leq N_i y_i$$

$$a_i \leq c_k \quad \forall k \text{ s.t. } K_k \in S_i$$

$$a_i \leq (1 - a_{i'}) \quad \forall i' \text{ s.t. } S_i \subset S_{i'}$$

$$c_k \leq 1$$

$$\mathbf{A_1 x + A_2 I \geq b}$$

$$x \in R^n, \mathbf{I} \in B^m, \mathbf{y} \in B^{n_p}, \mathbf{c} \in B^{n_K}, \mathbf{a} \in B^{n_\alpha}$$

Example

The problem is based on an example from Crooks and Macchietto (1992). The batch recipe and plant are shown in figures 1 and 2. The production of G and FI is maximized, while guaranteeing a reliability of 85% (SR>0.85). The reliability for tasks relying on pumps is 0.99, except for task Store&Cool_G. This relies on pump P6 and utility C2 for which the reliability is 0.95. Task Distillation_FGI requires also C2, thus its reliability is 0.95. The reliability for all other tasks is one.

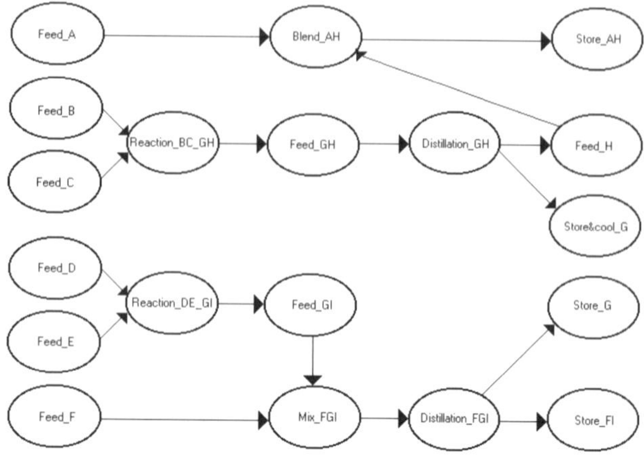

Figure 1. Example task network.

First, we find the two policies for the production of G and the single policy for the production of FI. These are shown in figure 3. The reliabilities of these policies are:

$$SR_1 = SR_3 = 0.99^6 \text{ x } 0.95 = 0.89$$

$$SR_2 = 0.99^3 \text{ x } 0.95 \text{ x } 0.99 = 0.91$$

Then, Model_1 is solved using CPLEX (1993). This results in a plan relying on policy Y_2 and producing 75 units of G and no mass of FI. Producing FI would demand the activation of Y_1 and Y_3, leading to a violation of the reliability constraint: $SR = 0.81 < 0.85$. Now, observing that Y_1 and Y_3 share seven task nodes, the *SR* bound is improved. This by reducing them to the modular structure shown in figure 4 and expressing *SR* as:

$$\log SR = y_2 \log 0.92 + c_1 \log 0.9 + c_2 \log 0.99 + c_3 \log 0.99$$

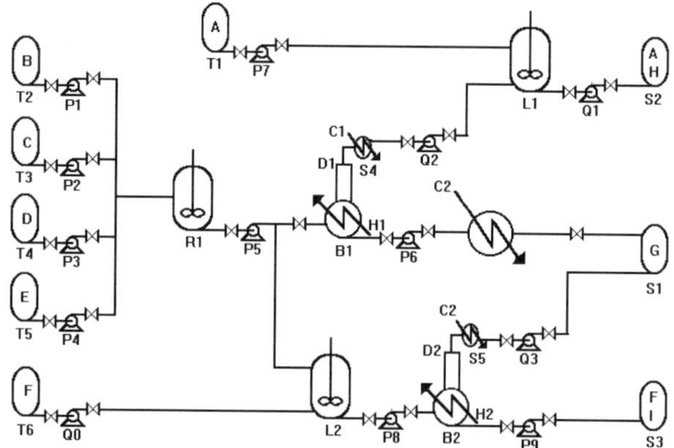

Figure 2. Example batch plan.

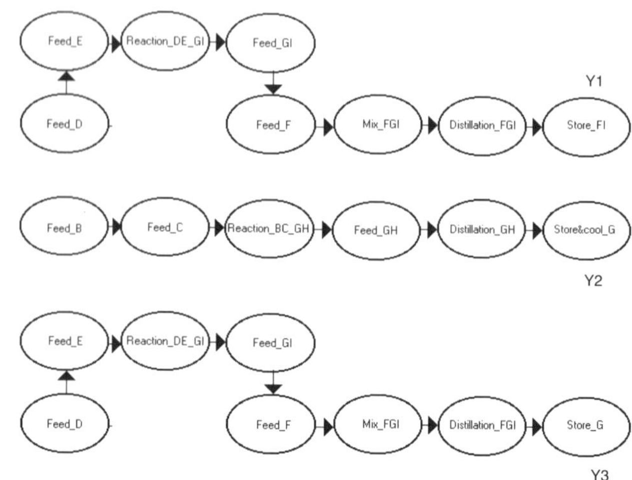

Figure 3. Production policies for G and FI.

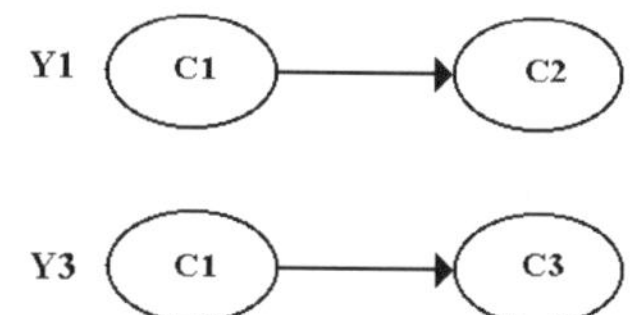

Figure 4. Reduced modular structure.

Now, the effect on the plan reliability of cut is considered only once. The solution of Model_2 resulted in a plan in which 40 units of G and 60 units of FI are produced. Due to the better SR estimation (SR=0.88) one can see that now the reliability constraint is not violated by the simultaneous activation of and. We can finally observe that share utility C2. There unique potential a value (α =1/0.95) becomes significant when all the policies are simultaneously activated

Conclusions

The synthesis of batch schedules resilient to process equipment failures is considered. The approach relies on a network representation for batch processes. The production routes for each product are aggregated into policies, whose reliability is estimated by established network reliability tools. A schedule stochastic reliability index is then calculated within a multiperiod MILP formulation. Even though the aggregation of routes simplifies the problem formulation it can result in a conservative reliability index. This will be the case if the production policies share process equipment. Hence, model reformulations are proposed, decomposing the production policies into their corresponding minimal cuts, and the cuts into networks of process units. Thus, a better estimation of the reliability index is obtained for the cases in which there are cuts common to the various production policies and when the cuts are associated (i.e. share process equipment). These cases are very common in batch processes. Therefore, the improvements obtained in the SR bounds using the proposed methodology can be very significant.

Acknowledgments

The comments provided by Dr. Daniel Lewin and Prof. Ram Lavie, both at the Technion, Israel, are gratefully acknowledged. Financial aid was provided by a Foreign and Commonwealth Office-Clore Foundation Scholarship.

References

CPLEX (1993). CPLEX Optimization Inc.

Gertsbakh I.B. (1989). *Statistical Reliability Theory*. Marcel Dekker, Inc.

Pistikopoulos E.N. and T.V. Thomaidis (1990). AICHE Annual Meeting, Florida.

Rotstein G.E., R. Lavie, and D.R. Lewin (1993). Submitted to *Com. & Chem. Eng.*

Straub D.A. and Grossmann, I.E. (1990). *Comp. & Chem. Eng.*, **14**, 967-985.

DESIGNING MOLECULES WITH GENETIC ALGORITHMS

Venkat Venkatasubramanian[†], King Chan, Anantha Sundaram and James M. Caruthers
School of Chemical Engineering
Purdue University
West Lafayette, IN 47906

Abstract

Designing new molecules possessing desired properties is an important and difficult problem in the chemical, material and pharmaceutical industries. The traditional approach to this problem consists of an iterative formulation, synthesis, and evaluation cycle that is long, time-consuming, and expensive. Current computer-aided design approaches include heuristic and exhaustive searches, mathematical programming, and knowledge-base methods. While all these methods have a certain degree of appeal, they suffer from drawbacks in handling combinatorially large and nonlinear search spaces. Recently, a genetic algorithm-based approach has been shown to be quite promising in handling these difficulties. In this paper, we review some recent results using this approach for the design of polymers and refrigerants.

Keywords

Computer-aided molecular design, Evolutionary design, Genetic algorithms, Structure-property relationships.

Introduction

The design of new molecules and composites with desired properties is an important and difficult problem encompassing the design of polymers, blends, drugs, pesticides, refrigerants, solvents, paints and so on. The goal of molecular design is to find a solution or a set of solutions in terms of the structure of the molecule given a set of target macroscopic properties or property constraints for the molecule. Traditional molecular design consists of an iterative formulation, synthesis, and evaluation cycle that is a long, time-consuming and expensive process. Thus, there exists considerable motivation for exploring computer-based molecular design (CAMD) approaches to the design problem.

Computer-based molecular design generally proceeds by combining pre-defined chemical building blocks, called base groups, along mainchain and sidechain positions. This construction process is difficult in general as the design space is large, complex, nonlinear and poorly understood. Current design approaches include heuristic and exhaustive searches, mathematical programming, and knowledge-base methods. However, the combinatorial complexity of the search space makes the heuristic and exhaustive searches less useful for large-scale molecular design problems. Mathematical programming methods require the solution of a difficult nonlinear optimization problem. The solution procedure is also susceptible to local minima traps. Lastly, the nonlinear structure-property relationships cannot be easily quantified for creating expert system design rules.

Recently Venkatasubramanian et al. (1992, 1994) proposed a new approach using genetic algorithms to tackle these difficulties. Genetic algorithms (GAs) are stochastic search procedures based upon the principle of Darwinian theory of natural selection and evolution. In this short paper, we discuss some recent results using this GA-based framework. After a brief introduction to the framework, we present results for polymer and refrigerant design case

[†] Author to whom all correspondence should be addressed

studies. The paper concludes with some thoughts on future directions.

Genetic Algorithms for CAMD

Genetic search approach allows a population, which represents potential solutions, to reproduce and create new individuals under competitive evolutionary laws (Holland, 1975). The interplay between all these components, in general, leads to gradual improvement and ultimately to solutions or near solutions to the design problem. The architecture of genetic algorithmic framework is generally divided into six components: (1) representation, (2) initialization, (3) fitness evaluation, (4) parent selection, (5) genetic operators, and (6) genetic parameters. These components are summarized in the CAMD context below. Detailed discussions on genetic algorithm fundamentals and applications can be found in De Jong (1975), Goldberg (1989), Davis (1991), Koza (1992), Michalewicz (1992) and Androulakis and Venkatasubramanian (1991).

Potential molecular design candidates are constructed from chemical building blocks called base groups. The GA is initiated randomly with feasible designs (i.e., fitness > 0) to provide good starting points. The parent selection procedure mimics the biological evolution process by governing the extent to which chromosomes survive to influence future generations. The most common form of selection uses a technique known as "fitness proportionate" selection. This selection is random but weighted in proportion to the normalized population fitness. The genetic operators appropriately modify the structure of the selected strings, to produce members of the next generation. The classical GA operators are one-point crossover and mutation (Goldberg, 1989). In one-point crossover, the two mating individuals are each cut at a randomly selected point, and the sections after the cuts exchanged. The offsprings produced, combine characteristics of both parents. In mutation each gene on the chromosome can be altered with a small probability. Mutation facilitates small-scale local searches in the design space while crossover facilitates large-scale global exploration. In our framework, the standard genetic operators have been modified to accommodate the rich chemistry of molecular interactions. Furthermore, new genetic operators were developed to facilitate the molecular design process. These include crossover at two positions, insertion or deletion of molecular groups, blending two designs into one, and repositioning base groups on the mainchain. These operators are called two-point crossover, insertion, deletion, blend and hop, respectively. The GA requires parameter values, such as population size, genetic operator probability rates, and fitness function constants which are normally chosen by a trail-and-error procedure. The adaptation of classical genetic algorithms to molecular design is described in greater detail in Venkatasubramanian et al. (1994).

Molecular Design Case Study

Polymer Design

This section presents an extension of the polymer design case study presented in Venkatasubramanian et al. (1994). The previous investigation utilized 4 mainchain (>C<, –⌬–, -C=OO-, -O-) and 4 sidechains groups (-H, $-CH_3$, -F, -Cl). The GA was required to design molecular structures given property values of known polymers, namely: (i) Polyethylene terephthalate (PET), (ii) Polyvinylidene propylene copolymer (PVP), and (iii) Polycarbonate of bisphenol-A (PC). The GA discovered all three target polymers in a fraction of the 200 generations allowed. The average generation number for locating the first target design and the success rate (in parenthesis) for an initial population initiated with random mainchain and sidechain groups having lengths 2-7 are: (i) PET 11.3 generations (100%), (ii) PVP 28.2 generations (100%), and (iii) PC 41.0 generations (100%). The paper also showed that the success rate of the GA search was in general better than that of random search.

In the follow-up case study, the base group and side group pools were increased to 17 mainchain and 15 sidechain groups as shown in Table 1. This causes the search space to explode from about 1.4×10^5 design candidates for the previous study to about 1.1×10^{13} for the current study for design lengths of 2-7 base units. Thus, this search space is about 100 million times larger than our earlier search space. As in the previous case study, the goal is to have the genetic algorithm design for property constraints of known engineering polymers to within ±0.5% of the desired property values. A weighted "gaussian" fitness function (Eq. 1) shown below was utilized:

$$F = \exp\left(-\alpha\left[\sum_{i=1}^{n} \frac{\left(P_i - \bar{P}_i\right)^2}{\left(P_{i,max} - P_{i,min}\right)^2}\right]\right) \qquad (1)$$

where P_i is the ith property value, $\bar{P}_i$, is the average of the maximum and minimum acceptable property values, $P_{i,max}$ and $P_{i,min}$, respectively and, α, is the fitness decay factor. The property constraints are density, glass transition temperature, thermal expansion coefficient, specific heat capacity, and bulk modulus. Physical property values are calculated by the van Krevelen (1990) group contribution methods. The initial population of all GA runs are initiated with random mainchain and sidechain groups. The results are compiled after 25 runs of 1000 generations each. The population size is 100, the maximum design length is 7, and the elitist policy keeps 10% of the previous best population members. The genetic operator probability values are as follows: i) crossover = 0.20, ii) mainchain mutation = 0.20, iii) sidechain mutation = 0.20, iv) insertion = 0.00, v) deletion = 0.10, vi) blend = 0.10, and vii) hop = 0.20.

Table 1. Polymer Design Base Groups.

17 Mainchain Groups

>C< -S- $-SO_2-$ -O- -C=O- -O-C=O-

-O-C=O-O- -C=O-O-C=O- -NH- -C=O-NH-

-O-C=O-NH- -NH-C=O-NH-

15 Sidechain Groups

-H $-CH_3$ $-C_2H_5$ $-mC_3H_7$ $-iC_3H_7$

$-tC_4H_9$ -F -Cl -Br -OH

$-O-CH_3$ $-O-C=O-CH_3$ $-O-C=O-O-CH_3$

-CN

Table 2. Polymer Design Search Results.

Target Polymer	Avg Gen No. for Locating Polymer Target	Success Rate
P1: Polyethylene terephthalate (PET)	184	**12%**
P2: Polyvinylidene propylene (PVP)	411	**36%**
P3: Polycarbonate of bisphenol-A (PC)	N/A	**0%**

Results and Discussion

The results are presented in Table 2. Column 2 gives the average generation when the target was first located. Column 3 gives the percentage of runs that were successful in locating the target (in **bold** text).

As one might have expected, the genetic search was not as successful as it was in the smaller case study. For the larger case study, the genetic search took much longer to find the target molecule, as seen from the larger generation numbers. Similarly, the genetic search was not successful in finding the target in every run as it was in the smaller case study. In fact, the success rates are considerably lower as one can see (in the case of polycarbonate the target was not found at all). However, the most important observation here is that the genetic search still succeeded in finding thc target molecule in two out of the three cases, though with a smaller success rate, even though the search space had exploded by a factor of 100 million. Given the tremendous increase in the search space size, it would not have been surprising at all if the genetic search had totally failed in all three cases. However, it is indeed quite surprising that the genetic search succeeded in some of the cases. This result offers a promising insight about the efficiency and the usefulness of the genetic search as a viable methodology for large, complex, molecular design problems. As in the earlier study, the GA was also able to determine many alternative structures with fitness greater than 0.90.

Refrigerant Design

Joback and Stephanopoulos (1989) have reported a detailed replacement refrigerant molecular design case study subjected to target property constraints using their heuristic-enumeration methodology. The present case study solves the same problem with the GA approach. The target property constraints as specified by Joback and Stephanopoulos are given below and summarized in Table 3. The current work also examines designing refrigerants which offer high thermal efficiency as well as satisfying property constraints. An ideal vapor-compression refrigeration system was used since it takes into account the thermodynamic properties of the refrigerant. Group contribution property estimation procedures of Joback and Reid (1988) and Joback (1989) were employed.

Table 3. Target Property Constraints for the Refrigerant Design Problem.

Property	Temperature	Value
1. $P_{V\text{-}low}$	272.05K	≥ 1.4 bar
2. $P_{V\text{-}high}$	316.45K	≤ 14 bar
3. ΔH_V	272.05K	≥ 18.4 kJ/g-mol
4. C_p	294.25K	≤ 32.2 cal/g-mol

The fundamental unit or base groups for constructing a refrigerant molecule is a variant and subset of those used in Joback's case study. The current GA does not consider ring compounds as many are toxic and thus environmentally unfavorable. Furthermore, the base groups are redefined such that all group combinations result in feasible compounds provided that the base groups appear in their appropriate positions in the design that are based on the categories defined below. This removes the enumeration step to determine chemically feasible designs as is required by Joback's method. The base groups are divided into 3 parts: i) mainchain, ii) sidechain, and iii) endgroups for cases when Lengths =1 and ≥2. This division is necessary since some mainchain groups can act only as endgroups or as non-endgroups. Moreover, the number of sidechains for a particular mainchain depend on the chain length as well as group location. The refrigerant design mainchain and sidechain groups are illustrated in Table 4.

Table 4. Refrigerant Design Base Groups.

10 Mainchain Groups

>C<	>C=C<	>C=C=C<	-C≡C-	>N-
-O-	-S-	-C=O-	>C=C=O	-C=O-O-

13 Sidechain Groups

-H	$-CH_3$	-F	-Cl	-F
-Br	-I	-OH	-CN	-SH
$-NH_2$	$-NO_2$	-COOH		

The genetic operators are slightly modified to account for the creation of infeasible molecules. Infeasible designs appear since some operators may result in inconsistent sidechain numbers. Chemical feasibility is maintained by repairing infeasible designs into feasible ones.

The property fitness function brackets off a lower and upper bound in which no constraints are violated and the fitness =1. The properties which fall outside these bounds have their fitness reduced by a "sigmoidal" shaped fitness function. The fitness functions employed for the refrigerant design study are given below:

$$Fi = \begin{cases} 1, \text{ if } Pr,min \le Pi \le Pr,max \\ Sig\,i, \text{ if } Pr,min > Pi \text{ or } Pi < Pr,max \end{cases} \quad (2)$$

$$Sig_i = \exp\left(-\alpha i\left[\frac{(P_i-P_{i,min})^2}{(P_{r,max}-P_{r,min})^2}\right]\right) \quad (3)$$

$$F_{prop} = \frac{1}{n}\sum_{i=1} F_i \quad (4)$$

were $P_{r,min}$ and $P_{r,max}$ are the absolute minimum and maximum property values to design for. Their difference also normalizes the property value to remove possible bias from any single property. $P_{i,min}$ is the minimum property value which yields a fitness of 1. The function F ranges from 0 to 1, 1 when P_i is within $P_{r,min}$ and $P_{r,max}$. The fitness decay rate is controlled by α_i. For example, α_i for a Cp value greater than 32.3 is 100 since it gives a rapid decay in fitness for constraint violations. On the other hand, α_i for Cp values less than $P_{cp,min}$ is 10 which allows a more gradual decay of fitness. The parameters for equations 3 and 4 are listed in Table 5.

Table 5. Fitness Function Design Constants.

Property	$P_{i,min}$	$P_{r,min}$	$P_{r,max}$	$\alpha_{i,rmin}$	$\alpha_{i,rmax}$
1. $P_{V\text{-}low}$	1.4	0	10.0	10	100
2. $P_{V\text{-}high}$	5.0	0	20.0	100	10
3. ΔH_V	18.4	0	60.0	10	100
4. C_p	17.0	0	40.0	100	10

The thermal efficiency fitness is defined as the ratio of the coefficient of performance (COP) of the ideal vapor-compression cycle vs. the COP of the Carnot cycle ($F_{eff} = COP_{vap\text{-}comp}/COP_{Carnot}$). For thermal efficiency calculations, the refrigeration system must maintain an environment at 260K for available cooling water at 295K. The refrigeration coils and condenser are of sufficient size that a 5K approach can be realized. More detailed thermodynamic theory on the refrigeration process is located in (Cerepnalkovski, 1991). The overall fitness when considering both property and thermal efficiency constraints is given by the linear combination of F_{prop} and F_{eff}.

$$F_{total} = \beta_1 F_{prop} + \beta_2 F_{eff} \quad (5)$$

where, β_1 and β_2 are the weight factors for F_{prop} and F_{eff}, respectively. The sum of β_1 and β_2 is unity so that F_{total} ranges from 0 to 1. Equal weight is given to F_{prop} and F_{eff} as both β_1 and β_2 are equal to 0.50.

The genetic search is required to design for refrigerants subjected to property and thermal efficiency constraints. The population size is 100, the elitist policy keeps 10% of the previous best population members. The initial population consist of refrigerants with random mainchain and sidechain groups and random design lengths of 1 to 7. For statistical significance, results were compiled after 25 GA runs of 500 generations each. The genetic operator probability values are as follows: i) crossover = 0.30, ii) mainchain mutation = 0.20, iii) sidechain mutation = 0.20, iv) insertion = 0.10, v) deletion = 0.10, vi) blend = 0.10, and vii) hop = 0.00. The hop operator is not required as the order of the base groups do not affect the predicted physical properties.

Results and Discussion

The number of solutions which satisfied all property constraints is 19. 15 of 19 design candidates had efficiency fitness values of 0.6 to 0.7. The other four were between 0.7 and 0.8. Table 6 provides an illustrative example of designs which satisfy all property constraints and have high thermal efficiencies. Refrigerants molecules #6, 8 and 9 in Table 6 are used in current practice. The GA was able to identify the correct refrigerant molecule length as no designs exceeded three base groups even though longer designs were allowed. Refrigerant designs with efficiency fitness values over 0.70 had the -C=O- group in common. Since the program does not consider cost, stability, toxicity or flammability constraints, etc., the proposed design must be experimentally tested for these properties. However, such additional constraints can be easily integrated into the genetic search. Furthermore, the GA also found numerous near-optimal solutions with high property and efficiency fitness values (71 refrigerants with $F_{tot} \geq 0.80$).

Table 6. Illustrative Refrigerant Design Results for Property and Thermal Efficiency Constraints.

Refrigerant Molecule	Pv-low	P_V-high	ΔH_V	C_p	F_{eff}
1. $H_3C\text{-}CH{=}C{=}O$	1.51	6.67	27.77	28.97	.820
2. $CH_3\text{-}C{=}O\text{-}CH_3$	1.73	7.34	27.14	30.17	.818
3. $H_3CCF{=}C{=}O$	1.57	6.97	26.96	31.39	.789
4. $FH_2C\text{-}CH{=}C{=}O$	1.58	7.12	26.88	31.83	.774
5. $HC{\equiv}CCH_3$	1.69	6.35	21.44	24.18	.690
6. CH_3Cl	1.61	6.25	21.58	20.10	.663
7. $H_3C{=}C{=}CH_3$	1.55	5.97	20.85	25.03	.657
8. CH_2FCl	1.67	6.66	20.71	23.00	.630
9. $H_3C\text{-}CH_3$	2.74	9.55	18.67	22.58	.622
10. $H_3C\text{-}N\text{-}FCH_3$	1.83	7.33	20.42	30.15	.610

Summary and Conclusion

In this work, some recent results from the design of polymer and refrigerant molecules are presented. The studies investigated the performance of the genetic search algorithm for molecular design for large search spaces. It was found in the polymer study that despite the tremendous increase in the search space size, the genetic search was generally able to find the target molecules though with a much less success rate and much more slowly. For the refrigerant study, the GA rapidly and efficiently comes upon a list of refrigerants which satisfy property and thermal efficiency constraints. Some of these designs may be promising refrigerant alternatives as they do not contain CFCs. Furthermore, an appealing feature of genetic search is that it finds a collection of nearly-optimal candidates whenever it fails to find the exact target. Another useful property is the ability to incorporate a wide variety of constraints such as higher-level chemical/biological knowledge, environmental and toxicology constraints and so on. Primary limitations of this approach are its heuristic character which does not guarantee an acceptable or optimal solution and the trial-and-error determination of the various search parameters.

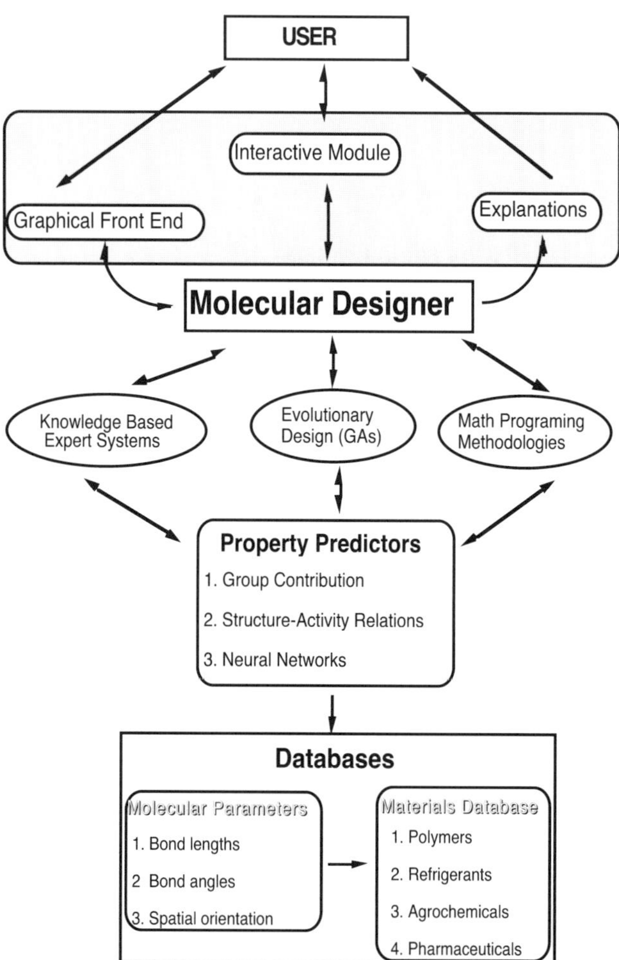

Figure 1. Proposed Integrated Framework for CAMD.

Speculating about the future directions of this approach, one realizes that the real-life CAMD problems are so difficult and complex that a successful approach would require a combination of several techniques integrated into a unified, interactive, framework. Figure 1 shows one such possible framework. In such a framework, the GA-based design engine would be coupled with expert systems as well as nonlinear programming methodologies to improve the effectiveness of the overall design process. We also envision such a system as an interactive one, where the molecular designer actively participates in the design process.

References

Androulakis, I.P., and V. Venkatasubramanian (1991). A Genetic Algorithmic Framework for Process Design and Optimization, *Computers Chem. Engng.*, **15**(4), 217-228.

Cerepnalkovski, I. (1991). *Modern Refrigerating Machines*. Elsevier, Amsterdam.

Davis, L. (Ed.) (1991). *Handbook of Genetic Algorithms*. Van Nostrand Reinhold.

De Jong, K.A. (1975). *An Analysis of the Behavior of a Class of Genetic Adaptive Systems*, Ph.D. Thesis, University of Michigan, Michigan, Ann Arbor.

Goldberg, D.E. (1989). *Genetic Algorithms in Search, Optimization, and Machine Learning*. Addison-Wesley.

Holland, J.H. (1975). *Adaptation in Natural and Artificial Systems*. The University of Michigan Press, Ann Arbor.

Joback, K.G., and R.C. Reid (1987). Estimation of Pure-Component Properties from Group Contributions, *Chem. Eng. Comm.*, **57**, 233-243.

Joback, K.G. (1989). *Designing Molecules Possessing Desired Physical Property Values*, Volume 1 and 2. Ph.D. Thesis in Chemical Engineering, MIT.

Joback, K.G., and G. Stephanopoulos (1989). Designing Molecules Possessing Desired Physical Property Values, *Proc. FOCAPD '89*, 363-387, Snowmass, CO.

Koza, J.R. (1992). *Genetic Programming: On the Programming of Computer by Means of Natural Selection*. MIT Press, Cambridge, Mass (1992).

Michalewicz, Z. (1992). *Genetic Algorithms + Data Structures = Evolution Programs*. Springer-Verlag, Berlin.

van Krevelen, D.W., and P.J. Hoftyzer (1990). *Properties of Polymers, their Estimation and Correlation with Chemical Structure*. Third Edition, Elsevier Scientific.

Venkatasubramanian, V., K. Chan, and J.M. Caruthers (1992). Designing Engineering Polymers: A Case Study in Product Design, *AIChE Annual Meeting*, 140d, Miami, FL, November.

Venkatasubramanian, V., K. Chan, and J.M. Caruthers (1994). Computer-Aided Molecular Design Using Genetic Algorithms, *Computers Chem. Engng*, **18**, 9, 833-844.

A MIXED INTEGER (NON) LINEAR PROGRAMMING APPROACH TO SIMULTANEOUS DESIGN OF PRODUCT AND PROCESS

Aydin K. Sunol
Chemical Engineering Department
University of South Florida
Tampa, FL 33620

Abstract

Product and process design are integrated through mixed integer (non) linear programming. The 0-1 variables are for discrete processing options and raw material/additive choices while continuous decision variables are for processing conditions and flows. Product design involves relating product mix and processing conditions to product characteristics. The multi-objective approach called for this type of problems is addresses through relative weighing of the product attributes in the objective function. The lumped parameters used in the model are derived from detailed distributed models using a two tier approach that includes heuristic and algorithmic elements. The example used is a porous matrix-polymer composite and its process design through supercritical fluid aided impregnation and surface treatment.

Keywords

Mixed integer nonlinear programming, Product design, Process design.

Introduction

How to define and set boundaries for Chemical Product Engineering is naturally subject to debate. One view point is as follows: A chemical product contains several components each of which have their own attributes; Although the individual components are usually separate substances, they could very well be groups of atoms making up a molecule. In this context, simple chemicals are not regarded as a chemical product. Differentiation of chemical products from other manufactured goods may be appropriate to focus as well. In chemical products, the formulation is important and its form is not. This last statement naturally brings a debate on whether chemical product attributes are affected by its form and whether that form is attained through chemical processing or not.

In product engineering, physical chemistry and transport processes is more important than the unit operations. The relative importance of marketing as well as research and development in product design is significantly more that in process design. Boundaries of product engineering and its interfaces with other sciences is depicted in Fig. 1.

It is no doubt that in educating the chemical engineers of the future more emphasis to product design has to be given (Wei, 1988). Depending on the volume of production and the particular chemical processing industry, the relative importance of product engineering, process engineering, and chemistry may change (Wesselingh, 1991). This relationship is depicted in Fig. 2. Regardless of the relative contribution, the need for an integrated effort for product and process design is important. It is not too surprising that most integrated approaches to simultaneous product engineering and process engineering is for composites (Olsen and Vanderplaats, 1989; Hajela and Shih, 1989).

In process design, satisfaction of multiple objectives is important and is increasingly becoming more important. The incorporation and meeting of the multiple objectives in product design is almost essential. Multiple objectives could be handled explicitly in the objective function or implicitly as constraints. If the product attribute measures are incorporated in the objective function, they are not bounded as one would have with their implicit incorporation as constraints. There is ever increasing concern in environmental acceptability of the process and the impact of the product. Therefore, the approach taken here is hybrid. While performance measures such as strength, durability, etc. are incorporated within the objective function in a weighted fashion, environmental restrictions are addressed as constraints or in pre-processing.

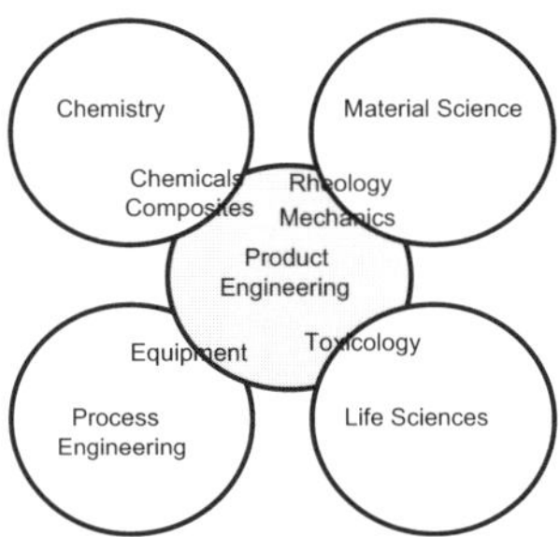

Figure 1. Boundaries and interfaces of chemical product engineering.

A Generalized Modeling Framework

The tasks involved in product and process design involve algorithmic and heuristic processing of symbolic and numeric data. Therefore, a hybrid approach that interweaves numerical and heuristic approaches is warranted. Only algorithmic component of the approach will be discussed here.

Traditionally, unit operations models of varying complexity have been used in flowsheet simulators or algorithmic synthesis procedures. These unit models are somewhat inadequate for product design purposes where the physical chemistry and transport phenomena become more important. Identification of the phenomena involved in product performance and techniques useful in design are the initial steps taken. This is followed a detailed mechanistic model (and/or experimental prototype) and a simpler lumped parameter model that could be used in optimization (and experimental design). A two tier approach that updates the rigorous and lumped parameter models is used here. The lumped parameter model is optimized using a MINLP approach.

For recent advances in the field, reader should refer to Grossmann and Floudas (1994).

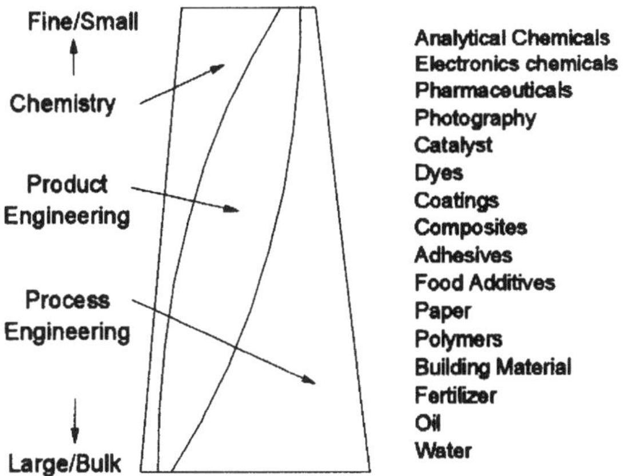

Figure 2. Relative contribution of process engineering, product engineering and chemistry in various CPIs.

Mathematical Modeling Formulation

The general formulation of the MI(N)LP problem

$$\max_{\underline{y}\,\underline{x}} \quad OBJ = (w)^T \underline{A} - (\alpha)^T \underline{y} - (\beta)^T \underline{x} \quad MILP$$
$$[-f(\underline{x})] \quad MINLP$$

$$\text{s. t.} \quad H(\underline{x},\underline{y}) = 0$$
$$G(\underline{x},\underline{y}) < 0$$
$$(\text{ the form used is: } \quad d^L \le D_1 \underline{y} + D_2 \underline{x} \le d^U$$
$$E_1 \underline{y} + E_2 \underline{x} = e)$$
$$y_j = 0,1$$
$$x_i \ge 0$$

The elements of the objective function includes the different product attributes, cost of product components, and the capital cost. Linear model with an intercept ($\alpha + \beta X$) is applicable for both the capital cost and product attributes while the raw material cost does not have an intercept. Each product attribute is weighted, w_i. Objectives (product performance characteristics) are a function of its components, their relative amount, and the processing conditions

$$A_l = \alpha^T \underline{y} + \beta^T \underline{x} \tag{2}$$

with appropriate activation of continuous decision variables.

$$X_j^L Y_j \le X_j \le X_j^U Y_j \tag{3}$$

Since the product design problem is primarily a blending problem, the material balance relations is sufficient and energy balance considerations are neglected. The batch vs. semi-continuous vs. continuous considerations could be incorporated. However, here, since we are primarily dealing with techniques/phenomena, a pseudo steady-state treatise is used.

$$\sum_{m\in I_n}\sum_{k\in K_m} S_{cmk}F_{cmk} - \sum_{m\in O_m}\sum_{k\in K_n} F^{m}_{cmk} = 0 \qquad \begin{matrix} c \in C \\ n \in N \end{matrix} \tag{4}$$

here you have C components and N techniques (phenomena).

The lower and upper bounds on continuous decision variables are very important both from the computational point of view as well as modeling. For example, use of the basic product rate as lower limit ensures meeting the target production. Similarly, care in setting the product attribute bounds is essential.

The types of logical constraints used are:

1. To activate the component flows associated with each unit.

$$\sum_{m\in O_n} F^{m}_{cmk} - Uy_{nk} \leq 0 \qquad k \in K_n, n \in N, c \in C \tag{5}$$

2. Co-existence of number of units (techniques), or components.

$$\sum_{k\in K_n} y_{nk} < Constant \qquad (also \leq \; or =) \qquad n \in N \tag{6}$$

3. Technique m can only exist in condition k only if unit n exists in condition k.

$$\sum_{n\in N} y_{mk} = y_{nk} \qquad k \in K_n \tag{7}$$

In the implementation stage of the formulation, disaggregation of the variables is required. The following variables and terms are used.

X_i^c = Component amount (or F for rates)
X_i^P = Product components used (or F for rates)
X_i^R = Amount of Raw materials used (or F for rates)
X_i = Intermediate components amounts (or F for rates)
X_{jk} = Operating condition value of technique k for technique j
y_{jk} = Operations condition of unit j at condition k
S_{ijk} = Split (yield) fraction of each component i in unit j for condition k
y_j^P = existence/non existence of component i in product
y_j = existence/non existence of unit (process function or concept) j
I_j = { j | unit j has input (rate/amount) from j}
0_j = {j' | unit j has output rate/amount) from j'}
w_l = Attribute weight
α_j^c = Fixed cost coefficient of unit j
β_j^c = Variable cost coefficient of unit j
β_i^R = Variable cost coefficient for raw material i
A_l = Attribute Value
$\alpha_{l,i,j(k)}$ = Fixed contribution coefficient of component i at condition of k of unit j
$\beta_{e,i,j(b)}$ = Variable contribution coefficient of component i at condition k of unit j

Description of the Example

A novel two stage approach for manufacture of custom-made products with superior mechanical and chemical properties is used as an example (Sunol, 1991). There is an existing pilot plant to gather experimental data as well as accumulated know how. (Ward et. al, 1990). The principle of the process is rather simple and involves combination of supercritical impregnation followed by supercritical fluid aided surface treatment. Solvent-impregnant mixture penetrate into a porous wood matrix in the supercritical state. The desired chemicals are than deposited inside the porous material while the supercritical solvent is flashed off to be recycled. The possible impregnants that include monomers may be subsequently further treated, e.g. polymerized. The second stage involves surface treatment where the coatings are sprayed from a supercritical phase. The coatings may be either liquid or solid upon release from the supercritical phase.

Implementation Issues and Computational Aspects

There are a number of possible impregnants, solvents, and entrainers that could be used to impart different product performance characteristics (namely aesthetic, fire-proofing, bioactivity resistance, mechanical strength, and machinability) to a variety of species (i.e. different woods or reconstituted woods such as fiberboard, flakeboard, or particleboard). Each combination will bring its corresponding processing implications such as possible use and extent of vacuum and extraction, continuous vs. batch impregnation, direction (traverse, radial, and/or tangential) and relative amounts of permeation and diffusion, the polymerization technique (e.g. in-situ temperature swings for polymerization, radiation curing, catalyst-heat curing) and the conditions used. Thus, the design and eventually the plant has to be quite flexible and multi-purpose.

A mechanistic distributed permeation model was developed to determine the monomer uptake of the wood. The model was coupled with algebraic set of equations for thermophysical parameters such as viscosity and density. The model successfully represents the experimental findings and provides parameters such as permeability constant and effective diffusivity useful for scale up.

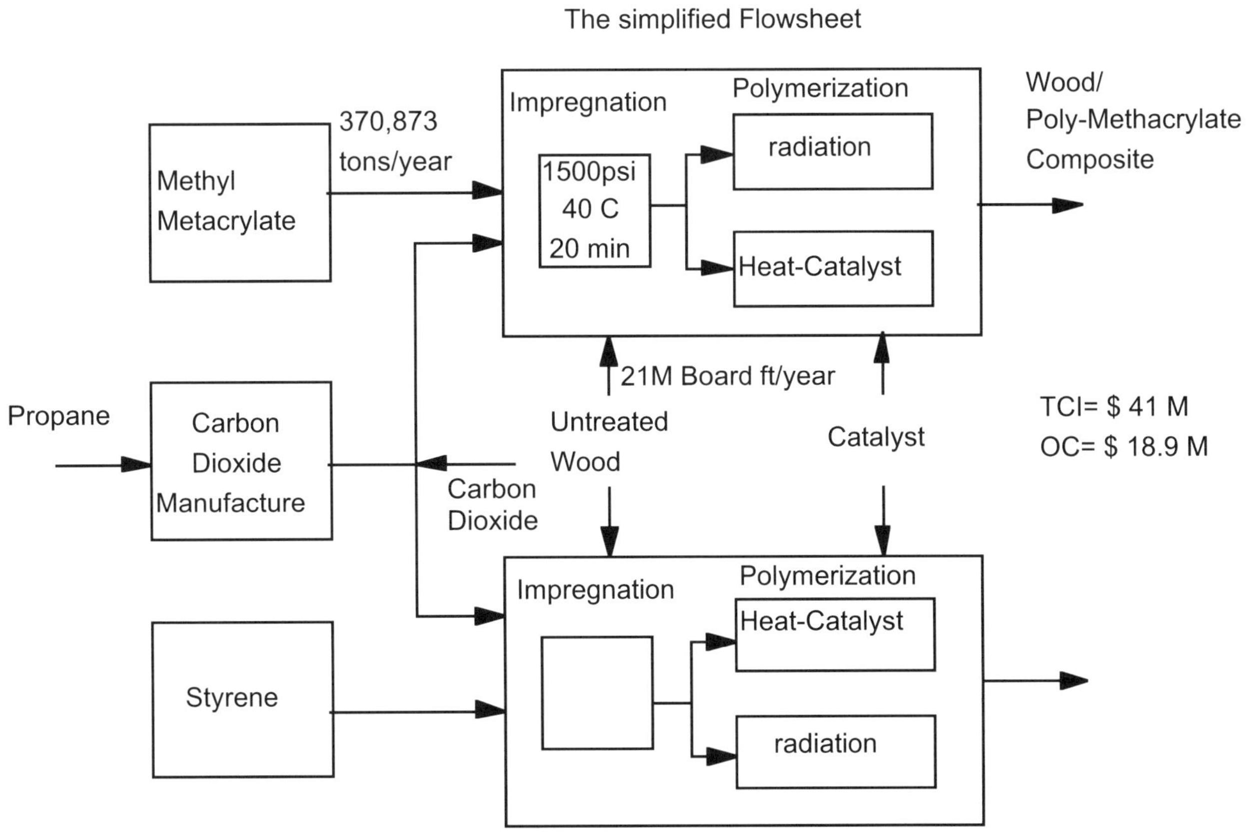

Figure 3. An example with the case study.

The permeation model provided the parameters used in a less detailed process simulation model that in turn provided the parameters for the MINLP model.

The MI(N)LP model is tested for various scenarios, one of which is shown in Figure 3. In this formulation, carbon dioxide manufacturing/purchasing options were explored as well. The continuous lines show the selected path and corresponding operating conditions. The weight and strength were the product attributes used in the particular example. The problem was solved using GAMS/DICOPT.

Conclusions

The utility of MI(N)LP in simultaneous design of processes and products is quite encouraging. The knowledge of the product and process is of probably the most important element in the success of the MINLP based modeling. Incorporation of product scheduling aspects would further enhance the utility of the approach. The nonlinear formulation is naturally much more powerful especially in representing the multiple objectives as a function of the operating condition history. Variable transformation/constraint reformulation that would convexify the system and special algorithms for bilinear MINLPs (Adams and Sherali, 1993) all contribute in enhancing its power.

Acknowledgments

The partial support provided by the Florida High Tech Counsel and the Advanced Separation Technology Inc. is gratefully acknowledged

References

Adams W.P., and H.D. Sherali (1993). Mixed Integer Bilinear Programming Problems. *Mathematical Programming,* **59**, 279-305.

Hajela P., and C.J. Shih (1989). Optimal Design of Laminated Composites using a modified Mixed Integer and Discrete Programming Algorithm. *Computers and Structures*, **32**, 213-221.

Grossmann, I. and C. Floudas (1994). Algorithmic Approaches in Process Synthesis: Logic and Global Optimization, FOCAPD-94, Colorado.

Sunol, A.K. (1991). Supercritical Fluid Aided Treatment of Porous Materials. U.S. process patent, #4992308.

Ward, D., T. Dinatelli, and A.K. Sunol (1990). Supercritical Fluid Aided Wood-Polymer Composite Manufacture. AIChE National Meeting, Orlando.

Wei, J. (1988). Educating Chemical Engineers for the Future. In Sandler and Finlayson (Ed.), *Chemical Engineering Education in a Changing Environment*. Engineering Foundation.

Wesselingh. Chemical Product Engineering. An article on a course at Delft University, Holland.

USING DISTRIBUTED COMPUTING TO SUPPORT INTEGRATED BATCH PROCESS SCHEDULING, PLANNING AND DESIGN UNDER MARKET UNCERTAINTY

Matthew. H. Bassett, Gautham. K. Kudva, Joseph. F. Pekny and Sriram Subrahmanyam
School of Chemical Engineering
Purdue University
West Lafayette, IN 47907-1283

Abstract

Mathematical programming approaches to engineering problems involve the solution of Mixed Integer Linear Programs. In the case of Design, Planning and Scheduling problems, the problem size may be quite formidable. In this paper, a formulation for the design and planning of batch chemical plants is used to show the value of distributed computing for accelerating the location of good feasible solutions for such problems.

Keywords

Distributed computing, Branch and bound, Design, Planning, Scheduling, Batch plants.

Introduction

Process scheduling, planning, and design decisions encompass order of magnitude differences in time scale in the sense that the duration over which choices remain in effect and period at which decisions must be made varies widely. For industrial processes that we have studied, typical scheduling time scales range from one to twenty-four hours corresponding to the length of time required to conduct basic production activities (e.g. batch processing times, equipment cleanout, assay times, shipment frequency, etc.). Process planning time scales typically range from weeks to months corresponding to various supply chain considerations, business cycle and/or marketing projections. At the upper extreme, design decisions typically address a time scale as short as several weeks or months (retrofit applications) to years or decades (grassroots design) corresponding to how frequently equipment can be installed to provide new capabilities as well as the demand forecasts. Following these differing time scales, a traditional separation exists between scheduling, planning, and design functions. This separation makes solution of underlying problems more tractable but is somewhat artificial in that effective design and planning choices cannot be made without considering how a plant can be operated. In a general sense, aggregate decisions must be consistent with the capabilities implied by detailed decisions made at the shortest time scale. One way to achieve this consistency is to impose conditions on operation which simplify design and planning. For example, one can assume that a facility will operate in a campaign mode with a certain cycle time or equipment can be dedicated to particular tasks. Clearly the validity of these types of assumptions depends on the application, but ideally any operating mode should be suggested as a result of the physics of the application rather than an artifact of the design or planning methodology. In the next section, we summarize a modeling approach that integrates scheduling, planning, and design decisions without imposing any conditions on operation. Computational results are given later for industrial test problems.

Integrating Scheduling, Planning, and Design

Theoretically, a comprehensive model could be built based on the scheduling time scale and used for planning and design but such a model would be impossibly large to solve for almost any application. Practically, such a model is unnecessary since detailed operating projections several weeks, months, or years into the future will certainly not be implemented as they are projected because of changes in conditions. As an alternative to a monolithic detailed model for integrated scheduling, planning and design, we have extended the hierarchical model of Subrahmanyam et al. (1994) that divides the time axis over the horizon of interest into a number of segments as in Figure 1. An aggregate Mixed Integer Linear Programming (MILP) model, the **Design SuperProblem** (DSP), taking into account resource constraints, production capacity, demands, costs, and material balances is built using such a non-uniform discretization of time. For a one year planning horizon, there may be twenty-six, two week time segments. For a twenty year design horizon there may be eighty, three month segments.

The actual number of time segments is controlled by the size and detail needed in the aggregate model, the frequency with which decisions must be made, and the available computational resources to be applied to solving it. Within each time segment a detailed scheduling problem is solved to determine if the production plans of the aggregate model can be met. If this detailed scheduling problem is still too large, the aggregate model is applied recursively until tractable scheduling problems are obtained. If the detailed scheduling model determines that the aggregate model was overly optimistic, information is fed back to the aggregate model in the form of constraints which further limit production allocation within the offending time segment(s). Subsequent solution of the aggregate model will then require more realistic production targets or equipment purchases and can adjust production over the entire planning horizon accordingly.

Thus our approach ultimately produces a number of detailed scheduling problems that are coordinated by the aggregate model. Because the scheduling problems only address a relatively small amount of time their solution is readily accomplished using our existing scheduling tools. Note that the scheduling problems are not solved to determine detailed operation over the entire planning and design horizon, rather they simply verify that the aggregate model solution can be implemented. When solved using the techniques described in Pekny and Zentner (1993), the scheduling problems will indicate a natural operating mode for the facility that is consistent with the overall planning and design in an uncertain market. If schedule regularity is a natural consequence of the problem, then it will be a result of scheduling problem solution rather than an imposed constraint. If, for some reason, simple or complex schedule regularity is a requirement, then additional constraints can be imposed on the scheduling problem formulations.

Uncertainty can be incorporated into scheduling, planning, or design models by making parameters random variables and imposing an objective function that seeks to accommodate uncertainty, e.g. maximize expected net present value, etc. Unfortunately such direct treatment greatly complicates solution efforts by introducing severe non-linearity. With current solution technology and foreseeable computational capability such non-linearity greatly compromises the size of problems that can be addressed and eliminates possible consideration of many practical issues.

As an alternative to such direct treatment of uncertainty, we have developed the concept of scenarios in the context of batch plant design under uncertainty (see Subrahmanyam et al. (1994)). Intuitively, a scenario is a set of estimates of demand and corresponding prices for products during a given time segment along with an associated probability. In a given time segment, the sum of scenario probabilities must be one. A user is free to specify scenarios in different time segments independently. Continuous probability distributions can be approximated by using scenarios and the quality of approximation scales with the total number of scenarios.

Computational expense also scales with the number of scenarios but the aggregate model remains a Mixed Integer Linear Program (MILP). The objective function of the aggregate model can reflect direct economic considerations (e.g. minimize expense, maximize expected net present value, etc.) or product based objectives (e.g. plan to meet the most probable demands for products). For example, the model of Subrahmanyam *et al.* (1994) uses maximization of expected net present value as the objective function.

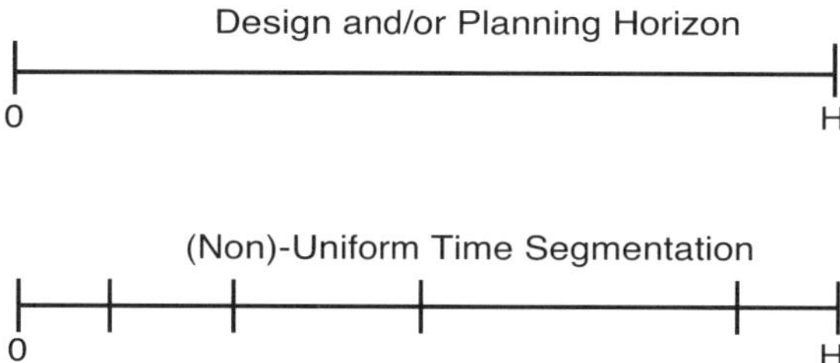

Figure 1. Decomposition Based on Segmenting Time.

Role of Distributed Computing

The above modeling approach provides several opportunities for exploiting parallelism. In particular, the branch and bound process for solving the Design SuperProblem MILP can be executed in parallel. In addition, the scheduling sub-problems implied by the Design SuperProblem (DSP) are independent and can be trivially solved in parallel. Finally, assuming each individual scheduling sub-problem is solved using exact or approximate branch and bound, then the same sort of parallel solution opportunities are available as with the DSP. There are three key issues associated with exploitation of parallelism in routine solution of integrated scheduling, planning, and design problems: (1) the computational details must be transparent to the end user, (2) most calculations are done in a "what-if" mode where obtaining good, but not necessarily optimal results, in a short time is more important than optimal results in a relatively long time, and (3) additional algorithm development to support specific properties of a particular problem must be implemented easily and rapidly.

The Distributed Control Architecture for Branch and Bound Calculations (DCABB) System provides facilities for addressing these issues in a parallel/distributed computing environment (Kudva, 1994, Kudva and Pekny, 1994). In particular DCABB supports parallel/distributed search of branch and bound trees, competing search of multiple trees that result from the use of different algorithms/formulations (Pekny, 1989), and cooperating branch and bound trees that result when a single thread of control spawns multiple branch and bound processes which must be completed before execution of the thread can continue. Upon its solution, the DSP implies such a thread of control for which the multiple scheduling subproblems must be solved before the next iteration can continue.

The DCABB system assumes full responsibility for conducting all branch and bound calculations on a locally or wide area distributed network of computers, some of which may possess multiple processors. Furthermore, because DCABB provides layered functionality ranging from pure communication issues at the low level to solution of general purpose MILPs at the high level, users are free to tailor algorithms as much as is needed/warranted for the task at hand. Finally by specifying that branch and bound trees may be terminated after a certain wall clock time or number of nodes, DCABB allows the user to apply the appropriate amount of computational effort depending on whether calculations warrant speed or precision.

Design/Planning Computational Results

To show the advantage of distributed computing in design/planning three problems are discussed (see Table 1). All the problems are based on industrial data for a three stage process. Within this process, all of the isolated intermediates, dirty solvents, and raw materials are storable. The only resources not storable are the mixtures obtained prior to filtration. In the first two cases, there are a total of 13 tasks to be performed and 22 resources used. A total of 9 unit types are available to choose from 5 different equipment families. Sales of the product are during the second quarter of each year for the life of the plant with 10 scenarios each quarter. See Subrahmanyam (1994) for details.

The first two problems are for plant designs covering 3 1/2 and 4 1/2 years respectively with a single product. The third problem is an extension of the first two with one additional product and covering 10 years. Table 1 details the dimensions of the problems. Since this paper details the advantage of using distributed computing in the design framework context, only the solution of the Design Super-Problem (DSP) is considered here. CPLEX-MIP (CPLEX Optimization, Inc., 1993) was unable to find any solution for Example 1 within 60 mins of CPU time using an HP 9000/755 computer. Example 2 was solved to the first feasible solution which was within 30% of the optimal solution in under 2 hours of CPU time using the same computer. In the case of Example 3, CPLEX-MIP was unable to find any solutions after 8 hours.

DCABB was used to perform both serial and distributed branch-and-bound searches for these problems. The serial (single processor) runs were conducted on a Hewlett-Packard 9000/755 workstation. One HP 9000/755 workstation and 7 HP 9000/715 workstations were used for the distributed runs. The rounding heuristic described by Subrahmanyam et al (1994) was applied at each node in the tree to obtain integer feasible solutions. These solutions allowed the search to quickly prune many parts of the tree, thereby narrowing the search space.

The ability to distribute the work among a number of processors led to significant improvements in the time taken to obtain incumbent solutions. In Example 1, neither method (serial and distributed) was able to close the bound gap significantly (solution within 5% of optimal), but using eight processors we were able to achieve this equivalent solution value about five times faster (serial case solution time was 550 CPU secs. whereas the distributed case took 110 CPU secs.). For Example 2, the bound gap was almost closed (to within 0.1% of optimal) with the distributed solution being obtained eight times faster than the serial one (serial case solution time was under 14000 CPU secs. and the distributed case solution time, under 2000 CPU secs.), whereas CPLEX-MIP could only manage a solution which was within 30% of optimal. For Example 3, both methods achieved some bound gap closure (8.35% for the distributed case in 4000 CPU secs., and 8.4% for the serial case in 20000 CPU secs.), with the distributed algorithm able to converge to a better incumbent than the sequential algorithm.

In most design and planning cases, obtaining good feasible solutions quickly is more important than obtaining a single optimal solution, especially in cases where feasibility studies are being conducted. In these situations, the ability to rapidly generate feasible solutions provides the engineer with the flexibility to explore a large number of designs, and choose between designs that are economically equivalent but differ radically from an implementation standpoint.

Note that the distributed MILP solver used in the above experiments has not been customized to exploit the structure inherent in the design problems. A number of enhancements, including pivot-and-complement heuristics for incumbent generation, use of warm-start basis, and variable fixing based on physical implications can be used to improve the performance of the MILP algorithm.

Table 1. Data for Example Problems.

Problem No.	Time Intervals	No. of Units	No. of Tasks	No. of Resources	No. of Products	Total Variables	Integer Variables	Constraints
1	14	9	13	22	1	1712	126	1370
2	18	9	13	22	1	2193	162	1760
3	60	6	27	41	2	14845	2760	11058

The next phase of the research will focus on integrating the solution of the design super-problem and the scheduling sub-problems in the distributed framework. The capability of DCABB to solve multiple branch-and-bound trees simultaneously can be used to solve the scheduling sub-problems in parallel. Development of an algorithm to automatically update the design-superproblem based on scheduling sub-problem solutions will enable the integration of the entire design-scheduling hierarchy.

Conclusions

The above examples illustrate the power of distributed computing in the context of solving large scale optimization problems. The speedup times and bound gap reductions are encouraging and hence, further research in this area is warranted. The application in the area of design, planning and scheduling is critical since they involve the solution of large MILPs. The quick generation of feasible solutions is important from the designer's standpoint to perform a detailed analysis of the problem.

References

CPLEX Optimization, Inc. (1993). CPLEX-MIP, Suite 279, 930 Tahoe Blvd., Bldg. 802, Incline Village, NV 89451-9436.

Eckstein, J. (1993). *Parallel branch-and-bound algorithms for general mixed integer programming on the CM-5*, Technical Report TMC-257, Thinking Machines Corporation.

Kudva, G.K. (1994). *DCABB: A Distributed Computing Architecture for Branch and Bound Calculations*. Ph.D. Thesis, Purdue University, West Lafayette, IN 47907-1283.

Kudva, G.K. and J. F. Pekny (1994). DCABB: A Distributed Computing Architecture for Branch and Bound Calculations, *Computers and Chemical Engineering*, in press.

Pekny, J.F. (1989). *Exact Parallel Algorithms for Some Members of the Traveling Salesman Problem Family*. Ph.D. Thesis, Carnegie Mellon University, Pittsburgh, PA 15213.

Pekny, J.F. and M.G. Zentner (1993). Learning To Solve Process Scheduling Problems: The Role of Rigorous Knowledge Acquisition Frameworks, *Foundations of Computer Aided Process Operations*, CACHE.

Subrahmanyam, S., J.F. Pekny, and G.V. Reklaitis (1994). Design of Batch Chemical Plants Under Market Uncertainty, *Industrial and Engineering Chemistry*, in press.

Subrahmanyam, S. (1994). *Design of Batch Chemical Plants Under Uncertainty*. Technical Report, Purdue University.

SIMULATION, DESIGN AND ANALYSIS OF SUPERCRITICAL EXTRACTION PROCESSES

Glen Hytoft, Lucie Coniglio, Kim Knudsen and Rafiqul Gani
Department of Chemical Engineering
Technical University of Denmark
DK-2800 Lyngby, Denmark

Abstract

Important aspects related to simulation, design and analysis of supercritical extraction (SCE) processes are highlighted through the use of appropriate computational tools. In process simulation, the question of correct simulation is addressed. In process design, the choice of condition of operation and process alternatives are investigated. In analysis, the question of consistent simulation is addressed and the dynamic behavior of a liquid-liquid extraction column is analyzed. Finally, as test examples, SCE-processes from chemical, biochemical and food industries have been investigated.

Keywords

Supercritical extraction, Simulation, Design, Phase equilibria.

Introduction

Supercritical extraction (SCE) employs as a solvent, compounds that are at or near their critical region at the condition of operation. The most commonly employed solvent is carbon dioxide although light hydrocarbons have also been proposed (Brignole et al. (1987)). A typical feature of SCE-solvents is that their critical pressure is high and their critical temperature is moderate. A SCE process operates at elevated pressures and temperatures around T_C. SCE processes have been established as a promising alternative to conventional separation techniques such as distillation, absorption and extraction. Biochemical industries and food industries are principal users of SCE-based processes.

Near its critical region, a solvent increases the solubility of the solute significantly. This phenomenon can be noted from the relationship between the partial molar volume of the solute in the fluid phase and the derivative of the fugacity coefficient with respect to pressure.

$$\left(\frac{\partial}{\partial P}\ln\varphi_2\right)_{T,x} = \frac{v_2}{RT} - \frac{1}{P} \qquad (1)$$

In the above equation, it can be noted that the partial molar volume of the solute (v_2) in the fluid phase increases significantly due to small changes in pressure and/or temperature near the critical region because the pressure derivative of fugacity coefficient is large at this condition. Also, for correct and consistent values, it is necessary to have an appropriate thermodynamic model since the fluid behavior is complex.

Increasing interest in SCE-based processes in the last decade has produced new methods for prediction of properties (Dahl et al. (1991)) and computational tools for simulation/design of SCE processes (Brignole et al. (1987), Cygnarowicz and Seider (1989)). The main problem until now has been the unavailability of experimental/plant data and the inability of thermodynamic models to predict the phase equilibria adequately. As a result, most design/analysis tools are empirical and most simulation systems are problem specific.

The objective of this paper is to present through a computer program package, the important aspects related to simulation, design and analysis of SCE processes. Specifically, this paper addresses issues related to correct and consistent simulation of a SCE-process, choice of condition of operation and process alternatives and finally, modelling aspects related to the dynamic behavior of the liquid-liquid extraction column. Examples are taken from three different processes involving chemical, biochemical and food industries.

Process Simulation

A typical SCE process essentially includes a fluid-liquid extraction column operating at elevated pressure. The solvent recovery operation occurs at lower pressure and therefore, since the extraction and recovery operations are connected by the solvent-rich stream in a closed-loop, there is need for expansion and compression.

Simulation of a SCE process requires the simulation of a fluid-liquid extraction column at elevated pressures and a (solvent) recovery operation at lower pressures besides non-mass transfer operations such as expansion, compression and heat exchange. The fluid-liquid operation is typically modelled as a liquid-liquid extraction operation. The recovery operation is typically a distillation operation at pressures lower than the extractor but may still be significantly higher than atmospheric pressure. An appropriate thermodynamic model therefore must predict the phase equilibria (liquid-liquid and vapor-liquid) for a range of temperatures and pressures. Most often, the chosen thermodynamic model also need pure component properties which may not be available for the solute. Also, correct determination of the phase identities is important since they affect the energy balance.

Computational Tools

The essential elements of a computer program package for study of SCE processes must have a collection of thermodynamic models coupled to steady state and dynamic simulation models for extractors, flash operations and distillation columns. Also, the program package must have models for heat exchangers, compressors, valves which employ multi-phase stream identification procedures to ensure consistent determination of stream enthalpies. More details (modelling, method of solution, etc.) can be found in Hytoft and Gani (1993). Finally, the program package must have capabilities for computation of various forms of phase diagrams.

Simulation Strategy

Simulation of any process also requires, (a) estimation of pure component data, and (b) correlation or adaptation of thermodynamic model parameters to known phase equilibrium data. Although in many applications, steps a and b may not be needed, in most SCE processes, they are usually necessary. In this work steady state simulation is performed with a modified modular approach (use of simplified/linear models), dynamic simulation with a structured equation oriented approach and the steady state simulator is coupled to a SQP-algorithm for process optimization. More details on these topics can be found in Hytoft and Gani (1993).

Application Example (Case 1)

Fatty esters (methyl and ethyl esters) obtained from fish oils after esterification with ethanol can be extracted by CO_2 (Staby, 1993). Staby (1993) has also presented experimental VLE-data for binary and multi-component mixtures involving heavy fatty esters (C12-C22), heavy fatty acids (C12-C22) and CO_2. The best correlation (step b) was found to be a modified version of the MHV2-mixing rule (Dahl et al. (1991)) coupled to the modified UNIFAC-method (see Dahl et al., 1991) with quadratic mixing rule for the “b” parameter and considering $l_{ij} = 0.4$ and $l_{ij} = 0$ as interaction coefficients for respectively fatty ester-CO_2 and fatty ester-fatty ester (see Coniglio et al., 1994). Since a group contribution approach has been used, the same group parameters can be employed for all types of fatty esters which can be described by the new groups. The fatty ester-CO_2 and fatty acid-CO_2 binary data have been correlated with an average absolute error of 9.7% in pressure for data covering a temperature range of 313-473 K and a pressure range of 10-330 atm. More details of the correlation and prediction results is given by Coniglio et al. (1994).

In the simulation step, a process involving two single equilibrium stage operations has been simulated (design for this process has not been found in the open literature). The performance of CO_2 as a SCE-agent is investigated in this simulation example. An equimolar mixture of C18-ethyl ester and C22-ethyl ester is to be purified in terms of C22-ethyl ester by gradually removing the C18-ethyl ester with supercritical CO_2. In the first operation, the lighter ester is removed with CO_2 and in the second operation, CO_2 is recovered. The two single stage operations are modelled as a multi-phase flash process with specified temperatures and pressures. The first flash operates at 323.15 K and 125 atm while the second flash operates at 323.15 K and 60 atm. Although, the extraction per stage is low, the relative composition of the C18-ester and the C22-ester is 10 to 1. Therefore, there is a good incentive to proceed further with the design for the process. The total computational time is 0.02 CPU seconds on a 486-based PC. Attempts to reduce the computational times further by adapting the SRK equation of state did not give predictions of acceptable accuracy.

Application Example (Case 2)

Design and simulation of β-Carotene recovery from water by CO_2 has earlier been reported by Cygnarowicz and Seider (1990). In this paper, we have attempted to repeat the simulation with a different thermodynamic model (MHV2-model). Since the pure component data used by Cygnarowicz and Seider (1990) do not seem to be consistent (they do not satisfy the relationship of carbon number versus critical properties and boiling points). The method of Constantinou and Gani (1994) has been used to predict the critical properties (T_C = 905.3 K, P_C = 6.95 atm and acentric factor ω = 1.461). The available experimental solubility data were then used to adapt the modified UNIFAC-method (see Dahl et al. (1991)) group parameters between the CH main group and CO_2 interactions for the MHV2-mixing rule.

In the simulation step, the process as described by Cygnarowicz and Seider (1990) was simulated. Not surprisingly, different energy cost for the same separation was obtained. The total energy cost for the two compressors (9.27 kJ/hr plus 0.44 kJ/hr) is determined to be 902 $/year (using the same cost data as Cygnarowicz and Seider (1990)). This value is more than twice as much as that reported by Cygnarowicz and Seider (1990) for case 2. (Note, however, that operating costs are quite low).

To be sure that the same solubility has been simulated, the separator is simulated separately as a single stage multi-phase flash (at 335 K and 235 atm) using the MHV2-model. Three phases (a fluid phase, a liquid phase and a solid phase) with similar compositions as reported by Cygnarowicz and Seider (1990) were obtained.

Design Aspects

In this section, two design aspects are addressed briefly: a) condition of operation, b) choice of operation.

Condition of Operation

It is clear that in SCE-based processes, the extraction operation has to be performed at elevated pressures. Brignole et al. (1987) have shown that in the case of ethanol production, the cost of production can be reduced by using a light hydrocarbon solvent. The optimum choice of the condition of operation depends on the SCE-agent as well as the design specifications (loss of solute and solvent in the raffinate and purity of product). Through simultaneous simulation and optimization, we have observed that while the pressure effects the solubility (and therefore, the amount of solvent needed), the temperature effects the above mentioned design specifications. Most often, the optimum appears to lie near the lowest allowable temperature and the highest allowable pressure (not considering equipment costs).

Choice of Operation

Aspects related to the choice of operations are described through two dehydration processes, ethanol dehydration and acetone dehydration. Both processes are characterized by the binary azeotropes being formed with water. While Brignole et al. (1987) have recommended the use of a SCE-based process for ethanol dehydration, Cygnarowicz and Seider (1989) concluded that SCE-based process was not suitable for acetone dehydration (with CO_2 as the SCE-agent). In this paper, the existence of another promising alternative made feasible by the disappearance of the binary azeotropes both at low pressures and at high pressures is illustrated. This observation is based on computation of relative volatility as a function of pressure using the MHV2-model. In Table 1 $1/K_2$ is the relative volativity at infinite dilution of component 2 in component 1. (K_1 is the relative volativity at infinite dilution of component 1 in component 2). Table 1 shows the results of computation of relative volatilities and it can be noted that the azeotropes indeed disappear (when $1/K_2$ and $K_1 > 1$) at both high and low pressures. For ethanol-water, the disappearance occurs at very low (near 0.001 atm) pressure and near 33 atm while for acetone-water disappearance occurs at around 1 atm and at the critical point.These results have been compared with experimental data and good agreement has been noted. The disappearance of azeotropes as a function of pressure can also be represented through T-x,y phase diagrams and therefore, from a design point-of-view, use of phase diagrams is recommended particularly when considering the choice of operations. Since in the case of ethanol dehydration, the solvent recovery operation occurs at pressures between 20-30 atm, a single distillation operation (without any SCE-agent) at 33 atm appears to be an alternative worth considering. For similar reasons, a single distillation operation at 0.8 atm appears to be a reasonable alternative for acetone dehydration.

Table 1. Relative Volatilities versus Pressure.

Pressure	Ethanol(1) - water(2)		Acetone(1) - water(2)	
(Bar)	$1/K_2$	K_1	$1/K_2$	K_1
0.001	1.270	6	-	-
0.10	0.820	14	2.10	67
1.00	0.750	18	1.06	45
10.0	0.820	15	0.70	19
30.0	0.997	11	0.76	10
40.0	1.050	9	0.84	8

Simulations have shown that approximately similar products can be obtained for the same energy requirements (in the case of ethanol dehydration) and significantly lower energy requirements for acetone dehydration. The simulation results (obtained by using the MHV2-model) for the two processes are given in Table 2. Note that in these results, optimization with respect to design of the distillation operation has not been attempted. A large saving in equipment costs appears to be feasible.

Analysis

In this section, simulation results which may not be consistent are analyzed and the dynamic behavior of a liquid-liquid extraction column is analyzed.

Consistent Energy Balance

In the case of simulation results, we have chosen to concentrate on the state of a stream with respect to energy balance. Again, we have used the two dehydrogenation processes that have been studied earlier. In the case of ethanol dehydration, by repeating the simulations at the condition reported by Brignole et al. (1987), we have found that the stream entering the recovery column is not a single liquid phase while the stream entering the compressor/pump is a fluid phase. Therefore, the enthalpy computations with assumed phase identities (liquid in both cases), cannot be correct. Also, in the case of acetone dehydration, the stream entering the distillation column has been found to have three-phases (one vapor and two liquid). Assuming that the MHV2-model is more correct than the earlier GCEOS-model, it is likely that inconsistent energy costs have been reported by Cygnarowicz and Seider (1989). Also, it is important to note that at the condition of operation, specific heat correlations for the solvent are not valid and therefore, extrapolation is necessary.

The reboiler heat duty for the recovery column in the case of ethanol dehydration (extractor operating at 70 atm and 380 K, recovery column feed pressure 25 atm, 99% ethanol recovery and < 0.1% solvent loss) has been found to be, in this work, 13800 kJ/hr. In the case of acetone dehydration (extractor operating at 103 atm and 310 K, recovery column feed pressure 65 atm, 99% acetone recovery and < 1% solvent loss) the reboiler heat duty for the recovery column has been found to be 2631 kJ/hr. In each case, the simulated stream conditions have been verified with phase envelope computations and multi-phase computations to avoid any inconsistencies. Values of temperatures and pressures have been adjusted to match the same state as those reported earlier.

Table 2. Distillation Column Details.

	Ethanol(1)-water(2)	Acetone(1)-water(2)
No. stages	50	30
Reflux ratio	30	3
Pressure	33.0	0.8
Composition feed (mol/hr)	21.74 500.00	10.0 90.0
Composition top (mol/hr)	19.587 1.067	9.869 0.131
Composition bot. (mol/hr)	2.153 498.93	0.131 89.89
Temp top (K)	479.5	323.0
Temp bot (K)	510.6	365.1
Q_c (J/hr)	$1.36*10^7$	$1.21*10^6$
Q_h (J/hr)	$2.02*10^7$	$6.94*10^5$

Liquid-Liquid Extractor Dynamics

Dynamic models for liquid-liquid extraction have been previously reported by Ramachandran et al. (1992) and Hytoft and Gani (1993). While the model of Ramachandran et al. is more simple (assumption of constant total hold-up), the model of Hytoft and Gani is more general. Also, the model of Hytoft and Gani allows simulations for a wide range of thermodynamic models. Therefore, low pressure processes (where usually, the extract is the heavy phase) as well as high pressure processes (where the extract is the light phase) can be simulated. In this paper, we show the dynamic behavior of the liquid-liquid extractor for the separation of iso-propanol (IPA) from water with CO_2 (at the same condition as reported by Ramachandran et al. (1992)). Specifically, we show the response due to a disturbance (-5%) in the feed flowrate. The simulations have been carried out with the MHV2-model. Figure 1 show the dynamic response (for the four plates) of the total liquid hold-ups and the IPA hold-ups. It can be clearly seen that the component hold-ups and the total hold-ups do change with time, however, the response for the total flowrates are much faster than the component flowrates.

Conclusions

The significance of accurate pure component data and, correct/consistent prediction of phase behavior for SCE-processes has been highlighted through the use of computational tools especially suited for studies of SCE-processes. Since SCE-processes operate near the phase boundary, the correct location of this boundary is important for computation of energy requirements for the process. The condition of operation of SCE-processes has been found to be influenced by the choice of the solvent (affects the pressure) as well as the product specification (affects the temperature). Finally, the simulation results show that an alternative to SCE-processes for dehydration of alcohols and ketones is distillation at a pressure where the binary azeotrope does not exist.

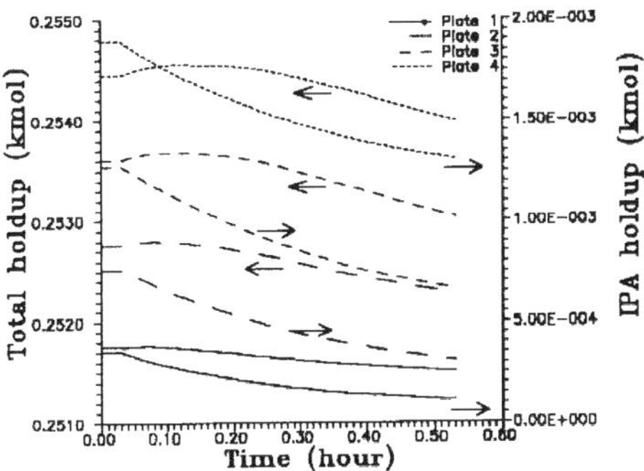

Figure 1. Total molar hold-up response (upper curves) and IPA hold-up response.

References

Brignole, E.A., P.M. Andersen and Aa., Fredenslund (1987). Supercritical fluid extraction of alcohols from water, *I&EC Research*, **26**, 254-261.

Constantinou, L. and R. Gani (1994). A new group contribution method for the estimation of properties of pure compounds, *AIChE J.* (in press).

Coniglio, L., K. Knudsen and R. Gani (1994). Model prediction of supercritical fluid-liquid equilibria for carbon dioxide and fish oil related compounds. *I&EC Research* (submitted).

Cygnarowicz, M. and W. Seider (1989). Effect of retrograde solubility on the design optimization of supercritical extraction processes, *I&EC Research*, **28**, 1497-1503.

Cygnarowicz, M. and W. Seider (1990). Design and control of a process to extract β-Carotene with supercritical carbon dioxide, *Biotechnol. Prog.*, **6**, 82-91.

Dahl, S. Aa. Fredenslund and P. Rasmussen (1991). The MHV2-model — A UNIFAC based equation of state model for prediction of gas solubility and VLE at low and high pressures, *I&EC Research*, **30**, 1936-1945.

Hytoft, G. and R. Gani (1993). Simulation tools for design and analysis of supercritical extraction processes, AIChE Annual Meeting, paper 9c, Nov. 1993, St. Louis, USA

Ramachandran B., J.B. Riggs, H.R. Heichelheim, A.F. Seibert and J. R. Fair (1992). Dynamic simulation of a supercritical fluid extraction process, *I&EC Research*, **31**, 281-290.

Staby, A. (1993). *Application of supercritical fluid techniques on fish oil and alcohols*. Ph.D. thesis, Technical University of Denmark, Lyngby, Denmark.

THE INFLUENCE OF DIFFERENT PARAMETERS ON PROCESS HEAT INTEGRATION

Majda Krajnc and Peter Glavic
Chemical Engineering Department
University of Maribor
62000 Maribor, Slovenia

Abstract

The influence of **minimum approach temperature** ($\Delta_{min}T$) and **temperature contributions** ($\Delta_{cont}T$s) on the grand composite curve (GCC) is decribed. Different approaches influence the design of processes with small differences between the lowest and the highest temperatures of the process. $\Delta_{cont}T$s in such processes are very low. **Different integration methods** (pinch, dual-temperature and nonlinear programming) affect process designs of a normal chemical process and a heat pump fractionator, as presented in the second part. The methods used can give quite different energy targets and structures. Varying **investment and utility prices** in different decades cause differences in the process structure, too. Utility prices, which are based on capital and fuel prices change with time. Structures as the result of the mentioned changes were obtained in three different decades with pinch and NLP methods.

Keywords

Heat integration, Chemical process design, Economic analysis, Utilities, Optimization.

Introduction

Chemical and process industries are significant energy consumers and extended studies have been made on how to utilize the energy more efficiently.

Pinch analysis was one of the first systematic synthesis methods. It was developed by Linnhoff (1983). The use of areic heat flow rate instead of $\Delta_{min}T$ as a basic parameter to vary for optimization purposes has been suggested by Fraser (1989). Trivedi (1989) specified two approach temperatures (network **H**eat **R**ecovery **A**pproach **T**emperature and individual **E**xchanger **M**inimum **A**pproach **T**emperature) as well as additional utility heat flow rate consumption, Φ, prior to design. Yee, Grossmann and Kravanja (1990) proposed a nonlinear programming, NLP, formulation which simultaneously targeted energy and area costs of a heat exchanger network.

We have compared the methods mentioned to see which one is the most suitable and efficient for process design and which one gives the optimum configuration. The procedure for chemical process design which includes the proper location of units and utilities in the process and the subsequent optimization of the obtained structure, supposedly suitable for permanent usage, is presented in three steps. **First** we would like to show the differences among the four different approaches to determine $\Delta_{cont}T$s of $\Delta_{min}T$ for streams. **Second**, different structures that can be deduced from the same starting points with different design methods will be described. **Further**, different prices of utilities and investment costs, which also influence the structure, have been tested with pinch and NLP methods over three different decades.

Influence of Different Temperature Contributions

In utility/process exergy analysis it is not only important to determine heat flow rates of utilities but also the kind of utilities. A good resource for choosing them is the GCC which shows the energy level relation between process streams and utilities. Temperatures of streams are lowered or raised by $\Delta_{cont}T$ which represents the two contributions to $\Delta_{min}T$ between process streams and utilities. In simple models, $\Delta_{cont}T$ equals $\Delta_{min}T/2$. But a detailed analysis shows that different fluids give different contributions to $\Delta_{min}T$ in a heat exchange unit. The heuristic, developed by Townsend and Linnhoff (1983), uses the following temperature contributions for different fluids: 20 K for gases, 10 K for liquids, 5 K for boiling liquids or condensing vapours while Glavic and Kravanja (1988) suggest 0 K for directly contacting streams in reactors, furnaces and compressors.

But further investigations have shown that it might be very convenient to take $\Delta_{min}T$ values as dependent variables. Fraser (1989, 1989b) has replaced $\Delta_{min}T$ with the areic heat flow rate, i.e. a heat flow rate divided by the heat exchanger area. $\Delta_{min}T$ for each stream was derived from its minimum areic flow rate and its stream surface coefficient of heat transfer. Values of minimum areic heat flow rates in the range 1000-3500 W/m^2 were recommended.

We wanted to compare the results obtained with process stream analysis of the two theories on two examples.

Example 1. Hydrodealkylation Process

Example 1 presents a HDA process (Douglas, 1988) with diphenyl recycled, shown in Fig. 1. The stream data were obtained by process simulation. After simulation, a thermodynamic analysis was carried out. At the beginning we checked $\Delta_{min}T$s of all heat exchangers, coolers and heaters so that we could estimate which approach would be appropriate for stream analysis. Only two process units had $\Delta_{min}T$ values lower than defined by Townsend's theory, and this theory would perhaps be suitable for analysing our process, as it will be shown next.

Utility Analysis (Townsend's and Fraser's Approaches)

Townsend's and Fraser's temperature contributions were used for the construction of GCC. When we compared both GCCs, the available utilitics (water, refrigerant, steam, fuel) were correctly located against pinch temperature. Different $\Delta_{cont}T$s changed the process heat exchange and utility targets. It was obvious that contribution values defined by Townsend's and Fraser's theories did not cause much difference in pinch temperatures, energy targets or in process structure when high ΔT in heat exchangers existed and the methods used gave satisfactory results.

Example 2. Heat Pump Fractionator

Then the analysis was carried out for a process shown in Fig. 2 where a three-component mixture was separated by distillation and energy was supplied with a heat pump; the process fluid was a working fluid (Barrett, 1980; Krajnc and Glavic, 1992). The operating parameters were determined by simulation again. It was assumed that two utilities would be necessary, a hot one (chemical utility) and a cold one (cooling air). We wanted to find out if the available utilities were appropriately located in the GCC. First the stream analysis was carried out with Townsend's and Fraser's approaches.

Utility Analysis (Townsend's and Fraser's Approaches)

The results showed quite different energy targets. The reasons were different $\Delta_{cont}T$s of streams which are shown in Table 1. They had a strong influence in this case. The process is not very large. Components, which were separated by distillation, had their boiling points close together, so the lowest temperature of the process was 16.7 °C in the flash drum, F, and the highest one was 61.7 °C after compression, C1. It is a very rare occasion that $\prod\Delta T$ of the whole process is only 45 K. In such cases it can happen that with Townsend's method energy targets are higher than they are in the "base process"; too high values of $\Delta_{cont}T$s cause this phenomenon. This conclusion was also evident from the location of utility streams (i.e. cooling air was erroneously placed above the pinch point and the chemical utility, i.e. reactor, below it).

Better results were obtained by Fraser's approach. Minimum areic heat flow rate values of 500 and 1000 W/m^2 were selected, and as it is evident from Table 1, $\Delta_{cont}T$s were very small. In both cases the energy target represented was a treshold case without any hot utility need but cooling air was inappropriately located above the GCC instead below it. It was higher at 1000 than at 500 W/m^2.

Table 1. $\Delta_{cont}T$s of Streams Determined with Different Approaches for Example 2.

Str. Type	Proc. Unit	Towns (K)	Fraser (K)		Individ. Contr. (K)	
			500 (W/m^2)	1000 (W/m^2)	Towns. Prop.	Fraser Prop.
Proc.	**Str.**					
hot	HE	5	0.72	1.45	2.89	2.89
hot	CD	5	0.72	1.45	1.56	0.52
cold	HE	5	0.72	1.45	2.89	2.89
cold	RH	20	10	20	12.4	17.36
Util.	**Str.**					
hot	RH	10	0.72	1.45	6.2	1.24
cold	CD	20	10	20	6.26	7.31

Individual $\Delta_{cont}T$ Approach

After obtaining the above mentioned results, we wanted to find an approach which could be useful also in processes with small ΔTs. We tried to solve the problem with previous approaches again, but using individually calculated $\Delta_{cont}T$s. The approach determines $\Delta_{cont}T$ of each stream individually from simulation results (individual simulation method). The rules of Townsend's and Fraser's methods were taken into account for partitioning both contributions of $\Delta_{min}T$. Calculated $\Delta_{cont}T$ values are presented in the last two columns of Table 1. With both approaches the GCCs were similar, i.e. treshold cases without a hot utility need. The cold utility was properly located below the GCC. In this case it was obvious that the individual approach gave the best and logical results.

The Influence of Different Optimization Methods on Process Structure

After stream analysis, where utilities and process units are properly located in the process, the structure optimization follows. Different optimization design methods can be classified into two categories: mathematical programming methods and heuristic evolutionary ones. While mathematical programming methods (e.g. NLP and MINLP) are very complex and require considerable mathematical expertise, the heuristic evolutionary methods (pinch, dual-T) are simple to comprehend and implement. Structures that can result from the methods mentioned will be shown next.

Energy and cost comparisons between all possible structures of the HDA process and the process with a heat pump are summarized in Table 2. It is evident that the process structure, obtained with the NLP method, gives the lowest total annual costs when observing Example 1. Structures were quite different. The last column in Table 2 shows relative costs in % when comparing configurations against the base structure. Cost reduction ranges from 11% (pinch configuration) to 16% (NLP configuration).

Table 2. Energy and Cost Comparison of Different Structures (Prices in 1992).

Method	I_h (kW)	I_c (kW)	C_{tot} (M$/a)	C_{rel} (%)
Example 1				
Base	2725	5830	2.542	100
Pinch	1117	4228	2.256	89
Dual-T	1663	4772	2.193	86
NLP	**2616**	**5722**	**2.148**	**84**
Example 2				
Base	733	3786	1.784	100
Pinch (no split.)	-	3053	1.794	101
Pinch (split.)	**-**	**3053**	**1.747**	**98**
NLP	-	3053	1.794	101

With a heat pump fractionator the NLP method gives the same energy targets (treshold case without hot utility needs) and the same total annual costs as the pinch analysis, but they are higher than the costs of the base configuration. But total costs, obtained with the pinch method, where stream splitting is considered, are lower than costs of the base structure. Cost reduction of the optimal solution was 2% as evident from Table 2 in the last column. Optimal process configurations of both examples are presented on Figs. 1 and 2.

The Influence of Utility Prices and Investment Costs on Process Structure

If we want to design a process structure which will be suitable for permanent use then the influence of time should be taken into consideration. The influence of time (e.g. inflation) will be presented with both investment and utility costs considered. As it was discussed by Ulrich (1992), utility prices depend on the base fuel price. In 1980 the base fuel price (e.g. No. 6, fuel oil) was higher than in 1970 or 1992, and this change influenced the utility prices and costs.

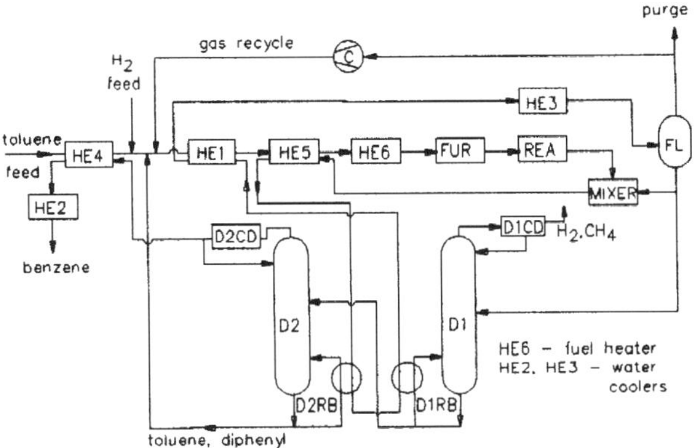

Figure 1. Optimal process configuration for HDA process (prices in 1992).

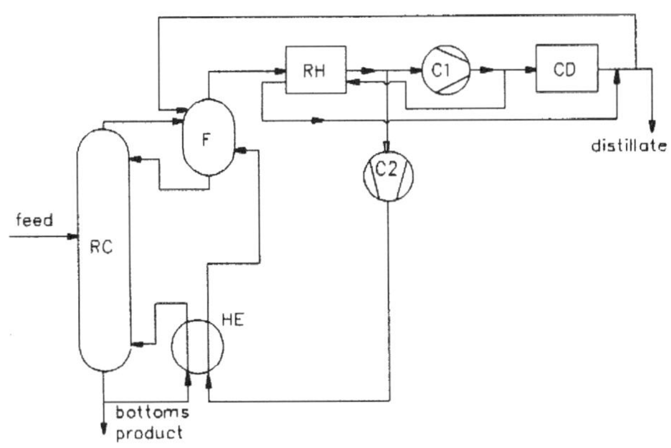

Figure 2. Optimal process configuration for heat pump fractionator (prices in 1992).

It is interesting to make a comparison between structures obtained in different decades with pinch and NLP methods. Economic results are summarised in Table 3.

Pinch Results

When results obtained with the pinch analysis are compared, the following conclusions are important: energy targets in the 1980 structure are lower in comparison with the 1992 and 1970 structures because energy was more expensive; utility costs are the highest in 1980 because steam became much more expensive; as investment and total costs increased with time the structures would be, consequently, different.

Table 3. Influence of Different Utility Prices on HDA Process Structure in Three Different Periods of TIme.

Method	I_h (kW)	I_c (kW)	C_{tot} (M$/a)	C_{tot} (1992) (M$/a)
1970				
Base	2725	5830	0.766	2.539
Pinch	1220	4330	0.725	2.291
NLP	**2616**	**5722**	**0.695**	2.282
1980				
Base	2725	5830	2.472	4.304
Pinch	976	4080	1.965	3.149
NLP	**1641**	**4747**	**1.801**	2.920
1992				
Base	2725	5830	2.542	2.542
Pinch	1120	4230	2.256	2.256
NLP	**2616**	**5722**	**2.148**	**2.148**

NLP Results

When the results obtained with the NLP method are compared, the conclusions are similar to those of the pinch analysis, except that utility costs are increasing regularly with time.

It is evident from Table 3 that structures obtained with the NLP method are economically the most suitable. The last column shows total annual costs recalculated to present time (year 1992). Present total costs differ a lot from each other. The cheapest structure is the one from 1992, costructed with the NLP method.

Conclusions

The influence of different $\Delta_{cont}T$s on GCC was studied. The values were determined or calculated by four different approaches: Townsend's, Fraser's and two individual simulation ones. We have found that in processes with small differences between maximum and minimum process temperatures only individually calculated $\Delta_{cont}T$s of streams from simulation results give logical results. In processes with large differences Townsend's (and Fraser's) method gave satisfactory results.

Further, systematic evolutionary methods based on heuristics (pinch, dual-T) and mathematical programming (NLP) method were used and studied on the HDA process and on the process with a heat pump. The minimum total annual costs were obtained with the pinch analysis, where stream splitting was considered, when the process with the heat pump was analysed, and with the NLP method when analysing the HDA process. The energy needs and economic results showed that different methods gave different structures.

The influences of utility prices and investment costs were studied on the HDA process with the classical pinch analysis and with the NLP method. The comparison between configurations in three different decades was carried out. It was found that at present it would be economical to use the structure which was designed in time periods when the fuel had lower prices.

It could be concluded that the pinch analysis is still the best for inventing optimal superstructures and that it gives optimal results. It is also appropriate for final optimization of processes which have few process streams. In processes with more than eight or ten streams the searching for the best solution becomes very tedious and in such situations the pinch analysis is used for conceptual design, but the superstructures have to be optimized with the NLP method to obtain the best results.

References

Barrett, S.J. (1980). First-prize-winning solution. *AIChE Student Members Bulletin 1980*, New York. 27-52.

Douglas, J.M. (1988). Cost diagrams and the quick screening of process alternatives. In B.J. Clark and J.W. Bradley (Ed.), *Conceptual design of chemical processes*, Part 2. McGraw-Hill Book Company, New York. 289-315.

Fraser, D.M. (1989). The use of minimum flux instead of minimum approach temperature as a design specification for heat exchanger networks. *Chem. Engng. Sci.*, **4**, 5, 1121-1126.

Fraser, D.M. (1989b). The application of a minimum flux specification to the design of heat exchanger networks. *DECHEMA Monography 116*, 253-260.

Glavic, P., Z. Kravanja, and M. Homsak (1988). Heat integration of reactors — I. Criteria for the placement of reactors into process flowsheet. *Chem. Engng. Sci.*, **43**, 3, 593-608.

Krajnc, M., and P. Glavic (1992). Energy integration of mechanical heat pump with process fluid as working fluid. *Trans. IChemE*, **70**, Part A, A4, 407-420.

Linnhoff, B., and E.C. Hindmarsh (1983). The pinch design method for heat exchangers networks. *Chem. Engng. Sci.*, **38**, 5, 745-763.

Linnhoff, B., H. Dunford, and R. Smith (1983). Heat integration of distillation columns into overall processes. *Chem. Eng. Sci.*, **38**, 8, 1175-1188.

Townsend, D.W., and B. Linnhoff (1983). Part II: Design procedure for equipment selection and process matching. *AIChE J.*, **29**, 5, 748-771.

Trivedi, K.K., B.K. O'Neill, and J.R. Roach (1989). A new dual-temperature design method for the synthesis of heat exchanger networks. *Computers Chem .Engng.*, **13**, 6, 667-685.

Ulrich, G.D. (1992). How to calculate utility costs. *Chemical Engineering*, **99**, 2, 110-113.

Yee, T.F., I.E. Grossmann, and Z. Kravanja (1990). Simultaneous models for heat integration — I: Area and energy targeting and modeling of multi-stream exchangers. *Computers Chem. Engng.*, **14**, 10, 1151-1164.

PROCESS ANALYSIS OF SULFOLANE PRODUCTION: A CASE STUDY IN APPLICATION OF PROCESS SIMULATION AND OPTIMIZATION FOR WASTE REDUCTION

C. Shyamkumar, M. Jayagopal, C. L. Czekaj and K. A. High
School of Chemical Engineering
Oklahoma State University
Stillwater, OK 74078

Abstract

Process simulation is an important tool for process improvement and optimization, especially in the retrofit scenario, which involves investigation of a large number of alternatives. In this paper, we examine the use of a simulated process model to propose waste reduction alternatives in the sulfolane process of Phillips Petroleum Company. The approach used in this study consists of process modeling, economic analysis, and generation and optimization of alternatives. The lack of quantitative data in this paper reflects the proprietary concerns of the industrial sponsors.

Keywords

Process modeling, Waste reduction, Sulfolane.

Introduction

Escalating waste disposal costs, increasingly stringent waste reduction regulations, and heightened public awareness are the primary motivations for pursuing waste minimization measures (NEWMOA, 1992). Along with the environmental considerations of processing, the changing nature of market conditions, the availability and prices of raw materials and utilities place uncertain demands on existing processes. Thus, there is a need to apply a consistent approach for retrofitting existing processes.

Process modeling as a tool can be used to analyze the possibility for reducing industrial pollution at the source. The potential benefits include synthesis, analysis, control, and optimization of environmentally friendly processes. Retrofit, especially for reducing the environmental impact of existing processes involves investigating a large number of potential alternatives. The simulation of a process model combined with an economic analysis aids in the generation and optimization of alternatives. A comparison of these alternatives can then be performed to select the techno-economically feasible options. In this work, an approach involving process simulation and economic analysis has been used to study improvements in the sulfolane process of Phillips Petroleum Company.

Process Description

Currently, the sulfolane process is based on chemistry which was first described in the early 1900's. The sulfolane process (Fig. 1) can be broken down into three basic blocks: synthesis, purification, and waste treatment.

After the sulfolane synthesis block, the product is purified in a set of downstream processes. The synthesis block consists of three steps carried out in stirred batch reactors. In the first reaction, 1,3-butadiene reacts with sulfur dioxide to form sulfolene. The products of this reaction are sent to a separator where a solvent is added and the unreacted reactants are removed and treated by conventional methods (Willis, 1971). Any untreated gases are burned in the flare. Sulfolene is then hydrogenated in the presence of a catalyst to form sulfolane. Side reactions leading to the formation of unwanted byproducts are known to occur in both of the reaction steps. The wastes generated from the synthesis steps include gaseous emissions, wastewater discharges, and spent catalyst. In this work, the synthesis block of the process was analyzed for waste reduction. The details of the process are available in U.S. Patent 3,622,598 (Willis, 1971).

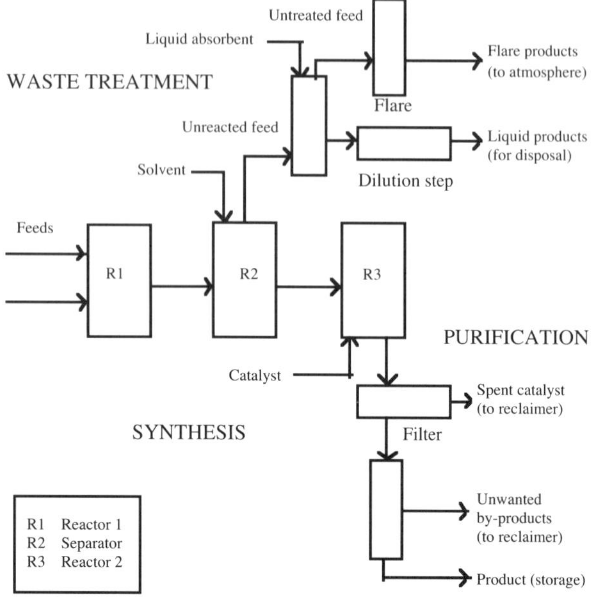

Figure 1. Schematic of sulfolane production process.

Approach

The approach used in this work has three basic elements:

- Process modeling
- Economic analysis
- Generation, optimization, and selection of alternatives

Process modeling involves categorization of the process, physical and kinetic data gathering, and iterative development of the model until the criterion for validation are satisfactorily met. The economic analysis consists of obtaining a base-case profitability for the existing process based on total product costs. Costs for waste treatment and disposal should be calculated as direct production costs. The magnitudes and sensitivity of the economics to specific process parameters can be analyzed. Based on this analysis several alternatives can be generated. These alternatives can be optimized (assuming annual profit as the objective) and compared to select the most beneficial alternative. Further details of the general approach can be found in Shyamkumar et al. (1993).

Process Modeling

The steady-state simulator ASPEN PLUS has been used for developing a model of the sulfolane process. This package contains a batch reactor module RBATCH which can be integrated into a continuous flow sheet. In addition, ASPEN PLUS contains physical property estimation, sensitivity, and optimization modules.

The physical properties of the intermediate product sulfolene are neither available in the open literature nor the ASPEN PLUS property databases. Therefore, these were either estimated or approximated with sulfolane properties (Table 1).

Table 1. Estimated Sulfolene Properties.

Property	Value
Normal boiling point	424 K
Critical temperature	648 K
Critical pressure	52.7 atm
Heat of formation	-2.59E+08 J/kmol.
Ideal gas heat capacity @ 350 K	1.13E+05 J/kmol•K
Vapor pressure @ 350 K	0.07 atm

The kinetic data for the first reaction were obtained from Phillips Petroleum Company. Reliable kinetic data for the hydrogenation reaction are not available. Proprietary operating condition data was provided by Phillips Petroleum Company. A process model of the sulfolane process was developed based on the approximate physical property and kinetic data. The model is described below.

First Reaction Step

The first reactor was modeled as a series of ideal batch reactors to incorporate the effect of imperfect mixing based on work done by Tanner, et al. (1985). The sizes of the ideal batch reactors were determined by a detailed theoretical study of the mixing phenomena in the reactor.

The main assumptions in developing the model were (a) a constant temperature profile, and (b) the mixing flow patterns in the reactor. The temperature profile in the reactor was determined by heat balance. The mixing patterns and flowrates were derived from correlations developed for single impeller tanks by Nagata et al. (1959) and Sachs and Rushton (1954). These correlations were extended for use in multi-impeller tanks. The model was used to study ways of improving the throughput and exit conversions from the reactor. The heat transfer capacity of the reactor was also analyzed.

Treatment Step

This vessel was modeled as a series of ideal equilibrium stages to account for the pressure transients of the system. The simulation of this step was found to be sensitive to the properties of sulfolene and the thermodynamic model used for predicting vapor liquid equilibrium.

An assumption made in developing this model was to neglect the decomposition reaction of sulfolene. Using this model the effect of operating conditions on the separation efficiency of the unreacted feed from the reaction mixture, and on the freezing point of the sulfolene mixture were studied.

Hydrogenation Step

The only data available for the hydrogenation step were heat removal profiles of the process plant reactor. This step was represented with a stoichiometric model to overcome lack of kinetic data. A study of the relative rates of the heterogeneous processes, i.e. gas absorption and external diffusion, was performed, assuming only the hydrogenation reaction is taking place. These were compared with approximate "process rates" or "observed rates" of hydrogenation (1.7 - 2.7 E-06 gmol/cc•sec) which were calculated based on the heat data.

The main assumption in the analysis of this reactor was that the reaction calorimetry can be related to an overall rate of reaction. In order to calculate the rate of absorption and diffusion, several transport properties of the reaction mixture were estimated. The model was used to study the rate limiting step of hydrogenation.

Results and Sensitivity Analysis

Using the process model the following variables were predicted and compared against process plant data. The selection of these variables was based on the availability of process information.

1. Exit composition from first reaction step
2. Sulfur dioxide concentration in treatment tank outlet
3. Overall material balance of the process

The sensitivity of the results to the assumptions made were studied. Further, the sensitivity of the above mentioned process variables to the model inputs were analyzed. The results of these studies justified the assumptions and approximations made in developing the model. Specifically the following assumptions and inputs were validated:

1. Size of ideal batches in the first reaction step.
2. Use of series of steady state equilibrium stages to represent the treatment step.
3. Approximation of sulfolane properties for sulfolene in the treatment step.

Economic Analysis

An economic analysis of the sulfolane process was carried out. The cost data used in this analysis are shown in Table 2. The main costs associated with the process are the raw material costs, followed by waste treatment costs. The benefits of waste reduction are often hidden because the environmental costs are combined under an overhead account (Freeman, 1992). For sulfolane production, the catalyst cost affects the profit levels the most. This reveals a potential for either conserving or improving the utilization of the reactants.

Table 2. Assumed Cost Data.

Category	Cost
Raw materials	
sulfur dioxide	\$0.20/lb.
1,3-butadiene	\$0.115/lb.
catalyst	\$7.15/lb.
nitrogen	\$0.70/MSCF
hydrogen	\$0.131/lb.
sodium hydroxide	\$0.165/lb.
Utilities	
steam	\$3.08/Mlb.
electricity	\$0.0345/kWhr
Waste treatment/disposal	
liquid disposal	\$0.01/gal.
solid waste reclamation	\$750/ton
gaseous waste treatment	\$0.165/lb.

Generation of Alternatives

The following alternatives were considered based on the integrated process and economic modeling approach.

1. Continuous operation of the first reactor.
2. Improving the separation of reactants in the treatment step.
3. Optimization of hydrogenation solvent.
4. Overcoming the rate limitation of the hydrogenation reactor.

Continuous Operation of the First Reactor

The first reactor was analyzed for the possibility of increasing the throughput of the reactor by operating it continuously. The main advantage of this alternative is that it distributes the initial heat of reaction evolved over a period of time. This ensures that the temperatures do not go over allowable levels even at relatively high flow rates. The reactor was modeled as a series of two reactors, a CSTR followed by a PFR. This configuration enables a reduction in reaction time and an increase in the conversion. The maximum throughput for this configuration was calculated and was found to be substantially higher than the current production rates.

Improving the Separation of Reactants in the Treatment Step

A reduction of sulfur dioxide concentration levels in the treatment step is important since SO_2 is a known poison for the hydrogenation catalyst (Willis, 1971). The separation of the raw materials from the sulfolene-solvent mixture can be improved by increasing the area of mass-transfer. A reduction of the gaseous emissions from the process can be achieved by coupling the enhanced separation with recycling the reactants. However, the recycling option is infeasible in the current operating scenario. This option might become more attractive if sulfur dioxide emission standards become more stringent or butadiene emissions are regulated (Bryant, 1992).

Optimization of Hydrogenation Solvent

Since the side reactions in the hydrogenation reactor involve the solvent and the unreacted sulfur dioxide, optimizing the solvent is a potential waste reduction alternative. In this study, three solvents, namely, water, sulfolane, and isopropyl alcohol have been compared with respect to the following criteria:

1. Polarity
2. Volatility
3. Solubility of reactants and hydrogen
4. Heat capacities

An ideal solvent would be highly polar, moderately volatile, able to dissolve reactants and hydrogen, have a high heat capacity and be chemically inert under hydrogenation conditions. The solvent properties were calculated either through ASPEN PLUS or were manually estimated. Based on the results of the study it can be concluded that isopropyl alcohol is an appropriate solvent.

Overcoming the Rate Limitation of the Hydrogenation Step

The rate of external diffusion of hydrogen in the reaction slurry was found to limit the sulfolene hydrogenation rate. This was based on the assumption that the Langmuir-Hinshelwood type of heterogeneous mechanism applies to the sulfolene hydrogenation reaction. A discussion of the alternatives for overcoming the external diffusion limitation follows. Increasing the catalyst charge while retaining the present hydrogenation solvent will increase the hydrogenation rate but also increase the amount of by-product formation. Decreasing the catalyst particle size is not a feasible alternative because of the increased cost of obtaining such a catalyst. Increasing the concentration of hydrogen in the liquid represents the most attractive alternative. This can be accomplished by either increasing the pressure of the reactor or changing to a hydrogenation solvent in which hydrogen solubility is higher.

Optimization and Comparison of Alternatives

The technical feasibility and incremental benefits of the suggested modifications were considered for their evaluation. The first criterion is often easy to assess quantitatively. The cost benefits, however, are often harder to measure due to the uncertainty and unavailability of data. For the case of the sulfolane process a retrofit option has been considered beneficial if it affects process profitability by reducing the byproducts generated from the process. This reflects the high likelihood of further regulations and the benefits of early compliance. Using the above mentioned criteria the following retrofit alternatives have been selected:

1. Modification of the first reactor to operate it as a continuous process.
2. Optimizing the hydrogenation solvent to achieve desirable characteristics (e.g. isopropyl alcohol).

Results and Conclusions

An approach involving process modeling and economic analysis was applied to the sulfolane process of Phillips Petroleum Company. A simulated process model was developed using ASPEN PLUS. Physical property approximations and equipment non-idealities were incorporated in the model. An economic analysis isolated catalyst cost as being the most significant variable affecting profitability. Several alternatives targeted towards waste reduction were then developed and optimized. A comparison of these alternatives was carried out to select the most beneficial of these.

Acknowledgments

This work was made possible through National Science Foundation grant CTS-9216834. Additional sponsors include Phillips Petroleum Company and University Center for Energy Research at Oklahoma State University. The Phillips Petroleum Company is acknowledged for process information and collaborative efforts.

References

Bryant, C.R., Weinberg, Bergeson, and Neuman (1992). EPA releases draft hazardous organic NESHAP. *Pollution Engineering*, **24**,25-26.

Freeman, H., T. Harten, J. Springer, P. Randall, M. A. Curran, and K. Stone (1992). Industrial pollution prevention: a critical review. *Journal of Air and Waste Management Association*, **42**, 619-656.

Nagata, S., K. Yamamoto, K. Hashimoto, and Y. Naruse (1959). *Mem. Fac. Eng. Kyoto Univ.*, **21**, 260.

Northeast Waste Management Officials Association (NEWMOA) (1991). *Total Cost Assessment: An Overview of Concepts and Methods*.

Sachs, J.P., and J.H. Rushton (1954). Discharge flow from turbine-type mixing impellers. *Chem. Eng. Progr.*, **50**, 597-603.

Shyamkumar, C., M. Jayagopal, and K.A. High (1993). Process design for waste minimization in the manufacture of the specialty chemical sulfolane. Presented at the AIChE Annual Meeting, St. Louis, MO.

Tanner, R.D., I.J. Dunn, J.R. Bourne, and M.K. Klu (1985). The effect of imperfect mixing on an idealized kinetic fermentation model, *Chem. Eng. Sci.*, **40**, 1213-1219.

Willis, J.L. (1971). to Phillips Petroleum Company, *U.S. Patent*. 3,622,598.

THE MODELLING AND SIMULATION OF REACTIVE DISTILLATION PROCESSES

Jianhua Zhu and Fu Shen
University of Petroleum
Changping, Beijing 102200, PRC

Abstract

A generalized model of the reactive distillation processes and a software (prototype) are developed via rate-based approach. The homotopy-continuation method is selected to solve the complicated nonlinear model equations efficiently. The simulation on the reactive distillation processes is carried out, and the stage temperature profile, composition profile and flow rate profile for both vapor and liquid phase are also obtained. From a further insight to the results, the pitfalls in experimental design are discussed.

Keywords

Model, Simulation, Reactive distillation process, Rate-based approach, Homotopy-continuation method.

Introduction

The reactive distillation process is a hybrid process. Due to its benefit in reduction of capital investment and lowering of operating cost, it has been highly interested by many researchers and engineers since 1970's. Now, the reactive distillation technology has been successfully applied in esterification process. Driven by the Clean Air Act, The Chemical Research & Licensing Co. began to develop catalytic distillation technology for the production of MTBE in 1978, and applied successfully this technique to the industrial production in 1981 (Smith, 1982). The reactive distillation technique is in further developing.

Since the reactive distillation process combines reactor and distillation column in a same vessel, the reaction and mass-transfer affect simultaneously on the process, this hybrid process is therefore much more complex than that in single reaction process or separation process. The tiny changes in process parameters, such as location of the feed plates, number of plates, residence time, catalyst, by-product concentrations and ratios of the reactants in feeds, will lead complex influences on process operation. It is hardly possible to determine influence of each parameter on the process operation solely by experimental studies.

Accompanying with the developing computer technology, the process simulation began to be effective. Now, the simulation is not only used to predict the results of the experiments, but also can be applied to determine the relationship of the variables and their relative importance. Therefore, to set up the mathematical model of the process, and to do the process simulation are good approach to understand the process. For the design of the reactive distillation processes, the modelling and simulation will also be very important. Here, a trial of generalized modelling and simulation for the reactive distillation processes is presented.

Mathematical Models

In practical multicomponent distillation operation, stages rarely operate at equilibrium. Usually the departures from equilibrium is expressed by incorporating stage efficiency into the equilibrium relation. Mass transfer in multicomponent mixtures is more complicated than in binary system because of the possible coupling between the individual concentration gradients. One of the interesting consequences of this interaction effect is that the individual point efficiencies of different species might be at anywhere in the range from $-\infty$ to $+\infty$. Further, it poses question on the validity of plate efficiency concept (Krishnamuthy, 1985). Thus, it will lead to the use of the non-equilibrium stage model to replace the equilibrium-efficiency model for the modelling of the reactive distillation processes.

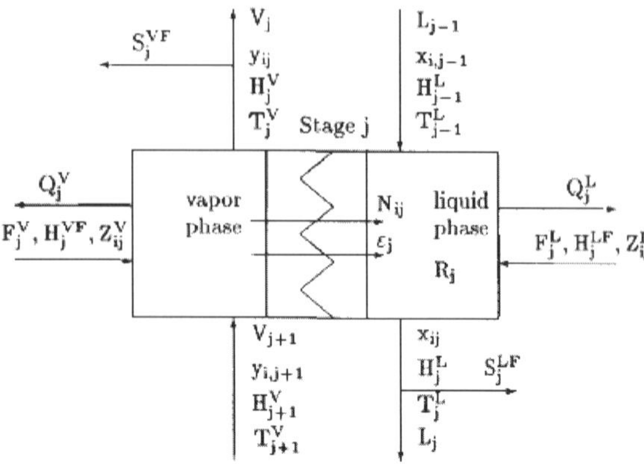

Figure 1. Schematic representation of a non-equilibrium stage with reaction.

A schematic illustration of a non-equilibrium stage with reaction is shown in Fig. 1. The non-equilibrium stage may represent one plate or section consisted of several plates in plate column, or section of packed or wetted-wall column, and the stage is numbered from top to bottom. Assuming the operation is at steady state; the reactions occur in liquid phase or on the surface of catalyst immersed in the liquid phase, no reaction in vapor phase; the stage is at mechanical equilibrium, $P_j^L = P_j^V = P_j$; the liquid phase is perfectly mixed and vapor phase is mixed well before arriving to top stage; and, the interface between vapor and liquid phase is uniform throughout the dispersion, the transfer rates and the reaction rate Rj respectively have the same value in anywhere on the stage. Provisions are made for vapor and liquid feed streams, side stream drawoffs of vapor and liquids, and for the addition or removal of heat. For a reactive distillation system with N non-equilibrium stages and C component mixtures, the equations used to model the behaviors of any stage in the reactive distillation column are catalogued as follows (Zhu, 1994):

- Material Balance Equations
  It includes vapor, liquid phase and interface material balance equations, totally 2C+1 for one normal non-equilibrium stage.
- Equilibrium Equations
  At the phase interface, there are C equations of phase equilibrium that relate the mole fractions on each side of the interface.
- Summation Equations
  For the vapor or liquid bulk phase, and the vapor and liquid on the each side of the phase interface, the sum of the mole fraction should be equal to one. Therefore there are four summation equations.
- Heat Balance Equations
  There exist 3 equations, represents respectively the energy balance of vapor phase, liquid phase, and the phase interface.
- Rate Equations
  Mass transfer is driven by concentration gradients. The mass transfer coefficients are the results to lump the effects of most of those variables besides compositions. With the more rigorous mass transfer models, the mass transfer coefficients depend on the mass transfer rates themselves. At the same time, the local energy flux may be represented in the general forms.

By imposing a particular shape on the bulk phase composition profiles, we know that the integrated total transport rates are equal to the average fluxes multiplied by the total interfacial area. Thus, the general transport rate, respectively for mass and heat transfer, can be expressed in 2C+2 equations.

Now, let us analyze the reaction rate. Assuming that the R_j is the apparent reaction rate on the stage j, thus R_j is the function of temperature, pressure, concentrations of reactants and catalyst, and liquid hold-up.

As shown above, we have illustrated all the equations in the generalized non-equilibrium stage model of the reactive distillation processes. From the analysis to this model, we know that there exist N(6C+8) equations for the reactive distillation process with N non-equilibrium stages and C component mixtures, but there are only N(5C+2) independent equations and the same number of independent variables is included in the model.

Vapor-Liquid Equilibrium Calculation

There is no adequate method for theoretical prediction or correlation of VLE data in reacting systems, since the equilibrium in such systems not only means concentration and thermal equilibrium, but also the reactions must be at equilibrium. However, the calculation method of VLE is important and limits the accuracy of the simulation results. In fact, for VLE of the same system, the different calculation methods were used by different researchers. Which method is better? we could not get the correct answer right now. Thus, an evaluation procedure on the calculation methods of VLE is developed in this work for the system of interest.

In most cases, the reactive distillation column operates at atmospheric pressure or at moderate pressures. Based on these facts, the truncated form of the Virial equation may be selected from various methods for calculating fugacity coefficients of vapor phase.

For the calculation of the activities or activity coefficients of components in liquid phase, a lot of the publications are available. Since the superiority of one method over the others is not always clear, the method choice must still rely on experience and analogy in practice. The most comprehensive comparison of the methods is made in the DECHEMA's VLE Collection (Gmehling, 1977). From the statistical analysis to the comparisons, the Wilson equation comes out the best, but there are also marked differences for particular classes of substances. By applying the evaluation procedure which is developed in this work, it is found that the Wilson equation is the best for the ethyl acetate-ethanol-acetic acid-water system. An artificial intelligence technique is also applied to help choice of the appropriate calculation method of VLE for a specific system.

Simulation Work

As another important part of this paper, the simulation is worked out for the reactive distillation process via non-equilibrium stage approach.

Process Description

For the esterification of ethanol and acetic acid using sulphuric acid as catalyst, the experimental data were determined in an Oldershaw column with diameter 0.05m, height is 1.5m, and equips 12 stage sieve trays (Kuang, 1991). In addition to 12 normal trays, there are the condenser at the top and the reboiler at the bottom of the column. Thus, the column has 14 non-equilibrium model stage. In general, the column is divided into fractionating, reacting and striping segment. Some pitfalls of the design of the experimental unit will be discussed below.

Process Model

According to the above statements, there are 14 non-equilibrium stages in the column for case study. The stages are numbered from top to bottom. Same specifications as in developed mathematical model were used. Therefore, the condenser is the first model stage, and the reboiler is the last stage, namely 14th stage.

For the condenser and reboiler, the mass balance and energy balance equations are significantly different from those for normal trays, or normal non-equilibrium stages. At first, the energy balance will be replaced by the specification equations. Secondly, the mass balance will be adjusted to describe the condenser properly. Since only the phase change occurs, the rate equations of mass and heat transfer are not considered in condenser.

For the specific system, the second-order reversible reaction kinetic equation is used to describe the reaction rate.

By assuming the vapor and liquid phase at the thermal equilibrium, the number of model equations can be reduced from 5C+2 to 5C for a normal tray. Thus the difficulty of solving model equations may be reduced.

Model Solver

Since the process model is developed via rate-based approach, the number of equations in nonequilibrium stage model is much more than that in equilibrium stage model. The introduction of transfer rate equations and reaction rate equations to the process model , makes the non-linearity higher than the equilibrium model. In order to solve the nonequilibrium model equations efficiently, we have evaluated some solution methods for nonlinear equations. On the basis of comparison, the Homotopy-continuation method is selected from various solution methods which are commonly employed.

Simulation Results and Discussion

The simulations on the stage temperature profile, composition profile and flow rate profile for both vapor and liquid phase are obtained, and they are quite satisfactorily agreed with the experimental data. Some comparisons are shown in Figs. 2-5.

Discussion

From an insight to above comparison, it is shown that the purity of comexperimental results for vapor composition profile at top and bottom of column is not high. The reason is that number of trays in fractionating and striping zone is not enough to separate the mixture. Thus, the pitfalls in the original experimental design are obvious. The conventional equilibrium approach is not adequate for reaction distillation processes. Therefore, the simulation on reactive distillation processes, can aid us to do process design and to optimize the operation.

A prototype software for the simulation of reactive distillation processes is developed.

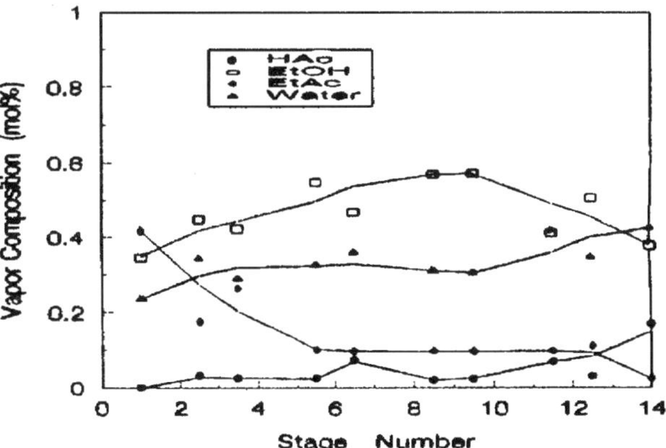

Figure 2. Comparison of simulation results with experiments for vapor composition profile.

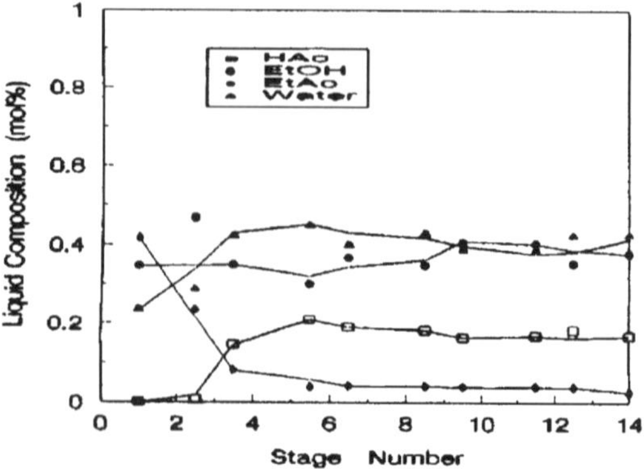

Figure 3. Comparison of simulation results with experiments for liquid composition profile.

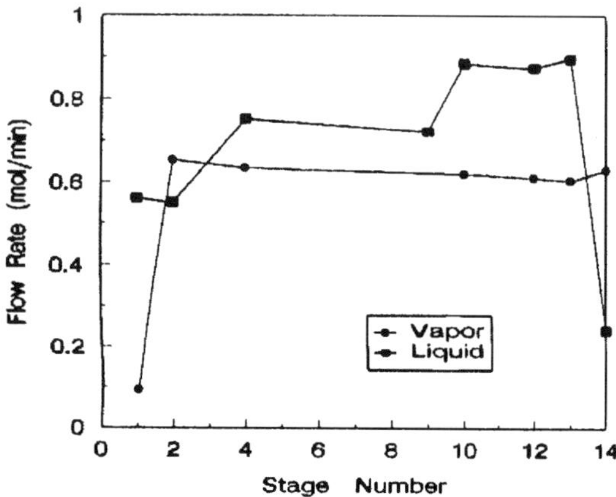

Figure 4. Comparison of simulation results with experiments for flow rate profile.

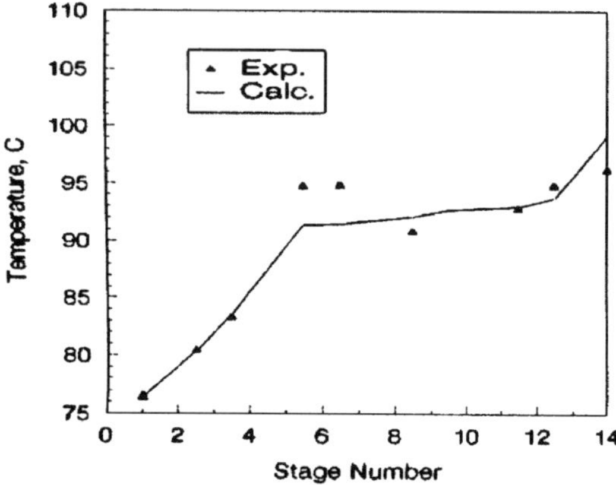

Figure 5. Comparison of simulation results with experiments for stage temperature profile.

Acknowledgments

This work is partly supported by SINOPEC, and the state-key project 29290504 of NSFC. The authors also wish to express their acknowledgment to Prof. Y.F., Xu for his kindly suggestions, to Dr. Kuang for his experimental collaboration.

References

Gmehling, J. and U., Onken (1979). Vapor-liquid equilibrium data collection. In D. Behrens and R. Eckermann (Ed.) *Chemistry Data Series*. Vol. I, DECHEMA.

Krishna, R. and G.L. Standart (1979). Mass and energy transfer in multicomponent systems. *Chem. Eng. Commun.*, **3**, 201.

Krishnamuthy, R. and R. Taylor (1985). A nonequilibrium stage model of multicomponent separation processes. *AIChE J.*, **31**, 449.

Kuang, J.W. (1991). *Studies on mass transfer model and tray efficiency of reactive distillation process*. Ph.D. Dissertation, Tianjin University, P.R. China.

Smith, L.A. and M.N. Huddleston (1982). New MTBE design now commercial. *Hydrocarbon Process.*, **61**, 121.

Zhu, J.H. and F. Shen (1994). Study on the Non-equilibrium Stage Model for Reactive Distillation Processes, *Computer and Applied Chemistry,* **11**, 167.

PINCHES IN REACTIVE DISTILLATION: CONDITIONS OF EXISTENCE

Martti Pekkanen
Department of Chemical Engineering
Helsinki University of Technology
02150 Espoo, Finland

Abstract

The conditions for pinches with reaction are derived. The results show that a pinch with net reaction is possible. In a reactive pinch the reaction effects and mass transfer or equilibrium effects between phases must cancel for each component and this poses very severe conditions to be fulfilled. In the special case of constant molar overflow it is shown that there can be no net reaction effects.

Keywords

Reactive distillation, Pinches, Conditions of existence.

Introduction

There has been a steady interest in reactive distillation (RD) in recent years, see e.g. Doherty and Buzad (1992) for a review. An important phenomenon to be studied in distillation as well as in reactive distillation, especially with respect to the design of RD units, is the possibility of pinches.

Definition 1

A pinch is a phenomenon where

a) the concentrations of both the liquid and the vapor phases are constant with respect to the column height in a steady state (reactive) distillation unit,
b) the liquid concentration is constant with respect to time in a batch (reactive) distillation unit.

The aim of this paper is to present the conditions for the existence of reactive pinches and to discuss the properties of both reactive and non-reactive pinches.

A pinch is a limiting phenomenon that cannot be realized (but can be approached) in the real physical world. Pinches are not useless entities, however, e.g. see Levy and Doherty (1986) and Kister (1992), but on the contrary are very useful e.g. in design. Here we distinguish three different kinds of pinches:

a) a tangent pinch (non-reactive non-azeotropic)
b) an azeotropic pinch (non-reactive)
c) a reactive pinch (non-azeotropic)

A pinch, where reaction occurs, is here named a reactive pinch instead of e.g. "reactive azeotrope" for reasons to be made apparent subsequently.

Reactions

Assumption 1

There are no reactions in the vapor phase.

This assumption could be easily relaxed, but this in turn would give no additional insight into the problem. Thus in this paper the hold-ups are liquid hold-ups in the respective units to be studied.

There are no restrictions with respect to the reactions that may be assumed to be present in the liquid phase. For convenience we define the net generation rate of moles per mole liquid, G, and the net generation rate of moles of component i per mole liquid, g_i, through reactions r_j, where r_j denotes the net reaction rate for reversible reactions, as follows

Definition 2

$$g_i = \sum_{j=1}^{J} \nu_{ij} r_j \; ; \qquad G = \sum_{i=1}^{I} g_i \tag{1}$$

Reactions and VLE

The basic assumption for the further development is that the vapor liquid equilibrium (VLE) and the reactions do not interact directly. The only interactions between the reactions and the VLE are changes in the concentrations caused by the reactions.

For two fluid phases the condition of physical equilibrium is the equality of the chemical potential of each component in the two phases, i.e.

$$\mu_i^L = \mu_i^V , \qquad \forall i \tag{2}$$

Assumption 2

The chemical potential of each component in any phase is independent of any reactions present in the system.

In practice the assumption 2. means that the fugacities and activities are not affected by the reactions and can thus be calculated using standard (non-reactive) correlations.

Stagewise Steady State RD Units

In a stagewise unit the numbering starts from the top. The treatment is based on balances, that can be expressed for reactive cases generally as

$$\text{ACC = IN - OUT + NET GEN} \tag{3}$$

It is assumed that there are no material side streams in a stage, that there may be heat streams, that the liquid hold-up inside a stage is known, that the liquid composition inside a stage equals the exit liquid composition throughout and that the relation between the exit compositions is known.

The steady state balances around the whole stage n are for total moles and for component i moles

$$0 = V_{n+1} + L_{n-1} - V_n - L_n + H_n G \tag{4}$$

$$0 = V_{n+1} y_{n+1,i} + L_{n-1} x_{n-1,i} - V_n y_{n,i} - L_n x_{n,i} + H_n g_i \tag{5}$$

The component i balance around the vapor phase in stage n is

$$0 = V_{n+1} y_{n+1} - V_n y_n + N_{n,i} \tag{6}$$

where the positive direction of mass transfer is from the liquid to the vapor phase. The enthalpy balance around the whole stage n is

$$0 = V_{n+1} I_{n+1} + L_{n-1} i_{n-1} - V_n I_n - L_n i_{n,i} + H_n \sum_{j=1}^{J} r_{nj} \Delta I_j + \sum_{k=1}^{K} \Phi_{n,k} \tag{7}$$

Now, in a pinch the concentrations are constant, and $y_{n+1,i} = y_{n,i} = y_i$ and $x_{n-1,i} = x_{n,i} = xi$ can be used in eqs. (5)-(6) and In+1 = In = I and in-1 = in =i in eq. (7). Eq. (4) is inserted in eqs. (5) and (7) and the results are manipulated to give

Condition 1

A pinch with net reaction can exist, if and only if

$$g_i = Gx_i - \frac{V_{n+1} - V_n}{H_n} (y_i - x_i), \quad \forall i \tag{8}$$

$$N_{n,i} = -V_{n+1} - V_n) y_i , \quad \forall i \tag{9}$$

$$V_{n+1} - V_n = \frac{1}{I-i} \left(H_n Gi - H_n \sum_{j=1}^{J} r_{nj} \Delta I_j - \sum_{k=1}^{K} \Phi_{n,k} \right) \tag{10}$$

It can be seen from eq. (8) that in a pinch reaction effects and mass transfer effects must cancel each other. If equilibrium stages are assumed, eq. (8) can be written as

$$g_i = Gx_i - \frac{V_{n+1} - V_n}{H_n} (K_i - 1), \quad \forall i \tag{11}$$

which shows that in a pinch kinetically controlled reaction effects and thermodynamical equilibrium effects must balance. This is even more clearly seen if a reaction scheme with G=0 is assumed. Eq. (8) also shows why the notion of reactive pinch is preferred to "reactive azeotrope" in this paper: an azeotrope is dependent on thermodynamics only whereas a reactive pinch depends on process phenomena (e.g. hold-up) also.

Eq. (10) shows that in a reactive pinch the flow rates need not be constant. Further, according to eq. (9) the direction of mass transfer must be the same for every component.

Batch RD Units

The application of the general balance equations for a batch RD unit gives

$$\frac{\partial H}{\partial t} = -V + HG \tag{12}$$

$$\frac{\partial}{\partial t}(Hx_i) = -Vy_i + Hg_i \tag{13}$$

The expansion of the derivative and the deletion of the concentration derivative in eq. (13) (because of the pinch), the insertion of eq. (12) into eq. (13) and manipulation gives

Condition 2

A pinch with net reaction can exist in a batch reactive distillation unit if and only if the following holds for each component i in the system.

$$g_i = Gx_i - \frac{V}{H}(y_i - x_i), \quad \forall i \tag{14}$$

Because xi is constant, it seems reasonable to assume that g_i and G remain constant also, and thus there follows that distillation must be carried out in a way that assures that the ratio V/H remains constant in time.

Constant Molar Overflow

Using the assumption of constant molar overflows in a pinch (i.e. zero differences of flow rates and concentrations) in eqs. (4) and (5), it immediately follows that G=0 and g_i=0 and thus

Condition 3

In a pinch with constant molar overflow G=0 and g_i=0 and there is no net effect due to the reactions.

Properties of Pinches from Operating Lines

In this section we assume that a pinch is reached. For simplicity the treatment is restricted to the rectifying section of a steady state stagewise distillation unit. The operating line for component i for the rectifying section is

$$y_{n+1,i} = \frac{L_n}{V_{n+1}} x_{n,i} + \frac{D}{V_{n+1}} x_{D,i} \tag{15}$$

For reactive distillation the operating line is

$$y_{n+1,i} = \frac{L_n}{V_{n+1}} x_{n,i} + \frac{D}{V_{n+1}} x_{D,i} - \frac{1}{V_{n+1}} \sum_{n=1}^{n} H_n g_{n,i} \tag{16}$$

Eqs. (15) and (16) alone cannot be used to perform stage to stage calculations. If ideal stages are assumed, the most common additional equation, i.e. an equilibrium relationship, can be used, e.g. as

$$y_{n+1,i} = K_{n+1,i}\ x_{n+1,i} \tag{17}$$

It should be noted that eq. (17) reflects thermodynamic restrictions only. Now, either of the eqs. (15) or (16) can be used with eq. (17) to calculate the concentrations in the rectifying section.

A Tangent Pinch

A tangent pinch (non-reactive non-azeotropic) occurs when xn calculated from the operating line eq. (15) and xn+1 calculated from the equilibrium relationship eq. (17) are the same for one yn+1. For such a case there will be no change in the liquid composition. This phenomenon can be visualized most clearly in binary distillation in a case where the operating line and the equilibrium curve intersect. It is thus seen that the liquid and vapor compositions need not be equal and that a tangent pinch can develop in the middle of a section. For a tangent pinch eqs. (15) and (17) give

$$\left(K_{n+1,i} - \frac{L_n}{V_{n+1}}\right) x_i = \frac{D}{V_{n+1}} x_{D,i} \tag{18}$$

For a tangent pinch to develop both mass balances (operating conditions) and equilibrium relations (thermodynamics) have to be taken into account. From this it also follows that a tangent pinch may be avoided by altering the operating conditions (e.g. by increasing the reflux)

An Azeotropic Pinch

An azeotropic pinch (non-reactive) occurs, if the distribution coefficient in eq. (17) become equal to one for all components i. In such a case the liquid and the vapor compositions are the same, no separation occurs and a pinch develops. For an azeotropic pinch eqs. (15) and (17) give

$$\left(1 - \frac{L_n}{V_{n+1}}\right) x_i = \frac{D}{V_{n+1}} x_{D,i} \tag{19}$$

from which is follows immediately, because $V_{n+1}=L_n+D$, that

$$x_i = x_{D,i} \tag{20}$$

which shows that in a rectifying section an azeotropic pinch must always lie at the top.

Thus for an azeotropic pinch to develop only equilibrium relations (thermodynamics) have to be taken into account. From this it also follows that an azeotropic pinch cannot be avoided by altering operating conditions (in given pressure).

A Reactive Pinch

A reactive pinch (non-azeotropic) occurs when xn calculated from the operating line eq. (16) and xn+1 calculated from the equilibrium relationship eq. (17) are the same for one yn+1. For such an case there will be no change in the liquid composition. For a reactive pinch eqs. (16) and (17) give

$$\left(K_{n+1,i} - \frac{L_n}{V_{n+1}}\right)x_i = \frac{D}{V_{n+1}}x_{D,i} - \frac{1}{V_{n+1}}\sum_{k=1}^{n} H_k g_{k,i} \qquad (21)$$

For a reactive pinch to develop both mass balances (operating conditions) and equilibrium relations have to be taken into account. From this it also follows that a reactive pinch may be avoided by altering the operating conditions (e.g. by altering the reflux or the hold-ups).

Azeotropes in a Reactive System

In a non-reactive system an azeotrope causes an azeotropic pinch to develop as seen from eqs. (15) and (17), when all distribution coefficients are equal to one.

In a reactive system, however, the use of eq. (16) with eq. (17) where all Ki=1 gives

$$x_{n+1,i} = \frac{L_n}{V_{n+1}}x_{n,i} + \frac{D}{V_{n+1}}x_{D,i} - \frac{1}{V_{n+1}}\sum_{k=1}^{n} H_k g_{k,i} \qquad (22)$$

which shows that a pinch does not necessarily follow from the presence of an azeotrope.

This phenomenon can be used in reactive distillation to overcome separation limitation imposed by azeotropes, if present.

Conclusions

The conditions for pinches in reactive distillation have been derived from balances. The results show that a pinch with net reaction is indeed theoretically possible. The conditions derived show that in a reactive pinch the reaction effects must cancel the mass transfer effects between the phases. Thus a reactive pinch is dependent on both the reaction and the mass transfer as described in the conditions 1.-3. above.

If equilibrium between the phases is assumed, kinetic reaction effects must balance thermodynamic equilibrium. The requirements for a reactive pinch are very severe indeed and it seems to be rather difficult to realize a reaction scheme that would fulfil the conditions presented here.

The condition 3. shows that a pinch with net reaction is not possible, if constant molar flow rates are assumed.

It is also shown that in reactive distillation a pinch does not necessarily follow from the presence of an azeotrope.

As a reactive pinch depends on operating conditions, on liquid hold-up and flow rates, it is a process phenomenon and not a thermodynamic phenomenon. Thus the term "reactive azeotrope" seems misleading and is not used here.

Notation

D = distillate flow rate, mol/s
g_i = net generation of moles of component i through reactions, mol/(s mol liquid)
G = net generation of moles through reactions = mol/(s mol liquid)
H = liquid hold-up, mol liquid
i = liquid enthalpy, J/mol
I = vapor enthalpy, J/mol
K_i = distribution coefficient for component i, $K_i=y_i/x_i$, dimensionless
L = liquid flow rate, mol/s
$N_{n,i}$ = mass transfer from liquid to vapor phase, mol/s
r_j = net reaction rate of reaction j, mol/(s mol liquid)
t = time, s
V = vapor flow rate, mol/s
x = liquid concentration, mol fraction
y = vapor concentration, mol fraction

Greek Letters

ΔIj = enthalpy generation of reaction j, J/mol
Φ = heat flow, J/s
μ_i^L = chemical potential of component i in liquid, J/mol
μ_i^V = chemical potential of component i in vapor, J/mol
$\nu_{i,j}$ = stoichiometric coefficient of component i in reaction j

Reference

Doherty, M.F., and Buzad, G. (1992). Reactive distillation by design. *Trans. IChemE* , **70**, 448-54.

Levy, S.G., Doherty, M.F. (1986), A simple exact method for calculating tangent pinch points in multicomponent nonideal mixtures by bifurcation theory. *Chem. Eng. Sci*, **41**(12), 3155-60.

Kister, H.Z. (1992). *Distillation Design*. McGraw-Hill, New York.

CRITICAL ANALYSIS OF EXPERIMENTAL DATA, REGRESSION MODELS AND REGRESSION COEFFICIENTS IN DATA CORRELATION

M. Shacham
Ben-Gurion University of the Negev
Beer-Sheva, 84105 Israel

N. Brauner
Tel-Aviv University
Tel-Aviv, 69978 Israel

M. B. Cutlip
University of Connecticut
Storrs, CT 06269

Abstract

Realistic modeling and accurate correlation of experimental data are essential to sound process design in the computer-aided design era. Many of the statistical techniques for analyzing the accuracy of correlations have been known for several decades but have not been convenient to utilize until the recent widespread availability of software on personal computers and workstations. This paper presents several examples where the use of interactive modeling and regression programs with convenient graphical output makes it possible to efficiently analyze and correlate experimental data, select the most appropriate among candidate models and estimate the most accurate model parameters for a selected model.

Keywords

Linear regression, Nonlinear regression, Parameter estimation, Model selection.

Introduction

Consistent and accurate correlation of experimental data is essential to accurate computer-aided process design. In the days when calculations were done by hand or with a calculator, inappropriate data could be immediately detected and discarded. When a correlation is included inside a large simulation program, it is very difficult to detect if it provides inaccurate or meaningless results in a particular range of temperature, pressure or composition. Thus, it is very important to critically examine a correlation using statistical and numerical considerations before the correlation becomes part of a large simulation program or data set.

Many of the statistical techniques for analyzing the accuracy of correlations have been known for several decades (Himmelblau (1970)). The emergence of interactive regression capabilities with graphical output, such as the regression program in the POLYMATH package (Shacham and Cutlip, 1994) makes the critical analysis of regression models an easy task that can be carried out in a short time. As in all statistical analyses, there is an inherent uncertainty involved with regression of data and furthering the analysis beyond a certain point can bring diminishing returns.

Ideally experimentation must be intermeshed with data analysis. Design of experiments to allow most effective model discrimination and most accurate parameter estimation is discussed, for example by Himmelblau, 1970 and Froment and Bishoff, 1990. In many instances, the analysis must be done on data that were taken in poorly designed experiments or utilizing data from the literature that were obtained before the statistical experimental design techniques were widely used.

Data Viability

Analysis of the data prior to the modeling and regression can reveal flaws that may significantly limit the applicability of the model being fitted. Consider the data used by Smith (1981) for the catalytic oxidation of sulfur dioxide.

Visual inspection of the data raises suspicion that the data do not represent true measured values. The reaction rate, which is the dependent variable is presented in round numbers, such as 0.02, 0.04, etc. While independent variable values can often be set to round numbers, which are more convenient for calculations, such numbers will rarely be obtained in measured values of the dependent variable. The original source of this data (Olson et al., 1950) reveals that the data were indeed extrapolated and smoothed.

A useful technique for analyzing the data is plotting one "independent" variable versus another "independent" variable as shown in Fig. 1.

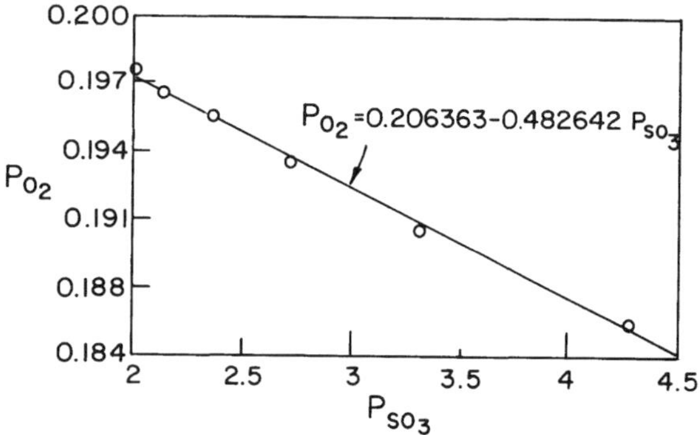

Figure 1. Plot of "independent" variable (P_{O_2}) versus "independent" variable (P_{SO_3}).

Figure 1 shows there is linear dependence between the presumed independent variables. Thus the contributions of the different independent variables to the variation of the dependent variable cannot be separated. Regression here will lead to nearly singular normal matrices and consequently to very large uncertainty regarding the calculated parameters (very wide 95% confidence intervals).

For instance, nonlinear regression of a power law rate expression

$$r = kP^a_{SO_3}P^b_{SO_2}P^c_{O_2} \quad (1)$$

to the data leads to the following parameter values (including 95% confidence intervals): k = 0.517 ± 113.3; a = -1.98 ± 7.02; b = -0.216 ± 4.556; and c = 6.078 ± 124.7. In practice most nonlinear regression algorithms will usually switch from the fast Newton-Raphson method to the slow (often very slow) steepest decent method. Our experience has shown that slow conversion often indicates inappropriate data in such cases rather than computational problems with the algorithm.

There are some techniques for handling data in order to alleviate the difficulties encountered due to linear dependence between the predictor variables used in a model (Draper and Smith, p. 258, 1981). For this example, these techniques did not improve the power-law correlation, probably due to the poor accuracy of the data.

Further analysis of the data can be performed in view of a particular model that is assumed to represent the data. Smith (1981) suggested the following model for the SO_2 oxidation data:

$$r = \frac{P_{SO_2}P^{1/2}_{O_2} - \left(\frac{1}{73}\right)P_{SO_3}}{\left(A + BP_{SO_3}\right)^2} \quad (2)$$

Nonlinear regression with the rate data yields the following parameter values: A = 0.1017 ± 0.0958 and B = 16.02 ± 4.33. These parameter values can be introduced into Eq. (2). Rearrangement of this equation yields:

$$73\left[r(0.1017 + 16P_{SO_3})^2 - P_{SO_2}P^{1/2}_{O_2}\right] = -P_{SO_3} \quad (3)$$

If the experimental data contain sufficient information to reflect the contribution of the reverse reaction, the plot of the left hand side of this equation versus P_{SO_3} should give a straight line with slope of -1. Figure 2 shows the plot of the left hand side of Eq. (3) (denoted Y) versus P_{SO_3} which indicates random distribution. Thus sufficient information is not available in this data set to reflect the contribution of the reverse reaction in the model.

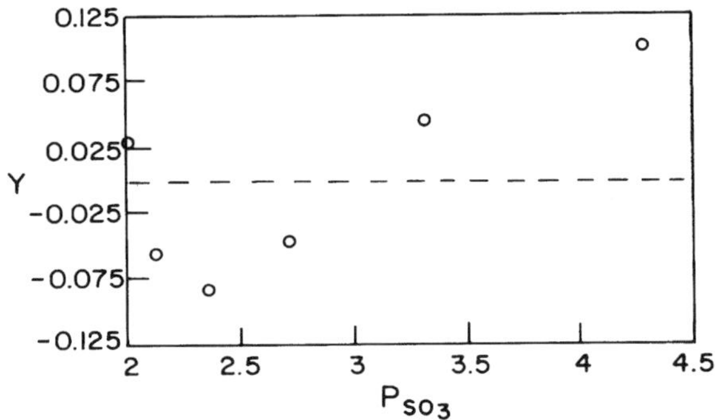

Figure 2. Plot of Eq. 3 as (Y) versus P_{SO_3}.

Selection Between Regression Models

The vapor pressure data for 3-methylthiacyclopentane (from Osborn and Douslin, 1966) is useful to demonstrate some principles in regression model selection.

There are several equations that can be used to correlate vapor pressure data. The most widely used are:

1. The two parameter Clapeyron equation:

$$\log P = A - B/T \quad (4)$$

2. The three parameter Antoine equation (Antoine, 1888):

$$\log P = A + B/(T+C) \quad (5)$$

3. The three parameter Cox equation (Cox, 1936):

$$\log(P/760) = 10^{(A+BT+CT^2)}(1 - T_{BP}/T) \quad (6)$$

where T_{BP} is the boiling point temperature.

4. The four parameter Riedel equation (Riedel, 1956):

$$\log P = A - B/T + C\ln T + DT^6 \quad (7)$$

Fitting the Clapeyron Eq. (4) to the vapor pressure data yields: A = 7.73562 ± 0.0323786, B = -1999.47 ± 12.4439 where s^2 and $R^2 = 0.999982$. The calculated curve of log (P) versus 1/T and the experimental points are presented in Fig. 3.

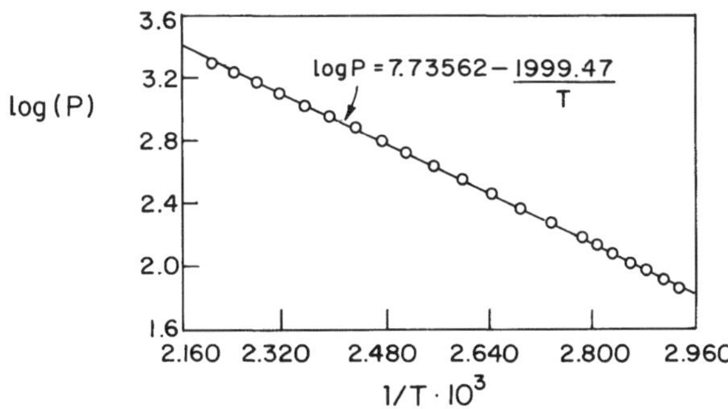

Figure 3. Correlation of vapor pressure data by the Clapeyron equation, calculated line and experimental data.

All the statistical and visual indicators show that the fit is excellent. The confidence intervals are narrow, the mean square about the regression is small relatively to the values, and the adjusted linear correlation coefficient is very close to one. It is always important to make a plot of the residuals defined by

$$\varepsilon_i = \log(P_{i,obs}) - \log(P_{i,calc}) \quad (8)$$

A plot of the residuals in Fig. 4 shows that they are not normally distributed but show a clear pattern. The experimental data exhibit a curvature which is not predicted by the Clapeyron equation; therefore this particular correlation is inadequate. Moreover, with such a definite trend in the residual distribution, the calculation of confidence intervals for this model violates the underlying statistical assumption of constant error variance for log P.

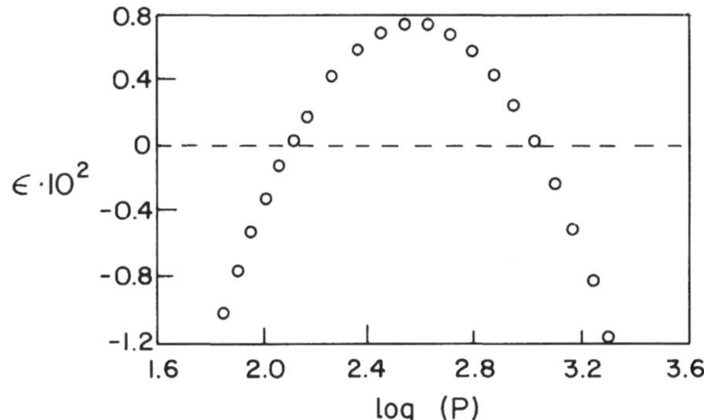

Figure 4. Residual plot for the Clapeyron equation.

The parameter values, the estimated variances (s^2), and the linear correlation coefficients (R^2) adjusted for the number of data points and number of parameters (Draper and Smith, p. 91, 1981) for all the vapor pressure correlations are summarized in Table 1.

Table 1. Parameter Values, Variance and Linear Regression Coefficient for Four Vapor Pressure Correlations.

	Eq.(4)	Eq.(5)	Eq.(6)	Eq.(7)
A	7.736± 0.0324	6.952± 0.00292	0.8404± 0.0019	45.426± 1.119
B	-1999.5± 12.44	-1433.6± 1.931	(-6.106± 0.097)*10^{-4}	-3528.7± 36.67
C	-	213.81± 0.220	(5.203± 0.123)*10^{-7}	-13.83± 0.4314
D	-	-	-	(5.315± 0.238)*10^{-2}
s^2	250.8	0.00516	0.00078	0.0084
R^2	0.999982	-	-	0.999997

The s^2 values in Table 1 indicate that the three and four parameter models are much better than the two parameter Clapeyron equation. For this particular case the Cox equation yields the smallest mean square value. The residual plot for this equation, given in Fig. 5, indicates that the residuals are randomly distributed without any particular trend. (Note that the data point representing the boiling point is excluded from this plot.)

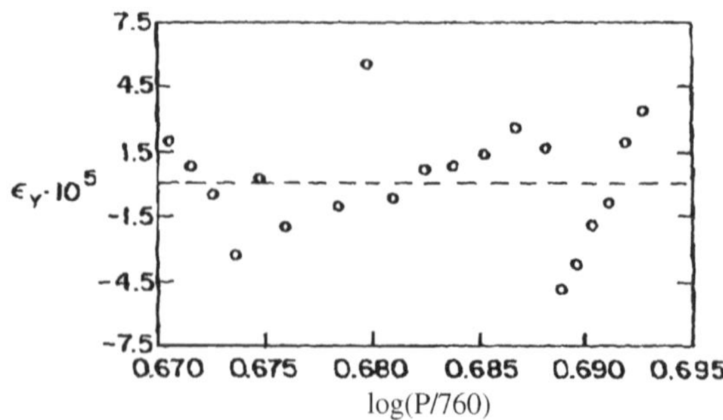

Figure 5. Residual plot for the Cox equation.

This example demonstrates a hazard in model selection when accepting the first model that is indicated as acceptable by one or even several statistical indicators. A residual plot will clearly indicate when the model is based on too few parameters while extremely wide confidence intervals will indicate existence of too many parameters. A comparison of the sum of squares of the deviations calculated on a common basis will allow selection of the best model from among several good candidates.

Calculation of the Model Parameters

There are several ways to calculate the numerical values of the model parameters. If the model is nonlinear with respect to the parameters, transformation of the model to obtain a linear expression is a very widely used technique. For example, the Antoine equation (Eq. 5), which is nonlinear in parameter C, can be linearized using the following transformation:

$$T \log P = (B+AC) + AT - C \log P \tag{9}$$

This is an example of a linearization which causes violation of the statistical assumption underlying least squares regression because logP on the right hand side of the equation contains experimental error where the independent variables are assumed to be free of error.

Linear regression is more widely utilized than nonlinear regression because software is readily available and because computational problems are occasionally encountered with nonlinear regression. For example, nonlinear regression algorithms may sometimes converge to a local minimum yielding incorrect results. Shacham et al. (1993) discuss in detail the pitfalls of linearization in the regression of activity coefficient data. They have shown that linearization may have contradictory effects on the accuracy of the calculated parameter values. The effect of linearization can be observed on the residual plots drawn using the transformed functions and data.

Conclusions

Interactive modeling and regression programs on a variety of computing platforms make it possible to efficiently correlate data with both linear regression and nonlinear regression.

The quality of the experimental data can be checked by plotting one independent variable versus another to detect linear dependency. If the model is composed of sums of terms of different order of magnitude (as in rate expressions for reversible reactions), plots can be derived which can indicate whether the data contain enough information to represent the less significant terms.

If there are several candidate models that can represent the same data, model selection can be accomplished by eliminating those with an insufficient number of parameters by using residual plots, detecting models with an excessive number of parameters by examination of confidence intervals, and by using the mean sum of squares of deviations on a common basis. The preferred method for calculation of the parameters (i.e., linear versus nonlinear regression) may be selected by making an error analysis of the transformation functions, residual plots, and a comparison of the mean sum of squares of the deviations on a common basis.

References

Antoine, C. (1888). *Compt. rend.*, **107**, 681.

Cox, E.R. (1936). *Ind. Eng. Chem.*, **28**, 613.

Draper, N.R. and H. Smith (1981). *Applied Regression Analysis.* John Wiley & Sons, New York.

Froment, G. and K.B. Bischoff (1990). *Chemical Reactor Analysis and Design.* John Wiley & Sons, New York.

Himmelblau, D.M. (1970). *Process Analysis by Statistical Methods.* John Wiley and Sons, New York.

Olson, R.W., R.W. Schuler and J.M. Smith (1950). Catalytic oxidation of sulfur dioxide, effect of diffusion. *Chem. Eng. Progr.*, **46** 12), 614-624.

Osborn, A.G. and D.R. Douslin (1966). Vapor pressure relations of 36 sulfur compounds present in petroleum. *J. Chem. Eng. Data*, **11**, 502.

Riedel, L. (1954). *Chem. Ing. Tech.*, **26**, 83.

Shacham, M. and M.B. Cutlip (1994). *Polymath 3.0 Users' Manual*, CACHE Corporation, Austin, Texas.

Shacham, M., J. Wisniak and N. Brauner (1993). Error analysis of linearization methods in regression of data for the Van Laar and Margules equations. *Ind. Eng. Chem. Res.*, **32**, 2820.

Smith, J.M. (1981). *Chemical Engineering Kinetics.* 3rd Ed., McGraw-Hill, New York.

TEMPERATURE OPTIMAL PROFILES IN REACTION SYSTEMS DESCRIBED BY TWO INDEPENDENT COMPOSITION VARIABLES

Tore Omtveit and Kristian M. Lien
Department of Chemical Engineering
University of Trondheim — NTH
N-7034 Trondheim, Norway

Abstract

This paper describes a method which extends the Attainable Region concept to cater for temperature profile optimization in cases where the reaction mixture composition is uniquely described by less than three independent composition variables. The method may be extended to also optimize profiles of pressure, pH and inerts.

Keywords

Reactor system synthesis, Attainable region, Optimal profiles, Targeting.

Introduction

The formation of byproducts in complex reaction systems may be controlled in several ways. One alternative is to focus on the impact of mixing patterns on the product distribution. Another alternative is to explore the impact of various profiles along the reactors. In this paper we will propose a method for finding temperature profiles and other profiles which will improve the overall selectivity of a reactor system. Optimal temperature profiles have been reported for batch reactors with many different types of kinetics. Rippin (1983) has reviewed works in this field. Chitra and Govind (1985) have presented a method for finding optimal recycle ratios and temperatures in a series of adiabatic intercooled recycle reactors. Kokossis and Floudas (1991) constructed a superstructure of interconnected adiabatic CSTR models and heat exchangers which they solved with an MINLP algorithm. Glasser and coworkers (Glasser 1987, 1992 and Hildebrandt 1990a, 1990b), have developed the **Attainable Region** approach. The main features of this approach are:

- The **limiting behaviour** of a reaction system can be **predicted prior to design**, given the kinetics of the system and its inlet conditions.
- The **structure** of optimal reactor systems may be **deduced** from the limiting behaviour.

The Attainable Region approach is graphically oriented, and may thus be visualized directly only for problems in two or three dimensions.

Balakrishna and Biegler (1992a) proposed a targeting approach for isothermal reactor networks. They use stepwise numerical optimization and are thus not restricted by dimensional limitations. Subsequent publications address nonisothermal reactors and heat recovery (Balakrishna, 1992b) and separation (Balakrishna, 1993).

Omtveit et al. (1994) have recently demonstrated cases where the dimensionality limitation of the Attainable Region approach may be overcome by exploitation of implicit constraints, using e.g. the Chemical Reaction Invariance principle to eliminate dependent variables from the problem.

Related to the optimization of other profiles than temperature, Van der Vusse and Voetter (1961) discussed the effect of pressure profiles. Ho and Humphrey (1970) demonstrated a numerical optimization of the temperature and pH profiles in an enzymatic biochemical reaction.

It should finally be mentioned that that the idea of profile optimization using the Attainable Region approach was recently presented in a conference paper by Glasser et al. (1993). Their approach was not described in a sufficient detail to be compared directly to the suggested approach, but the underlying ideas may seem quite similar.

The proposed method

The proposed method is a **targeting** procedure, i.e. we do not consider how the temperature profile is going to be realized until after it has been obtained. Mass- and heat-transfer resistance may thus prevent the obtained profiles from being attainable in practice. The scope of the method is further limited to 2-dimensional composition problems, i.e. the composition of the reaction mixture is assumed to be uniquely defined in a plane.

If the Attainable Region for an isothermal 2-dimensional reaction system is constructed, it may be possible that other temperatures along the boundary of the region lead to reaction vectors pointing out of the region, thus making it possible to extend the region. Here we ask whether it will be possible to construct a 2-dimensional region where the temperature is allowed to vary freely.

We will first consider a PFR:

- Find the optimal starting temperature; the temperature yielding the maximum initial selectivity.
- Take a small step in this direction.
- Repeat with the new point as starting point.

If sufficiently small steps are taken, a temperature optimal boundary of the attainable region will be constructed where no temperature exists at any point such that reaction vectors point out of the region.

PFR trajectories with *other temperature profiles*, coming from the *interior* of the attainable region, *cannot possibly cross such a temperature-optimized* boundary:Crossing would require that temperatures exist at the boundary where reaction vectors point out of the region. Temperature optimization along the boundary prevents the existence of such temperatures.

If the resulting temperature optimized PFR profile is not convex, then additional improvements may be made by mixing, and the procedure is continued analog to the construction of an isothermal Attainable Region: Construct a temperature optimized CSTR locus. If this is also not convex, then continue with a new temperature optimized PFR trajectory starting at the point on the CSTR locus where the overall selectivity is highest. In the latter case, the optimal system will consist of a CSTR followed by a temperature optimized PFR, and it is easily deduced that the optimal starting temperature for the PFR must be the same as the optimal temperature of the preceeding CSTR.

This procedure is based on construction of a *region*, described by a small number of *lines* (PFR trajectories, CSTR loci and mixing lines). In higher dimensions, PFR trajectories and CSTR loci will still be lines, but *surfaces* will be needed to describe a region. The procedure is thus limited to 2-dimensional problems.

Example

We use the van der Vusse reaction scheme using the parameters given by Balakrishna and Biegler (1993) as an illustrative example. The reaction scheme is:

$$A \rightarrow B \rightarrow C \tag{1}$$

$$2A \rightarrow D \tag{2}$$

The associated kinetic constants are

$k_{ab0} = 8.86e6h^{-1}$, $k_{bc0} = 9.7e9h^{-1}$

$k_{ad0} = 9.84e3\ Lmol^{-1}h^{-1}$, $E_{ab} = 15.00$ kcal/gmol

$E_{bc} = 22.70$ kcal/gmol, $E_{ad} = 6.920$ kcal/gmol

To better understand the character of the reaction system, we first analyse it isothermally. Figure 1, Fig. 2 and Fig. 3 show the comparison of isothermal PFR and CSTR reactors at temperatures ranging from 400K to 600K.

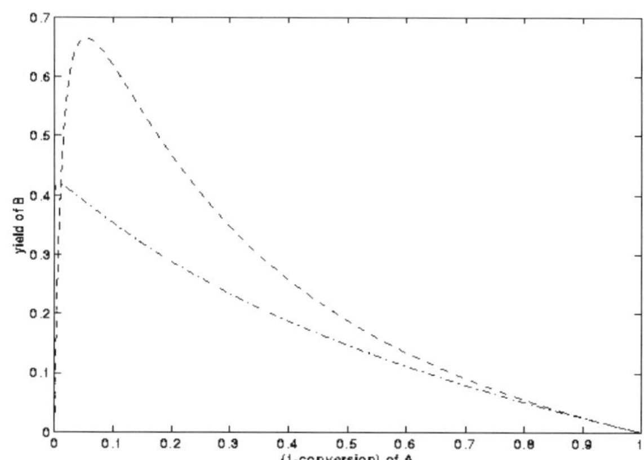

Figure 1. Isothermal reactor trajectories at 400K. (PFR: _._. CSTR: ---)

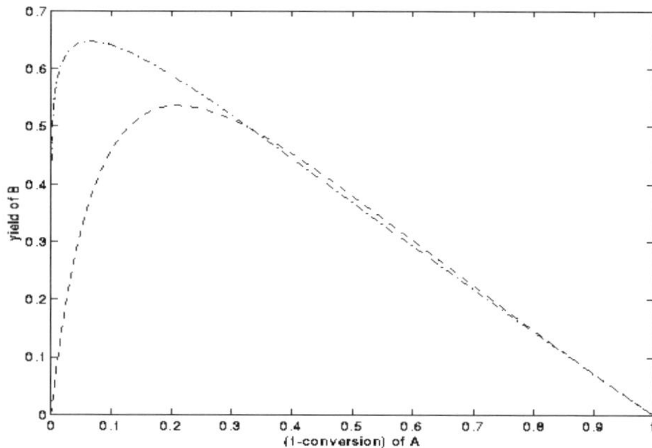

Figure 2. Isothermal reactor trajectories at 500K. (PFR: _._. CSTR: ---)

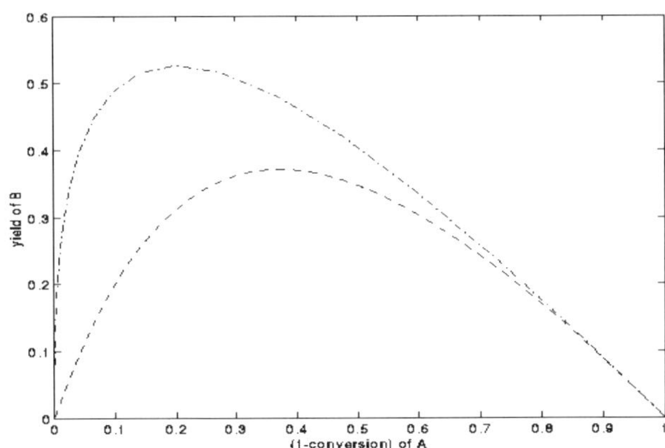

Figure 3. Isothermal reactor trajectories at 600K. (PFR: _._. CSTR: ---)

The shapes of the trajectories change significantly with temperature: At low temperatures the shapes are concave, indicating a CSTR reactor. At high temperatures the shapes are fully convex, and a PFR would then yield the best product distribution.

Fig. 4 shows an isothermal attainable region for the reactions at 450 K. At this temperature, the optimal isothermal reactor system is seen to be a CSTR followed by a PFR.

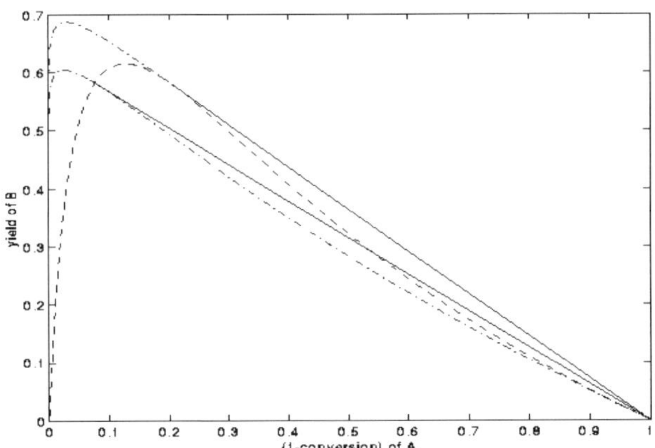

Figure 4. Isothermal Attainable Region at 450K.(PFR: _._. CSTR: ---)

Fig. 5 illustrates the temperature optimized trajectories for CSTR and PFR reactors. The temperature optimal PFR is convex, it gives a better product distribution than any CSTR reactor, and it outperforms any of the three isothermal PFRs.

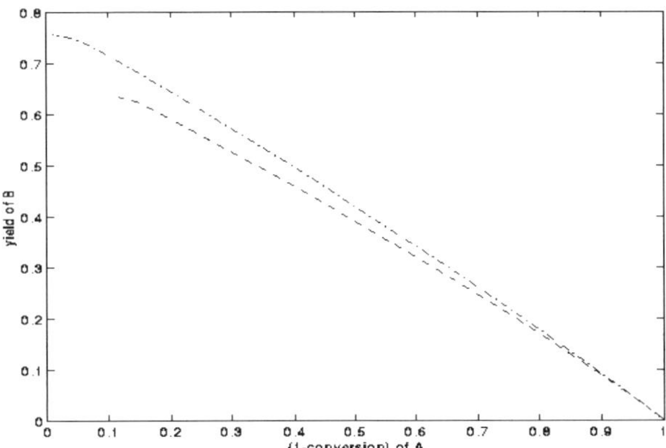

Figure 5. Temperature optimized reactor trajectories. (PFR: _._. CSTR: ---)

Fig. 6 displays the corresponding temperature profiles for the optimized reactor trajectories. It is seen that the reactors start at maximum temperature and approach the minimum temperature as conversion is increased.

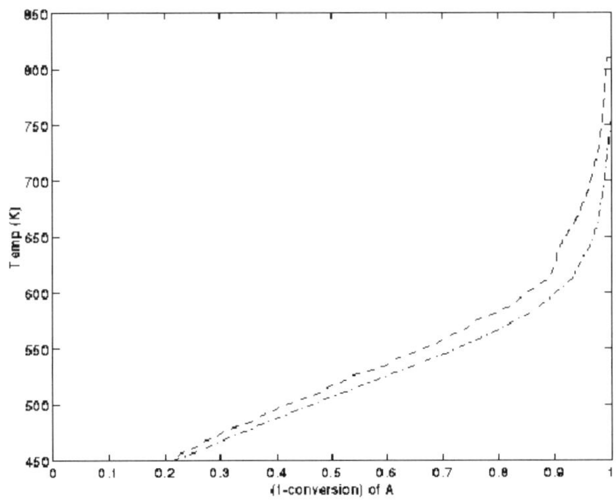

Figure 6. Optimized temperature profiles. (PFR: _._. CSTR: ---)

In general, it may be difficult to explain the trends of the optimal temperature profile directly from the kinetic model, but in this simple case with power law kinetics and Arrhenius type temperature dependence, some qualitative assessments may be given: At low conversions, i.e. at high concentrations of A, the 2nd order D-producing side reaction must be suppressed. The lower acivation energy of this side reaction than that of the main reaction indicates that such suppression could be be gained by keeping the temperature high at low conversion. At higher conversion, i.e. at lower concentrations of A and higher concentrations of B, the main reaction must compete with both side reactions, with suppression of the 1st order C-producing side reaction becoming increasingly important as the concentration of B increases. The C-producing side reaction has a higher activation energy than the main reaction, which indicates that the temperature should be reduced with increasing conversion.

Possible extensions

Similar to temperatures, the total pressure in a gas phase reactor can have a significant impact on the reaction rates, and on the equilibrium composition. The proposed method may be used to optimize the pressure profile, analog to the temperature profile optimization. With respect to practical implementation, decreasing pressure profiles will be easy and inexpensive to implement. Increasing pressure profiles pose more of an engineering challenge.

Addition of inerts has a dilution effect on reaction rates and equilibrium composition similar to reduction of the total pressure in the reactor. Note that a decreasing inert profile will require separation, and this may be difficult to implement at the reactor conditions.

In complex biochemical reactions, undesired reactions may be controlled by changing the pH level in the solution. The pH is adjusted by addition of acids or bases in amounts small enough that it may be assumed that these do not significantly alter the concentrations of other components in the reactor/fermentor.

Conclusions

We have proposed a method for optimizing the control profiles for a limited class of reaction problems; reactions with a fixed inlet composition, where the composition can be uniquely defined by two components. In combination with previous developments of the Attainable Region Concept, this is a step further towards a total concept for setting targets for reactor systems.

Acknowledgements

This work has been funded in parts by the Norwegian Science Foundation (NFR), Norsk Hydro, Statoil and the Nordic Energy Research Programme. We thank Magne Karlsen and Eirik B. Sund for their contributions to the computer programs used to generate the examples of the paper.

References

Balakrishna, S. and L.T. Biegler (1992a). A constructive targeting approach for the synthesis of isothermal reactor networks. *Ind. Eng. Chem. Res.*, **1**(31), 300.

Balakrishna, S. and L.T. Biegler (1992b). Targeting strategies for the synthesis and energy integration of nonisothermal reactor networks. *Ind. Eng. Chem. Res.*, **31**(9), 2152.

Balakrishna, S. and L.T. Biegler (1993). A unified approach for the simultaneous synthesis of reaction, energy and separation systems. *Ind. Eng. Chem. Res.*, **32**(7), 1372-1382.

Chitra, S.P. and W.R. Govind (1985). Synthesis of optimal serial reactor structure for homogeneous reactions, part ii: Nonisothermal reactors. *AIChE Journal*, **31**(2), 185-194.

Glasser, D., D. Hildebrandt and C. Crowe (1987). A geometric approach to steady flow reactors: The attainable region and optimization in concentration space. *Ind. Eng. Chem. Res.*, **26**, 1803-1810.

Hildebrandt, D., B. Glasser and C. Crowe (1990a). The geometry of the attainable region generated by reaction and mixing; with or without constraints. *Ind. Eng. Chem. Res.*, **29**, 49-58.

Hildebrandt, D. and D. Glasser (1990b). The attainable region and optimal reactor structures. *Chem. Eng. Sci.*, **2**, 2161-2168.

Glasser, B., D. Hildebrandt and D. Glasser (1992). Optimal mixing for exothermic reversible reactions. *Ind. Eng. Chem. Res.*, **31**, 1541-1549.

Glaser, D., D.Hildebrandt, S.Godorr and M.Jobsan (1993). A Geometric Approach to Variational Optimization: Finding the Attainable Region. Paper presented at the IFAC Conference, Sydney.

Ho, L.Y., and A.F. Humphrey (1970). Optimal control of an enzyme reaction subject to enzyme deactivation. i:batch process, *Ind. Eng. Chem. Res.*, **29**, 49-58.

Kokossis, A.C. and C.A. Floudas (1991). Synthesis of non-isothermal complex reactor networks. In 1991 AIChE Annual Meeting, Los Angeles, USA, 157f.

Omtveit, T., J. Tanskanen and K.M. Lien (1994). Graphical targeting procedures for reactor systems. *Comp. Chem. Eng.*, **18**(suppl), 113-118.

Rippin, D.W.T. (1983). Simulations of single- and multiproduct batch chemical plants for optimal design and operation. *Comp. Chem. Eng.*, **7**(3), 137-156.

Van der Vusse, J.G. and H. Voetter (1961). Optimum pressure and concentration gradients in tubular reactors. *Chem. Eng. Sci.*, **14**(25), 90-98.

PLANTWIDE DYNAMIC SIMULATION ON SUPERCOMPUTERS: MODELING A BAYER DISTILLATION PROCESS

Stephen E. Zitney
Cray Research, Inc.
Eagan, MN 55121-1560

Ludger Brull and Lothar Lang
Bayer AG
D-51368 Leverkusen, Germany

Robert Zeller
Cray Research, Inc.
D-80992 Munich, Germany

Abstract

With the goal of demonstrating the advantages of supercomputer simulations for process design and operations, Bayer AG recently teamed up with Cray Research to perform plantwide dynamic simulations of an entire Bayer production facility in Leverkusen. This paper presents project results and benefits from both a computational standpoint and a Bayer company perspective. Regarding computational aspects, we focus special attention on the performance of the process simulation software (SPEEDUP from Aspen Technology). SPEEDUP on Cray Research systems features new sparse matrix technology, vectorized residual calculations, and parallel asynchronous data transfer for exploiting the full capabilities of supercomputer architectures. From Bayer's perspective, the collaboration clearly demonstrated that plantwide dynamic simulation using SPEEDUP on a supercomputer provides a powerful and valuable extension to traditional process engineering activities, and most importantly, delivers tangible financial benefits to the bottom line.

Keywords

Supercomputing, Dynamic simulation, Plantwide modeling, Distillation, Sparse matrix methods.

Introduction

The future success of a chemical company depends on its ability to design and operate processes that achieve the highest possible output of the best quality using the smallest possible amount of raw materials and energy and generating as little waste as possible–and all of this with the highest possible safety for employees and environment. To meet these challenges, Bayer critically analyzes its production processes using state-of-the-art simulation and computing technology (*research*, 1993).

In Bayer's most extensive project to date, the Systems Process Technology Department developed a rigorous plantwide dynamic model for eight coupled distillation columns (nearly 1200 trays) representing one of the largest separation systems ever built in Europe. With some of the columns measuring four meters in diameter and towering to a height of nearly 80 meters, the heat-integrated plant distills eight high-value products from more than 40 components. The products are crude silanes, the basic materials for making silicones for sealants, resins, fluids, and rubbers. Plant operation is characterized by a very large total holdup (more than 500 m^3) and time-varying feed conditions. Depending on the yield in the upstream reaction, the composition of the feed changes by a considerable amount (changes of up to 100 percent in some components). Moreover, the plant has nearly no buffering between the single distillation columns, which leads to an unsteady continuous process. The incentives for Bayer to model this specific plant were to study process control problems and to look for ways to increase plant capacity, while at the same time improving quality and reducing the energy requirements.

SPEEDUP Case Studies and Performance

Over a period of about one year, a detailed mathematical model of the entire distillation plant was developed with SPEEDUP (Aspen Technology, 1993) using process flowsheet information and actual plant data. The model includes standard mass and energy balances and vapor-liquid equilibrium relations (Wilson equation). Off-line steady-state and dynamic simulations of the individual distillation columns and the whole plant were performed on a CRAY C90 supercomputer. Dynamic simulation of the eight coupled distillation columns required the repeated solution of a differential-algebraic system that contains more than 75,000 equations, making it the largest known industrial SPEEDUP application in production use in the world.

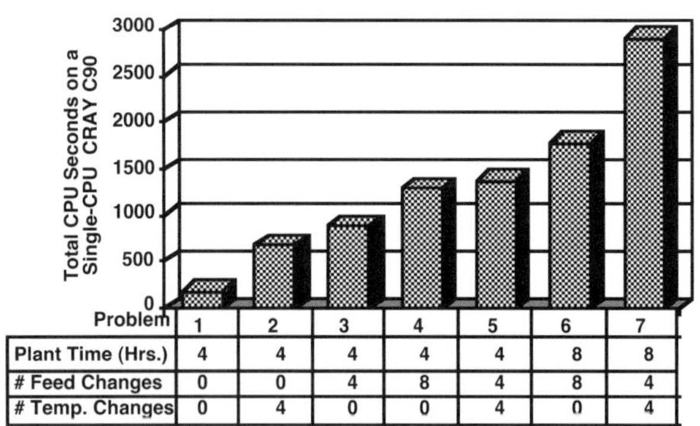

Problem	1	2	3	4	5	6	7
Plant Time (Hrs.)	4	4	4	4	4	8	8
# Feed Changes	0	0	4	8	4	8	4
# Temp. Changes	0	4	0	0	4	0	4

Figure 1. CRAY C90 single-CPU times for seven Bayer SPEEDUP case studies.

Figure 1 presents the CRAY C90 single-CPU times for seven Bayer case studies differing in the length of plant operation time simulated (four or eight hours) and the type (feed flowrate and/or temperature) and number of process disturbances. The results show that the solution time not only depends on the plant operation time being simulated, but is also strongly correlated to the number and type of external perturbations affecting the chemical process during the simulation. Since the relaxation times of several components in the Bayer process are relatively long, plant operation times of several days with many external perturbations must be simulated to obtain a realistic picture. As discussed below, the Cray Research version of SPEEDUP offers enhanced capabilities to make such simulations tractable.

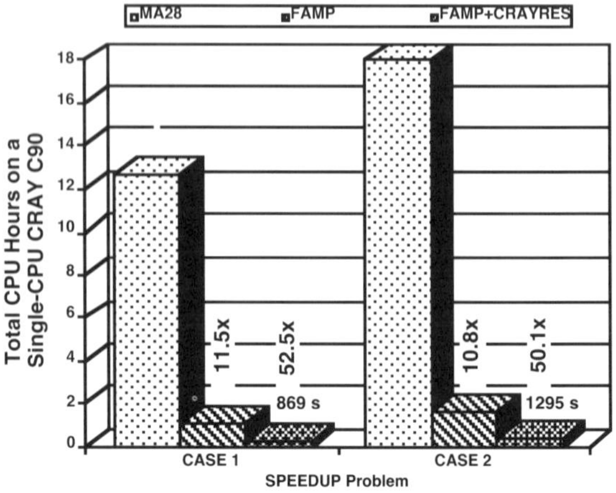

Figure 2. Performance improvements from using CRAY-optimized version of SPEEDUP on a single-CPU CRAY C90 supercomputer.

Figure 2 presents the results for two case studies simulating four hours of plant production time, with one (CASE 1) and two (CASE 2) disturbance(s) per hour in total feed flowrate. For each case, the bar on the left represents the CRAY C90 simulation using the standard version of SPEEDUP available on other computer platforms. The other two simulations for each case represent the value of using the CRAY optimized version of SPEEDUP which contains improved algorithms for sparse linear equation solving (FAMP) and residual evaluation (CRAYRES).

Compared to SPEEDUP's conventional sparse matrix techniques (e.g., Harwell's MA28), an improved out-of-core frontal solver, FAMP (Zitney and Stadtherr, 1993; Zitney et al., 1994), provides substantial savings in overall problem solution time, more than an order of magnitude for the Bayer simulations in Fig. 2. By relying on efficient dense matrix kernels, the frontal method factors a sparse matrix with a series of dense frontal matrices, each of which corresponds to one or more steps of the overall LU factorization. The frontal process avoids the problem of indirect addressing, which degrades the vector performance of conventional sparse matrix methods. Table 1 shows that FAMP is well over 250 times faster than the MA28 routine when solving a single linear system from the Bayer problem on a single-CPU C90 system. The middle bar for each case study in Fig. 2 shows the effect of this new solver on overall SPEEDUP performance. Performance gains of 11.5x and 10.8x are achieved using FAMP since slightly over 90 percent of the SPEEDUP solution time for each case is spent in the linear solver when using MA28.

Table 1. Single-CPU C90 Time Comparison of Sparse Matrix Solvers on a Single Sparse Linear System from the Bayer Problem.

			CPU Time (Sec.)	
Order	Nonzeros	% Sparsity	MA28	FAMP
75724	349716	99.994	854.8	3.2

After rethinking the sparse matrix methods used in SPEEDUP, a performance analysis shows that a substantial amount of computation time is spent in residual calculations. By invoking the CRAYRES command in SPEEDUP, a post-processor automatically modifies the SPEEDUP residual FORTRAN to generate DO loops and vectorized code, thereby improving vector performance when running dynamic simulations using either the DAE or SUPERDAE integrators. Figure 2 shows that the optimized residual calculations give nearly a factor of five performance increase on the Bayer simulations.

As model size and the length of simulation time increases, SPEEDUP stores more and more data to its database file for use in run-time and post-run plotting. Therefore, it becomes more important to perform as many input/output (I/O) operations as efficiently as possible. The Cray Research version of SPEEDUP provides an easy-to-use flexible file I/O library for parallel asynchronous data transfer to and from the SPEEDUP database. For the Bayer application, overlapping I/O and arithmetic computations in this way reduces the I/O wait time by more than an order of magnitude. This uses supercomputer resources efficiently, and lets large-memory SPEEDUP jobs complete and exit the system more quickly.

Providing more than a 50-fold increase in overall SPEEDUP performance in some cases, the software enhancements described above coupled with the fast I/O and CPU performance of the CRAY C90 hardware let Bayer and Cray Research engineers perform plantwide dynamic simulations that were previously intractable for mainframes and workstations. For example, simulating four hours of plant production time in CASE 2 now takes only 21 CPU minutes instead of the 18 CPU hours required with the standard version of SPEEDUP. As a result, Bayer engineers can run this plantwide simulation many more times per day and can even consider it for on-line use where real-time simulations are required.

Project Results and Benefits

Steady-state simulations were done with SPEEDUP to optimize on the basis of different feed concentrations and loads. The result of these optimizations is the plant characteristic, which reveals how much of the plant needs to be revamped to achieve a given increase in plant capacity. The plant characteristic found with the simulations indicated that to double capacity requires revamping only half of the plant. As a result, Bayer aims to expand production capacity for silicone products at its Leverkusen complex by more than twofold over the next few years.

Since plant characteristics only provide optimal operating points, a method must be developed to run the plant at these optimal operating conditions regardless of load changes and varying feed conditions. Such a process control strategy requires dynamic simulation of the plant. Our SPEEDUP dynamic simulations had the following objectives:

- Perform sensitivity studies to define the necessary process measurements (temperature, pressure, analyzers) for control purposes.
- Perform sensitivity analyses to define the control structure, i.e., which variable should be controlled by which manipulated variable. Determine if a conventional PID (proportional integral derivative) control can be used or if a more advanced control is required.
- Test the designed process control by simulation of load and setpoint changes of the complete plant, including controllers.

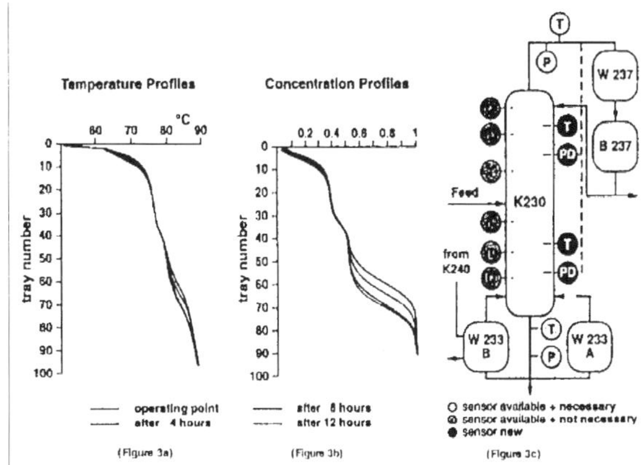

Figure 3. Sensitivity analysis and measurement selection.

Figure 3 shows part of the sensitivity analysis for a heat-integrated distillation column. We are looking at an increase of the reboiler duty in the column. Figures 3a and 3b show temperature profiles along the column and concentration profiles of the main component in this column. The profiles are moving from the operating point (lower curve) upward. The time difference for the profiles is four hours, which provides excellent insight into the dynamics within this unit. From these sensitivity studies for all the manipulated variables (approximately 30) in the plant, we can define the necessary process control measurements (Fig. 3c). The chief result was that one-third of the existing analyzers on the plant were no longer necessary.

On the basis of the newly defined process control measurements, we developed a completely new control strategy for the plant, which has shown potential for energy savings of three to five percent. To test the new controllers in simulation, we first had to design all the controllers by evaluating the necessary controller parameters. Then controllers and plant models are simulated together to check the controllers' performance. The major points of interest in this case are disturbance rejection, for example the ability to handle changing feed conditions and decoupling. By decoupling we mean that we want as few influences as possible by one control loop on another one.

Being able to test all the controllers before implementing them on the plant saves much troubleshooting and leads to a faster and more successful startup of the new controllers. Moreover, the well-defined experiments, which we can do in simulation but which are virtually impossible in the real plant, lead to a much better understanding of the process and may also yield decisive directions for further optimization and, ultimately, a competitive advantage.

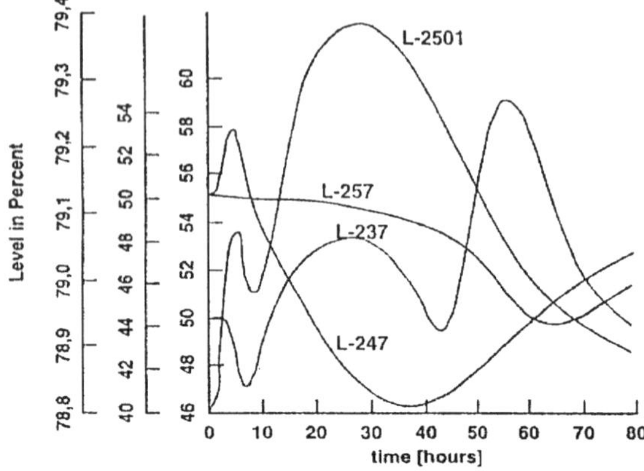

Figure 4. Dynamic plant behavior for a change in feed concentration.

The enormous time constants within the plant are another interesting outcome of the simulation experiments with the complete plant model. Figure 4 shows the responses of some level controllers for a simulation of 80 hours plant operating time for a change of feed concentration. Looking at these extremely long transients and keeping in mind that plant operators usually make some setpoint changes and perform some manual operations during their shifts, it became apparent that these operators never see the result of their manipulations, but a later shift does.

We concluded that plant operators should leave their process in automatic operation for as long as possible. This will require the operators to change their basic approach to work. Again, simulation can be helpful by showing possible reactions of the plant controllers to disturbances, thereby leading to a faster acceptance of the process controller as a member of the team.

Conclusions

The results described in this paper confirm that total plant simulation offers great potential for reducing operating and capital costs, leading to the optimization of chemical plant processes, such as Bayer's, and ultimately to a competitive advantage in quality, flexibility, and costs. The results of the joint project demonstrated to Bayer the effectiveness of dynamic simulation made possible by the high performance of supercomputers. To continue these efforts, Bayer AG has installed a CRAY C92 system at its Leverkusen facility. Bayer, Cray Research, and Aspen Technology plan to carry out further improvements in the capabilities and the performance of dynamic process simulation software on supercomputers.

Acknowledgments

The authors wish to thank James Goom, Phil Mahoney, and Peter Ward of Aspen Technology, Inc. for their responsiveness and support in using SPEEDUP. We also wish to express our appreciation to the Benchmarking and Corporate Computing and Networking groups at Cray Research for providing generous amounts of assistance and computer time on the supercomputer systems in Eagan, Minnesota and Chippewa Falls, Wisconsin.

References

Aspen Technology (1993). *SPEEDUP User Manual*, Vols. I and II, Version 5.4, Cambridge, Massachusetts.

research (1993). Optimizing chemical processes is the responsibility of research — sticking to the ideal course. *The Bayer scientific magazine*, Seventh Edition, 74-81.

Zitney, S.E. and M.A. Stadtherr (1993). Frontal algorithms for equation-based chemical process flowsheeting on vector and parallel computers. *Computers chem. Engng.*, **17**(4), 319-338.

Zitney S.E., K.V. Camarda and M.A. Stadtherr (1994). Impact of supercomputing in simulation and optimization of process operations. *Proc. Second International Conference on Foundations of Computer-Aided Process Operations.* (FOCAPO-II, Mt. Crested Butte, Colorado, July 18-23, 1993; D. W. T. Rippin, J. C. Hale, and J. F. Davis, Eds.), CACHE, Austin, TX, 463-468.

AUTOREGULATORY FEEDBACK IN INDUSTRIAL PROCESS DESIGNS

Richard D. Braatz, Babatunde A. Ogunnaike, James S. Schwaber and William Rose
DuPont Experimental Station
Wilmington, Delaware 19880

Abstract

It is of substantial industrial interest to modify processes in the design stage to improve controllability without decreasing profitability. With this goal in mind, we investigate improving controllability by introducing physical or chemical feedback *within the process design*. We refer to such feedback as being *autoregulatory*, and show that designs with autoregulatory feedback can simplify the plant-wide control design task, enhance process reliability, reduce capital costs, and reduce pollution. For a compressor system it is shown that autoregulatory feedback suppresses disturbances and reduces the tendency for the compressor to enter the undesirable stall or surge modes of operation.

Keywords

Process design, Controllability, Process control, Compressor instabilities, Self-regulation.

Introduction

One approach to process design and control is first to design the process based only on steady state profitability considerations, and then to try to control the plant once it is built. Although ignoring control considerations in the design stage makes the design engineer's job much easier, the resulting plant may be very difficult to control. If the controllability problems associated with a chosen design are not understood until after the plant is built, costly design modifications or convoluted control structures may be required to provide a plant which can be controlled with satisfactory performance (Fisher, Doherty, and Douglas, 1988).

A popular research topic over the last decade has been to include control considerations while developing the process design (Braatz, Lee, and Morari, 1994). When some measure of controllability indicates that a candidate design is difficult to control, then the design is modified to improve the controllability. Typical process design modifications include overdesigning process units and adding capacitances. These modifications increase capital cost, increase the potential for day-to-day chemical leaks due to the increased number of connections between process units, and increase the likelihood of catastrophic chemical release to the environment.

It is of substantial industrial interest to develop strategies for designing processes with improved controllability which do not have the disadvantages of the traditional design strategies. With this goal in mind, we investigate the use of physical or chemical feedback *within the process design*. We refer to such feedback as being *autoregulatory*, and show that these designs simplify the plant-wide control design task and may have the following additional benefits: enhanced process reliability, reduced capital costs, and reduced pollution. An autoregulatory flow-divider valve is described with such good open loop performance that feedback control is unnecessary. For compressor systems, a method of providing autoregulatory physical feedback is introduced in which disturbances are suppressed and the tendency for the compressor to enter the undesirable stall or surge modes of operation is reduced.

Traditional Process Design Strategies

The traditional industrial process design strategies for improving controllability include: arranging and selecting process units so that the overall plant design is self-regulating, placing holding and/or mixing tanks between process units, and overdesigning process units.

A process is said to be *self-regulating* if it is stable without the application of controls (Seborg, Edgar, Mellichamp, 1989). (Note that many authors have used the term *self-regulating* to also apply to *control structures* (Luyben, 1988). Though selecting among control structures is closely related to selecting among plant designs (Braatz, Lee, and Morari, 1994), the focus of this manuscript is plant design and the use of the term to refer to control structures will not be discussed further here.) Self-regulating processes are easier to control, and safer to start up and operate than unstable processes. In contrast, an open loop unstable process must be continually attended by an operator during power outages or any other abnormality in process operation which requires the controller to be taken off-line. A disadvantage of designing processes to be self-regulating is that the capital costs are usually higher, and that the range of operation and potential for optimizing operations may be limited (Downs and Ogunnaike, 1994).

One of the main traditional strategies for improving process controllability is to place holding or mixing tanks in streams to dampen disturbances (Buckley, Luyben, and Shunta, 1985; Fisher, Doherty, and Douglas, 1988). The disadvantages of adding capacitances (that is, tanks) between process units include increased capital cost, increased potential for day-to-day chemical leaks due to the increased number of connections between process units, and increased likelihood of catastrophic chemical release to the environment. The latter disadvantage should not be underestimated---the chemical accident in Bhopal, India which resulted in over 1000 deaths was due to accidental gaseous release from an unnecessary holding tank (Kletz, 1985).

Another traditional strategy for improving process controllability is to overdesign process units. This is one of the main strategies applied to distillation column design (Buckley, Luyben, and Shunta, 1985). One disadvantage of overdesigning process units is that it increases capital costs. In some cases (for example, in some homogeneous azeotropic distillation columns), overdesign can lead to an increase in operating costs (Laroche et al., 1992) or a *decrease* in controllability (Bekiaris et al., 1993).

Autoregulatory Feedback

Although new metrics for *quantifying* controllability have been developed over the last few decades (Braatz, Lee, and Morari, 1994), strategies for *improving* controllability have changed little over this time. It is of substantial industrial interest to develop such strategies which do not have the disadvantages associated with the traditional approaches.

With this goal in mind, Braatz et al. (1994) investigated improving controllability by introducing physical or chemical feedback *within the process design*. Such feedback is said to be *autoregulatory*, and we will illustrate the use of autoregulatory feedback with two examples: 1) a flow divider valve, and 2) a compressor system. The flow-divider valve is introduced as a design in which the introduction of autoregulatory feedback provides such good open loop performance that feedback control is unnecessary. This motivates the investigation of using autoregulatory feedback to provide good open loop performance characteristics for *industrial process designs*. A method of providing autoregulatory feedback in compressor systems is introduced in which disturbances are suppressed and the tendency for the compressor to enter the undesirable stall or surge modes of operation is reduced.

Flow Divider Valve

Flow divider valves are used to divide flow in a predetermined ratio independent of loading conditions. Fedoroff et al. (1992) have recently designed an autoregulatory divider valve which can maintain the ratio between outlet flow rates with less than 2% error over a wide range in inlet flow and downstream pressures (see reference for detailed design). The design is based entirely on physical feedback—differences between downstream pressures cause changes in orifice sizes internal to the valve, which in turn changes the flow resistance in each of the outlets so that the ratio of flows remains constant. This valve would be appropriate for cultivators which are lifted hydraulically on both sides (Fedoroff et al., 1992). If the magnitude of the flow to each side were not equal, then the cultivator would tip over. With the autoregulatory valve no ratio control loop is necessary, and the cultivator will not tilt even under a partial or complete power outage. The cost of the autoregulatory valve is negligible compared to the cost of the cultivator.

Compressor System

Much of the energy used in petroleum refining and chemical processing powers compressors (Shinskey, 1978). It is well-known that compressor systems can become unstable due to changes in flow rate, pressure, molecular weight of the gas, and inlet flow pattern. One of these flow instabilities, referred to as *surge*, involves one-dimensional rapid flow pulsations which can grow to full flow reversals. These pulsations can severely stress the compressor blades and casing. For large compressors, even a limited number of surge flow oscillations can destroy the compressor (Shinskey, 1978).

A generic compressor system is shown in Fig. 1, where a plenum is used to represents all capacitances downstream of the compressor. A simplified description of the physics behind surge oscillations can be found in the text by Shinskey (1978), and can be understood by looking at Fig. 2, which is plot of the head-flow characteristics for a typical centrifugal compressor. The surge line separates the stable and unstable regions of operation. In the stable region, the compressor behaves as would be normally expected---as flow is reduced the discharge pressure increases. In the unstable (or surge) region, the compressor characteristic has negative slope. If the flow is reduced in this region, then the discharge pressure falls, causing the flow and pressure to be further reduced. When the discharge pressure of the compressor falls below the pressure in the outflow line, a momentary flow reversal occurs which causes the line pressure to fall. Once the line pressure falls sufficiently, the compressor is able to force the flow to move forward again. A more detailed description of the fundamental physics behind surge phenomena is provided by Greitzer (1981).

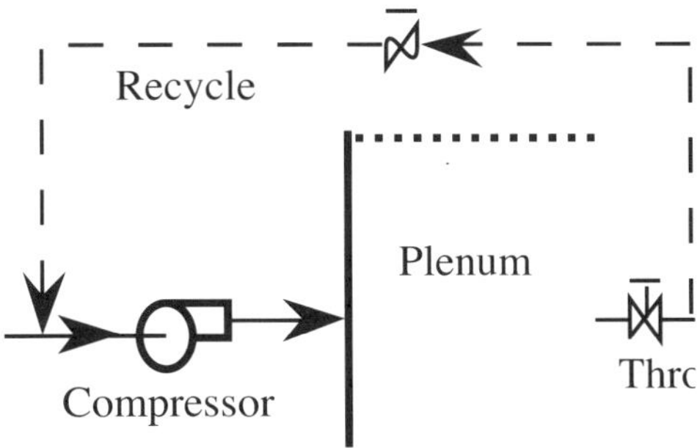

Figure 1. A generic compressor system.

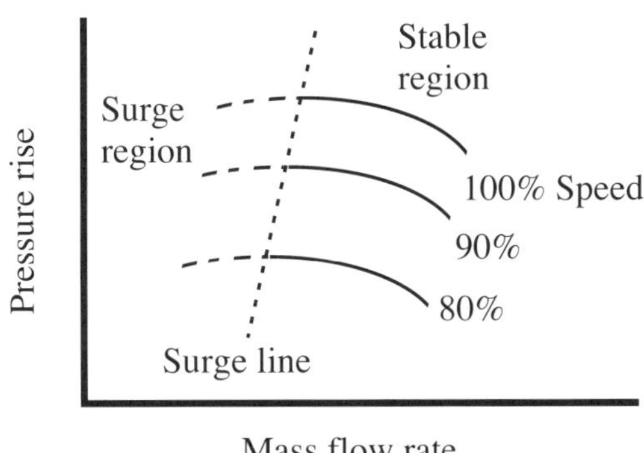

Figure 2. Typical compressor head-flow characteristics.

In the chemical process industries it is typical to operate compressor systems under conditions in which surge oscillations are unlikely to form. In Fig. 2 this corresponds to operating the compressor in the stable operating region, far to the right of the surge line. Many compressors are operated with a recycle flow (shown in Fig. 1) to keep the flow rate high enough to provide this safe stable operation. This flow circulation wastes energy. Literature by the Compressor Controls Corporation claims that a yearly energy savings of $1.2 million was achieved for a 40,000 HP compressor by reducing the amount of recirculation through improved controls (by operating closer to the surge line). The difficulty with operating near the surge line is that the location of this line is imprecisely known and can move to some degree. Furthermore, it can be difficult to prevent a flow oscillation which begins to develop from growing into a full surge oscillation (Greitzer, 1981). A design change which would remove the need for recirculation could lead to energy savings comparable to that reported above.

For brevity the ordinary differential equations which describe the compressor system are given elsewhere (Equations 1-7 of Abed, Houpt, and Hosny (1990)). Fig. 3 is a plot of the compressor mass flow rate response due to a disturbance in downstream flow resistance which moves the operating point to the unstable region. The mass flow oscillations are seen to be growing in magnitude, and eventually lead to full flow reversals (not shown to improve scale).

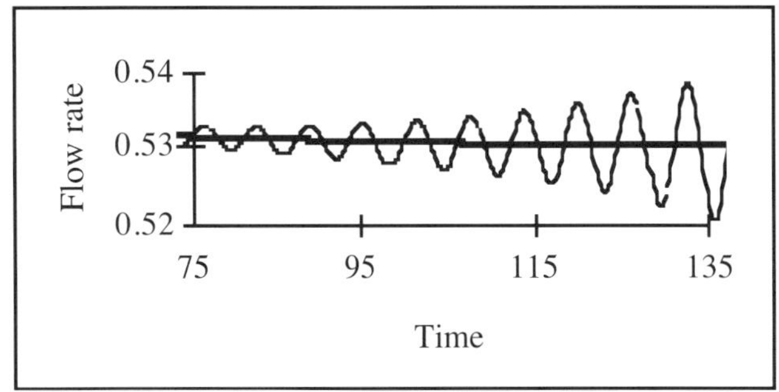

Figure 3. Compressor mass flow rate response with (-) and without (-) viscoelastic membrane.

A simple method for providing autoregulatory physical feedback would be to replace one of the walls of the plenum (represented by the short dashed line in Fig. 1) with a viscoelastic polymer membrane, where the properties of the membrane would be chosen to counteract mass flow oscillations with changes in plenum volume (in practice, this polymer membrane could replace a rupture disk in one of the downstream tanks, with the polymer membrane designed to break at the same pressure that the rupture disk would normally break). Since a purely elastic membrane would cause the volume changes to move with pressure changes, the viscoelastic membrane was designed to have a significant amount of damping.

The partial differential equations describing the membrane (which can be found in Chapter 9 of Wiley and Barrett (1982)) were discretized and the resulting ordinary differential equations combined with the equations describing the original design. The compressor mass flow response to the same disturbance as before is shown in Fig. 3 (as a thick solid line). At this point, we see that the design change has stabilized the compressor system at this operating condition. Further simulations (not shown due to space considerations) indicate that the design change causes disturbances to be suppressed in operating regimes which are stable for both designs. This design change allows the process to operate at lower mass flows, and may for some compressor systems lead to a removal of recirculation which would lead to a very large energy savings.

It will be described in a future manuscript how to optimize the disturbance suppression characteristics over the viscoelastic properties of the polymer membrane. Methods for manufacturing polymers to have specified physical properties are available in the literature (Akay, 1993). Other materials should also be considered in the development of a practical autoregulatory design.

Note that the increased capital costs of autoregulatory process modifications over the original design can be negligible, as can be seen from the flow divider and compressor systems.

Comparison Between Autoregulation and Traditional Design Strategies

Shinskey (1988) gives many examples of self-regulating processes. An example of special interest here is a holding tank (which is an integrating system) in which the level is stabilized by placing a valve on the outflow. Shinskey describes how this valve stabilizes the level by introducing feedback within the design. As shown by Shinskey, though it is possible for this autoregulatory feedback to keep the process output (the level) within a narrow range, the typical size of the process units would rarely provide such good steady-state performance.

Designing a process to be autoregulatory does not require adding capacitances or overdesigning process units. As can be seen from the compressor system example, the disadvantages of these traditional design strategies (potential increased chemical releases, capital costs, and operating costs) need not be shared by an autoregulatory design. However, the design of autoregulatory feedback mechanisms should be done with care, to ensure that the range of operation and potential for optimizing operations are not limited.

Conclusions

Two systems have been described in which autoregulatory feedback provides not only stability but also open loop performance. This leads to the conclusion that a wide range of controllability improvement is possible by introducing autoregulatory feedback into designs. Improving the open performance of the design has three related benefits: 1) for small units the performance may be good enough that control design is unnecessary (as for the flow-divider valve); 2) when control is required, the control system design and implementation will be less complex; 3) operational safety and product quality will deteriorate to a lesser degree under conditions which require the controller to be taken off-line. This leads to the conclusion that design engineers should actively search for autoregulatory mechanisms which provide good performance.

Designing successful autoregulatory designs requires both creativity and an intimate understanding of the underlying physics of the process. The use of *physical* feedback within the process to suppress disturbances was illustrated through two examples: 1) a flow-divider valve, and 2) a compressor system. Approaches for providing autoregulatory *chemical* feedback should be investigated. It is important to develop autoregulatory mechanisms for specific industrial processes (for example, distillation columns, chemical reactors), and to determine which classes of processes can benefit by the use of autoregulatory feedback.

References

Abed, E.H., P.K. Houpt, and W.M. Hosny (1990). Bifurcation analysis of surge and rotating stall in axial flow compressors. In *Proc. of the American Control Conference*, 2239-2246.

Akay, M. (1993). Aspects of dynamic mechanical analysis in polymeric composites. *Composites Science and Technology*, **47**, 419-423.

Bekiaris, N., G.A. Meski, C.M. Radu, and M. Morari (1993). Multiple steady-states in homogeneous azeotropic distillation. *Ind. Eng. Chem. Res.*, **32**, 2023-2038.

Braatz, R.D., J.H. Lee, and M. Morari (1994). Screening plant designs and control structures for uncertain systems. In *Proc. of the IFAC Workshop on the Integration of Process Design and Control*, 242-247.

Braatz, R.D., B.A. Ogunnaike, J.S. Schwaber, and W.C. Rose (1994). Autoregulation in industrial processes. In *Proc. of the IFAC Symposium on Modeling and Control in Biomedical Systems*, 127-128.

Buckley, P.S., W.L. Luyben, and J. Shunta (1985). *Design of Distillation Column Control Systems*. Instrument Society of America, Research Triangle Park, North Carolina.

Downs, J.J. and B.A. Ogunnaike (1994). Design for control and operability: an industrial perspective. *FOCAPD*, Snowmass Village, Colorado.

Fedoroff, M.R., T. Burton, G.J. Schoenau, and Y. Zhang (1992). Dynamic and steady-state analysis of an auto-regulator in a flow divider and/or combiner valve. *J. of Dynamic Systems, Measurement, and Control*, **114**, 306-314.

Fisher, W.R., M.F. Doherty, and J.M. Douglas (1988). The interface between design and control, Parts 1, 2, and 3. *Ind. Eng. Chem. Res.*, **27**, 597-615.

Greitzer, E.M. (1981). The stability of pumping systems — the 1980 Freeman scholar lecture. *ASME J. of Fluids Eng.*, **103**, 193-242.

Kletz, T.A. (1985). *What Went Wrong?: Case Histories of Process Plant Disasters*. Gulf Publishing Company, Houston.

Laroche, L., N. Bekiaris, H.W. Andersen, and M. Morari (1992). The curious behavior of homogeneous azeotropic distillation — implications for entrainer selection. *AIChE J.*, **38**, 1309-1328.

Luyben, W.L. (1988). The concept of "eigenstructure" in process control. *Ind. Eng. Chem. Res.*, **27**, 206-208.

Seborg, D.E., T.F. Edgar, and D.A. Mellichamp (1989). *Process Dynamics and Control*. Wiley, New York.

Shinskey, F.G. (1978). *Energy Conservation Through Control*. Academic Press, New York.

Shinskey, F.G. (1988). *Process Control Systems: Application, Design, and Tuning*, 3rd ed. McGraw-Hill, New York.

Wylie, C.R., and L.C. Barrett. (1982). *Advanced Engineering Mechanics*, 5th ed. McGraw-Hill, New York.

MATHEMATICAL PROGRAMMING TESTS FOR DYNAMIC STABILITY

Amy R. Ciric
University of Cincinnati
Cincinnati, OH 45221

Abstract

This paper introduces a set of tests for *practical* stability. A dynamic system is practically stable in a specified neighborhood *V* if there are no trajectories leading out of the neighborhood. A simplified Lyapunov function test is used as a basis for mathematical programming tests of operational stability. The proposed approach is quite general and can be formulated for constrained dynamic systems, systems with variable or uncertain parameters, and systems containing discretely different dynamic modes. Two examples illustrate the proposed approach.

Keywords

Dynamic stability, Lyapunov function tests, Practical stability, Discretely different dynamic modes, Distillation column dynamics.

Introduction

Understanding the dynamic stability of a chemical process is a key step toward developing safe and controllable chemical manufacturing processes. Instabilities and steady state multiplicities in reactor systems are well known, and a recent review has been provided by Razon and Schmitz (1987). Recent work suggests that instabilities and steady state multiplicities also occur in separation systems (ex., Widagdo et al., 1989).

Stability can be informally evaluated by performing a series of dynamic simulations, or formally assessed with eigenvalue or Lyapunov function tests. These formal tests can be difficult to apply to systems containing algebraic constraints, uncertain parameters, or discretely different dynamic modes. This paper presents an optimization based approach that can be readily applied to complex dynamic systems.

Testing the dynamic stability of a process *F* requires a dynamic model of the form

$$\frac{d\mathbf{x}}{dt} = f(\mathbf{x}, \mu) \tag{1}$$

$$h(\frac{d\mathbf{x}}{dt}, \mathbf{x}, \mu) = 0 \tag{2}$$

$$g(\mathbf{x}, \mu) \leq 0 \tag{3}$$

Here, $\mathbf{x}$ is a vector of process variables, such as temperatures, pressures, flowrates, etc.; $\frac{d\mathbf{x}}{dt}$ is the rate of change of $\mathbf{x}$ with respect to time; μ is a vector of model parameters, such as equipment sizes, controller settings, heat transfer areas, etc. The vector function f describes the rate of change of the process variables. The equality constraints may specify relationships that always hold, such as the sum of the mole fractions always equals one, or they may describe internal processes with very fast dynamics relative to the process described by equation (1), such as vapor liquid equilibria in a distillation column. The inequalities express fundamental relations, such as that all mole fractions must be positive.

A *steady state* $\mathbf{x}_o$ will satisfy

$$f(\mathbf{x}_o, \mu) = 0 \tag{4}$$

$$h(\mathbf{x}_O, \mu) = 0 \tag{5}$$

$$g(\mathbf{x}_O, \mu) \le 0 \tag{6}$$

In general, $\mathbf{x}_O$ is a *stable* solution of Equations (4), (5), and (6) if small perturbations away from $\mathbf{x}_O$ decay with time.

If the functions $f(\mathbf{x}, \mu)$ are differentiable at $\mathbf{x}_O$ and there are no constraints (Eq. (2) and (3)), then the asymptotic stability of (1) can be tested by computing the eigenvalues of the Jacobian matrix ∇f, evaluated at $\mathbf{x}_O$. If the real part of all of the eigenvalues are less than or equal to zero, then the system is stable, and small perturbations away from $\mathbf{x}_O$ will decay with time. Conversely, if the real part of one or more eigenvalues is greater than or equal to zero, then there are some perturbations that will not decay to zero, but will grow with time. In this case, the system is dynamically unstable.

Eigenvalue analysis is a popular and powerful technique for testing asymptotic stability. However, there can be a number of difficulties associated with this approach. First, the analysis is *local,* and consequently it is only valid for small perturbations. Second, eigenvalue analysis requires that the dynamic equations be *differentiable*. If the dynamic model contains discretely different dynamic modes, then eigenvalue analysis can be difficult to apply. Lastly, some of the parameters μ may be uncertain. Determining whether these variations will effect the stability of a steady state requires performing a bifurcation analysis over the range of the variations.

These complications can arise in a dynamic model of a chemical process. The dynamic model of a distillation column may contain algebraic constraints specifying that mole fractions are always positive and sum to one. The model may also contain discretely different dynamic modes. For example, each tray of a distillation column can have two discretely different dynamic modes. In one mode, the holdup volume is full of liquid: there is liquid flowing over the weir, and the volume of liquid on the tray is constant. In the other mode, the holdup volume is not full: there is no liquid flowing over the weir, and the amount of liquid on the tray can vary. Eigenvalue analysis cannot be directly applied to this overall system, and applying eigenvalue analysis to each of the 2^N dynamic modes of the overall column is prohibitively time consuming.

Alternatively Lyapunov function tests (Guckenhiemer and Holmes, 1983) can provide nonlocal information about the stability of a dynamic system. This test seeks a function $W(\mathbf{x})$ where

$$W(\mathbf{x}_O) = 0;\ W(\mathbf{x})>0 \tag{7}$$

A contour of $W(\mathbf{x})$ encloses a region that includes the point $\mathbf{x}_O$, and the gradient of $W(\mathbf{x})$ points out of this region. If a trajectory of the dynamic system is entering this region, then $\nabla W \cdot \frac{d\mathbf{x}}{dt} < 0$. If $\nabla W \cdot \frac{d\mathbf{x}}{dt}) < 0$ everywhere in the neighborhood V, then all trajectories in V lead inexorably to the point $\mathbf{x}_O$. If these conditions hold, then the system is stable.

Lyapunov functions are a powerful measure of stability. They are a nonlocal test that is valid for finite perturbations, and they can be applied to systems with discretely different dynamic modes. However, they can be difficult to use. Although the existence of a Lyapunov function gives positive proof of dynamic stability, failing to find one is not proof of dynamic instability.

Operational Stability

Operational or *Practical Stability* (Denn, 1975) is an alternative definition of stability that is easy to apply and retains the nonlocal features of a Lyapunov test function. Consider a neighborhood *V* in the R^n dimensional space of $\mathbf{x}$, defined by

$$V = \{\mathbf{x} : V(\mathbf{x}) \le 0\} \tag{9}$$

A system is *operationally stable in a specified neighborhood V* if all trajectories entering or originating in *V* always remain within *V*. Conversely, a system is *operationally unstable in neighborhood V* if there are any trajectories leaving *V*. Figure 1 illustrates two operationally stable systems (F1, F2) and two operationally unstable systems (F3, F4).

Three points should be noted about this definition:

- The neighborhood *V* is specified by the analyst, not by the definition of operational stability. A dynamic system may be operationally stable in one neighborhood and operationally unstable in another.
- The Lyapunov function test must hold for *some* neighborhood; this definition must hold for a *specific* neighborhood *V*.
- The test does not consider the ultimate fate of a trajectory leaving the bounding neighborhood. Thus, a system is considered operationally unstable if some trajectory leaves the neighborhood, even if the trajectory ultimately returns to the neighborhood.

Testing for Operational Stability

Notice that the boundary of the operational neighborhood *V* is described by $V(\mathbf{x}) = 0$, and that the vector ∇V is always perpendicular to the boundary and points of the neighborhood. The trajectory at a point $\mathbf{x}*$, on the boundary is moving in the direction $\frac{d\mathbf{x}^*}{dt}$.

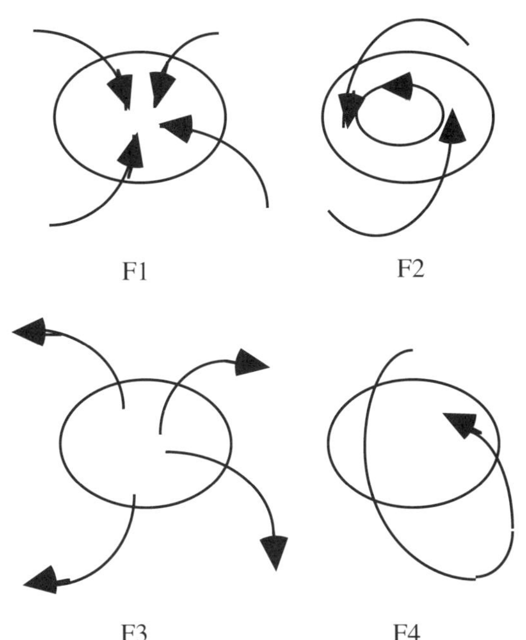

Figure 1. Operationally stable and unstable systems.

If the trajectory is leaving *V*, then $\frac{d\mathbf{x}^*}{dt}$ will point out of *V*, and the dot product $\nabla V \cdot \frac{d\mathbf{x}^*}{dt}$ will be greater than zero. If system F is operationally stable in neighborhood V, then $\nabla V \cdot \frac{d\mathbf{x}^*}{dt}$ will be less than zero everywhere on the boundary.

It is inefficient to compute the dot product at every point on the boundary. Solving the following optimization problem is a more efficient approach:

$$Z = \mathbf{max}\ \nabla V \cdot \frac{d\mathbf{x}}{dt}$$

subject to (**P**)

$$\frac{d\mathbf{x}}{dt} = f(\mathbf{x}, \mu)$$

$$h(\frac{d\mathbf{x}}{dt}, \mathbf{x}, \mu) = 0$$

$$g(\mathbf{x}, \mu) \le 0$$

$$V(\mathbf{x}) = 0$$

If F is operationally unstable, then there is at least one feasible solution of problem **P** where the dot product is greater than zero. Consequently, if F is operationally unstable in *V*, then Z will be greater than zero. If F is operationally stable in *V*, then Z will be less than zero.

It is important to note the special case where Z = 0. This can occur when (a) a steady state lies on the boundary of the neighborhood, (b) when the trajectory moves tangentially to the boundary, or (c) when then function V(**x**) has a stationary point on the boundary. These cases require additional analysis to determine whether the system is operationally stable or not.

Problem (**P**) is not necessarily a convex optimization problem; consequently, problem (**P**) may have more than one locally optimal solution. The problems in this paper were solved with a local optimizer. The validity of the solutions was checked by using several starting points.

Extended Tests for Special Cases

Uncertain parameters. The operational stability test (**P**) can easily account for uncertain parameters. If these parameters are treated as bounded variables within (**P**), then the optimization problem will search over the range of the uncertain parameters for conditions where the system trajectories can cross the boundary.

Discretely Different Dynamic Modes; Discontinuities in V. Suppose that the dynamic equations have a discontinuity at x=z:

$$\text{If } x < z \text{ then } \frac{d\mathbf{x}}{dt} = f_1(x,\mu) \tag{10}$$

$$\text{If } x > z \text{ then } \frac{d\mathbf{x}}{dt} = f_2(x,\mu) \tag{11}$$

This logical relationship is captured by introducing a binary variable Y. If Y=1, then x<z and $\frac{d\mathbf{x}}{dt} = f_1(x,\mu)$. If Y=0, then x>z and $\frac{d\mathbf{x}}{dt} = f_2(x,\mu)$. The following algebraic constraints express this relationship:

$$zY - U(1-Y) \le x \le z(1-Y) + UY \tag{12}$$

$$-UY \le \frac{d\mathbf{x}}{dt} - f_1(x,\mu) \le UY \tag{13}$$

$$-U(1-Y) \le \frac{d\mathbf{x}}{dt} - f_2(x,\mu) \le U(1-Y) \tag{14}$$

Replacing the first constraint in problem (**P**) with equations (12) to (14) leads to a mixed integer programming formulation for testing the operational stability of a dynamic system with discretely different dynamic modes. A similar approach can be used to include discontinuities in V(**x**).

Examples

Fold Bifurcation.

The fold bifurcation is one of the simplest dynamic stability problems (Guckenhiemer and Holmes, 1983):

$$\frac{dx}{dt} = \mu - x^2 \qquad \frac{dy}{dt} = -y.$$

If μ is greater than zero, then this system has two steady states: $(x_o, y_o) = (\sqrt{(\mu)}, 0)$ and $(x_o, y_o) = (-\sqrt{(\mu)}, 0)$. It is easy to show that the first steady state is asymptotically stable, while the second steady state is asymptotically unstable.

We will test the operational stability of this system within the neighborhood

$$V(x,y) = (x - x_o)^2 + (y - y_o)^2 \leq 1$$

Problem (**P**) becomes

$$\max 2(x - x_o)(\mu - x^2) + 2(y - y_o)(-y)$$

$$\text{s.t. } (\mathbf{P1}) \qquad (x - x_o)^2 + (y - y_o)^2 = 1$$

The stability of both steady states was tested for μ=4. When (x_o, y_o)=(2,0), the maximum of (**P1**) equals -6, and occurs when (x,y)=(1,0). When (x_o,y_o)=(-2,0), the maximum of (**P1**) equals 6, and occurs when (x,y)=(-1,0). Thus, the dynamic system is operationally stable when the neighborhood *V* is centered around (x_o, y_o)=(2,0), and operationally unstable when the neighborhood *V* is centered around (x_o, y_o) = (-2,0).

Binary Distillation

In this example, a 100 mol/hr steam containing a 50/50 mixture of two components is to be separated in a 9 tray distillation column. The vapor-liquid equilibria is ideal, and the relative volatility α_{AB} = 4. The heat of vaporization is λ_A = 25 kJ/mol for component A and λ_B = 60 kJ/mol for component B. Operational stability concepts are used to determine if tray drying will occur during a perturbation of this column.

The base case design specifies 0.026 m^3 of liquid holdup on each tray. The liquid density equals 108 mol/m^3. The reflux ratio and the boil up ratio are set to 1.75 and 0.545, respectively. The feed enters the column on the 5th tray. The column produces a distillate stream containing 96.4% A and a bottoms stream containing 90.5% B. These compositions are maintained by proportional controllers that adjust the reflux ratio and the boil up ratio. The gain on the distillate control is 0.5, while the gain on the bottoms controller is 10. Thermal transients are assumed to be very fast relative to composition profile transients.

The amount of liquid held on each tray is taken as a dynamic variable, and as a result, each tray has two distinctly different dynamic modes. In one mode, the tray is full of liquid, the amount of liquid on the tray is constant, and any fluctuations in the overall material balance are reflected in fluctuations in the liquid reflux flowing off the tray. In the other mode, the tray is not full, there is no liquid reflux flowing off the tray, and the amount of liquid on the tray varies with any fluctuations in the overall material balance for the tray. This dynamic discreteness was modelled with integer variables, as discussed earlier.

In the operating neighborhood, the volume of liquid held on each tray is at least 10% of the design holdup volume. The operational stability of the column within this neighborhood was tested by solving a mixed integer nonlinear programming problem, constructed from (a) dynamic material balances, (b) integer relations capturing the discretely different modes on each tray, (c) steady state energy balances, (d) VLE relationships, (e) the control terms, and (f) equations defining the operational neighborhood. This MINLP was solved with Generalized Benders Decomposition (Geoffrion, 1974) using GAMS (Brooke et al., 1988). The problem converged in 13 iterations of the GBD algorithm, and consumed 131 CPU seconds on a Sun Sparc 1+ workstation computer.

The objective at the optimum solution equalled zero. Closer inspection of the solution showed that all trays were operating in the second mode, where the holdup volume is varying and there is no liquid runoff. There is no vapor or liquid flow anywhere within the column. The holdup on trays 1-4 and on trays 6-9 are constant; the holdup on tray 5 (the feed tray) is *increasing* at the rate of 0.926 m^3/h. This solution describes a column at startup: there is no significant holdup anywhere on the tray, and the feed entering the column is beginning to fill the liquid holdup on the feed tray.

Conclusions

The concept of operational or practical stability provides a flexible test evaluating dynamic stability in a wide variety of systems, including constrained dynamic systems and systems containing discretely different dynamic modes. This paper demonstrated that operational stability can be tested with an optimization problem. Two examples illustrated the approach.

References

Brooke A., D. Kendrick, and A. Meerhaus (1988). *GAMS: A User's Guide*, The Scientific Press, San Francisco.

Denn M. (1975). *Stability of Reaction and Transport Processes*, Prentice-Hall, Englewood Cliffs NJ.

Geoffrion A.M. (1974). Generalized Benders Decomposition. *JOTA*, **10**, 237-260.

Guckenhiemer J. and P. Holmes (1983). *Nonlinear oscillations, dynamical systems, and Bifurcations of Vector Fields*, Springer-Verlag, New York NY.

Razon L.F. and R.A. Schmitz (1987). Multiplicities and Instabilities in Chemically Reacting Systems — A Review. *Chemical Engineering Science*, **42**, 1005-1017.

Widagdo S., W.D. Seider, and D.H. Sebastian (1989). Bifurcation Analysis in Heterogeneous Azeotropic Distillation. *AIChE J.*, **35**, 1457-146.

PRACTICAL IMPLEMENTATION OF ENGINEERING DESIGN INTEGRATION TECHNOLOGY

John R. Cassata, Paul S. Odom and Chris Santner
The M.W. Kellogg Company
Houston, Texas 77210-4557

Design Team: Louis G. Archuleta, Byron L. Hardy, Steve A. Kalota,
Ralph F. Pascoe and Jatin T. Shah

Abstract

Engineering Design Integration Technology, EDIT, has been in development at M.W. Kellogg for the past four years. The goal of this effort was to develop a prototype system for Fluid Catalytic Cracking Technology that integrates engineering design calculations and plant operating performance evaluation into a single work process. The system that has resulted addresses all the design life cycle issues, allowing the engineer to deal with grassroots design, revamps and operating data with equal ease in the same environment. At the same time, this environment addresses the issues of quality assurance, compliance with ISO 9000 standards and communications. The elements required to implement this system included database concepts based on dynamic database structures, application of intelligent interface concepts, information management by an expert system and the use of AspenPlus® as a calculation engine. The system has validated new concepts in information management and path independent interactive calculations with an intelligent interface.

Keywords

Process design, Design evaluation, Design work process, Data management, Integration, Life cycle.

Introduction

Over the past four years M.W. Kellogg has been working on integrating the design work process by using new data management concepts, intelligent interfaces and artificial intelligence to drive complex calculation engines. For convenience, the concepts used in this approach are called the **E**ngineering **D**esign **I**ntegration **T**echnology (EDIT).

In early work applied to Kellogg's ethylene technology, the plant wide heat and material balance was computed using an object-oriented interface concept which passed information to an expert system that created a batch AspenPlus® job. Though powerful, this system did not directly address the issues of data management, work process, quality assurance, or plant performance measurement. In the current effort, a prototype system that addresses these issues has been developed for Kellogg's Fluid Catalytic Cracking Technology (FCC). This work has integrated into a single environment the historical data, design and rating calculations and the administrative issues involved in the design work process.

EDIT concepts can be better understood by examining the scope of the problem that has been solved for the FCC Technology.

Design "Life Cycle" Problem

Typically the design problem is viewed from the standpoint of process synthesis and heat and material balance. EDIT takes a broader view of the design in what could be called the design "life cycle." The proposed definition of the design life cycle starts with the initial grassroots design and installation. Then the process proceeds through cycles of evaluation of plant operating data and subsequent plant modification (revamp), and eventually ends with the abandonment of the design in favor of new technologies (Fig. 1). The evaluation of plant operating data can be focused on troubleshooting, optimizing operations, or evaluating performance, but the results of these evaluations can also be used to improve the design. EDIT concepts allow the engineer to deal with the design "life cycle" in a single work environment.

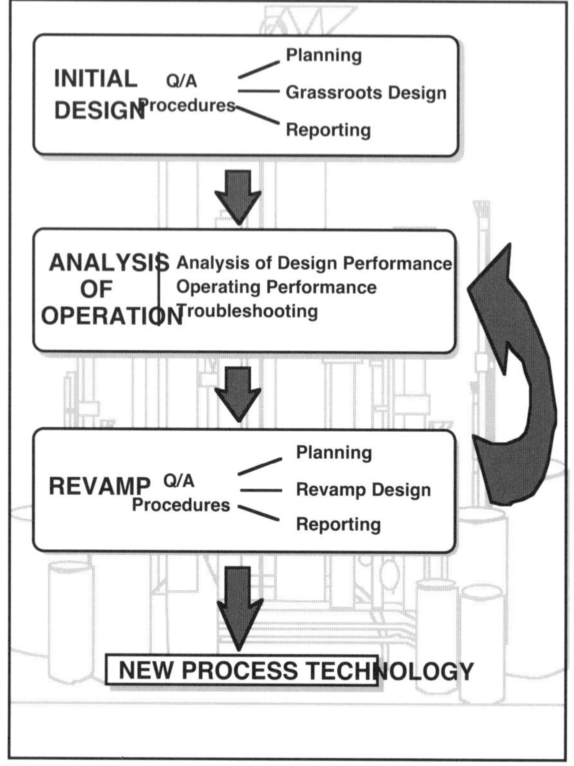

Figure 1. Design "Life Cycle" concept.

System Description

Functionality

The system is designed to reflect the engineering work process. This includes: administrative planning, design, rating, quality assurance, compliance with ISO 9000 standards and publishing. Almost without exception, definable attributes are table driven, i.e. they can be maintained by the user. A partial list of examples of definable attributes are quality assurance(Q/A) authorizations, dimensional units, work process procedures, design practice constraints, report content and E-Mail notification.

The database is designed to hold roughly 10 years worth of design and operating data at any stage of development. The data can be queried by the user community and portions can be downloaded onto a portable computer for field work.

The administration of the design process, which is executed in accordance with ISO 9000 standards, is controlled by the Process Design Manager. The design quality control procedures are selected when the work process is initiated.

The engineer can generate a design or rate performance of an existing facility within the same environment. The system calculations are order independent. Where appropriate the user determines which variables are independent and dependent within each calculation. This feature allow the engineer to easily meet specifications since the specification can be defined as an independent variable. It also allows the design problem and rating problem to be solved in the same solution environment. The user controls convergence at three levels: within a calculation group, in a calculation region and for the global solution. There is a graphic display of the FCC Unit in which key design values are dynamically updated so the engineer can see the impact of local variable changes on the global solution.

Dimensional units are defined by the Process Design Manager, but can be converted to other defined sets by making a simple system request. Any individual data item can be viewed and/or entered in any units. This feature allows the engineer to deal with plant data without making manual conversions.

Quality assurance is monitored continuously. Through color coding and messaging the engineer is immediately informed of deviations from accepted design practice and job specific constraints. From a single central screen, the system allows the user to visually scan the entire design for constraint violations or missing input.

Reports are generated after the computations have completed the Q/A processes. Operating data can be imported and averaged with outliers automatically removed.

Architecture

The EDIT system components for the FCC Technology are shown in Fig. 2. The application makes extensive use of third party software. The interface component uses MAGIC®, by Magic System Enterprises. The expert system shell is ART-IM®, by Inference Corporation, the database is ORACLE®, 7.0, and the calculation engine is AspenPlus®, by Aspen Technology Inc. The system is setup in a client server mode. A SUN Sparc 10 is presently being used as the server and a SUN Sparc 2 is used as the client.

The MAGIC® interface, which is controlled by the user, directs the operation of the system. The interface automatically reads and updates information in the database as required. When the user requests a calculation, the interface passes high level directives to ART-IM® which converts the directives into a series of AspenPlus® interactive commands. The AspenPlus® results are collected by ART-IM® and are made available to the interface through function calls.

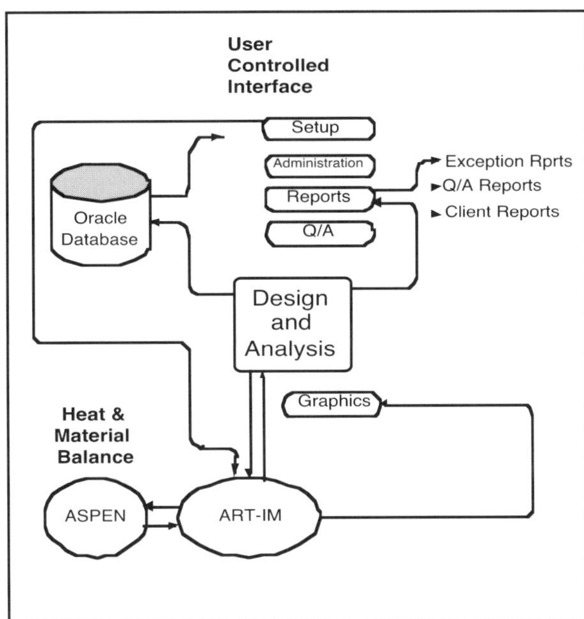

Figure 2. FCC System architecture.

EDIT's Key Concepts

Database

EDIT has to deal with massive amounts of data and a significant number of relationships. The object-oriented analysis of the problem identified over 2,000 objects per case having a total of roughly 40,000 attributes. The maintenance of historical data yields requires storage of 12,000 instances(cases). Classical relational database implementation would give unacceptable performance because of the high number of large table joins that would be required to execute the calculations. Object-oriented databases might be able to cope with a problem of this size, but like relational technology, there is an assumption that the data schema is constant. In reality, at no time during the life of the software is the schema static. This is especially true during software development. EDIT uses object-oriented concepts with dynamic relationships. The objects are accessed using a relational engine. This approach almost completely eliminates the need to define the data schema apriori. Changes to the data schema can now be accomplished in a very short time with no effect on the system since the object relationships are automatically updated. Performance appears to be the same or better than object-oriented databases.

Calculations

Every calculation in the system is defined by its set of variables and the degrees of freedom in the calculation.

Each calculation is related to its associated group by an equation set. In turn, the group of equations is related to other groups by convergence within a region. And finally, regions are related by global convergence. The result is a path independence and a user controlled problem solution. In this environment, grassroots design, revamp design and rating calculations can be done within any individual calculation, since any variable can be selected as the dependent variable.

Work Process Integration

The FCC design process is a complex interaction of yield, process, mechanical and equipment requirements. This process has been reduced to a logical progression of computational events, but still retains the option to do the calculations in any order. This allows the technology expert the freedom to do case studies for optimization.

Quality control and ISO 9000 requirements have been reduced to a set of steps that are defined by engineering management at the beginning of the project. The system controls access and publishing of results based on this work process by required the appropriate electronic approvals. An automated notification process communicates to the appropriate individual(s) when their electronic sign off is required.

Conclusions

The success of a design is as much a product of accumulated knowledge as it is fundamental calculations. For most petrochemical processes the complexity makes it impractical to do calculations that integrate all aspects of chemical reactions, fluid dynamics and mechanical design. The ability to relate operating experience directly to the design calculation environment offers the engineer greater insight into the design's performance.

The limitation of Kellogg's work on EDIT concepts is that it has been applied to individual technologies with a relatively narrow scope of design. The modeling concepts will have to be generalized so that the design process can overlay any process technology. In addition, an improvement in information modeling is required. The present system, though flexible, has a set definition of what information is contained within a task and how information is passed from task to task. True dynamic information/task modeling is required to make EDIT more general.

Information modeling, degrees of freedom, structural equation solutions and quality control are all key concepts that have been addressed in the FCC System. The present application of EDIT demonstrates how these key concepts can become part of the design environments of the future.

MODELLER — AN OBJECT-ORIENTED COMPUTER-AIDED MODELLING TOOL

Heinz A. Preisig
School of Chemical Engineering and Industrial Chemistry
University of New South Wales
Sydney, New South Wales, Australia 2033

Abstract

This new tool is the result of a decade of research and experimenting with a suitable system representation and separation from other elements of systems engineering problem solving. It is the realization of a modelling kernel providing an easy-to-use graphical and textual interface to define and modify process models. It supports hierarchical modelling, inheritance on all levels and minimizes on structural modelling errors by implementing a number of modelling paradigms.

Keywords

Modelling, Computer-aided design, Dynamic systems.

Introduction

The design and development of plants involves a wide range of computer-based tools all of which use a process model in one or another form. The models vary in nature and complexity depending on the use of the model which led to the definition of the term "multifaceted" tasks and associated problems are defined for the same plant component. Thus different models of the same plant coexist in the same project.

Computer-aided modelling addresses the problem of generating and modifying process models efficiently. Different models can be defined in two different ways (I) they are either generated directly by applying alternative theories for the description of a process or (ii) existing models may be modified by the means of mathematical methods such as model reduction or linearization (Preisig, 1994b).

Defining the scope of computer-aided modelling is not a trivial matter. Different projects made different contributions to the subject. Other groups involved are Herzberg and Balchen (N), Perkins and Sargent (UK), Åstroem and coworkers (S), Marquardt (BRD), Stephanopoulos (USA), Westerberg (USA). Leaving the scope wide, several of these groups defined a formal language suitable to capture process modelling knowledge such as ASCENT III (Piela, 1991), or Model.la (Stephanopoulos et al., 1990), Omula (Anderson, 1989). A data base oriented approach is being suggested by Marquardt (Marquardt, 1992).

Scope of the MODELLER

This project took a different approach. Whilst there were attempts to define a formal language early in the project, later preference was given to the construction of a modelling kernel. For this purpose process engineering problem solving was subdivided into a number of smaller problems. Not too surprisingly, we found that some of the identified problems were already solved (knowledge database handling), some are still waiting to be solved (comprehensive problem specifier, comprehensive solution analyzer), others are rather straightforward to realize (algebraic manipulations, compilation and splicing), and a modelling kernel, which eventually was subject of this research (Preisig, 1991c). The resulting tool solves the core modelling problem but does not solve any of the other problems, though it provides the necessary interfaces.

The MODELLER presented here is the second prototype. The first version (Lee, 1991), which was used to justify the suggested split, established most of the concepts (Preisig, 1990a, 1990b). The scope of the MODELLER, the model designer tool, was limited to generating symbolic equations (dynamic equations) and definitions giving a complete mathematical representation of the modelled plant in a state-space form.

Concepts

The model designer, our MODELLER, implements a special purpose editor supporting hard-core mechanistic modelling. It constructs plant models using two principal building blocks, namely systems and connections.

Systems

Systems represent capacities, that is anything with mass. *Systems* are dynamic elements and are defined as consisting of a single phase or a pseudo phase (an average of several phases such that it appears as a single phase). The term *system* is very much used as it is defined in axiomatic thermodynamics. The equations descrbing the dynamic behaviour of a *system* are the conservation principles. State variable transformations are introduced as definitions.

Connections

Connections represent communication paths between parts of the overall system. They always connect two systems. A connection being attached to only one system is not allowed and is completely disabled by the MODELLER. The consequence is that the global system is always closed. All sources and sinks must be included in the description even if they are only represented as simple reservoirs.is constraint was introduced to meet one of our main objectives, namely: *The tool must eliminate as many modelling errors as possible*. One ended streams are common errors showing in process models. Inconsistencies in the conservation principles are the consequences (Rodukas, 1989).

Connections are defined as separate objects. They are not part of either of the two connected systems. *Connections* also do not represent the dynamics of the process but describe the transfer of extensive quantities through a boundary assuming pseudo-steady state for the physical system associated with the actual transfer. The transport law, which is the main component of the mathematical representation of a connection, is generally a function of the states of the two systems, again a fact supporting the use of a state-space representation as the basic internal representation. *Connections* are interpreted similarly to idealized walls in thermodynamics, though with the extension that they represent the behaviour of finite capacity elements at pseudo-steady state. It is important to recognize this difference. If one finds that the pseudo-steady state assumption is not good enough to satisfy the requirements of the application, this assumption must be revised. The modification would then involve replacing the connection in question by one or several systems and connections representing a more refined dynamic description of this part of the plant.

Connections also incorporate all effects associated with system surfaces, such as change of phase across the boundary. Also each *connections* introduces a reference coordinate system against which the direction of the flow is defined. Therefore, the transfer laws must indicate a direction.

Networks

Process design draws extensively from flowsheeting with the flowsheet being the graphical representation of the plant. With the two building blocks of systems and connections and the conditions on how connections are being introduced, the traditional flowsheet representation was put into the background and was replaced by a network of systems and connections. This network, which now carries the label *physical topology,* is the basic structure on which all the other parts of the model build, for the time being excluding information processing systems (control). A digraph serves the purpose of a graphical representation of these basic networks, with the vertices being the systems and the arcs representing the connections.

Hierarchical Networks

Complexity is handled by introducing hierarchical grouping of networks. Such a grouping requires a definition of a set of systems which together are given a name and represent a composite system. This grouping leads to a classification of connections. *Internal connections* are connections between systems being member of the defined group and *external connections* being connections of members of the group with systems that are not included in the group and are thus part of the environment of the composite system (Preisig, 1991b).

Species Topology

The latest development is associated with exploring the different mechanisms and policies of defining the species topology in the containment of the physical topology. The species topology defines which chemical or biological species is present where in the physical system. A number of concepts were introduced for this purpose.

Firstly, a *species sets* defining a finite number of species. Taken from a base set, the data base, in the process of defining the species distribution, gradually a subset is being assembled which then is associated with the plant model.

Secondly, a *reaction set* was defined. It introduces the reactions between the various species in the species set relevant to the process. The species set and the reaction set must be compatible, that is the union of all species in the reaction set must be a subset of the associated species set. With the base species set, a reaction set is associated defining the basic reactions that may be considered. The reactions selected in the process of defining the species topology are then associated with the plant.

Thirdly, a property *permeability* was introduced for mass connections. Permeability is a species set defining which of the species that may be present on either side of the connection may pass through the connection. Permeability is not a directional quantity but it only defines a kind of "gate mechanism". The species included in the species set permeability may pass through the connection associated with it. All others may not. Since it is often easier to define which species may not pass, the complement of permeability may be defined. Permeability thus constraints the exchange of species between systems and can be used to describe the behaviour of a semipermeable membranes or constrains imposed by the species' state of aggregation. Chemical reactions introduce new species in a system as the consequence of a reaction that may potentially take place. The species topology must be defined before the component mass balances can be generated.

Mechanistic Details

Mechanistic details incorporates all remaining bits of information such as

- state variable transformations provide the link between the internal state and the state variables used in transfer laws, physical property models, kinetic laws and other such relations. They allow to view the process state through different glasses, so-to-speak.
- Transfer laws describe how extensive quantities are being transferred between systems.
- Production term describe mass conversion in a system by the means of a chemical or biological reaction.
- Physical properties.
- Geometrical properties.

Some of this information is of empirical nature. Examples are kinetic laws and physical property relations. Others are strictly mathematical such as state variable transformations. Others are somewhere in-between such as transfer laws. All of this knowledge can be supplied through interfaces to knowledge bases. The necessary tools are essentially available, reason for which this project has not been involved in any of this development up to now.

Equation Generation

The current version generates a textual output with general information about the structure of the system and equations for all parts of the plant. The equations and definitions are written in the syntax of an algebraic manipulator.

Implementation

The MODELLER is a special purpose data editor. Written in Modula-2 and using the API of MS Windows, the program provides quite a professional interface. All dialog is context dependent, that is, only those activities are shown in a menu that are possible and applicable to the object pointed to when requesting a dialogue. This technique removes a large portion of documentation and the user finds quickly his way into generating and manipulating process models.

Hierarchical Network of Systems and Connections

Two levels are shown at the time, an internal node of the tree and its children, the subsystems of the system represented by the internal node. This view into the tree can be moved around by simple point-and-klick operations.

All tree nodes except the top one (root of the tree), have an environment, the components of which are shown below the parent window. The environment components can be moved about in this outer frame, but are not allowed inside the parent window. The child systems can be arranged in the usual grab-and-drop fashion.

Connections

Connections are shown as lines from subsystem to subsystem. In the middle of the connection a box shows the type of direction (m :: mass, h :: heat, w :: work), the stream number and the reference direction against which the transfer of extensive quantity through this connection is measured. Multiple connections are shown in the same way, with the exception that the box changes to a list box, Fig. 1.

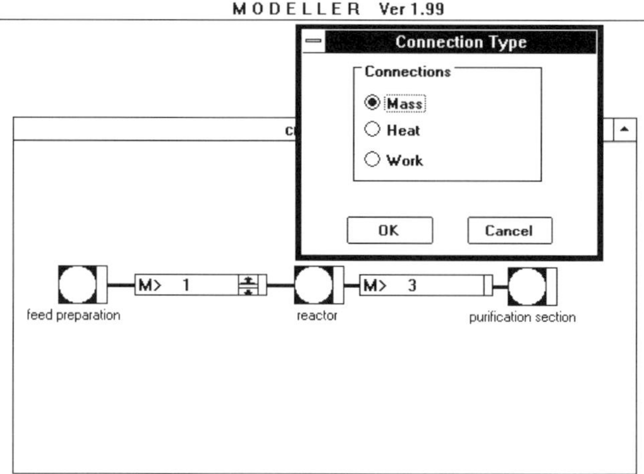

Figure 1. Establishing a connection.

Species Distribution

The computation of the species topology must be initiated by seeding species into systems. Dependent on the policy the species then propagate into the environment of the seed systems, meet other species and consequently undergo reactions. Two policies are implemented in the current program. The first is a maximal solution, which allows species to move along any mass connection in both directions if the transfer is not explicitly prohibited by the means of defining the connection non-permeable for the species.

The second policy is called the constraint approach in which the directionality of the mass flow can be constraint to be always positive and constraining the reactivity for individual systems.

Interfaces to Data Bases

Simple textual lists are presently substituting for more advanced interfaces to data bases. A project is currently under way to study the particularities of the data bases associated with the MODELLER.

Inheritance

One of the desired properties of such a tool is the ability of inheriting existing model elements. This may be only the containment, the physical topology of a plant unit, such as a heat exchanger or a distillation tray or a containment together with a full chemistry or biology. The MODELLER allows saving of any portion of the plant at any stage of the definition process. If an internal node, a composite system, is chosen, all external connections are cut and the sub-tree with the selected system as the root is saved. The saved portion represents an independent plant, a thermodynamic universe. If this plant is imported again into another plant as a component, the connections must be re-established.

Output

The current version generates a textual output file with the equations and definitions using the syntax of Mathematica. The file first lists the structure of the model, that is, the hierarchical grouping and the connections as source || sink where both systems are defined in an object-oriented notation. The plant species and reaction sets are in textual form together with the associated indices, providing the map between the species names and the indices used in the equations. The same applies to the reactions and the connections. The next section then lists the total mass balance, the species balances and the energy balance for all systems in the hierarchy.

Conclusions

The MODELLER demonstrates the implementation of a computer-aided modelling tool, which is based on a physical-chemical view of process models. The computer science part of the project, whilst often results in a quite painful exercise, is of secondary importance for the basic concepts of modelling. An appealing and easy to use user interface is essential for the user and improves productivity immensely. The knowledge based part, namely tasks such as selecting transfer laws, equations of states (state transformations) or kinetics can be solved independently and must not be clustered into the modelling tool. An output suitable to be processed in an algebraic manipulator opens the possibilities to process the symbolic models further with linearization and model reduction being two of the most frequent algebraic operations.

Acknowledgments

The work of Mr. Mehrabani, who implemented this first version of the MODELLER as part of his Ph.D. Thesis is acknowledged.

References

Andersson M. (1989). *Omola – An Object-Oriented Modelling Language*, Internal Report Dept. of Automatic Control, LUTFD2/(TFRT-7417)/1-018/(1989). Lund Institute of Technology, Lund (Sweden).

Lee T.Y. (1991). *The Development of an Object-Oriented Environment for the Modelling of Physical, Chemical and Biological Systems*, Doctoral Thesis, Texas A&M University, College Station, TX, USA.

Marquardt, W. (1992). An Object-Oriented Presentation of Structured Process Models, ESCAPE-1 Elsemore, Danmark, May 24-28.

Piela P.C., T.G. Epperly, K.M. Westerberg and A.W. Westerberg (1991) ASCEND: An Object-Oriented Computer Environment for Modelling and Analysis: The Modelling Language, *Comp. & Chem. Eng.*, **1**(1), 53-72.

Preisig, H.A., T.Y. Lee and F. Little (1990a). A Prototype Computer-Aided Modelling Tool for Life-Support System Models, 20th Intersoc. Conf. on Env. Systems, SAE Tech. Pap. Series No 901269, Williamsburgh, 1-10.

Preisig, H.A.,T.Y. Lee and F. Little (1990b). Computer-Aided Modelling of Physical-Chemical-Biological Processes, AIChE Annual Meeting 1990, Chicago.

Preisig, H.A., D.-Z. Guo, A.Z. Mehrabani (1991a). Computer-Aided Modelling: A New High-Level Interface to Process Engineering Software, *Proceedings of CHEMECA-91*, Newcastle, Australia, 954-960.

Preisig H.A. (1991b). On Computer-Aided Modelling for Design, AIChE Meeting 1991, Los Angeles, 138e.

Preisig H.A. (1991c). View on the Architecture of a Computer-Aided Process Engineering Environment, AIChE Annual Meeting 1991, Los Angeles, 137b.

Preisig, H.A. (1994a). Computer-Aided Modelling -- Species Topology, *ADCHEM'94*, Kyoto, Japan.

Preisig, H.A. (1994b). Components of a Computer-Aided Process Engineering Environment, in preparation.

Rodukas M.R., E.R. Cantwell, P.I Robinson and T.W. Shenk (1989). DAWN (Design Assistant Workstation), 19th Intersoc. Conf. on Env. Systems, SAE Technical Paper Series No 891481, San Diego, 10 pages.

Stephanopoulos, G. Henning and H. Leone (1990). MODEL.LA. A Modeling Language for Process Engineering--I The Formal Framework, *Comp. Chem. Eng.*, **8**(14), 813-846.

Zeigler, B.P., (1984). *Multifacetted Modelling and Discrete Event Simulation*, Academic Press, New York.

EXTENDING A PROCESS DESIGN SUPPORT SYSTEM TO RECORD DESIGN RATIONALE

René Bañares-Alcántara, Josh M. P. King and Geoffrey H. Ballinger
Department of Chemical Engineering
University of Edinburgh
Edinburgh EH9 3JL, Scotland

Abstract

KBDS (Bañares-Alcántara, 1994, Bañares-Alcántara and Lababidi, 1994) is a prototype design support system for conceptual design of chemical processes. This paper describes an extension to KBDS that enables the recording of design rationale. The extension makes use of an IBIS representation (Rittel and Webber, 1973) and is integrated to the history of the design process maintained by KBDS. An example is given of how design rationale is kept. This shows potential advantages derived from the maintenance and further use of design rationale in the design process.

Keywords

Design support systems, Design rationale, Issue-based information systems, Conceptual design, Design history.

Introduction

According to Mostow (1985), a comprehensive model of design should support the representation of the

1. State of design (description of the design object).
2. Goal structure (goals are prescriptions of how the descriptions of the artifact should be manipulated).
3. Design decisions (choices between alternative design paths).
4. Rationale for design decisions (justifications for goal selection).
5. Control of the design process (selection of the best goal to work on and the best plan with which to achieve it).
6. Learning in design (both, of general knowledge about the domain and of specific knowledge about the problem).

KBDS (Knowledge Based Design System), as described in Bañares-Alcántara and Lababidi (1994), can maintain design alternatives and design constraints, which are adequate for the representation of item 1 and part of item 2. We have extended such a representation with an IBIS (Issue-Based Information Systems, Rittel and Webber (1973)). This extension can account for items 2 through 4, and provide a solid base for items 5 and 6.

Next section provides a brief introduction to the core of KBDS. A description of the extensions to record design rationale is then presented and illustrated through an example. Finally we present a list of expected advantages from recording design rationale and future work proposed to achieve them.

KBDS: a Design Support System

KBDS is a prototype computer-based support system for integrated and cooperative chemical plant design. Its task is to assist a group of cooperating designers during the course of the design process. Design is an evolutionary process, i.e. it generates new design alternatives as transformations of existing alternatives. Given the complexity of chemical process design, such transformations are incremental, that is, each design step affects only a small part of the current design state. Thus, KBDS is based on a representation that accounts for the evolutionary, cooperative and exploratory nature of the design process, covering design alternatives, constraints, rationale and models in an integrated manner. A design process is represented in KBDS by means of three interrelated networks that evolve through time: one for design alternatives, another one for models of these alternatives, and a third one for design specifications and constraints.

In KBDS each partial or alternative design is linked to other design alternatives .Furthermore, KBDS maintains a more detailed "understanding" of a design state, i.e. it maintains a representation of an alternative's constituent unit operations (or black-box representations of plant sections), their relation to each other, and to those in other design alternatives (e.g. refinement). Design constraints can be used to evaluate the adequacy of a design alternative. The prediction of the behaviour of design alternatives is done by means of models.

Next section describes how KBDS maintains the designers' rationale. Like in the rest of KBDS, the information is represented in a prescriptive fashion, i.e. in a form that is amenable to computer processing.

Design Intent Representation

Recent work in our group has focused on the maintenance of the designer's intent, for the recording and use of design deliberation, argumentation and rationale. This is done by means of IBIS networks (Rittel Webber, 1973) which store every decision along with its competing alternative decisions and the arguments used in the selection. A network is maintained for every problem or issue discussed; networks can be interrelated. An IBIS network consists of *issues* (questions or problems relating to a design, item 2 of Mostow's list), *positions* (possible alternative solutions to the issue raised, item 3 of Mostow's list), and *arguments* (reasons or justifications for selecting the suggested solutions, item of Mostow's list).

The IBIS methodology has been used to model deliberation in software development (Conklin and Yakemovic, 1991), but only in a declarative fashion. Extensions were required to make these ideas prescriptive and applicable to Process Engineering, for example, by connecting the IBIS network to the design alternatives history maintained by KBDS. This was achieved by means of two new objects (Ballinger et al, 1994): *steps*, transformations of a design alternative suggested by a selected *position* , and *tests*, constraints placed on the design alternatives and associated to the *arguments* .

Design revisions or alternative design paths may be added at any point in the IBIS network should the designer wish to back-track and are indicated in the network by a *step* from a *position* other than that currently selected.

As an example, Fig. 1 shows the evolution of the design alternatives for the separation section of the HDA plant considered in Chapter 7 of Douglas (1988). This part of the design spans from the **separation-block** to the **separation4** design alternatives. The purpose of the **separation-block** is to separate the benzene product from the impurities: toluene, diphenyl and light-ends.

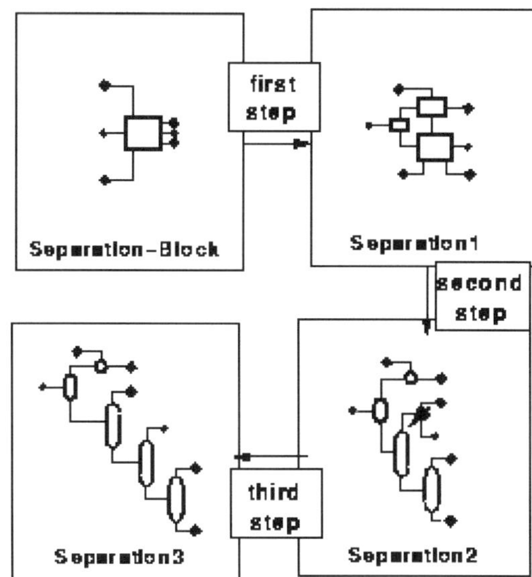

Figure 1. The design history of the HDA plant separation section.

The contents of the IBIS network can be examined, manipulated and generated through the *Intent Tool (Fig. 2)*. The window shows the first (and simplest) network in the example case. It deals with the **general-separation-structure?** *issue* **(I)** associated with the **separation-block** design alternative. This *issue* raises the question of the choice of the general structure of the separation system. Also shown are three possible *positions* or design paths **(P's)**, the *arguments* or justifications related to each choice **(A's)**, and a set of tests that can be applied to **separation block** alternative in order to evaluate the adequacy of each choice **(T's)**. Douglas (1988) considers three choices depending on which phases are present in the reactor effluent. In the figure, the *position* **initial-phase-split** was selected as the best choice because the reactor effluent is a VL mixture. The selection of this *position* gives rise to a step that transforms **separation-block** into **separation1**. The other positions considered were to have a liquid separation system only (for liquid reactor effluents) or to partially condense the reactor effluent and then have a phase split (if the reactor effluent is a vapour).

Using Design Intent Records

This section discusses the possible uses of design intent records and illustrates how they may be used to support the design process. From the declarative point of view, an IBIS structure can be used to keep track of the issues that have been discussed, which design alternative suggested them, and, if they are resolved, which alternative they affect directly. It is also possible to find whether an issue has been resolved , the choices that have been selected, and the reasons for their selection. Since part of the information kept in the network is in a prescriptive form, functions have been created that rank positions responding to an issue in order of desirability and suggest the most appropriate position for a given issue. Also, parts of the design object or process can be re-used in other projects with a fuller understanding of the assumptions and implications of the action, i.e. re-use can be *context-sensitive*.

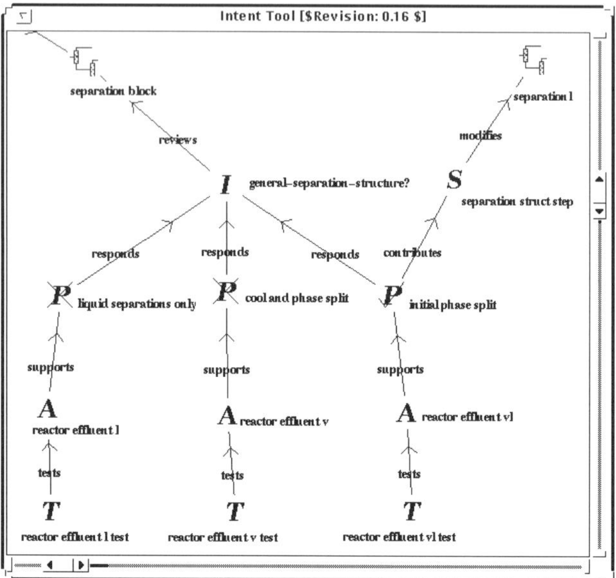

Figure 2. The Intent Tool window.

There are three immediate ways in which the design rationale can be used to the designer's advantage, all of them provided by KBDS's *Intent Tool* :

1. Storage of the design rationale in a prescriptive fashion allows the design team to identify which parts of the plant must be re-designed when there has been a change in the internal assumptions, constraints or specifications of the plant, or any external factor affecting them. This is known as dependency-directed backtracking.
2. An automatic evaluation of positions which allows arguments to be given a weight. Weights can be assigned their currently displayed values or be reset to their previously stored values. Thus, the system supports "what-if" studies.
3. An automatic report generator that produces documents describing the evolution of the design alternatives and the argumentation that resulted in a given decision.

Next section contains a more detailed presentation of the last facility.

Automatic Report Generation

After associating a text-based description to each IBIS node it is possible to request the automatic generation of reports. There are two types of reports that KBDS can generate: *Steps reports* which describe the design rationale for the evolution of one design alternative to another, and *Argumentation reports* describing the deliberation resulting in a particular decision. The designer is able to select the level of detail of each report.

An extract from an automatically generated *Steps report* can be seen in Fig. 3. It describes the evolution of **separation-block** to **separation4**, and contains information about the creator and creation time, the flowsheets for the starting and finishing design alternatives, a description of each design *step* with its selected *position*, its supporting *arguments*, and the alternative *positions* considered.

A design *step* may be the result of the selection of various *positions* (each *position* related to a different *issue*). The *argumentation report* is centred on an *issue* and can list the *arguments* that were not included in the reasoning process but that have the potential to affect the validity of the results.

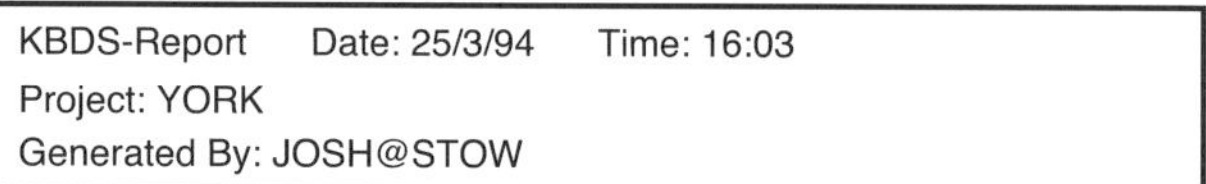

KBDS-Report Date: 25/3/94 Time: 16:03
Project: YORK
Generated By: JOSH@STOW

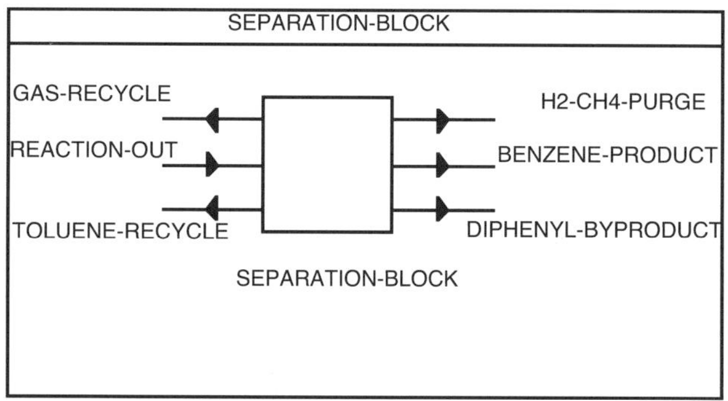

The evolution of **separation-block** to **separation4** involved four steps:

The first step modified **separation-block** to **separation1**
It was decided

Position: to have a phase split as the first unit operation
argument: because the reactor eF-fluent stream contains a vapour liquid mixture

Alternatives considered were:

Alternative: to have only liquid separation unit operations.

Figure 3. An extract from an example report.

Conclusions and Future Work

There are a number of reasons that suggest that the information supporting design decisions should be explicitly recorded. This information can be used to improve the documentation of the design process, verify the design methodology used and the design itself, and provide support for analysis and explanation of the design process. KBDS is able to do this by recording the design artifact specification, the history of its evolution and the designer's rationale in a computable form.

In the future, we envisage the application of knowledge-based systems to the information recorded by KBDS-IBIS to achieve the following:

1. Automatic generation of documents, such as summaries of achieved results, lists of pending tasks, etc.
2. During retrofit, in the generation of an "artificial history" of the design process for those plants for which no computable record of their design has been kept.
3. Automatic generation of design alternatives for those situations where it is possible to recognise an identical situation in the records of a previous design history.
4. Case-based reasoning, drawing analogies from similar situations in the past.
5. Application of design "critics" from early stages of design, thus spreading the consideration of downstream issues, such as controllability, safety and environment, along the whole design process time span.
6. Selection of competing application packages, e.g., physical property prediction methods, flowsheeting packages, user interfaces.

References

Bañares-Alcántara, R. (1994). Design support systems for process engineering. I. Requirements and proposed solutions for a design process representation. Accepted in *Computers and Chemical Engineering*.

Bañares-Alcántara, R. and H.M. Lababidi (1994) Design support systems for process engineering. II KBDS: an experimental prototype. Accepted in *Computers and Chemical Engineering*.

Ballinger, G., R. Bañares-Alcántara, and J.M.P.King (1994). Using an IBIS to record design rationale. ECOSSE Technical Report 1993-17, Department of Chemical Engineering.

Conklin, E.J. and K.B. Yakemovic (1991). A process-oriented approach to design rationale. *Human Computer Interaction*, **6**, 357-391.

Douglas, J.M. (1988). *Conceptual Design of Chemical Processes*. McGraw-Hill.

Mostow, J.(1985). Toward better models of the design process. *The AI Magazine*, **6**(1), 44-57.

Rittel, H.W.J.,and M.M. Webber (1973). Dilemmas in a general theory of planning. *Policy Sciences*, **4**, 155-169.

A HEURISTIC-NUMERIC CONSULTING SYSTEM FOR THE CHOICE OF CHEMICAL REACTORS

Gerhard Schembecker, Thomas Dröge, Ulrich Westhaus and Karl H. Simmrock
Department of Chemical Engineering
University of Dortmund
D-44221 Dortmund, Germany

Abstract

A new type of computer based consulting system for the choice of chemical reactors during the first step of process design is presented. Because neither expert systems nor numerical calculations alone are able to solve the problem of proper reactor choice, the heuristic-numeric consulting system READPERT (Reactor Development, Choice and Design Expert System) was developed. On the one hand it uses expert system technology and knowledge representation forms like rules and frames, on the other hand numerical calculations. By that way advice for the design of a chemical reactor can be given even in an early stage of process design. READPERT can be used like a stand alone system as well as part within PROSYN (*Pro*cess *Syn*thesis), an environment of cooperating distributed heuristic-numeric systems, concerning computer aided process synthesis.

Keywords

Reactor choice, Reactor design, Expert system, Process synthesis, Heuristic-numeric consulting system.

Introduction

Choice and design of a chemical reactor is the authoritative step in process synthesis. Inlet and outlet streams of the reactor determine the structure of the raw material preparation and product separation systems. Due to this, the costs of the whole chemical plant are strongly related to the reactor design. Unfortunately the proper design of a reactor is not an easy task, because there are a lot of parameters influencing the problem. Fitting, e. g., mode of operation, type of reactor, catalyst and process variables like temperature and reactant concentration just experimentally or by calculations is a difficult task. The idea to use computers and heuristic knowledge to solve this problem has led to the development of the heuristic-numeric consulting system READPERT (*R*eactor- *E*valuation *A*nd *D*esign Ex*pert*system).

At the beginning of a chemical process design usually no definite values concerning the kinetics of a reaction system are known. Therefore READPERT is starting with mainly qualitative criteria derived from first experiments and reasonable assumptions to find a proper reactor and its relevant operating conditions. In addition a set of numeric calculations is included. The whole task is divided into a variety of subproblems, which can mostly be solved independent from each other. In order to get handy and clearly arranged knowledge bases the subproblems are bundled in four different modules (Fig. 1).

Module 1: General Reactor Type

The first step after starting READPERT is the definition of the reaction scheme. Multiple reaction schemes are defined by combination of more than one single reaction step. For each single reaction step READPERT provides five kinetic models:

- irreversible reaction
- reversible reaction
- reaction depending on catalyst
- autocatalytic reaction
- inhibited reaction

Free choice and combination of any of these types is possible. Afterwards the user has to specify the aim of optimization during reactor development. Possible terms are either conversion, selectivity or yield.

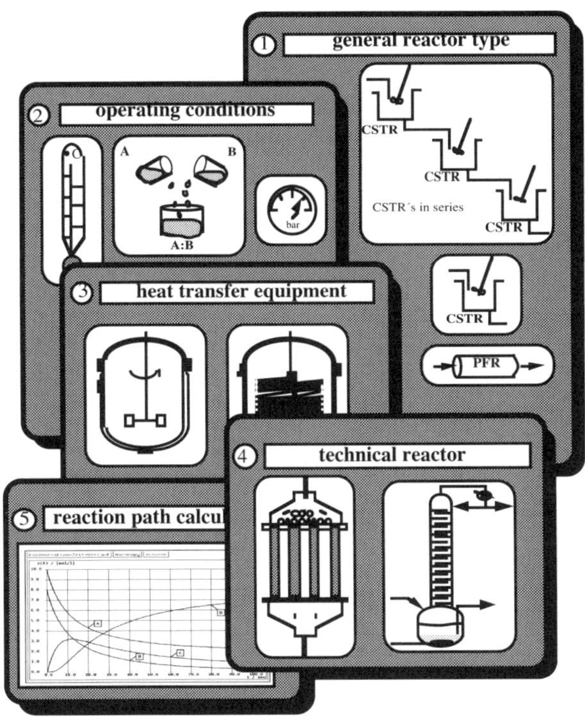

Figure 1. Main modules of READPERT.

The problem definition is followed by the determination of a general reactor type. If side reactions occur, READPERT needs some information about activation energies and the reaction orders. Informations about certain relationships between the main reaction and the side reactions (bigger, equal or smaller) are sufficient. Based on reaction orders and the structure of the reaction system READPERT gives advice about favourable conditions like the concentration levels for the reactants, the residence time distribution, the contacting pattern (equal flow or cross flow) and the degree of conversion. The results are presented in a dialog and have to be confirmed by the user. Afterwards these results are used to determine the optimal degree of backmixing for the reactor.

In a similar way READPERT gives advices for the choice of temperature profile and the temperature levels within the reactor. During this step informations about activation energies, structure of reaction system, heat of reaction and qualitative hints about reaction rates are used. If some data are unkown, READPERT tries to evaluate them by alternative ways (in the case of unknown activation energies READPERT derives the temperature profile due to the temperature dependency of undesired byproduct formation).

Two additional constraints of technical nature are considered during selection of a general reactor type. The first one is the mode of operation (batch, semi-batch, continuous), the second one is the phase of reaction mixture. Usually the user will know them. If not, READPERT can estimate the phase of reaction from the physical data of the pure substances, temperature and pressure of reaction and the composition of the reaction mixture. Some advice for the choice of operation mode is possible, too.

Module 2: Operating Conditions

Based on the results and the input data from module 1 (Fig. 1) module 2 gives recommendations for the most important operating conditions of the reactor. The following parameters are determined:

- temperature profile within the reactor
- necessity of recycle and cyclestreams
- qualitative temperature levels at beginning and end of reactor
- qualitative concentration levels for educt components
- necessity of inerts
- degree of conversion
- rates of reactants
- location or manner of reactant input
- location or manner of product withdrawal
- pressure level

An isolated consideration of these points seems to be desirable but fails in most cases. There are many links between the parameters. The correct choice of the degree of backmixing depends on the optimal concentration levels for the reactants. This point depends on the structure of the reaction system, the reaction orders of reactants in main and side reactions, the aim of optimization and last but not least on the shape of the reaction rate vs. conversion plot. The shape of that curve is influenced by the temperature profile and the heat effects during reaction. That means the degree of backmixing and the temperature profile cannot be choosen independent from each other.

The result of these dependencies is a complex network instead of a hierarchical structure. To prevent logic cycles within the net was an important task during the evaluation of READPERT therefore.

Module 3: Heat Transfer Equipment

A step by step procedure is used in the third module. Starting point is an equipment list of all conceivable heat transfer elements. The first step is to check which kind of equipment is suitable for the special problem. Helical coils cannot be used for an agitated vessel, if the reaction mixture has a high viscosity and tends to build incrustions on the inner surface of the reactor. The second step is the calculation of the specific heat flow for each of the elements comming into consideration. To achieve quick and easy screening short cut calculations are used to determine the heat flows. If the maximum flow of an element is higher than the needed one, it may be used. The choice of the best elements among the remaining possibilities of heat transfer is the last part of this procedure. Equipment costs are estimated to get hints for a proper choice.

For a plug-flow reactor additional conditions have to be considered. Probably a hot spot has to be suppressed or a special temperature profile has to be realised. A combination of several elements may be useful in such a case. The corresponding graphical flowsheet is generated by READPERT (Fig. 2).

• **feed stream as heat transfer medium**

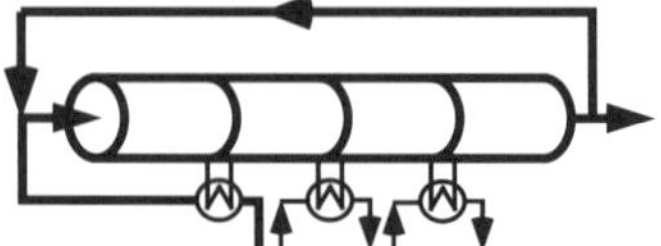

• **backmixing at beginning of reactor**

• **heat transfer between feed and output**

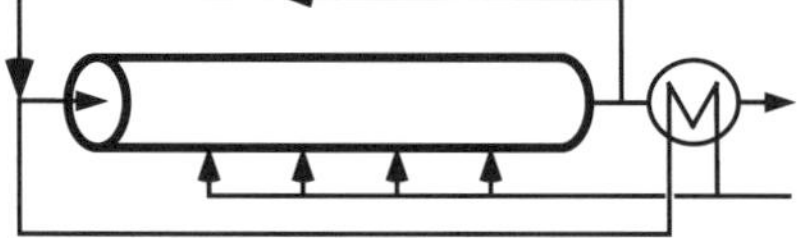

Figure 2. Graphical flowsheets generated by READPERT in the third module.

The choice of heat transfer elements terminates the basic design of a reactor.

Module 4: Technical Reactor

After solving the above mentioned fundamental questions, READPERT tries to find a technical reactor (module 4). A appropriate technical reactor has to satisfy the proposals developed in the previous modules and some further criteria of technical relevance. Heterogeneous catalysed reactions or gas/liquid-reaction systems, e. g., need to deal with reaction engineering questions as well as mass and energy transfer. Therefore, criteria like solidity and coarseness of catalyst particles, pressure loss or mechanical abrasion influence the selection of a technical reactor. The choice of a technical reactor represents the step between basic engineering ideas and final reactor design. In difference to other modules of READPERT using production rules and algorithmic programming, the technical reactor module uses a frame based structure.

The structure is the result of a classification of technical reactors primary using the phase of reaction mixture as main criterion. Other criteria like catalyst or viscosity allow further differentiations. Multiple inheritance and additional knowledge lead to the efficient choice of a proper reactor. A knowledge acquisition component allows to change properties of predefined reactors or to insert new objects (reactors), classes or properties into the knowledge base. By this way the user can build up his personal knowledge base for technical reactors.

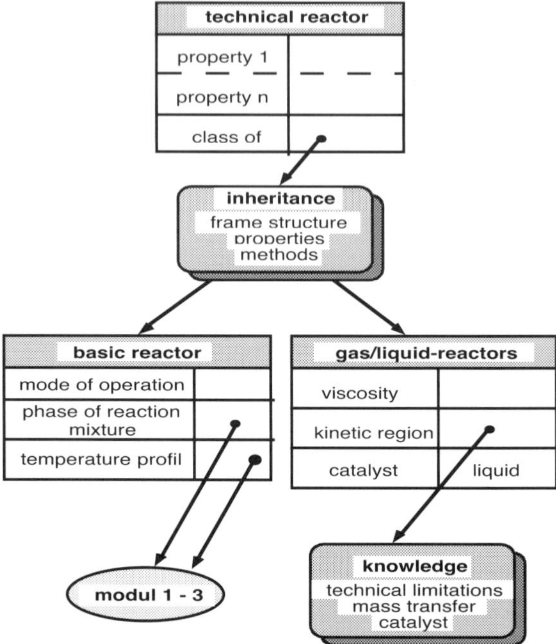

Figure 3. Universal frame of a technical reactor.

Additional Functions

READPERT is completed by some tools and interfaces. A calculation module (5) allows the determination of reaction data for several basic reactor types. The results can be plotted in various forms, e.g. as concentration-time or selectivity-conversion plots. Physical data of pure components and mixtures can be determined and managed by a blackboard system. If a session has been finished, the results are stored in a data base. By this way an overlook to previous sessions similar to the actual one is guaranteed. To have an overview about the actual task of determination and also the general context hypertext and textretrieval are implemented. The dialog and explanation component of READPERT answers questions like "How was this result generated?" or "Why is this question of importance?" even during a session. Last but not least a "What-if"-function allows an quick and easy change of, e. g., input data, constraints and subtargets.

Example: Production of 1,2-Dichloro-Ethane

Reaction system

1. $C_2H_4 + Cl_2$ -> $CH_2Cl\text{-}CH_2Cl$
2. $CH_2Cl\text{-}CH_2Cl + Cl_2$ -> $CH_2Cl\text{-}CHCl_2 + HCl$
3. $CH_2Cl\text{-}CHCl_2 + Cl_2$ -> $CHCl_2\text{-}CHCl_2 + HCl$

Additional information

Type of reaction:	catalytic
Phase of catalyst:	dissolved in liquid phase
Phase of reaction:	gas / liquid
Heat of reaction:	exothermic
Reaction orders RO:	$RO_1 < RO_2, RO_1 < RO_3$
Activation energies EA:	$EA_1 < EA_2, EA_1 < EA_3$
Global reaction rate:	fast
Operation mode:	continuous

Results of module 1 (General reactor type)

Goal of optimization:	maximum of selectivity
Concentration level:	low
Temperature level:	low
Temperature profile:	isothermic
General reactor type:	Continuous stirred tank

Results of module 2 (Operating conditions)

Rate of reactants:	excess of C_2H_4
Pressure level:	high
Product withdrawal:	yes, if possible
Degree of conversion:	total conversion of Cl_2
Inert addition:	no necessity
Recycle streams:	unconverted C_2H_4

Results of module 3 (Heat transfer equipment)

External reflux condenser is cheapest possibility of heat transfer. The combination with a distillation column is recommended.

Results of module 4 (Technical reactor)

A profile for a suitable technical reactor is derived from the known data like temperature profile, mode of operation, degree of backmixing, etc.. Further questions like backmixing of the different phases, corrosion, kinetic region and flow rates have to be derived or answered by the user. Then the following reactors are recommended:

- Stirred tank reactor with distillation column above
- Stirred tank reactor with external heat exchanger
- Bubble column
- Packed column
- Jet stream reactor with loop

Hardware and Software

READPERT requires a computer with UNIX operating system and X-Windows environment. This concept guarantees the use of READPERT on a wide hardware platform together with commercial programs such as ASPEN PLUS™ which is directly connected to READPERT

READPERT in Computer Aided Process Synthesis

READPERT is part of an environment of cooperating distributed heuristic-numeric systems, called PROSYN (*Pro*cess *Syn*thesis), concerning computer aided process synthesis (Simmrock, 1989). An overall manager system called PROSYN-Manager (Schembecker et al., 1994, Wolff, 1994) is responsible for the development of total chemical flowsheets with two main tasks: the design of unit operations and the choice of an appropriate reactor. In case of a reactor problem PROSYN-Manager activates READPERT.

Conclusions

A new type of computer based consulting system for the selection of chemical reactors in an early stage of process synthesis has been presented. The heuristic-numeric system READPERT on the one hand uses expert system technology and knowledge representation forms like rules and frames. On the other hand several numerical calculations, linked via interfaces like a subroutine, are used by READPERT too. This combination allows to find a proper reactor with a minimum of knowledge about the actual reaction system.

Acknowledgements

The authors would like to thank the BASF AG, Ludwigshafen, for their support of the work reported here.

References

Fried, B. (1990). Regelbasierte Auswahl grundlegender Reaktortypen mittels wissenbasiertrer Programmierung, Ph.D. Thesis, Universität Dortmund. *Fortschritt-Berichte VDI-Verlag*, Reihe 3, **247**.

Simmrock, K.H., Fried, A., Funder, R. and Schüttenhelm, W. (1989). Cooperating Expert Systems in Process Synthesis. Computer Applications in the Chemical Industry, Erlangen 23.-26. April 1989, *DECHEMA-Monographie*, VCH Verlagsgesellschaft, **110**, 135-144.

Schembecker, G. , Simmrock, K.H., Wolff, A. (1994). Synthesis of chemical process flowsheets by means of cooperating knowledge integrating systems, *IChemE Symposium Series*, **133**, 333-341.

Wolff, A. (1994). *Heuristisch-numerisches Managersystem zur Synthese chemischer Verfahren*, Ph.D. Thesis in preparation, University of Dortmund.

Westhaus, U. (1994). *Auswahl von Reaktoren mittels heuristisch-numerischer Verfahren*, Ph.D. Thesis in preparation, University of Dortmund.

Dröge, Th. (1994). *Beratungssystem zur Auswahl technischer Reaktoren*, Ph.D. Thesiss in preparation, University of Dortmund.

OPPORTUNITIES IN PC AND MAC NUMERICAL SOFTWARE FOR PROCESS ENGINEERING

Lee R. Partin
Eastman Chemical Company
Kingsport, TN 37662

Abstract

Numerical software packages for personal computers have been applied extensively by process engineers. Spreadsheet software is very popular for day-to-day tasks such as numerical calculations, data analysis, and plotting. Recent advances are making spreadsheets and other numerical tools even easier to apply and more powerful. For example, Excel®[1] has a built-in MINLP equation solver. Add-in packages are available to incorporate capabilities like Monte Carlo simulation and neural networks. Excel has a powerful Visual Basic® macro language (VBA) for writing your own functions, macros or add-in applications. Other mathematical packages complement the capabilities of spreadsheets and they link well in the window's environment. Mathcad® software provides an appealing graphical environment for dealing with model derivation, calculation and plotting. Maple V® software is a system for symbolic, numeric and graphic computation. Its strengths are in symbolic manipulations during model development, in its powerful programming language and in its graphical tools. These developments offer the process engineer with new opportunities in applying the personal computer. As an example, a flowsheet was modeled within Excel using a process simulator add-in called SimTools™ written with VBA. The software was linked with pinch technology analysis programmed within Maple V. The windowing environment makes it easy to exchange data, switch tasks and review results.

Keywords

Process engineering, Modeling, Personal computing, Spreadsheets, Process simulation, Numerical software.

Introduction

Process engineers were quick in exploiting personal computers. Spreadsheets are applied extensively for process calculations and design activities. Schmidt and Upadhye (1984) and Julian (1985) show the application of spreadsheets for material balances. Turner (1992) describes spreadsheets for the design specifications of hydraulics, pumps, vessels and heat exchanges. The process simulator firms have written special versions of their products for the personal computer with links to spreadsheets. There are several other numerical software packages that can be applied to process engineering tasks. The windowing operating system makes it easier to link the software for increased productivity. The goal of this paper is to highlight areas where process engineers can further exploit the capabilities of personal computing.

Spreadsheet Advances

Spreadsheets are the most widely used numerical tool in personal computing. Recent advances are making them even easier to use and more powerful. The Excel program is a good example of the new advances.

[1] EXCEL and Visual Basic are registered trademarks of Microsoft Corporation. Mathcad is a registered trademark of MathSoft, Inc. Maple is a registered trademark of Waterloo Maple Software. SimTools is a trademark of L R Partin Enterprises.

Solving Equation Sets Including Inequalities and Optimization

Excel has several built-in functions to assist with solving sets of equations as documented in its manuals.

For linear algebra, matrix functions solve sets of linear equations ($Ax = b$) via spreadsheet formulae. Therefore, a linear mass balance model of a process can be programmed within the spreadsheet. One option is to solve a linear equation set of flows for each chemical within the flowsheet. The solution to a given set contains the flows of the specified chemical in all process streams.

Excel's Solver routine provides a wide range of functionality. In one option, it handles optimization of linear programming (LP) models. In its full application, it handles nonlinear equations as well as integer constraints (MINLP) for a small number of equations or variables. Solver has numerous potential applications for the process engineer. Here are some examples from the author's programming:

1. Perform nonlinear regression analysis for heat of vaporization parameters to simultaneously fit data on heat of vaporization and liquid heat capacity.
2. Solve the Underwood equation for minimum reflux ratio.
3. Regress for Wilson activity coefficient parameters using binary activity coefficient data.
4. Calculate Wilson activity coefficient parameters of a binary set given the infinity dilution activity coefficients.

Excel also has a Goal Seek command for simple applications of manipulating an input cell to obtain a desired value in a target cell.

Macro Programming

Macro programming features are critical in allowing users to tailor the software to their applications. Turner (1992) applies macros in the automation of data transfer from process modeling through the equipment design phase. Macros are also a great way to add functionality. Excel 5.0 has a Visual Basic Application language for writing subroutines and functions. Subroutines are the means of automating tasks (i.e., passing data, converting data, entering information, searching databases, printing reports). Functions provide a way to add spreadsheet functions that can be called within the cells of the spreadsheet. Functions take inputs via entry or cell references and calculate a value or array of values.

The power of a macro language was demonstrated by the author through the development of a process simulator called SimTools as an Excel add-in package. Physical property data are stored within a sheet of the workbook for access by the functions. The add-in contains standard functions needed to perform energy and material balances:

- property calculations (molecular weight, density, heat of vaporization, vapor and liquid enthalpy and vapor pressure)
- vapor-liquid K values by various methods
- vapor-liquid flash routines (bubble point, dew point, T P flash, P V flash)
- calculate the stream temperature from its enthalpy
- calculate simplified unit operation balances

The functions are available in the same manner as the standard functions (e.g., SUM, AVERAGE, etc.). A subroutine for tear stream convergence via the Wegstein method completes the toolkit. Visual Basic provides a tool for writing extensive applications for the process engineer.

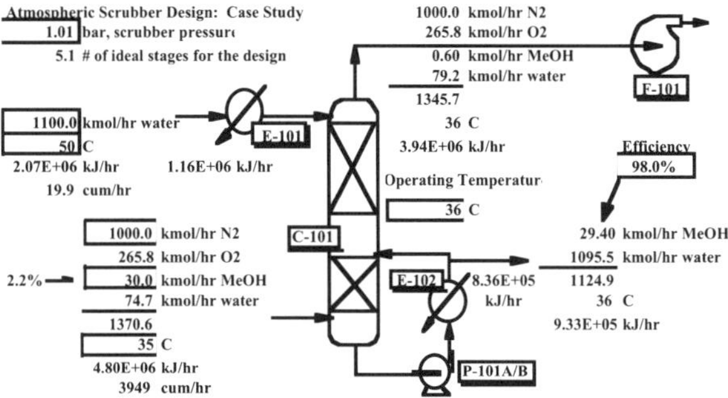

Figure 1. Example Excel spreadsheet for process design.

Figure 1 presents a simple example of applying spreadsheets to the design of a scrubber system for absorbing methanol from air with water. The Excel drawing toolbar was applied to sketch the process flowsheet on top of the cells that calculate process conditions. The cells containing input parameters are highlighted by boxes. They are also highlighted in a light blue color on the monitor. The engineer can easily change an input parameter and see the results quickly appear on the diagram. Other screens in the spreadsheet perform the rest of the analysis including equipment designs, equipment purchase costs, plant installed cost and net present value analysis. Excel's Scenario Manager and Sover routine provide the needed tools for optimizing the process parameters for maximum value.

Figure 2 shows another application built with SimTools. The graph is based on research in distillation system synthesis. Distillation lines or residue maps are a graphical technique for representing the capabilities of distillation systems (Doherty and Perkins, 1979; Stichlmair et al., 1989; Partin, 1990). Distillation lines are calculated as a distillation tower operating at total reflux and fixed pressure. For a given distillation line, the user picks a liquid composition. A bubble point flash calculates the vapor in equilibrium with it. The vapor composition becomes the liquid composition for the next stage up the tower at total reflux. The ethanol-water azeotrope in the methanol-ethanol-water system introduces a boundary in the diagram. Distillation lines are constrained to one of the two regions within the diagram. This is an important observation for the synthesis of distillation schemes since it limits the potential products from distillation towers.

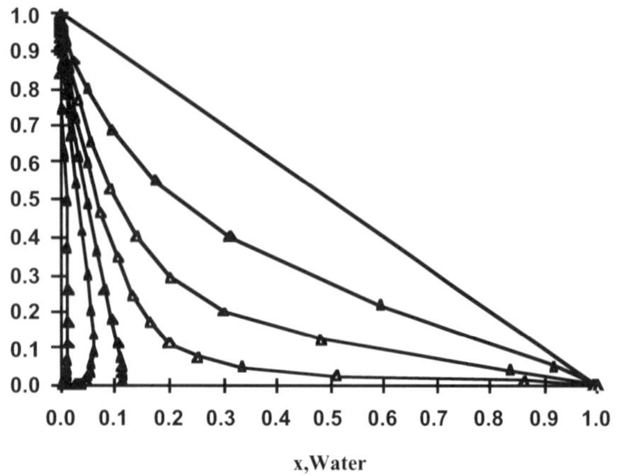

Figure 2. Distillation lines for Methanol-Ethanol-Water system at 1.013 bar.

Third-Party Add-Ins

Several add-in packages to Excel are sold by third-party vendors. Once installed, the packages work within the Excel framework. New menus are added to the standard toolbars. Two of the add-ins are documented for this report.

The Crystal Ball®[2] add-in provides Monte Carlo simulations within Excel. Key model parameters are defined as assumptions and described through distribution functions (e.g., flow = 1,100 kmol/hr with a standard deviation of 55 kmol/hr). Key spreadsheet results are defined as forecasts and the software monitors their values during the simulation. Numerous trials (perhaps 1,000) are calculated through the spreadsheet with random values taken from the distributions of the assumptions. The distribution functions of the forecasts are tabulated from the results. This technology lets the engineer quickly document the impact of parameter uncertainties on the model outcomes. For example, the design value of the area for exchanger E-101 from Fig. 1 is uncertain due to uncertainties in the design basis and design methods. Figure 3 plots the resulting heat exchanger area distribution. For an introduction to uncertainty analysis for chemical engineering applications, see Frey and Rubin (1992).

Another add-in called Braincel™ [3] incorporates neural network features into Excel. Neural networks are a modeling tool for fitting a wide range of models including traditional models, highly nonlinear models, discrete functions, and dynamic models. Its availability as an add-in makes this technology more accessible to the engineer.

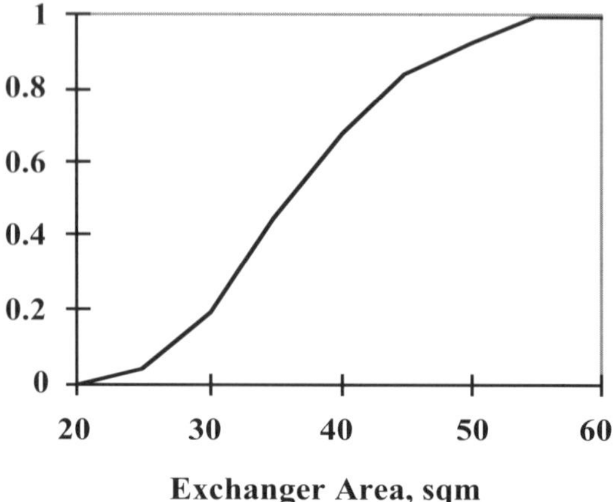

Figure 3. Cumulative probability distribution on Exchanger E-101 size.

Phiroz (1990) gives an introduction to neural networks for chemical applications.

Oher Numerical Packages

Other software packages are available for personal computers to handle more intensive mathematical needs.

Inputs: $p := 2$ bar $\mu := 1 \cdot 10^{-8}$ bar-sec

$C := 0.06$ $n := 0.4$ $k := 6 \cdot 10^{12}$ $R_m := 6 \cdot 10^{10}$ 1/m

Provide solver for filtrate flow rate, q:

$q := 0.003$ initial guess at q

$$g(v) := \mathrm{root}\left[\left(p - \mu \cdot q \cdot R_m\right)^{(1-n)} - \mu \cdot q \cdot C \cdot k \cdot v, q\right]$$

Solve the differential equation for volume vs time:

$V_0 := 0$ initial filtrate volume

$D(t, V) := \mathrm{Re}\left(g\left(V_0\right)\right)$ dv/dt or q

$Z := \mathrm{rkfixed}(V, 0, 7200, 100, D)$ function to solve it

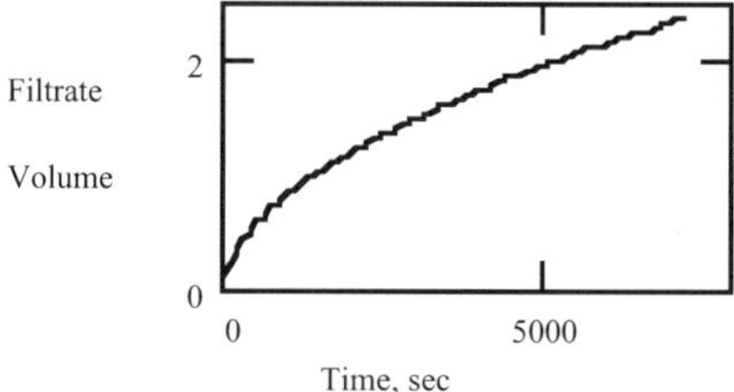

Figure 4. Example Mathcad document.

Mathcad

Figure 4 was programmed with the Mathcad software to solve for the filtrate flow during cake filtration at a fixed supply pressure using the method from Tiller (1990). Mathcad has an appealing interface that combines text, mathematical expressions, computation routines and graphics. Expressions are entered in typeset form. Model development is self-documenting and interactive. Results are presented as information is entered or updated. Numer-

---

2. Crystal Ball is a registered trademark of Decisioneering, Inc.

3. Braincel is a trademark of Promised Land Technologies, Inc.

ical functions handle the solution of equation sets, differential equation sets and partial differential equations. In Fig. 4, root solves an implicit equation for the filtrate flow, q, given the value of the filtrate volume, n, and rkfixed integrates over the range of 0 to 7,200 seconds in 100 intervals.

Maple V

Maple V is a software system for symbolic, numeric and graphic computation. Its advantages are in its symbolic capabilities and programming language. The knowledge from college mathematics courses has been programmed into its functions. For example, a solve routine performs the algebra in solving a set of equations. Figure 5 demonstrates its ability to solve the set of equations of a heat exchanger for expressions of the outlet temperatures in terms of the other parameters.

```
  Entering equations:
> eqnset:={q=CPhot*(T1-T2),q=CPcold*(T4-T3),
     q=UA*((T1-T4)/(T2-T3))/ln((T1-T4)/(T2-T3))}:
  Solve for q,T2 and T4 in terms of the other values:
> SolveSet:=solve(eqnset,{q,T2,T4}):
  Define two dimensionless values,
     B = UA*(CPhot-CPcold)/CPhot/CPcold and
     R = CPcold/CPhot:
> subs([exp(-UA*(CPhot-CPcold)/CPhot/CPcold)=B,
     CPcold=R*CPhot],SolveSet[2]):
  Simplify the answer:
> simplify(");
```

$$T_2 = \frac{BRT_3 + RT_1 - T_1 - RT_3}{BR - 1}$$

$$T_4 = \frac{BT_1 - BT_3 - T_1 + BRT_3}{BR - 1}$$

$$q = \frac{CP_{hot}R\,(BT_1 - BT_3 - T_1 + T_3)}{BR - 1}$$

Figure 5. Example Maple V application.

The solve function performs the algebra to find the three variables in terms of the remaining five parameters. The result is a set of formulas for calculating T2, T4 and q. Without this solution, the engineer is faced with a difficult numerical convengence problem.

Maple's programming language is a powerful tool for developing process engineering systems. The author easily wrote a pinch technology application for generating composite, grand composite and supertargeting plots from stream table data. Other experiences include: (1) solving heat exchanger networks, (2) kinetic modeling, and (3) kinetic parameter fitting. Its VCR-style animation is an enjoyable way to fit kinetic parameters by watching a video of model results at varying kinetic constants.

Windowing Environment

The windowing environment makes it possible to combine the capabilities of the different software. You can quickly move between programs and exchange information. In one example, the author programmed the dehydration of toluene process within Excel using SimTools and linked it with pinch technology calculations in Maple V. The link allows the engineer to update the process model and then compare plots of old versus new composite curves.

Mathcad supports DDE exchange of data so it is possible for an Excel macro to pass parameters to a Mathcad document and obtain results from the calculations. This provides access to numerical software like differential equation solvers within the spreadsheet.

Conclusions

Recent advances in personal computing software offer several opportunities for applications in process engineering. It is time to exploit the capabilities.

References

Doherty, M.F. and J.D. Perkins (1979). On the Dynamics of Distillation Processes-III. *Chem. Eng. Sci.*, **34**, 1401-1414.

Frey, H.C. and E.S. Rubin (1992). Evaluate Uncertainties in Advanced Process Technologies. *Chem. Eng. Prog.*, **88**, 5, 63-70.

Julian, F.M. (1985). Flowsheets and Spreadsheet. *Chem. Eng. Prog.*, **81**, 9, 35-39.

Partin, L.R. (1993). Use Graphical Techniques To Improve Process Analysis. *Chem. Eng. Prog.*, **89**,1, 43-48.

Phiroz, B. (1990). An Introduction to Neural Nets. *Chem. Eng. Prog.*, **86**, 9, 55-60.

Schmidt, W.P. and R.S. Upadhye (1984). Material Balances on a Spreadsheet, *Chem. Eng.*, Dec. 24, 67-70.

Stichlmair, J.G., J.R. Fair and J.L. Bravo (1989). Separation of Azeotropic Mixtures via Enhanced Distillation. *Chem. Eng. Prog.*, **85**, 1, 63-69.

Tiller, F.M. (1990). Tutorial: Interpretation of Filtration Data, I. *Fluid/Particle* Sep. J., **3**, 2, 85-94.

Turner, J. (1992). Put Process Engineering in the Fast Lane, *Chem. Eng. Prog.*, **88**, 8, 59-62.

INTERACTIVE COMPUTER SIMULATIONS FOR INTEGRATED LEARNING

Nico van Klaveren
Consultant
Camino, CA 95709

Ahmet N. Palazoglu and Karen A. McDonald
Department of Chemical Engineering and Materials Science
University of California at Davis
Davis, CA 95616

A. Terrell Touchstone
Chevron Research & Technology Comp.
Richmond, CA 94802

Abstract

Interactive, game-like simulators are emerging educational tools for industry and academia alike. As an example, the SimRefinery™ prototype (Fig. 1), a dynamic simulator of a petroleum refinery has been developed and tested in several education environments. In concept, SimRefinery™ is similar to the highly successful SimCity™ computer game produced by Maxis, Inc. This paper describes the prototype simulator, its development, and its implementation in industry (at Chevron) and in education (at the University of California, Davis).

Keywords

Computer games, Simulators, Computer instruction, Refinery operations.

Introduction

Imagine for a moment that you are in a classroom, seated in front of a computer monitor. On the screen is an aerial view of a refinery stretched out across a landscape. The screen is alive - oil tankers are docking, railcars are pulling into the loading terminal, and vehicles are moving about.. Then, remembering your instructors suggestion, you reach for the mouse and click on the button labeled "crude unit" and learn that you are feeding 80,000 barrels of a light crude to the refinery. By clicking on the button labeled "operating margin report" you find out that your enterprise is losing $2 per barrel of crude oil processed, and that you're overproducing on regular gasoline. Worse yet, you can't sell the gasoline because it does not meet product specifications since the octane rating is too low!

You decide to lower the crude feed rate to correct the gasoline production. By adjusting the components blended into the gasoline you are able to just meet the octane requirements of your batch. You are now almost breaking even financially.

Now that the operation is running smoothly, you decide to explore. You click on the landscape over several of the processing units to find out details about their function and status. During your investigation, you suspect that the catalytic reformer unit is underutilized and decide to increase its "severity". This results in more octane for the pool. After adjusting the gasoline blending recipe you find you are now earning ten cents on the barrel.

Then unexpectedly, a flange fire is reported in the alkylation unit. The unit will be shut down for two shifts and, as a result, a main gasoline feedstock is suddenly unavailable. A hint button tells you that the incident was preventable had you appropriated more of the budget to training. It further adds that you could have continued production despite the fire if you had built up an inventory in the component tank.

Soon the class is over and you are amazed at how quickly the time flew. You save the status of the game to continue to explore and experiment tomorrow.

SimRefinery

The "game" described above is called SimRefinery™ and is adapted from the highly successful SimCity™ produced by Maxis Inc. SimRefinery™ is basically a dynamic, open ended system simulation which mimics the processes, their interactions and responses to present conditions. Besides refinery processes the simulation also integrates financial and marketing aspects, training, the environment, safety, labor, etc. Also, in certain circumstances, random events, such as fires or natural disasters, can occur in the refinery.

A prototype SimRefinery™ was developed by Chevron in cooperation with Maxis. The goals of the project were:

- to demonstrate the effective use of new computer technologies for training on complex systems/processes and for information transfer
- to replace current training materials which teach employees about what a refinery is
- to generate interest and ideas for other business simulations which would be of value to the company

Advantages

The narrative presented above shows many of the advantages of an interactive game-like simulator such as SimRefinery™ over traditional learning processes:

- The intuitive, graphical interface encourages learning by exploration and experimentation which is motivating to most people and accelerates learning.
- The "player" can use the learning mode most effective for him/her: exploring the help screens, analysis of the effects of changing operating conditions, or trial and error experimentation. Consequently, a larger group of trainees benefit from using this tool. It is reported that only 20% of the population learns well using the traditional, passive mode (lectures, books, etc.). Others (35%) learn more effectively through visual input (demonstrations, models), and the largest group (45%) learns best through experimentation (Schneider, 1990).
- The simulation teaches "systems thinking", covering interactions among diverse areas such as technology, labor, economics, the environment, R&D, and training. Harwit (1994) points out the drawbacks of "compartmentalization of knowledge in our educational system" which can lead to technical specialists who lack insight and knowledge about the real world.

SimRefinery Development

The design and development of a SimCity™-like simulation requires a dedicated team of experts. The SimRefinery™ project team consisted of a refinery expert, two process engineers with software development backgrounds, a team of Maxis' software developers and a potential customer: an instructor enthusiastic about integrating SimRefinery™ in one of her classes at Chevron.

The learning design must be guided by a clear understanding of the way in which the end product will be used. The highly technical learning needs of refinery operators differ greatly from staff employees who need mainly to learn terminology and see the interdependencies of the organization.

Our objective was to develop a "generic" simulation whose main purpose was to demonstrate the advantages of SimCity™-like simulations for training and education. The resulting prototype is a low-fidelity simulation which uses very simple steady state representations of the refinery processes and economics obtained from a variety of sources (e.g. textbooks, plant data, experts). The fidelity of the correlations should be tailored to meet the requirements of the intended audience. The development of the SimRefinery™ prototype took 6 months; we estimate that a complete package can be developed in less than a year.

Implementation

Chevron

The SimRefinery™ prototype was demonstrated to management, training professionals and prospective trainees at Chevron. The response was exceedingly favorable, however, the next stage of development of SimRefinery™ has been delayed for several reasons:

- established training philosophies/approaches are hard to change,
- hardware has to be made available
- instructors need to be trained
- high development cost and limited market potential.

To assure the required level of technical accuracy and applicability for site-specific training in a high quality game-like simulator, a multidisciplinary development team is needed. As a result, the development cost is very high and the final product may not be of interest to the refining industry in general.

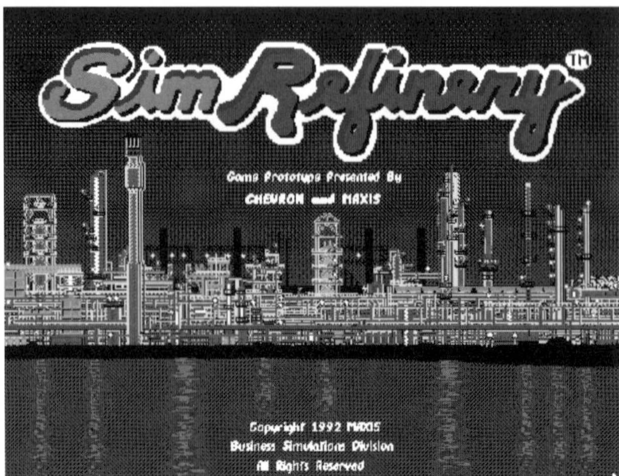

Figure 1. SimRefinery™ title screen.

University of California at Davis

The SimRefinery™ prototype has been used at UC Davis for several purposes. It was adopted in ECH 1, Introduction to Chemical Engineering, a seminar course for first year students which provides an overview of the various subdisciplines of chemical engineering. Unlike previous seminars in ECH 1 which use overhead transparencies or traditional lecture styles of presentation, we used computer presentation software in conjunction with the simulator to teach the students about refining. The simulation was used to give students an overall view of what goes on in a refinery, very much like the walkthrough narrative described in the introduction. We had several objectives for the presentation:

- To familiarize students with a simplified overall flow diagram and the unit operations involved in refining
- To familiarize students with terminology used in the oil refining industry
- To point out the interdependencies between the technical aspects of the process, economics, market demands, quality specifications, safety etc.
- To challenge the students to explore SimRefinery™ on their own (the program was available on campus in computer laboratories and disks were distributed to interested students) to try to improve operation. A "challenge problem" was defined (with a copy of SimCity' as the prize for the winner) to encourage students to further explore SimRefinery™.

As expected, SimRefinery™ facilitated the introduction to oil refining tremendously. It is safe to say that it would be impossible to cover the subject any other way as clearly in the allotted time (50 min.). In course evaluations students stated that the simulator was fun to use, encouraged their interest in the chemical engineering profession and increased their understanding of refinery processes.

Although only a few of the students completed the "challenge problem", a significant fraction did explore SimRefinery™ on their own. Since ECH 1 is graded on a pass/not passed basis and students are normally taking a heavy load of coursework, it is not surprising that only a few of the most enthusiastic students completed the challenge problem. Further refinements to make the simulator easier to use (e.g. faster on networks in computer classrooms), and more complete would make it more effective as a teaching tool.

SimRefinery™ has also been used at UC Davis to provide prospective students with a glimpse of what chemical engineers do. SimRefinery™ was demonstrated to visiting high school students interested in majoring in chemical engineering. It was also demonstrated on Picnic Day, the annual "open house" for the UC Davis campus, where students, families, alumni and the general public visit the campus to learn more about the campus, activities, and various departments. In both cases, SimRefinery™ proved to be a very useful, exciting and fun way to show people with a variety of different backgrounds and levels of technical understanding about what goes on in a refinery.

The Future

Chevron is actively trying to bring together the right combination of people to further develop SimRefinery™. Proposals under consideration at Chevron are to join with other industry sources.

The use of game-like simulators for higher education, particularly manufacturing, is very promising. With initial funding from the Teaching Resources Center at UC Davis and the UNOCAL Foundation, we plan to:

- Develop a new, more complete module to be used in an introductory course (ECH 1: Introduction to Chemical Engineering) and an upper division course (ECH 158A: Chemical Engineering Economics and Optimization) for Chemical engineering students. Not only will it provide students with a realistic overview of how a refinery works but it will also teach engineering students, at an early stage in their career, that technical issues are not the only ones that dictate the overall profitability of complex manufacturing processes but that issues such as environmental regulations, equipment maintenance, scheduling, market economics etc. have a significant impact as well.
- Develop a new General Education course for non-science students on Issues and Options in Manufacturing which uses the refinery simulator. The objective of the course is to bridge the gap between scientists/engineers and students in the humanities/liberal arts. For non-

science students the course will provide a better understanding of the technical aspects of a manufacturing plant (using the refinery as an example throughout the course) as well as the tradeoffs involved in meeting competing objectives. For science/engineering students the course will provide an integrated view of manufacturing in its societal context.

The potential impact of this educational project is very broad. First of all, the simulator modules and course designs can easily be transferred to other universities for use in courses similar to the ones we have described above. Secondly, it could be a model for development of new modules for other industries such as SimChemical (chemical plant), SimBiotech (a biotechnology plant), SimChip (this could be either integrated circuit manufacturing or potato chip production, depending on the target audience) etc. We are currently evaluating existing development software which would allow faculty to develop their own modules for their classes. In addition, the modules may be useful at lower levels, such as in high schools. Finally, further extensions using multimedia could make these teaching tools even more effective.

Conclusions

During the last five years simulator-based computer games have become very popular. An example is SimCity™, developed by Maxis Inc. in 1989. The SimRefinery™ prototype is an adaptation of SimCity™ built initially to test the effectiveness of these types of simulators as teaching/training tools for industry. More recently, SimRefinery™ has been introduced into university courses.

The development of a prototype SimRefinery™, modeled after SimCity™, and its introductory use in industry and academia have taught us that

- Simulators significantly improve the learning process and facilitate teaching about complex systems. They accommodate diverse learning styles and thus reach more people. The interactive nature of the simulators generates excitement and a desire to explore further, all without financial or operating risk. Students gain a hands-on feel for the issues and variables affecting the refining business.
- The technology for developing these simulations is readily available.
- As is true for any delivery system, the learning design is the major part and challenge of the development project.
- Implementation needs a new educational paradigm. Without this, there are barriers for the introduction of these simulators.

Acknowledgments

We are grateful to Chevron Research and Technology Company and Maxis Inc. for the permission to use the SimRefinery™ prototype, and the UNOCAL Foundation and UC Davis Teaching Resources Center for funding the educational project at UC Davis.

We thank Bill Wanat and Susan Gustin (Chevron) for their participation in the SimRefinery™ project, for their perseverance and unwavering belief in the idea. We would also like to thank David Mills and James Wong at UC Davis for their assistance in the implementing the simulator in ECH 1.

References

Harwit, M. (1994). Fostering a Nation of Science Generalists. *Technology Review*, **97**, 3.

Schneider, D. (1990). High Noon for High Tech Training - Preparing America for the 21st Century, Keynote at the 9th Annual Computer-Based Training Conference, San Antonio, TX, May, 1990.

ACTIVITY MODELING FOR PROCESS ENGINEERING

Kenneth A. Debelak
Department of Chemical Engineering
Gabor Karsai, Samir Padalkar, Janos Sztipanovits
Department of Electrical Engineering
Vanderbilt University
Nashville, TN 37235

Frank DeCaria
E. I. Dupont DeNemours, Old Hickory Works
Old Hickory, TN 37138

Abstract

Activity modeling is a new functionality which seeks to integrate information which will be provided from different process monitoring and control functions and describes exactly what is done with this data. Activity modeling allows the process, monitoring, and control system to function according to the needs of different users, operators, control engineers and plant managers. The Intelligent Process Control System with its rich modeling paradigm for plant and activity modeling and automatic program synthesis capabilities offers a powerful new tool for process engineering.

Keywords

Activity modeling, Model-based systems, Fault diagnosis, Process modeling and simulation.

Introduction

Plant operation requires a diversity of problem-solving activities which must occur in an information-rich environment. Current and historic plant data, static and dynamic models of processes, characteristics and state of equipment and other information must be available and integrated at the points of decision. Efficient support for accessing relevant information and integrating it into problem-solving activities is critical for economical and reliable plant operation.

The Intelligent Process Control System (IPCS) is a domain-oriented, model based programming environment for developing monitoring, control, simulation, diagnostic, and recovery applications. Process engineers with little or no software engineering can easily specify models of the plant. IPCS then creates executable applications from them automatically. The important point here is that the models are domain specific, i.e., built from concepts characteristic of the field. IPCS has the following capabilities:

- Collect, maintain and provide access to key plant information typically generated in design time. This information provides a plant-specific context for decision procedures.
- Model and execute high-level activities that use the available process monitoring and control (PM&C) functionalities and provide direct and relevant support for decisions.

The first capability provides support in building and managing *plant models*. Models are presented in the form of various diagrams, and are stored in an object-oriented data base (OODBMS). The basic plant-related information, which is typically expressed in process flow sheets, P&I diagrams, and in a variety of models, is explicitly represented. The second capability provides support for building *models of activities* that combine various PM&C functionalities, associate the activity models with plant models and automatically synthesize executable systems in a real-time environment from the activity models. In its simplest manifestation, modeling activities might involve the gathering and interpretation of data from one of the functional modules such as the event historian, comparing the data to the standard operating conditions that are expressed in plant models (solutions of material and energy balances, outputs from a process simulator, e.g. ASPEN) and making a decision as to the possible faults in the plant. In a broad sense, the modeling of activities would describe what plant information is to be retrieved (historical data, on-line data, recipe data, etc.), what will be done with the plant information (comparison against plant models, display of the data for an operator, statistical analysis of the data, etc.), and what conclusions will be reached from an analysis of the data (determination of faults, change in control strategies or parameters, determination of production schedule, etc.).

Model Categories

IPCS offers a broad selection of *basic modeling concepts* which can be used to model various properties of a plant and related operation support activities. The basic modeling concepts are obviously not enough to model all possible plants or activities. IPCS provides a set of *model organization principles* which can be used to build more specialized, or more complex models using the basic concepts as building blocks.

There are two basic categories of IPCS models:

- *PLANT models*, which describe what is known about a plant, and
- *ACTIVITY models*, which describe the models of operation support procedures.

Users can build various models of the above categories using IPCS' graphical model builder. Once they are created, the PLANT models serve as a rich source of information about the plant, which can be easily maintained or modified. PLANT models can be built in design time as a documentation of the design process. In the case of existing plants, the models can be created retroactively as required by the ongoing need of plant operation. ACTIVITY models represent arbitrary operations, processing activities that are built from basic activity blocks supplied with IPCS. The ACTIVITY models are transformed into executable programs, called the IPCS APPLICATIONS, which execute the modeled activities (as they were specified in the models). ACTIVITY models may have multiple references to PLANT models and are in close relationship to them. During design time, ACTIVITY models can be used to specify simulation studies using the actually available PLANT models. During plant operation, ACTIVITY models may represent a variety of operation support systems, such as real-time simulators to predict the behavior of processes, plant model verification procedures, or fault recovery procedures that use the IPCS diagnostic system. These activities are also closely linked to the PLANT models, since they generally use plant information. IPCS can also model the relationship between PLANT models and ACTIVITY models, therefore directly supporting their consistency during the lifetime of the plant.

Plant Models

The basic plant modeling concepts are divided into three categories:

- Process models
- Stream models
- Equipment models.

Process models represent interactions in the plant in terms of processes and various material energy and information streams. Processes perform some transformation on input streams, and produce output streams. Streams are characterized by their components and their attributes (e.g. pressure, temperature). Equipment models describe the physical structure of the plant, which implement the processes, i.e. the hardware.

Activity Models

Activity models are divided into two basic categories, SYSTEM ACTIVITIES and USER ACTIVITIES. SYSTEM ACTIVITIES exist in one unique copy in an IPCS activity configuration, which may comprise a large variety of activities. The models of SYSTEM ACTIVITIES are part of the IPCS model library, and are used to model their connections to USER ACTIVITIES. SYSTEM ACTIVITIES are automatically generated by the IPCS program synthesis system using the information in the PLANT MODELS. USER ACTIVITIES may exist in many copies in an activity configuration, and are defined by the users.

Currently supported SYSTEM ACTIVITIES in the IPCS are:

- *Diagnostic Activity* refers to the IPCS real-time diagnostic system, which receives and processes alarms and generates hypotheses for the underlying fault causes and future fault events in the plant.
- *System State Identifier Activity* follows the state transitions in the processes and process equipment and generates events for other activities that are influenced by the states.

Currently supported USER ACTIVITIES in the IPCS are:

- *Algorithmic Activities* are simple data processing activities which can execute a piece of code (their script) written in some high-level language (currently C).
- *Timer Activities* can be used to generate timed delays and timed event sequences for triggering other activities.
- *Finite State Machine (FSM) Activities* are for

building simple state machines. They can be used to define FSM controllers in terms of states and transitions which can be triggered or can generate triggers for other activities.

- *Simulation Activities* include algebraic and differential equation solvers that can compute the static and/or dynamic behavior of processes using their equation models.
- *External Interface Activities* model communication channels to the outside world of an IPCS applications. They are typically used to configure the plant data acquisition system, retrieve data from a process historian, or start a simulation (e.g. ASPEN or SPEEDUP) on a remote node.
- *Operator Interface Activities* model operator interactions. They describe what is shown to the operator and how the operator can interact with the activities.
- *Compound Activities* are for building hierarchies of activities: they can contain any of the above activities and other compound activities.

The IPCS modeling methodology supports a variety of model organization principles for managing complexity. Among the various organizational principles used are (1) multiple aspect to capture a particular kind of relationship or abstraction, (2) hierarchical composition to build deep hierarchies, and (3) connections in which the meaning of objects is determined by the objects they couple. Further reading on modeling and model organization principles in Multigraph-based systems can be found in Karsai et al. (1992), Padalkar et al. (1991), Karsai et al. (1991), Abbot et al. (1993), and Wilkes et al. (1993).

Implementation

A graphical model editor provides a visual programming environment for building models. The use of graphics is not exclusive: whenever reasonable text is used (e.g. equations). Models are stored on an OODBMS, which supports sophisticated data structures and network access. A system integrator tool facilitates the creation of executable programs from the models. The executables contain: (1) model interpreters to create run-time objects, (2) special run-time support modules, e.g. equation solvers, diagnostic reasoning engines, (3) user-defined subroutines compiled and linked, (4) the Multigraph Kernel to implement low-level scheduling, (5) external data interfaces for plant data and other external packages, e.g. ASPEN, and (6) a graphical operator interface. When an executable is started, the model interpreters create and configure the computations implementing the required functionalities. The models are read from the database. After interpretation, the database is not needed, since all required information is in the running code.

Two types of applications will be presented to illustrate the use of plant and activity models. The first is a sensor fault diagnosis application. The second demonstrates the retrieval of plant data, the simulation of a distillation column with the data, and the display of the information on an operator interface.

Example one diagnoses faulty sensors (level or flow) and predicts the actual readings from the faulty sensors. The process model consists of a set of coupled material balances. An example would be the material balances around the distillation column in Fig. 1. Three material balances can be derived:

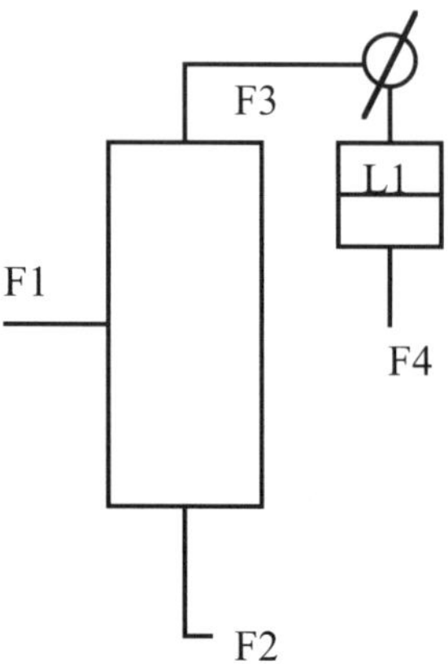

Figure 1. Material balances around a distillation column.

1. Balance B1: F1 = F2 + F3
2. Balance B2: F1 = F2 + ΔL1 + F4
3. Balance B3: F3 = ΔL1 + F4

The flow sensors F1, F2, F3, and F4 and level sensor L1 are modeled such that each has a fault state. Violation of balances B1, B2, and B3 are modeled as process failure modes (BV1, BV2, and BV3). Fault propagations are modeled from the *faulty* fault-state of each sensor to each of the balance violation failure modes, as long as that sensor participates in that balance. For example, the *faulty* failed state of sensor F1 propagates to the balance violation failure-modes BV1 and BV2. Each fault propagation is weighted with a probability of 100 and a [0, sampling interval] time interval.

Padalkar et al. (1991) developed a generic fault modeling and diagnosis methodology. The basic fault modeling and diagnostics methodology has been used to generate diagnostics for power generation plants (Karsai et al. 1992), chemical plants (Karsai et al. 1991), and aerospace vehicles (Carnes et al. 1991). Briefly, the fault modeling decomposes the plant functions to generate a process hierarchy , and a plant structure to generate an equipment hierarchy. A process in the process hierarchy has failure-modes that represent violation of normal behavior. A fault propagation graph can be specified which represents a causal link among process failure-modes. A fault propagation is weighted with a probability and a fault propagation time interval. Equipment in the hierarchy can have failed states. Fault propagations can also be specified between equipment failed-states and process failure-modes. Alarms, which are generated from sensor values using suitable algorithmic activities, can be associated with process failure modes. An ON alarm triggers the diagnostic reasoning engine. The reasoning engine uses current ON alarm information and structural, probabilistic, and temporal constraint enforcement upon fault models to generate a diagnosis. In this example instances of the same type of algorithmic activity are used to monitor each balance, and flag an alarm if a balance is violated. These alarms are associated with their respective balance violations failure-modes. Once a sensor is diagnosed to be faulty, a prediction of its true reading is made in terms of a range based on associated balance violation residuals and reading from other non-faulty sensors.

The current plant problem has 10 material balances and 35+ sensors. Each sensor is modeled as having failed-states *read-hi*, *read-lo*, *stuck-hi*, *stuck-lo*, and *stuck*. The results are communicated to an operator interface, and eventually will communicate with the plant's DCS. Calculations based on these sensor readings are modified to include a correction factor determined from predicted values. The application has detected 3 faulty sensors. A set of generic algorithmic activities that can be instantiated from a set of material balances and can configure themselves from a specification file containing relevant information about the material balances has been prepared.

Example two is a simulation activity. An ASPEN+ simulation activity (ASA) provides an interface to an ASPEN+ simulation. At modeling time one creates an ASA model, which embeds and specifies an ASPEN+ simulation. The simulation is executed when all the input conditions for the ASA activity are satisfied. From the user's point of view, it is similar to an algorithmic activity: it executes a subroutine (i.e. an ASPEN+ simulation runs) when the required input data are present. After execution, the results are propagated through the output ports of the activity to an operator interface. The ASA model configures an ASPEN+ simulation using an already existing ASPEN+ simulation file prepared with Model Manager, and then modifies that file based on the incoming process data. The ASPEN Summary File Toolkit is used to retrieve data for display. The current application models a DMT column in polyester a intermediates plant. Process data for the simulations are retrieved from a process historian running on a VAX. A server program running on the VAX support this data retrieval. IPCS is currently running on an HP 710 workstation. After the data are received, a request is made from another server running on a CRAY or VAX to execute the ASPEN simulation. After the simulation is completed, the data are retrieved and displayed on an operator interface. All communication is over TCP/IP connections. The operators can use the information from the simulation to determine the composition of the feed for which there is no on-line measurement, and several of the controlled and manipulated variables. This is a steady-state simulation. However, a dynamic simulation could also be configured in this same manner. Fig. 2 shows a schematic of the simulation activity. Both of these applications are beyond the development stage, but are not yet "commercialized" (commercialized meaning on-line 24 hours a day and being used by operators to make process decisions). They are, however, being used by operators and engineers to routinely evaluate process operations.

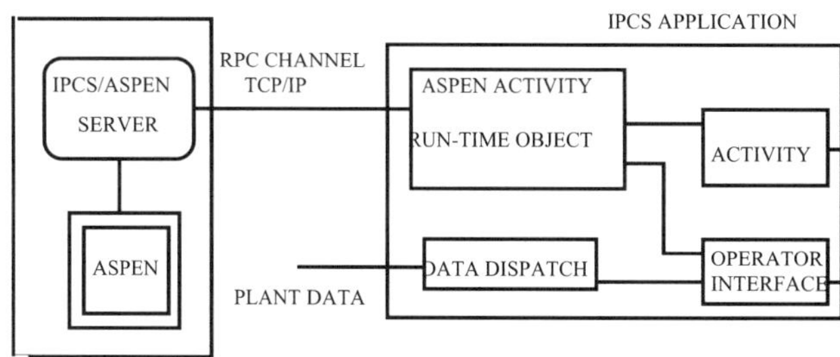

Figure 2. IPCS/ASPEN Interface: Run-time.

REFERENCES

Abbott, B. A. et al. (1993). Model-based approach for software synthesis. *IEEE Software*, May, 42-52.

Carnes, J., W. Davis, C. Biegel, G. Karsai (1991). Integrated Modeling for Planning, Simulation and Diagnosis. IEEE Conference on AI Simulation & Planning in High Autonomy Systems, April.

Karsai, G., J. Sztipanovitz, S. Padalkar, C. Biegel, K. Debelak, S. Droes, N. Miyasaka, K.Okuda, F. DeCaria, M. Lopez (1991). A model-based approach for plant-wide monitoring, control, and diagnosis. *Proc. AIChE Annual Meeting*, Los Angles.

Karsai, G. et al. (1992. Model based intelligent process control for cogenerator plants. *Journal of Parallel and Distributed Computing*, June, 90-103.

Padalkar, S.J. et al. (1991). Real-time Fault Diagnosis. *IEEE Expert*, June, 75-85.

Wilkes, D.M., et al. (1993). The multigraph and structural adaptivity. *IEEE Transactions on Signal Processing*. August, 2695-2717.

METHODOLOGY FOR DATA MODELING FOR THE PROCESS INDUSTRIES

James J. Fielding, Neil L. Book, Oliver C. Sitton, Michael R. Blaha,
Barbara L. Maia-Goldstein, John L. Hedrick and Rudolphe L. Motard
Consortium for Advanced Process and Control Engineering Technologies
Department of Chemical Engineering
University of Missouri — Rolla
Rolla, MO 65401

Abstract

Reliable information sharing has become of critical importance to all industries including the chemical process industries. In the modern industrial environment, information is shared between tasks, many of which are automated, both within and across organizations. Data modeling provides the tools to allow the development and implementation of information sharing in an efficient, effective, and reliable manner.

Data modeling is a set of techniques used to analyze information based on the structure and relationships of data in a given domain. The structures and relationships described can be used to provide a clear and unambiguous context for all data that must be shared. Implementation methods for the automation of data sharing are being created that are independent of the content of models but instead rely only on their forms. This allows implementation to be independent of the domain of the information being shared, thereby promoting implementation reusability. Data modeling provides reliability and and extensibility implementation reusability provides efficiency and effectiveness.

The development of a methodology for the sharing of data in the process industries domain will be described. This methodology includes the establishment of the scope and contents of the domain, the techniques and tools used to develop data models, and the use of implementations for the data models. The **P**rocess **D**ata e**X**change **I**nstitute's project provides a large scale example. The domain of the **PDXI** project is process engineering data to and from simulators, selected equipment design programs, and selected equipment specification sheets.

Keywords

Data modeling, Information modeling, Information sharing, **PDXI**, **STEP**.

Introduction

Complexity is the one overwhelming common thread in modern engineering problems whether they be from the traditional engineering fields or from software engineering. Chemical engineering is no exception. This explosion in complexity is caused by the profusion of more and better tools many of which are automated. Managing the complexities of the problems and the data generated in finding solutions requires new techniques and tools. Information or data modeling is a paradigm which holds many of the answers to such problems.

One of the fundamental tasks in managing the complexity of problems is to provide information sharing between the tools employed. An engineering problem may require several organizations, e.g. an operating company, an AE&C company, and a number of process equipment design and manufacturing companies, each running a number of different software tools to share information between tasks and organizations. Reliable information sharing is critical. Information modeling provides tools to allow the development and implementation of information sharing in an efficient, effective, and reliable manner.

Common approaches to information sharing

There are five approaches used to exchange information between tasks (or systems) (Doty, 1992). They are:

- manual re-entry of data
- standardization on a single system
- direct or point-to-point translators
- translators to a common neutral exchange format
- standardization on a common data format.

Manual re-entry of data is very common today, especially in company-to-company exchanges. It, however, is very expensive and cannot be done without introducing errors.

Standardization on a single system is not a realistic solution in the process industries. The evolution of systems today nearly guarantees that any system selected as a current standard will be a poor economic choice in the near future.

Point-to-point translators are a solution many organizations in the process industries currently use internally. The major draw back is the number of translators which must be created and more importantly maintained is for systems exchanging data. Any time a system is modified, all translators to and from that system must be maintained. Since software systems evolve rapidly, this puts an enormous maintenance burden on an organization.

Using translators to a common neutral exchange format reduces the number of translators required to only two translators per system, to and from the common format. When a single system is changed, only the two translators directly involved with it must be maintained. However, when the common neutral format is changed all translators must be maintained. Thus, extreme care must be taken in the design of the neutral format to minimize later changes.

Standardization on a common data format is a step beyond a common neutral file format. The structure of the data used by all processes is the same, so data exchange is not lossy. Data can be exchanged directly between processes or stored in a repository that is accessible by the processes. A common data format promotes reusability of code. Interfaces can be specified and code developed that all processes can use.

Data Modeling

Two approaches discussed above, a common neutral file format and a common data structure, can be said to be data driven. A data driven approach is one where an understanding of the nature and relationships of the data forms the initial analysis of the problem.

It is important that the analysis of the data be in the language of the domain of the problem rather than a computer or implementation view. Experts in the domain are the only people who can determine if the right data is being described and if the descriptions are correct.

The structures and relationships described by data models are used to provide a clear and unambiguous context for all data that must be shared. Often a graphical technique is used to establish a reviewable data model and then that model is translated to a textual data specification language in order to remove any ambiguities and to rigorously establish constraints.

After an information domain is defined, the basic structure of the data is captured in a conceptual model. The relationships and finer structures of the data are added to this model until the relationships between the data and critical attributes of the data are captured. These form an analysis model of the data. This basic structure and relationships of data form the foundation for all later efforts.

Development of design data models follows the analysis model. Details are added to the analysis model until a complete version of the data is modeled. Influences from the implementation(s) chosen begin to enter the models. At this point a graphical model still works well, but as the remaining implementation decisions are made a translation to a text based model is advisable.

Developing Implementations

Ideas about how the final product will be implemented are generally in place before a data modeling project is undertaken. These ideas should be filed away and forgotten until the analysis stage is completed. When implementations creep into analysis model, the true structure of the data is be obscured. When an analysis is skewed towards a particular implementation, it is difficult to evolve workable designs that are based on some alternative implementation.

STEP - ISO 10303

The manufacturing industries have taken the lead in the development of information sharing technologies. A need for a standardization effort for how discrete parts are described and specified has been recognized. There is an ongoing international effort for the standardization of product data through out its life-cycle. This effort is the International Standards Organizations (**ISO**) Product Data Representation and Exchange project, commonly known as **STEP** (ISO, 1992). **STEP** provides mechanisms for the description of product data and a neutral file format for the exchange of that data. These and other portions of **STEP** have been released as International Standards.

The chemical process industry has many facets which although not normally thought of as products have the same sorts of life-cycle data that are captured by **STEP**. The issues of data exchange may be even more critical to the chemical process industries than other industries due to provisions in some regulatory requirements (OSHA part 1910 & CAAA part 68) that such data be accessible by governmental agencies at all times.

Methodology for Data Modeling

In the creation of information models and implementations for the **PDXI** project, a methodology emerged that makes use of the technologies from **STEP**, the pecullarities of the process engineering domain, and the lessons learned from this project. This methodology can be viewed as a number of steps:

- domain identification,
- scope definition,
- object modeling and review
- EXPRESS modeling,
- API specification
- API implementations.

The original view was that such steps could be done in a sequential fashion, but experience quickly demonstrated that a some iteration is required.

Domain Identification

A common method of defining a domain for modeling is to use a knowledge domain. A knowledge domain is a domain defined by one field of knowledge at some level of specialization. The knowledge domain for the **PDXI** project is process engineering. The advantage of using a knowledge domain is it makes it easier to identify domain experts. Most experts in the domain will use the same terminologies. **STEP** is organized on knowledge domains providing context for application protocols.

The boundary of a domain is all data which is supplied from or to activities outside the domain. Often such data is transformed in either format or terminology as it passes from one domain to another. A clear description of the data on both sides of the boundary can alleviate this data impedance.

A technique that is useful in defining a knowledge domain and its boundaries is activity modeling. Activity modeling is a way of representing the activities that occur in an enterprise and the flows of information and entities between these activities. Activity models are created using largely graphical syntaxes with few constructs and rules (Book et al., 1993). Activity models are effective for defining the scope of the body of data associated with a knowledge domain. The **PDXI** activity model is registered with **ISO** as **ISO** TC184/SC4/WG3N272. Figure 1 shows as portion of this document.

Scope Definition

Defining the actual scope of the data to be included in the PDXI project proved to be the most difficult task of the project. An important lesson learned is that scope definition must be made early in the project. While it is not possible to decide on every bit of data to be included at the outset, it is possible to decide the boundaries that the scope will fall within. These boundaries must be strictly adhered to or the scope will continue to creep outward and the project will lag.

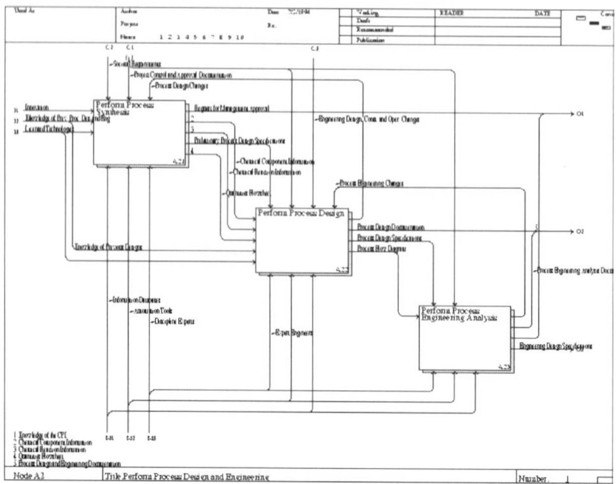

Figure 1. Activity model — Perform process design and engineering.

We recommend initially developing a planning level model for any project. This is an abstract model that defines the relationships between the major areas of the project. Fundamental modeling decisions will be made at this level which affect the entire scope.

Object Modeling and Review

The graphical object modeling language used in the PDXI project was the Object Modeling Technique (OMT) (Rumbaugh et al., 1991). OMT is a graphical method allows modelers to quickly modify data models. A data model will undergo a great deal of revision. A modeler should not become too attached to an initial model.

Our approach to data modeling was to allow one modeler to make an initial draft of a model. These models tended to be fairly abstract. Then, the other modelers on the project team would make a brief written review of that model. It is important to get a model down on paper at an early stage. The initial team reviews were incorporated into the model. The team would then meet and review the model in more detail. Sometimes consensus could not be reached in one review. When this occurred, specific examples were be constructed against both viewpoints; examination of specific examples at a second review meeting usually resolved any conflicts. Figure 2 is two OMT models from the planning level showing the relationship between equipment and simulation.

The top part of the figure shows the original configuration; the second part is the final form which separates equipment specifications from simulation. This second viewpoint was adapted in the team meetings on the argument that the unit operations hierarchy in current simulators is very dissimilar to any organization of the modeled equipment. This simple organizational decision

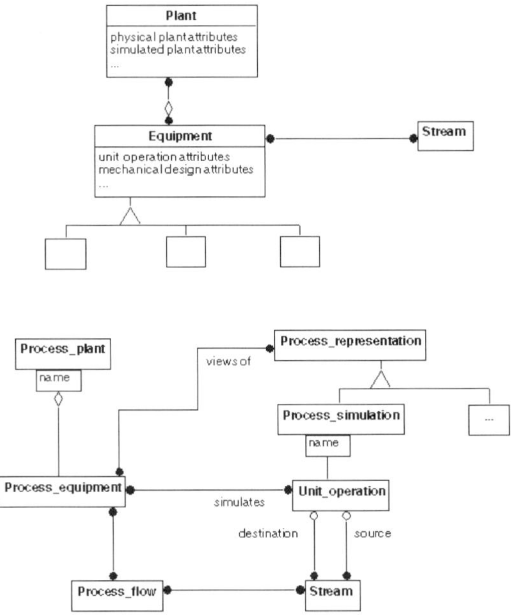

Figure 2. OMT models of the relationships between equipment and simulation.

took place over three team review meetings; as simple as this change is, the impact over the structure of the rest of the data is enormous. It is an indication of the importance of the basic structural decisions..

Once the structure was set, details were filled in and the model distributed to the PDXI membership for review. This level of review is necessary to make sure the data models reflect the domain. Good review by experts in the domain is critical to the success of a data model. The first wide review is used to make sure the structure and scope of the models is correct. If the structure is not correct, whether all the details are in and correct does make any difference. Once the structure is set, a detailed level of review is required. This level of review is done on small pieces of the model.

EXPRESS modeling

EXPRESS is the data definition language used in STEP. It is a text based language with a number of object-oriented concepts. As such the mappings from OMT to EXPRESS are mostly mechanical. However, constraints, especially usage and referential, can be rigidly specified and enforced with EXPRESS. Such constraints are appropriate to the design model. A detailed explanation of the OMT to EXPRESS mappings can be found in Fielding et al. (1993).

Application Programming Interface Specification

An interface between layers of software architecture is often termed an Application Program Interface (API). An API specification can consist of the functional description of the interface and a specification of any calls which can be made to the API. It may also contain one or more specific language bindings (subroutine library calls) which provide the functionality and comply with the specifications. Having a compliant API provides two large advantages: 1) the applications are isolated from the details of the data exchange and 2) the data exchange software is reusable.

Implementations

Having the PDXI project layered on top of STEP technologies provides a great deal of leverage in implementation. STEP is a large effort and has much more resources committed to developing implementations than any one industry can easily match. This allows individual industries to spend their efforts on modeling their own data.

Any modeling effort is evolutionary. Actual usage of the models in an implementation will quickly point out both the strengths and weakness of the underlying data models. We anticipate that all the models in the PDXI project will undergo some revision after an initial period of use.

Conclusions

Data modeling provides the tools to accomplish reliable and efficient information sharing for the process industries. The PDXI project has shown that this approach is workable. While this project is just a start, the tools, methods, and lessons presented here will be valuable for ourselves and all others who will undertake the same task.

References

Book, N.L., O. Sitton, R. Motard, B. Goldstein, J. Hedrick, and J. Fielding (1993). PDXI Activity Modeling With the IDEF0 Methodology presented at the American Institute of Chemical Engineers national Meeting, Houston, Texas.

Doty, R., (1992). *An Introduction to STEP*. Digital Equipment Company, Chelmsford, MA.

Fielding, J., N. Book, and J. Hedrick, (1993). From OMT Data Model to Neutral File Format for Process Data Exchange. Presented at the American Institute of Chemical Engineers national Meeting, Houston, Texas.

ISO, (1992). ISO 10303 — 1 Industrial Automation Systems -- Product Data Representation and Exchange -- Part 1: Overview and Fundamental Principals, Draft International Standard. ISO TC184/SC4.

Rumbaugh, J., M. Blaha, W. Premelani, F. Eddy, and W. Lorenson, (1991). *Object Modeling and Design*. Prentice Hall, Englewood Cliffs, NJ.

ROBUST PHASE STABILITY ANALYSIS USING INTERVAL METHODS

Mark A. Stadtherr and Carol A. Schnepper
Department of Chemical Engineering
University of Illinois
Urbana, IL 61801

Joan F. Brennecke
Department of Chemical Engineering
University of Notre Dame
Notre Dame, IN 46556

Abstract

Conventional equation solving and optimization techniques for solving the phase stability problem may fail to converge or may converge to an incorrect result. A technique for solving the problem with mathematical certainty is needed. One approach to providing such assurance can be found in the use of interval methods. An interval Newton/generalized bisection technique is applied here to solve the phase stability problem. Results for two models of liquid-phase systems, using several different feed compositions, indicate that the technique used is reliable and very efficient.

Keywords

Phase stability, Interval computations, Nonlinear equation solving, Interval Newton method, Tangent plane criterion, NRTL equation.

Introduction

The determination of phase stability is a recurrent problem in the computation of phase equilibria, and thus is especially important in the analysis and design of separation operations such as distillation and extraction. The problem is basically to determine whether a phase of given composition, temperature, and pressure will split into multiple phases. This problem is frequently formulated in terms of the tangent plane condition (Baker et al., 1982). Minima in the tangent plane distance are sought, usually by solving a system of nonlinear equations for the stationary points (Michelsen, 1982). If any of these yield a negative tangent plane distance, indicating that the tangent plane intersects (or lies above) the Gibbs energy of mixing surface, the phase is unstable. The difficulty lies in that, in general, given any arbitrary equation of state or activity coefficient model, most computational methods cannot find with complete certainty all the stationary points, and thus no guarantee of stability can be provided. What is needed are robust techniques that can find *all* solutions to a system of nonlinear equations, and do so *with certainty*, or techniques for finding the *global* optimum of a function, and do so again *with certainty*. Recent advances in the field of interval mathematics make possible just such techniques.

Interval mathematics involves computation with intervals as opposed to real numbers. Interval Newton methods, when combined with a generalized bisection approach, provide the power to find with confidence all solutions of a system of nonlinear equations (Neumaier, 1990; Kearfott and Novoa, 1990), and to find with total reliability the global minimum of a nonlinear objective function (Hansen, 1992), provided only that upper and lower bounds are available for all variables. Efficient techniques for implementing interval Newton/generalized bisection are a relatively recent development, and thus such methods have not yet been widely applied. Schnepper and Stadtherr (1990) suggested the use of this method for solving chemical process modeling problems, and recently described an implementation (Schnepper and Stadtherr, 1994). The technique proved very efficient on a number of small to moderate size problems, including a vapor-liquid equilibrium problem with multiple roots, and was made even more efficient by taking advantage of the fact that the technique is amenable to parallel computing.

For the phase stability problem various approaches have been proposed recently. For example, Sun and Seider (1992,1993) apply a homotopy-continuation method, which will often find all stationary points, but may be initialization dependent and provides no theoretical guarantees that all solutions have been found. Also, McDonald and Floudas (1993,1994) show that for certain activity coefficient models, the problem can be reformulated to make it amenable to solution by global optimization techniques, which do guarantee the correct answer. However, in general there appears to be a need for an efficient general-purpose method that can perform phase stability calculations with complete reliability for any arbitrary equation of state or activity coefficient model.

In this paper, we demonstrate the use of interval methods for phase stability computations. These methods can be applied in connection with any equation of state or activity coefficient model, and when properly implemented are completely reliable.

Phase Stability Analysis

The determination of phase stability is often done using tangent plane analysis (Baker et al., 1982; Michelsen, 1982). A phase at specified temperature, pressure, and feed composition $\mathbf{z}$ is unstable if the Gibbs energy of mixing versus composition surface $m(\mathbf{x}) = \Delta g^M(\mathbf{x})/RT$ ever falls below a plane tangent to the surface at $\mathbf{z}$. That is, if the tangent plane distance

$$D(\mathrm{x}) = m(\mathrm{x}) - m_0 - \sum_{i=1}^{n} \left(\frac{\partial m}{\partial x_i}\right)_0 (x_i - z_i) \qquad (1)$$

is negative for any composition $\mathbf{x}$, the phase is unstable. The subscript zero indicates evaluation at $\mathbf{x} = \mathbf{z}$, and n is the number of components. A common approach for determining if D is ever negative is to minimize D subject to the mole fractions summing to one. It is readily shown that the stationary points in this optimization problem can be found by solving the system of nonlinear equations:

$$\begin{gathered} \left(\frac{\partial m}{\partial x_i}\right) - \left(\frac{\partial m}{\partial x_i}\right)_0 - \left(\frac{\partial m}{\partial x_n}\right) + \left(\frac{\partial m}{\partial x_n}\right)_0 = 0, i = 1{:}n-1 \\ 1 - \sum_{i=1}^{n} x_i = 0 \end{gathered} \qquad (2)$$

The $n \times n$ system (2) has a trivial root at $\mathbf{x} = \mathbf{z}$, and frequently has multiple nontrivial roots as well. Thus conventional equation solving techniques such as Newton's method may fail by converging to the trivial root or give an incorrect answer to the phase stability problem by converging to a stationary point that is not the global minimum of D. This is aptly demonstrated by the experiments of Green et al. (1993), who show that the pattern of convergence from different initial guesses demonstrates a complex fractal-like response for even the simple liquid phase model used as an example below.

We demonstrate here the use of an interval Newton/generalized bisection method for solving the system (2). The method requires no initial guess, and will find *with certainty* all the stationary points of D.

Interval Computations

A real *interval*, X, is defined by $X = [a,b] = \{x \in \Re \mid a \le x \le b\}$, where $a,b \in \Re$ and $a \le b$. A real interval vector $\mathbf{X} = (X_i) = (X_1, X_2, ...X_n)^T$ has n real interval components X_i, and since it can be interpreted geometrically as an n-dimensional rectangle, is frequently referred to as a *box*.. Several good introductions to computation with intervals are available, including recent monographs by Neumaier (1990) and Hansen (1992).

Of particular interest here are interval Newton methods. Consider the solution of the system of real nonlinear equations $\mathbf{f(x)} = \mathbf{0}$, where it is desired to find all solutions in an specified initial box $\mathbf{X}^{(0)}$. The basic idea in interval Newton methods is to solve the linear interval equation system

$$\mathrm{F}'(\mathbf{X}^{(k)})(\mathbf{N}^{(k)} - \mathbf{x}^{(k)}) = -\mathbf{f}(\mathbf{x}^{(k)}) \qquad (3)$$

for the interval $\mathbf{N}^{(k)}$, where k is an iteration counter, $\mathrm{F}'(\mathbf{X}^{(k)})$ is a suitable interval extension of the real Jacobian $\mathrm{J}(\mathbf{x})$ of $\mathbf{f(x)}$ over the current box $\mathbf{X}^{(k)}$, and $\mathbf{x}^{(k)}$ is a point in the interior of $\mathbf{X}^{(k)}$, usually taken to be the midpoint. It can be shown (Moore, 1966) that any root $\mathbf{x}^* \in \mathbf{X}^{(k)}$ of $\mathbf{f(x)}$ is also contained in $\mathbf{N}^{(k)}$. This suggests the iteration

$$\mathbf{X}^{(k+1)} = \mathbf{X}^{(k)} \cap \mathbf{N}^{(k)}. \qquad (4)$$

There are various interval Newton methods, which differ in how they determine $\mathbf{N}^{(k)}$ from equation (3) and thus in the tightness with which $\mathbf{N}^{(k)}$ encloses the solution set of (3).

While the iteration scheme given by equations (3) and (4) can be used to tightly enclose a solution (e.g., Shacham and Kehat, 1973), what is of most significance here is the power of (3) to provide an existence and uniqueness test. For several techniques for finding $\mathbf{N}^{(k)}$ from (3), it can be proven (Neumaier, 1990) that if $\mathbf{N}^{(k)} \subset \mathbf{X}^{(k)}$, then there is a *unique* zero of $\mathbf{f}(\mathbf{x})$ in $\mathbf{X}^{(k)}$, and furthermore that Newton's method with real arithmetic *will converge* to that solution starting from *any* point in $\mathbf{X}^{(k)}$. Thus, if $\mathbf{N}^{(k)}$ is determined using one of these techniques, the computation can be used as a root inclusion test for any interval $\mathbf{X}^{(k)}$. If $\mathbf{X}^{(k)} \cap \mathbf{N}^{(k)} = \varnothing$, then there is no root in $\mathbf{X}^{(k)}$; if $\mathbf{N}^{(k)} \subset \mathbf{X}^{(k)}$, then there is exactly one root and Newton's method with real arithmetic will find it; otherwise, no conclusion can be drawn. In the last case, one could then repeat the root inclusion test on the next interval Newton iterate $\mathbf{X}^{(k+1)}$, assuming it is sufficiently smaller than $\mathbf{X}^{(k)}$, or one could bisect $\mathbf{X}^{(k+1)}$ and repeat the root inclusion test on the resulting intervals. This is the basic idea of interval Newton/generalized bisection methods. If $\mathbf{f}(\mathbf{x}) = \mathbf{0}$ has a finite number of real solutions in the specified initial box, a properly implemented interval Newton/generalized bisection method can find *with mathematical certainty* any and all such solutions to a specified tolerance, or can determine *with mathematical certainty* that there are no solutions in the given box (Kearfott, 1987,1990).

The technique used here for computing $\mathbf{N}^{(k)}$ is the preconditioned Gauss-Seidel-like technique developed by Hansen and Sengupta (1981) and Hansen and Greenburg (1983). Neumaier (1985,1990) has proven the existence and uniqueness test for this method of determining $\mathbf{N}^{(k)}$.

Since all variables are mole fractions, the initial box $\mathbf{X}^{(0)} = [\mathbf{0},\mathbf{1}]$ is suitable. In practice the initial lower bound is set to an arbitrarily small positive number ε (10^{-10} was used) to avoid taking the logarithm of zero in subsequent calculations. Since it is known (Michelson, 1982) that the roots sought lie in the interior of $[\mathbf{0},\mathbf{1}]$, this can be done without the loss of reliability providing a sufficiently small value of ε is used.

Our implementation of the interval Newton/generalized bisection method for the phase stability problem is based on appropriately modified routines from the packages INTBIS (Kearfott and Novoa, 1990) and INTLIB (Kearfott et al., 1993). To demonstrate the potential of the technique we use two models of liquid-phase systems, one a simple model of excess Gibbs energy, the other the NRTL model.

Simple Model

This model is used by Green et al. (1993) to demonstrate the difficulties in using Newton's method for phase stability problems. It is a three component system with the excess Gibbs energy given by $\Delta g^E/RT = 3x_1x_2$. This may be regarded as a special case of the two-suffix Margules equation. We solved the system (2) for this model for the three different feed compositions used by Green et al. The roots found and the corresponding value of D are shown in Table 1, and are the same as those reported by Green et al. The results indicate phase stability for only the third feed composition. Performance data, including the number of root inclusion tests required, the depth reached in the binary tree generated in the bisection, and the computation time on an HP 9000/735 workstation, is given in Table 2.

NRTL Model

This example is used by McDonald and Floudas (1993,1994) to demonstrate a global optimization technique for solving the phase stability problem when the NRTL model for excess Gibbs energy is used. There are two components: n-butyl acetate (1) and water (2). The parameters in the NRTL model for this system are G_{12} = 0.30794, G_{21} = 0.15904, $\tau_{12} = 3.00498$, and τ_{21} = 4.69071. The roots of system (2) found for five feed compositions are shown in Table 3, and corresponding performance data is shown in Table 4. There are as many as five roots for some feed compositions and phase instability is indicated for all five cases.

Table 1. Simple Model: Roots Found.

Feed: (z_1,z_2,z_3)	Roots: (x_1,x_2,x_3)	D
(0.45,0.45,0.10)	(0.8509,0.0857,0.0634)	-0.0671
	(0.0857,0.8509,0.0634)	-0.0671
	(0.45,0.45,0.10)	0.0
(0.60,0.18,0.22)	(0.1185,0.6842,0.1973)	-0.0284
	(0.4675,0.2923,0.2401)	0.00152
	(0.60,0.18,0.22)	0.0
(0.90,0.06,0.04)	(0.3447,0.5820,0.0733)	0.1664
	(0.1193,0.8277,0.0530)	0.1479
	(0.90,0.06,0.04)	0.0

Table 2. Simple Model: Performance.

Feed: (z_1,z_2,z_3)	Root Inclusion Tests	Level Reached in Binary Tree	CPU Time: HP 9000/735 (sec.)
(0.45,0.45,0.10)	111	14	0.04
(0.60,0.18,0.22)	119	15	0.03
(0.90,0.06,0.04)	117	14	0.03

The CPU times indicate performance that is significantly faster than a GAMS implementation of the model-specific procedure of McDonald and Floudas (1993), and that is comparable to a more recent C implementation of their technique (McDonald and Floudas, 1994). For example, for the last feed composition (z_1 = 0.65) McDonald and Floudas (1993,1994) report CPU times on a HP 9000/730 of 10.74 s for the GAMS implementation of their technique and 0.11 s for the C implementation of the method, versus 0.06 s on an HP 9000/735 for the interval Newton/generalized bisection method used here. (Note that, based on the well-known LINPACK-100 benchmark, the Model 735 used here is about 70% faster than the Model 730 used by McDonald and Floudas.) In addition to being very efficient, the technique used here has the advantage of being model-independent; it can be easily used to solve the phase stability problem for other models of excess Gibbs energy, and can also be used in connection with equation of state models.

Table 3. NRTL Model: Roots Found

Feed: (z_1,z_2)	Roots: (x_1,x_2)	D
(0.50,0.50)	(0.1602,0.8398)	0.0278
	(0.00421,0.99579)	-0.0325
	(0.50,0.50)	0.0
(0.10,0.90)	(0.96346,0.03654)	-0.2142
	(0.00599,0.99401)	-0.0291
	(0.10,0.90)	0.0
(0.20,0.80)	(0.3922,0.6078)	-0.00607
	(0.00379,0.99621)	-0.0743
	(0.20,0.80)	0.0
(0.15,0.85)	(0.5423,0.4577)	-0.0388
	(0.8549,0.1451)	-0.0260
	(0.9216,0.0784)	-0.0267
	(0.00438,0.99562)	-0.0557
	(0.15,0.85)	0.0
(0.65,0.35)	(0.7593,0.2407)	0.00063
	(0.9413,0.0587)	-0.00671
	(0.1346,0.8654)	0.0632
	(0.00471,0.99529)	0.0150
	(0.65,0.35)	0.0

Table 4. NRTL Model: Performance.

Feed: (z_1,z_2)	Root Inclusion Tests	Level Reached in Binary Tree	CPU Time: HP 9000/735 (sec.)
(0.50,0.50)	206	13	0.05
(0.10,0.90)	116	11	0.02
(0.20,0.80)	204	14	0.04
(0.15,0.85)	245	15	0.06
(0.65,0.35)	268	17	0.06

Conclusions

The interval computation method demonstrated here provides an efficient and completely reliable means for solving the phase stability problem. The method is model-independent, straightforward to use, and requires no problem reformulation. Though we have treated the problem here as one of solving a system of nonlinear equations for all its roots, the method can readily be extended to try to take advantage of the fact that this can be considered a global optimization problem, and that to determine phase instability all that really needs to be found is an interval value of D with a negative upper bound. Since this requires the additional work of doing an interval function evaluation of D for each box tested, it is not clear that this will always result in any overall savings, though for problems in which the phase is unstable, savings are likely and potentially substantial.

Acknowledgement

This work was funded in part by the National Science Foundation under grant DDM-9024946.

References

Baker, L.E., A.C. Pierce, and K.D. Luks (1982). Gibbs energy analysis of phase equilibria. *Soc. Petrol. Engrs. J.*, **22**, 731-742.

Green, K.A., S. Zhang, and K.D. Luks (1993). The fractal response of robust solution techniques to the stationary point problem. *Fluid Phase Equilib.*, **84**, 49-78.

Hansen, E.R. (1992). *Global Optimization Using Interval Analysis*. M. Dekkar, New York.

Hansen, E.R. and R.I. Greenburg (1983). An interval Newton method. *Appl. Math. Comput.*, **12**, 89-98.

Hansen, E.R. and S. Sengupta (1981). Bounding solutions of systems of equations using interval analysis. *BIT*, **21**, 203-211.

Kearfott, R.B. (1990). Interval arithmetic techniques in the computational solution of nonlinear systems of equations: Introduction, examples, and comparisons. *Lectures in Applied Mathematics*, **26**, 337-357.

Kearfott, R.B. (1987). Abstract generalized bisection and a cost bound. *Math. Comput.*, **49**, 187-202.

Kearfott, R.B. and M. Novoa III (1990). INTBIS, A portable interval Newton/bisection package. *ACM Trans. Math. Softw.*, **16**, 152-157.

Kearfott, R.B., M. Dawande, K. Du, and Ch. Hu (1993). *INTLIB: A portable FORTRAN 77 interval standard function library*. Submitted for publication.

McDonald, C. and C.A. Floudas (1993). Global optimization for the phase stability problem. Presented at AIChE Annual Meeting, St. Louis, November.

McDonald, C. and C.A. Floudas (1994). Global optimization for the phase stability problem. *AIChE J.*, to appear.

Michelsen, M.L. (1982). The isothermal flash problem. Part I: stability. *Fluid Phase Equilib.*, **9**, 1-19.

Moore, R.E. (1966). *Interval Analysis*. Prentice-Hall, Engelwood Cliffs, NJ.

Neumaier, A. (1990). *Interval Methods for Systems of Equations*. Cambridge University Press, Cambridge.

Neumaier, A. (1985). Interval iteration for zeros of systems of equations. *BIT*, **25**, 256-273.

Shacham, M. and E. Kehat (1973). Converging interval methods for the iterative solution of a non-linear equation. *Chem. Eng. Sci.*, **28**, 2187-2193.

Schnepper, C.A. and M.A. Stadtherr (1990). On using parallel processing techniques in chemical process design. Presented at AIChE Annual Meeting, Chicago, November.

Schnepper, C.A. and M.A. Stadtherr (1994). Robust process simulation using interval methods. Submitted for publication.

Sun, A.C. and W.D. Seider (1993). Global minimization of the Gibbs free energy. Presented at AIChE Annual Meeting, St. Louis, November.

Sun, A.C. and W.D. Seider (1992). Homotopy-continuation algorithm for finding globally stable phase equilibria. Presented at AIChE Annual Meeting, Miami Beach, November.

EXPERIMENTS WITH PATH-TRACKING ALGORITHMS FOR HOMOTOPY CONTINUATION METHODS

D. A. Harney and N. L. Book
University of Missouri — Rolla
Rolla, MO 65401

Abstract

Homotopy continuation methods are now widely accepted as both theoretically valid and implementable tools for obtaining the roots of systems of nonlinear equations. An important part of their implementation is the method used to trace the homotopy path. A quantitative analysis of the most commonly used path-tracking algorithms is presented, along with examples of their utilization in large-scale process flowsheeting design and optimization problems of degree up to 917.

It has been proved (den Heijer and Rheinboldt, 1981) that the local convergence radius of the continuation path is impossible to calculate a priori, but most of the better algorithms make estimates of its bounds using available properties such as previous curvature, contraction ratios and/or step-sizes. It is shown here that careful monitoring of the determinant of the augmented jacobian is also necessary to prevent any algorithm from segment jumping in the continuation space, without significantly reducing its efficiency.

Keywords

Homotopy continuation, Predictor-corrector, Determinant monitoring.

Introduction

The problem of finding solutions to the system of equations

$$\mathbf{f}: D \subset R^n \rightarrow R^n, \qquad \mathbf{f}(\mathbf{x}) = \mathbf{0} \tag{1}$$

has been made considerably more robust by the introduction of homotopy continuation methods. A homotopy, or deformation $\mathbf{h}: R^n \times R \rightarrow R^n$ can be defined such that,

$$\mathbf{h}(\mathbf{x},0) = \mathbf{g}(\mathbf{x}), \qquad \mathbf{h}(\mathbf{x},1) = \mathbf{f}(\mathbf{x}) \tag{2}$$

where $\mathbf{g}$: $R^n \rightarrow R^n$ is a smooth map with known (or easily obtainable) zeros and $\mathbf{h}$ is also smooth. A convex linear homotopy can then be defined as

$$\mathbf{h}(\mathbf{x},t) = t\mathbf{f}(\mathbf{x}) + (1-t)\mathbf{g}(\mathbf{x}) = \mathbf{0} \tag{3}$$

By following the solution path of $\mathbf{h}(\mathbf{x},t) = \mathbf{0}$, from $t = 0$, to $t = 1$, the problem is continuously deformed from the "easy" problem, $\mathbf{g}(\mathbf{x}) = \mathbf{0}$, to the original system, $\mathbf{f}(\mathbf{x}) = \mathbf{0}$. The theorem of Leray and Schauder (1934) states that there is indeed a continuation path connecting the starting point $\mathbf{x}^o$ with a solution vector $\mathbf{x}^*$, if

a. $\mathbf{h}(\mathbf{x},t)$ has no solution on the boundary of D and with $t \in [0,1]$

b. $\mathbf{g}(\mathbf{x}) = \mathbf{0}$ has a unique solution $\mathbf{x}^o$.

Also, the Parameterized Sard's theorem states that if $\mathbf{0}$ is a regular value of $\mathbf{h}$ with the dependence on $\mathbf{x}^o$ taken into account, then $\mathbf{0}$ is a regular value of $\mathbf{h}(\mathbf{x},t)$, for all $\mathbf{x}^o$ except for a set $\mathbf{x}^o$ of measure zero. This theorem essentially prohibits bifurcation in the real domain while traversing the continuation path. However, bifurcation points can be encountered when complex domain calculations are initialized by a starting point lying wholly in the real domain. See Allgower and Georg (1990, 1993) for a description of the implementation of these theorems.

The more commonly used homotopies are as follows (Seader, 1989),

$$\mathbf{h}(\mathbf{x},t) = t\mathbf{f}(\mathbf{x}) + (1-t)\{\mathbf{f}(\mathbf{x}) - \mathbf{f}(\mathbf{x}^o)\} = \mathbf{0} \quad (4)$$

$$\mathbf{h}(\mathbf{x},t) = t\mathbf{f}(\mathbf{x}) + (1-t)(\mathbf{x} - \mathbf{x}^o) = \mathbf{0} \quad (5)$$

$$\mathbf{h}(\mathbf{x},t) = t\mathbf{f}(\mathbf{x}) + (1-t)\frac{\partial \mathbf{f}}{\partial \mathbf{x}}(\mathbf{x}^o)(\mathbf{x} - \mathbf{x}^o) = \mathbf{0} \quad (6)$$

where equations (7), (8) and (9) are referred to as the Newton, fixed point and scale-invariant affine homotopies respectively. It should be noted that the Newton homotopy does not necessarily satisfy condition (b) of the Leray-Schauder theorem, and the fixed point and affine homotopies easily violate condition (a).

Path Tracking

Most path tracking algorithms are of a predictor-corrector type, wherein the next point on the continuation path is first predicted along the tangent line,

$$\mathbf{w}_o^{k+1} = \mathbf{w}^k + \frac{d\mathbf{w}}{ds}(s)\Delta s_{k+1} \quad (7)$$

where $\mathbf{w}^k(s)$ is the (n+1) vector, $(\mathbf{x}^k, t^k)^T$ and s^k is the arclength. If the predicted point is sufficiently close to the continuation path, a locally convergent corrector step is available due to the fact that $\mathbf{w}$ satisfies $\mathbf{h}(\mathbf{w}) = \mathbf{0}$. This corrector step can be made to lie orthogonal to the tangent, i.e.,

$$\Delta \mathbf{w}_i^{k+1} \bullet \frac{d\mathbf{w}}{ds}(s^k) = 0 \quad (8)$$

or, by utilizing the Moore-Penrose inverse, one can find $\mathbf{w}^{k+1}$ such that it solves the optimization problem,

$$\left\| \mathbf{w}^{k+1} - \mathbf{w}_o^{k+1} \right\| = \min_{\mathbf{h}(\mathbf{w}) = \mathbf{0}} \left\| \mathbf{w} - \mathbf{w}_o^{k+1} \right\| \quad (9)$$

(Allgower and Georg, 1990).

The problem remains of choosing a step-length in equation (7) that most efficiently uses the contractive properties of the homotopy path without causing either segment jumping onto another (possibly nearby) path, or divergence of the corrector operator. Schwetlick (1984) has suggested a step-length near the maximal one accepted by the Kantorovich theorem, when uniform accuracy of all continuation points is required.

den Heijer and Rheinboldt (1981) have proposed three algorithms that estimate the safe local convergence distance, ρ^{k+1}, between the predicted point and the desired point on the homotopy path, based on an estimate of the convergence quality of the corrector. For the first two methods, they suggest models of the error behavior of the corrector derived from an analysis of the Newton-Kantorovich theorem. The third method is a slightly more empirical approach, applicable with any corrector type, which uses the local curvature and the ratio of corrector to step-size to estimate ρ^{k+1}, while requiring that certain parameters lie between calculated maximal and minimal values.

Corvalan and Saita (1991), using a somewhat similar analysis, have attempted to keep the number of corrector steps constant. They claim that in order to achieve convergence after the specified number of steps N, the error of the predictor must lie between an upper and lower bound calculated by,

$$\left\| \mathbf{w}^k - \mathbf{w}_o^k \right\| = [\varepsilon^{1/2(N-\theta)} / \lambda_k^{(1-1/2(N-\theta))}] \quad (10)$$

for some $\theta \in (0,1]$, where ε is the specified convergence tolerance of the corrector and λ is a function of the successive errors after each of the Newton corrector steps.

Allgower (1981) used upper and lower bounds α^1 and α^2, of the contraction ratios, to indicate both possible divergence, resulting in a predictor step with a reduced step-size, and when the rate of convergence is rapid enough that the subsequent step-size can be increased.

Allgower and Georg (1993) recently suggested step-length control using asymptotic expansion estimates for the step-size, which turns out to yield a similar control strategy as that derived by den Heijer and Rheinboldt (1981).

Determinant Monitoring

Any quantity $q(s)$, which is a continuous function of arclength and which equals zero when the augmented jacobian is rank deficient can be used to detect bifurcation points, or indicate the presence of near bifurcation points, (Kearfott, 1983). Both the determinant of the augmented jacobian and the reciprocal of its condition number have this property, whereas the determinant has the additional property that it will change sign when a simple bifurcation point is passed, or a higher order one where an odd number of eigenvalues of the augmented jacobian have changed sign.

It has been observed that it is necessary to monitor one of these continuation variables, in order to prevent an algorithm from segment jumping onto another solution path, regardless of the path-tracking algorithm used (Choi and Book, 1991, 1994) This monitoring is achieved by rejecting any continuation point that causes a sufficiently large change of $q(s)$, or one where the absolute value of $q(s)$ is very small, and then reducing the step-size. It is a relatively simple task to implement this monitoring phase in the algorithms, but care needs to be taken when implementing the monitoring phase into some of the algorithms. For example, the algorithm of Corvalan and Saita (1991), requires that the parameters ε and θ are temporarily updated when the monitoring phase causes a step-size reduction.

Numerical Results

Three of the above algorithms (Allgower (1981), method III of den Heijer and Rheinboldt (1981) and Corvalan and Saita (1991), henceforth to be referred to as algorithms (i), (ii) and (iii), respectively), were coded into DZINE[1] and a number of test problems were then solved using the homotopy equation solver HOMES (Choi, 1990), to solve the nonlinear partitions. The determinant of the augmented jacobian was the continuation variable monitored in all cases, as the calculation of the κ^1 condition number is equivalent to that of an extra LU decomposition at each step which makes it prohibitively expensive in the large scale problems, and the accuracy of the κ^1 condition number estimate (Cline et al., 1979) was not high enough for it to be used as a monitoring variable.

Problem 1: The Williams-Otto Process

The material balance equations for the Williams-Otto plant (Rijckaert and Martens, 1974) were generated by the DZINE flowsheeting system. This problem consists of five units, nine streams and six species, resulting in a non-linear partition of dimension 85. All three algorithms, using the Newton homotopy reached the desired solution from the given starting point. Table 1 lists the total number of LU decompositions required and the Equivalent Operations Count (EOC) (Armstrong et al., 1988), with what were found to be the optimal parameters for this problem. Algorithm (iii) performed best here, not only in terms of numerical efficiency, but also in the fact that reasonable alteration of its parameters does not cause excessively large intermediate step-sizes, as was the case with Algorithms (i) and (ii). These large step-sizes occasionally predicted to a point which lies outside the domain of definition of the vector **x**.

Problem 2: Distillation Tower

A multiple inputs, multiple outputs distillation tower (Lien, 1993), was modeled by the design flowsheeting system using the MESH equations, resulting in a nonlinear partition of dimension 917. Again, all three algorithms successfully tracked the Newton homotopy continuation path to the solution vector at t = 1, with Algorithm (iii) showing the highest efficiency (see Table 2 for results of path-tracking). This was due in part to the fact that it requires less restrictive upper bounds placed on the relative increase in step-size allowable, which allows it to reach the maximal step-size of approx. 2.5E+8 more rapidly. This is a good example of the poorly scaled problems that are regularly encountered in process modeling. This necessitates a careful choice of path-tracking algorithm parameters if one is to proceed efficiently.

Table 1. Operations and LUd Count for the Williams-Otto Problem.

Algorithm	Parameter Values	LUd	EOC
(i)	α^1= 0.05 α^2 = 0.5	64	2.1908E+5
(ii)	κ = 5.0 α = 0.05 δ^i =1.0E-4	56	2.0720E+5
(iii)	ε = 1.0E-3 θ = 0.15 N = 4	33	1.2680E+5

Table 2. Operations and LUd Count for the Distillation Tower Problem.

Algorithm	Parameter Values	LUd	EOC
(i)	α^1= 0.15 α^2 = 0.5	80	1.2350E+8
(ii)	κ = 3.0 α = 0.05 δ^i =1.0	55	8.4906E+7
(iii)	ε = 5.0 θ = 0.15 N = 3	48	8.0427E+7

Problem 3: Segment Jumping

Choi and Book (1991) suggested the following problem

$$\begin{aligned} f_1(x_1,x_2) &= x_1^2 + x_2^2 - 1.5^2 = 0 \\ f_2(x_1,x_2) &= x_2 - x_1^2 + 2 = 0 \end{aligned} \tag{11}$$

that creates an ill-behaved homotopy path when tracked by a fixed point homotopy in the complex space, with a starting point of (2.336, -2.5). This homotopy path contains near bifurcation points, which can lead to segment jumping and hence, a possible failure to locate all four roots. Without determinant monitoring, algorithm (iii) with N = 3, $\varepsilon = 10^{-3}$ and θ = 0.10, segment jumped twice, thereby locating all roots, but in the incorrect order. Algorithms (i) and (ii) similarly jumped onto another segment of the solution path, but did not make the second jump. This resulted in the homotopy paths returning to the starting point, which is not possible for a fixed point homotopy with a non-singular starting point. Monitoring the determinant prevented segment jumping in all cases and lead to a sig-

[1]. DZINE is an equation-oriented, process flowsheeting system developed at UMR.

nificant decrease in efficiency only in the region of the near bifurcation points.

The algorithms can also be made to follow the correct path by adjusting their parameters in such a way that the step-sizes calculated are consistently small enough. This however, is much less efficient than simply monitoring the determinant. For example, algorithm (iii) with the above parameters required 786 LU decompositions to find all roots with determinant monitoring, but without the monitoring phase and N set equal to three, the correct path was traced when ε and $1/\theta$ are less than 1.0E-08 and 2.0 respectively, thus giving rise to a LUd count of greater than 950, or an increase of more than 20%. Algorithm (ii) required an increase in the number of LU decompositions of over 100% in order to correctly trace the path without the monitoring phase.

Conclusions

Of the three algorithms tested here, the one proposed by Corvalan and Saita outperformed the other two on a large number of the problems tested including those presented here. This can be attributed mainly to the fact that it took the number of corrector steps at each continuation point into account explicitly, and therefore produces a more continuous variation of the step-sizes. In this regard, it is similar to the first two step-size control strategies of den Heijer and Rheinboldt (1981), but is not as conservative in its calculation of relative step-size changes. All the algorithms performed reasonably well in the distillation tower problem, supporting the claims of Wayburn and Seader (1983) and Lin et al. (1987), that homotopy methods are a viable tool for the analysis of large scale separation problems.

It is necessary to install a determinant monitoring phase in most algorithms if one wants to prevent segment jumping without significantly reducing the efficiency of the homotopy continuation method.

References

Allgower, E.L. (1981). A survey of homotopy methods for smooth mappings. In E.L. Allgower, K. Glashoff and H.D. Peitgen (Ed.), *Lecture Notes in Applied Mathematics*, No. 878, *Numerical Solution of Nonlinear Equations*, Springer-Verlag, New York, 1-29.

Allgower, E.L. and K. Georg (1990). *Numerical Continuation Methods: An Introduction*, Vol. 13 of *Series in Computational Mathematics*, Springer (Berlin, Heidleberg, New York), 14, 114-118.

Allgower, E.L. and K. Georg (1993). Continuation and path following. In E.L. Allgower (Ed.), *Acta Numerica*, Cambridge University Press (Cambridge, New York), 1-64.

Armstrong, J.P., M.F. Cummins, M.C. Whelan, O.C. Sitton and N.L. Book (1988). A chemical process flowsheeting system based on the functionality matrix. *Simulators V*, **19**, 4, 261-266, *Proceedings of the SCS Simulators Conference*, April 1988, Orlando, FL, The Society for Computer International, San Diego, CA.

Choi, S.H. (1990). *The application of global homotopy techniques to chemical process flowsheeting problems*. Ph.D. Thesis, University of Missouri-Rolla.

Choi, S.H. and N.L. Book (1991). Unreachable roots for global homotopy continuation methods. *AIChE J.*, **37**, 1093-1095.

Choi, S.H. and N.L. Book (1994). A robust predictor-corrector algorithm. *Fifth International Symposium on Process Systems Engineering*. Seoul, Korea.

Cline, A.K., C.B. Moler, G.W. Stewart and J.H. Wilkinson (1979). An estimate for the condition number of a matrix. *SIAM J. Numer. Anal.*, **16**, 368-375

Corvalan, C.M. and F.A. Saita (1991). Automatic stepsize control in continuation procedures. *Computers Chem. Engng.*, **15**, 10, 729-739.

den Heijer, C. and W.C. Rheinboldt (1981). On steplength algorithms for a class of continuation methods. *SIAM J. Numer. Anal.*, **18**, 925-948.

Kearfott, R.B. (1983). Some general bifurcation techniques. *SIAM J. Stat. Comput.*, **4**, 52-68.

Leray, J. and J. Schauder (1934). Topologie et equations fonctionelles. *Ann. Sci. Ecole. Norm. Sci.*, **51**, 45-78.

Lien, K.T. (1993). *Solving multicomponent distillation problems with an equation oriented chemical process flowsheeting system*. M.S. Thesis, University of Missouri-Rolla.

Lin, W., J.D. Seader and T.L. Wayburn (1987). Computing multiple solutions to systems of interlinked separation columns. *AIChE J.*, **33**, 6, 886-897.

Rijkaert, M.J. and X.M. Martens (1974). Analysis and optimization of the Williams-Otto process by geometric programming. *AIChE J.*, **20**, 4, 742-750.

Schwetlick, H. (1984). On the choice of steplength in path following methods. *Z. Agnew. Math. Mech.*, **64**, 9, 391-396.

Seader, J.D. (1989). Recent developments in methods for finding all solutions to general systems of nonlinear equations. *Proc. Conference on Foundations of Computer-Aided Process Design*, July, Snowmass, CO.

Wayburn, T.L. and J.D. Seader (1983). Solution of interlinked distillation by differential homotopy continuation methods. *Proc. 2nd Int. Conf. Found. Computer-Aided Process Design*.

APPLICATION OF INTERVAL-NEWTON METHOD TO CHEMICAL ENGINEERING PROBLEMS

G. V. Balaji, J. D. Seader, J. J. Chen, and S. Sharma
Department of Chemical and Fuels Engineering
University of Utah
Salt Lake City, UT 84112

Abstract

The interval Newton method in conjunction with generalized bisection, as implemented in the public domain software program INTBIS of Kearfott and Novoa, is applied to the solution of single and simultaneous nonlinear and transcendental equations. This method is used to solve 15 test problems from different chemical engineering application areas, one equation with known singularity, and one ill-scaled transcendental equation. The interval method as implemented in INTBIS is capable of finding all the real roots of an equation within a specified domain. The complex roots were also found by transforming the complex part into a real variable. In the case of polynomial equations, the computer time for INTBIS is compared to that of the parallel-path continuation method of Morgan as implemented in the public domain software program CONSOL8. The parallel-path continuation method could find all real and complex roots without specifying a domain, while INTBIS requires the specification of a domain. The effect of the size of the initial domain on the computer time for INTBIS was also studied for a set of multi-variable polynomial equations.

Keywords

Interval-Newton method, Nonlinear equations, Polynomial equations, Continuation, Software.

Introduction

There have been extensive efforts to develop techniques for obtaining roots of single nonlinear equations. Such equations are encountered in a variety of engineering fields. Recently efforts have been directed toward obtaining all roots, real and complex. This can pose a formidable challenge, especially when the equation involves transcendental terms. Nevertheless, it is becoming increasingly important for the method to obtain all the roots, real and complex, even when only some roots have any physical significance. This enables one to have an insight into the equation and the method for root-finding, which would otherwise be lost if some of the roots were not obtained. For example, an equation may be defined for which there is no root in a certain interval, or the roots are all complex. If one applies a method to find the root, or only the real root, within this interval, the root-finding procedure would fail to obtain any root. The user would not know whether to attribute this to the root-finding technique, or the algorithm, or the equation itself. If all the roots, both real and complex, are found, this question does not arise.

There are two general classes of root-finding methods for nonlinear equations. They are: (1) local methods, which require some knowledge regarding the location of the roots and find at best one root from each starting point, and (2) global methods, which can track down all the roots from a supposedly arbitrary starting point. Local methods, such as Newton's method, have been favored due to their speed of convergence. But they are not always robust. Furthermore, they are not guaranteed to find the root closest to the starting point. Therefore, interest has been growing in less efficient but more reliable global methods, such as the continuation method, in order to obtain all the roots all the time. Such continuation methods can require an order-of-magnitude more computation time, but with the speed of today's computers, this time is becoming less and less a factor in chosing a method.

Due to the inherent approximation in today's digital computers, because of the finite length of bytes and words, round-off errors are inevitable during computations. These round-off errors build-up and propagate, leading to inaccurate results. One way to avoid round-off and propagation errors is to use interval arithmetic, which treats the numbers as intervals that bound the number (Moore, 1979). Thus, when round-off error is the only error present in a computation, the width of the interval results tend towards zero as the length of the machine word increases. Due to this advantage, use of interval arithmetic to bound the exact solutions within intervals is finding applications in many numerical techniques. Finding the roots of a system of equations is one growing application.

Many engineering problems require finding roots of nonlinear equations. Application of interval arithmetic is an attractive choice because of the reduced round-off and propagation errors that are very common with finite numerical techniques that have been used. Often the bounds for the region of interest is known for engineering problems. This provides an added incentive to use local methods. One method for finding roots for a system of nonlinear equations with interval arithmetic is the multidimensional interval Newton method. The public domain software INTBIS (Kearfott and Novoa, 1990) is based on this method, in conjunction with the generalized bisection method. In chemical engineering, root finding is a very important and often used step during process design and flowsheeting. Commercial software that simulates process flowsheets employs root-finding techniques during each iteration while solving nonlinear design equations. The number of iterations may be large, even for nominal-sized problems. During the iterations, the errors due to finite machine word lengths propagate. The interval Newton method may be an attractive choice for solving nonlinear equations in chemical engineering.

Without modification, the current version of INTBIS is designed to handle only polynomial equations in the real domain. Nevertheless, with the availability of transcendental functions in the interval arithmetic library INTLIB (Kearfott, Dawande, and Hu, 1993), modifications to accommodate transcendental functions into INTBIS were easily made. With the simulation of the complex domain using two-dimensional real variables, the complex roots were also found using INTBIS.

As a first step for applying the interval Newton method, INTBIS was used on a set of 14 chemical engineering problems (Shacham, 1989). which involved single nonlinear equations typically encountered in chemical engineering. INTBIS was also applied to a transcendental function of purely mathematical interest (Watson, 1988) and to an equation involving terms in the numerator and the denominator (Gritton, 1991), which could lead to a singularity. In order to evaluate the routine on a system of linear and nonlinear equations, a two-stage reactor problem (Seader et al., 1990) involving a set of polynomials was also included.

Principle

The public domain program INTBIS is based on the interval Newton/generalized bisection algorithm, which combines a geometric bisection method with the interval Newton method. This algorithm is explained by Kearfott, (1978). Consider the system of nonlinear equations represented as:

$$\mathbf{f(x) = 0} \qquad (1)$$

The interval extension for the system of equations in (1) can be represented as:

$$\mathbf{F(X) = 0} \qquad (2)$$

Equation (2) is linearized as follows to apply the interval Newton method:

$$\mathbf{F'(X^k)\,(X^{k+1} - x^k) = - f(x^k)} \qquad (3)$$

where the superscript $\mathbf{k}$ represents the iteration counter, while $\mathbf{F'(X^k)}$ represents the interval extension of the Jacobian of $\mathbf{f}$ evaluated at $\mathbf{X^k}$, with $\mathbf{x^k}$ as the midpoint vector of $\mathbf{X^k}$. The combination of the interval Newton method, using the interval Gauss-Seidel technique, with the generalized bisection technique is capable of locating all the real roots to Eq. (1) in a designated domain.

Complex Roots

The interval Newton method is derived from the mean-value theorem, which does not hold in the complex domain (Hansen, 1978). One technique to find complex roots is to separate the real and imaginary parts of the variables, and treat them both as real variables. The multidimensional interval Newton method can then be applied to this two-dimensional system of equations involving the two real variables, one representing the real part, and the other the imaginary part. Of course, this doubles the number of variables and equations. This scheme was first tested with INTBIS on a single equation, problem number 1 of Shacham (1989), which was known to have real and complex roots. This problem involves a polynomial for computing the equilibrium fractional conversion, z, of nitrogen and hydrogen to ammonia, starting with stoichiometric quantities. At reaction conditions of 250 atm and 500 °C, the chemical equilibrium relation, in polynomial form,is:

$$z^4 - 7.79075\,z^3 + 14.7445\,z^2 + 2.511\,z - 1.674 = 0 \qquad (4)$$

From the analytical solution of a quartic equation, Eq. (4) is found to have the following roots, two of which form a complex conjugate pair: 0.278, -0.384, 3.949 ± 0.316i.

To compute the four roots with only real variables, Eq. (4) is modified as follows: Let $z = x + i\,y$. Substituting z in Eq. (4) and expanding, we obtain:

$$\begin{aligned}(&x^4 - 6\,x^2y^2 + y^4 - 7.79075\,x^3 + 23.37225\,xy^2 \\ &+ 14.7445\,x^2 - 14.7445\,y^2 + 2.511\,x - 1.674) \\ &+ (4\,x^3\,y - 4xy^3 - 23.37225\,x^2y + 7.79075\,y^3 \\ &+ 29.489\,xy + 2.511\,y)i\end{aligned} \quad (5)$$

For this equation to be equal to zero, the real and imaginary parts should individually and simultaneously be equal to zero. Therefore, we obtain two equations, one representing the real part, and the other the imaginary:

$$\begin{aligned}&x^4 - 6\,x^2y^2 + y^4 - 7.79075\,x^3 + 23.37225\,xy^2 \\ &+ 14.7445\,x^2 - 14.7445\,y^2 + 2.511\,x - 1.674 = 0\end{aligned} \quad (6)$$

$$\begin{aligned}&4\,x^3y - 4xy^3 - 23.37225\,x^2y + 7.79075\,y^3 \\ &+ 29.489\,xy + 2.511\,y = 0\end{aligned} \quad (7)$$

When INTBIS was used to solve these two equations simultaneously for x and y, the four roots for equation (4) were obtained as intervals. This procedure was also employed on those equations known to have complex roots amongst the remaining 13 chemical engineering problems of Shacham (1989).

It is important to note that by introducing the two variables, x and y, for the one variable z, we not only increase the number of equations from one to two, but also, by Bezout's Theorem (Morgan, 1987), the number of roots to the two equations is increased from four for the original equation to as many as 16 for the two equations. Of these, only four roots are real in x and y, representing the four original roots. The other 12 roots were located by continuation using CONSOL8 of Morgan (1987) and, as expected, were found to be complex in x, or y, or in both. As stated earlier, INTBIS can find only the four sets of real roots for x and y, corresponding to all the original roots.

Results

INTBIS was applied to : (1) 14 of the 15 chemical engineering problems due to Shacham (1989), (2) a CSTR problem due to Seader et. al. (1990), (3) an equation with numerator and denominator due to Gritton (1991), and (4) a purely mathematical transcendental equation suggested by Watson (1988).

Problems 1, 2, 4, 5, 6, 11, 12, 13 and 15 of Shacham (1989) were reduced to polynomial form. Problems 3, 7, and 10 involved transcendental terms, while problems 8 and 9 involved fractional powers of the dependent variable. Problem number 14 of Shacham was not included because it could be converted to a linear equation. The CPU times for solving the polynomial equations using INTBIS were compared with the times for CONSOL8 when using a SUN Sparc10 workstation. INTBIS and CONSOL8 found all real and complex roots for each problem. The results are given in Table I.

Seader et al. (1990) discuss the application of continuation methods to solve a system of nonlinear and linear equations resulting from the 1985 AIChE Student Contest Problem. The example involves an acid-catalyzed esterification reaction carried out continuously in two CSTRs in series. The resulting system of equations consists of five polynomial equations and two linear equations. Of the first five, one is third-degree and the other four are second-degree. The maximum number of roots as determined by Bezout's Theorem (Morgan, 1987) is 48. Only one root has any physical significance, while the others have either negative real values or complex values for some variables. CONSOL8 solved this system of equations for all 48 roots in about 134 s on a Sun Sparc10 workstation. When the initial range, with the only root of physical significance bounded by about 20% on either side, was given to INTBIS, it solved for that root in about 32 s. When the bounding range was increased to ± 50%, the CPU time increased to 153 s.

Gritton (1991) used a global-fixed point homotopy method to solve the problem 1 of Shacham (1989) for the calculation of chemical equilibrium in ammonia synthesis. The original equilibrium equation is:

$$\frac{8\,(4-x)^2x^2}{(6-3x)^2(2-x)} - 0.186 = 0$$

This equation has a singularity at x = 2, when the denominator becomes 0. When given the initial interval (-5, 2.5) INTBIS found both real roots, x = - 0.384 and 0.278, and also indicated the singularity at 2.

Watson [5] suggested the following transcendental equation :

$$\exp(-\,x^2) * \sin(x) = 0$$

The roots for this equation coincide with the roots for sin(x) = 0. But the exponential term introduces a large variation of the function in the interval (-25, 25). INTBIS found all the roots in this interval. Outside this interval the function produces underflow errors with double precision variables.

Conclusions and Recommendations

Unlike the single solution found by the classic Newton method, the exhaustive search for roots combined with the avoidance of round-off errors makes INTBIS an attractive choice for application in chemical engineering design programs. INTBIS can be easily incorporated into programs through its main driver routine GENBIS. In spite of its simulated interval arithmetic, its CPU times are comparable with those from CONSOL8, when solving polynomial equations. When only real solutions are desired, INTBIS may require less time than CONSOL8. INTBIS has the added advantage of being applicable to nonlinear equations with transcendental terms.

Table 1. Comparison of CPU Times for Solving the Polynomials from [4].

Problem		CPU Times (s)		Bounds for INTBIS	
		INTIBS	CONSOL8	Lower	Upper
1A	Ammonia synthesis	2.200 *	0.217	- 1 - i	4 + i
1B	Ammonia synthesis	2.717 *	0.183	- 3 i	5 + 3 i
2	Isothermal flash	1.9	---	- 12	7
4	Azeotropic point	0.033	0.133	0	6
5	Adiabatic flame temp	0.150 *	0.167	- 9000 - 9000 i	5000 + 9000 i
6	Beattie-Bridgeman EOS	1.267 *	0.200	- 5 i	2 + 5 i
11	Chemical Equilibrium	3.117 *	0.183	- 0.4 i	0.7 + 0.4 i
12	Virial EOS	1.050 *	0.133	100 - 100 i	300 + 100 i
13	Redlich-Kwong EOS	0.400 *	0.150	- 0.1 i	0.1 + 0.1 i
15	Sphere sinkage depth	0.033	0.167	- 1	3

** denotes representation of complex arithmetic with two real variables*

With the interval arithmetic being simulated using the currently available floating-point arithmetic compilers, each interval arithmetic operation takes more time compared to its countepart in floating-point operation. With the possible availability of interval arithmetic compilers, this time could be substantially reduced, making it competitive for use in design and simulation software. For multiple equations involving transcendental terms, the application of INTBIS in its present format is unwieldy because programming the functions and jacobians in interval arithmetic is very tedious. With a future modification for internal computation of the function and jacobian in interval arithmetic from the user input equations in floating-point arithmetic, the algorithm will be suitable for application to sophisticated simulation software.

Acknowledgements

We appreciate the assistance of Professor R. Baker Kearfott at every step of applying INTBIS to our equations. Financial support was provided by a graduate fellowship from the Phillips Petroleum Company.

References

Gritton, K. S. (1991). *Application of Global Fixed-Point Homotopy to Single-Nonlinear-Equation,* Ph.D. Dissertation, University of Utah.

Hansen, E. (1978). A Globally Convergent Interval Method for Computing and Bounding Real Roots, *BIT,* **18**, 415-424.

Kearfott, R. B. (1978). Abstract Generalized Bisection and a Cost Bound, *Math. Comput.*, **49**(179), 187-202 .

Kearfott, R. B.and M. Novoa III (1990). INTBIS, a Portable Interval Newton/Bisection Package, *ACM Trans. Math. Software,* **16**(2), 152-157.

Kearfott, R. B., Dawande, M., Du, K., and Hu, C. (1993). INTLIB: A Portable Fortran-77 Elementary Function Library, accepted for publication in *Interval Computations*.

Moore, R. E. (1979). *Methods and Applications of Interval Analysis,* SIAM, Philadelphia.

Morgan, A. P. (1987). *Solving Polynomial Systems using Continuation for Engineering and Scientific Problems*, Prentice-Hall, Englewood Cliffs, NJ.

Shacham, M. (1989). An Improved Memory Method for the Solution of a Nonlinear Equation, *Chem. Eng. Sci.*, **44**(7), 1495-1501.

Seader, J. D., Kuno, M., Lin, W.-J., Johnson, S. A., Unsworth, K., and Wiskin, J. S., (1990). Mapped Continuation Methods for Computing All Solutions to General Systems of Nonlinear Equations, *Computers and Chem. Engng.*, **14**(1), 71-85.

Watson, L. (1988). Personal communication.

AUTHOR INDEX

SUBJECT INDEX

T

U

V

W